Critical Issues in
CRIME
and JUSTICE

Second Edition

This book is dedicated to all local, state, and federal law enforcement officers, prosecutors, criminal investigators, forensic specialists, and emergency service workers who are committed to fighting terrorism, crime on the streets, and facilitating justice for all citizens of the United States of America.

Critical Issues in
CRIME
and JUSTICE

Second Edition

Editor
Albert R. Roberts
Rutgers University, Piscataway

SAGE Publications
International Educational and Professional Publisher
Thousand Oaks ▪ London ▪ New Delhi

For information:

Sage Publications, Inc.
2455 Teller Road
Thousand Oaks, California 91320
E-mail: order@sagepub.com

Sage Publications Ltd.
6 Bonhill Street
London EC2A 4PU
United Kingdom

Sage Publications India Pvt. Ltd.
B-42, Panchsheel Enclave
Post Box 4109
New Delhi 110 017 India

Printed in the United States of America

Library of Congress Cataloging-in-Publication Data

Critical issues in crime and justice / editor Albert R. Roberts. —2nd ed.
    p. cm.
Includes bibliographical references and index.
ISBN 978-0-7619-2686-3 (paper)
    1. Criminal justice, Administration of—United States. 2. Law enforcement—United States.
3. Corrections—United States. I. Roberts, Albert R.
HV9950 .C77 2003
364,973—dc21

                                    2002153871

This book is printed on acid-free paper.

09 10 11 9 8 7 6 5 4

Acquisitions Editor:	Jerry Westby
Editorial Assistant:	Vonessa Vondera
Copy Editor:	Elisabeth Magnus
Production Editor:	Denise Santoyo
Typesetter:	C&M Digitals (P) Ltd.
Indexer:	Kathy Paparchontis
Cover Designer:	Ravi Balasuriya

CONTENTS

LIST OF CONTRIBUTORS

Gaylene S. Armstrong, Arizona State University West
Mary Bosworth, Wesleyan University
Mark Blumberg, Central Missouri State University
Beau Breslin, Skidmore College
Timothy P. Cadigan, Administrative Office of the U.S. Courts
Norman B. Cetuk, Police Academy, Raritan Valley Community College
Meda Chesney-Lind, University of Hawaii at Manoa
Walter Edwards, Rutgers University
James Alan Fox, Northeastern University
Lisa A. Frisch, New York State Office for the Prevention of Domestic Violence
Kathleen M. Heide, University of South Florida
Vincent E. Henry, Pace University
Sean Hill, Sam Houston State University
Mary S. Jackson, East Carolina University
Michael Jacobson, John Jay College of Criminal Justice
David R. Karp, Skidmore College
Betsy Wright Kreisel, Central Missouri State University
Karel Kurst-Swanger, SUNY at Oswego
Jack Levin, Northeastern University
Charles Lindner, John Jay College of Criminal Justice
Doris Layton MacKenzie, University of Maryland
Daniel P. Mears, the Urban Institute, Justice Policy Center
Scott K. Okamoto, Arizona State University
Joan Petersilia, University of California at Irvine
Laura J. Praglin, University of Northern Iowa
Gina Pisano–Robertiello, Seton Hall University
Albert R. Roberts, Rutgers University
Joseph F. Ryan, Pace University
Margaret Ryniker, SUNY at Oswego
Lori Koester Scott, Maricopa County Probation Department
Donald J. Sears, Rutgers University
Cassia C. Spohn, University of Nebraska
Hung-En Sung, Kings County District Attorney's Office
Lisa Weber, American University
Michael Welch, Rutgers University
David B. Wilson, George Mason University
Katherine van Wormer, University of Northern Iowa
Richard H. Ward, Sam Houston State University

PREFACE

The overriding objective of this thoroughly revised second edition is to trace the tremendous progress achieved during the past decade toward resolving criminal justice issues, dilemmas, and controversies, while providing futuristic visions for the criminal justice field. Each chapter includes a comprehensive and up-to-date examination of the current critical issues and policy dilemmas within the system of criminal justice affecting local, state, and federal jurisdictions throughout the United States.

This book has been prepared for introductory and advanced courses on the U.S. criminal justice system and its interrelated components. Twenty-three of the 27 chapters are brand new to this second edition, and the remaining four are completely updated. Each contributing chapter author examines the current controversies and issues confronting the police, prosecutors, defense attorneys, judges, victim advocates, legislators, juvenile justice officials, and probation and correctional administrators. Most chapters are original and specially written with the goal of bridging theory to practice, promoting critical thinking, and applying new technology and strategic planning to criminal justice practice and management.

There is a growing realization that law enforcement and judicial agencies need to safeguard the legal rights of both the accused and the crime victim. While attempting to balance the rights of the accused with those of the innocent victim, the criminal justice system is plagued by a shortage of staff resources, a large number of violent crime victims, a low percentage of arrests for serious crimes, repeated cases of police use of excessive force, tremendous court backlogs, severely overcrowded jails and prisons, a high percentage of charges dismissed or reduced, and skyrocketing costs of terrorism as well as property-related crime. Several promising and cost-effective strategies to crime control, offender rehabilitation, legal reforms, domestic violence, and juvenile aftercare have been developed. This volume will provide an extensive background and discussion of the critical issues, policy reforms, legal remedies, and model program developments.

Critical Issues in Crime and Justice will be useful as the primary or supplementary text for Introduction to Criminal Justice, Critical Issues, Special Topics, Contemporary Issues in Justice and related overview courses. This book includes up-to-date chapters on the following important critical issues of our time:

- The recent expansion of law enforcement agencies, criminal justice personnel costs, victim costs, victimization rates, and criminal justice trends
- Case illustration of criminal justice processing of an alleged offender charged with illegal possession of a firearm, domestic violence, and murder

- The crisis induced by catholic priests committing numerous child sexual assaults and statutory rape
- Adult sex offender treatment programs that work
- Compstat and police management strategies
- The COPS federal grant program resulting in 100,000 newly hired police
- Violent youth gangs and law enforcement intervention programs
- International counter-terrorism strategies
- Myths and facts regarding serial rapists and mass murderers
- Forensic evidence and homicide investigations
- Probation and pretrial services automated case tracking-electronic case management system applications
- Justice response to woman battering
- Death penalty debates and the U.S. Supreme Court decisions
- Critical probation failures and promising innovations
- Prosecutor-based employment diversion programs for offenders
- Sentencing options and judges discretion
- Juvenile justice and delinquency prevention strategies
- The effectiveness of juvenile boot camps based on national data
- Content analysis of the prison overcrowding debate
- Federal and state prison rehabilitation and vocational training programs
- Restorative justice and advocating for offenders

In contrast to the traditional textbooks, this book includes detailed information and specific chapters on the above cutting edge topics as well as futuristic topics. In addition, this book includes the systematic application of timely and thought-provoking case illustrations and trend data in most chapters, policy and model program examples from many of the 50 states, and end-of-chapter summaries. This comprehensive and up-to-date book is recommended to all criminal justice practitioners and students.

Acknowledgments

First and foremost, I would like to acknowledge my editor at Sage, Jerry Westby, for practical guidance, editorial insights, and very useful brainstorming sessions related to the development of this thoroughly revised new edition. Second, I offer special thanks to Jerry's editorial assistant, Vonessa Vondera, for important technical assistance, and critical reminders in the preparation of the final manuscript. My sincere appreciation also goes to Denise Santoyo, my production editor, whose patience and care for detail was outstanding. The Sage team of Jerry Westby, Vonessa Vondera, and Denise Santoyo are to be congratulated for bringing out the best in me and in the individual chapter authors. Last, but most important, are the exceptionally talented scholar–practitioners who contributed substantial, original, and up-to-date chapters to this book. Without the dedication of my outstanding author team in meeting my deadlines in a timely fashion this book would never have been completed.

For further information go to: www.crisisinterventionnetwork.com

Albert R. Roberts, Ph.D.
Rutgers University, Livingston College Campus, Piscataway, N.J.

Part I

OVERVIEW OF CRIMINAL JUSTICE

1

CRIME IN AMERICA

Critical Issues, Trends, Costs, and Legal Remedies

ALBERT R. ROBERTS

This chapter provides an overview and orientation to the critical issues, trends, policies, programs, and intervention strategies of the criminal justice system. The critical issues surrounding the criminal justice system and its subsystems have evoked strong emotions and debate. Public outcry and legislative support for change are often prompted by the media coverage and the public's fear about youth violence, drug abuse, murder, and other victimizations. In an effort to respond to these pressures while also doing their job, the components of the criminal justice system—including law enforcement agencies, the court system, the juvenile justice system, and the correctional system—are constantly in flux, seeking ways to maximize their efficiency.

This text focuses on the critical issues within the organizations and agencies administering justice services in American society. Chapter authors were selected because of their expertise in a particular topical area. Most of the chapters are original and have been specially written for this book. This volume is divided into six main parts:

1. Overview of Criminal Justice

2. Understanding Crime and Criminals

3. Critical Issues in Law Enforcement

4. Critical Issues in Juvenile Justice

5. Critical Issues in the Criminal Court System

6. Correctional Systems and Alternatives

Each of the chapters within the above sections of the book examines the most timely critical issues and controversies confronting cities and communities throughout the United States.

The American criminal justice system consists of the formal governmental agencies and personnel who are empowered with the responsibility of enforcing the criminal code. These agencies or components of the system, which are present at federal, state, and local levels, include the police, the prosecutor's office, the public defender's office or legal aid, the courts, and corrections. Police officers detect crime, investigate citizen complaints, attempt to control crime by making arrests, and provide emergency services to crime victims and the community. The courts handle defendants. They have the legal responsibility of enforcing the state criminal code against defendants who have been indicted for specific criminal acts. They are also empowered to protect the same defendants from the violation of their constitutional rights by criminal justice practitioners. Jails and short-term county correctional institutions confine pretrial detainees and sentenced misdemeanants. Corrections professionals supervise convicted offenders in county, state, and federal prisons as well as community correctional centers. County and city probation officers are responsible for supervising juvenile and adult offenders in the community. State parole officers supervise juvenile and adult offenders in the community. In the federal system, the functions of probation and parole are combined. As a result, we have federal probation and parole agents monitoring, supervising, and checking up on offenders in the community.

To provide a context for interpreting the critical issues raised in this book's chapters, I shall present a review of current crime statistics and trends.

THE NATURE AND EXTENT OF CRIME

Crime is one of the most serious social problems facing American society. The two most comprehensive statistical sources of crime data in America are the annual Federal Bureau of Investigation (FBI) Uniform Crime Reports (UCR) and the National Crime Victimization Survey (NCVS). The UCR provides nationwide crime data based on "crimes known to the police" and police arrests in 20 crime categories made by almost 17,000 city, county, and state law enforcement agencies. The FBI provides all police agencies throughout the United States with standardized forms to report arrest data to the FBI. The NCVS is a massive victimization survey conducted by the Bureau of Justice Statistics (BJS), in cooperation with the U.S. Bureau of the Census, on approximately 45,000 to 60,000 households every 6 months. Household members are interviewed to determine whether they have been victims of the major crimes of rape, robbery, assault, burglary, personal and household larceny, or motor vehicle theft. The NCVS discloses the extent to which specific crimes are not reported to the police and the reasons victims give for not reporting crimes.

Figure 1.1 depicts a Crime Clock based on an aggregate representation of data from the UCR. The Crime Clock indicates that in 2000 there was one murder every 33.9 minutes, one

forcible rape every 5.8 minutes, one robbery every 1.3 minutes, one burglary every 15.4 seconds, one larceny-theft every 4.5 seconds, and one motor vehicle theft every 27.1 seconds. The FBI cautions that the Crime Clock should not be interpreted to imply regularity in the commission of crimes in the almost 17,000 police agencies contributing crime data; rather, it represents the annual ratio of crime to fixed time intervals.

The data reported by the FBI provide estimates of the extent of criminal activity known to law enforcement agencies. In 2000, there were over 1.4 million violent crimes reported to the police (Table 1.1). These included 90,186 forcible rapes, 910,744 assaults, and 15,517 murders. The murder rate for 2000, down 0.1% from the 1999 rate, was the lowest murder rate since 1965 (5.5 per 100,000 people). According to the supplemental UCR data, males composed 76.0% of the murder victims. Twenty- to 34-year-olds made up the largest portion of victims. Firearms were used in 7 out of 10 murders, and larger cities had a higher share of victims.

The frequency of major offenses has decreased considerably in the 10-year period of 1990 through 2000 (see Figure 1.2 and Table 1.2). The Crime Clock indicates that

- The frequency of murder declined from one every 22 minutes in 1990 to one every 33.9 minutes in 2000.
- The frequency of completed forcible rape declined slightly from one every 5 minutes in 1990 to one every 5.8 minutes in 2000.
- The frequency of robbery arrests decreased from one every 49 seconds in 1990 to one every 1.3 minutes in 2000.
- The frequency of reported burglary decreased from one every 10 seconds in 1990 to one every 15.4 seconds in 2000.
- The frequency of reported larceny-theft decreased slightly from one every 4 seconds in 1990 to one every 4.5 seconds in 2000.
- The frequency of motor vehicle thefts decreased from one every 19 seconds in 1990 to one every 27.1 seconds in the year 2000.

Nevertheless, both property crimes and violent crimes are still pervasive public health and social problems in American society.

The NCVS uses the panel survey method to collect information on personal and household

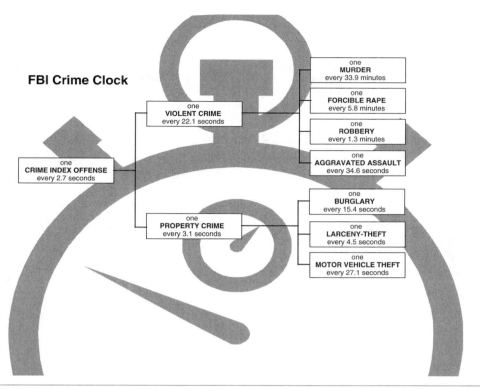

FBI Crime Clock

one
CRIME INDEX OFFENSE
every 2.7 seconds

one
VIOLENT CRIME
every 22.1 seconds

one
PROPERTY CRIME
every 3.1 seconds

one
MURDER
every 33.9 minutes

one
FORCIBLE RAPE
every 5.8 minutes

one
ROBBERY
every 1.3 minutes

one
AGGRAVATED ASSAULT
every 34.6 seconds

one
BURGLARY
every 15.4 seconds

one
LARCENY-THEFT
every 4.5 seconds

one
MOTOR VEHICLE THEFT
every 27.1 seconds

Figure 1.1   FBI Crime Clock

SOURCE: FBI (2001).

Table 1.1   Crimes of Violence: National Crime Survey and Uniform Crime Reports, 2000

Crime	*National Crime Survey*		*Uniform Crime Reports*	
	Number*	*Rate**	*Number**	*Rate**
Total Violent	6,323	27.9	1,424,289	52,300
Forcible Rape	261	1.2	90,186	3,200
Robbery	732	3.2	407,842	14,490
Aggravated Assault	1,293	5.7	910,744	32,360

*Rate is per 1,000 people
**The NCVS is based on interviews with victims and therefore cannot measure murder.
The total population age 12 or older was 226,804,610 in 2000.
SOURCE: Bureau of Justice Statistics (2001); FBI (2001).

victimizations. A representative sample of households is interviewed seven times at 6-month intervals. This method is called a "panel survey" because the same people are interviewed again. For example, 16,000 people over 12 years of age are interviewed in February. In the next month (March), and in each of the 4 successive months, an independent probability sample of the same sample size is interviewed. In August, the apartment dwellers and homeowners first interviewed in February are revisited and interviewed again. Similarly, the original March subsample is revisited in September, the April subsample in October, and so on.

The NCVS has documented a large amount of unreported crime. In some cities, the disparity

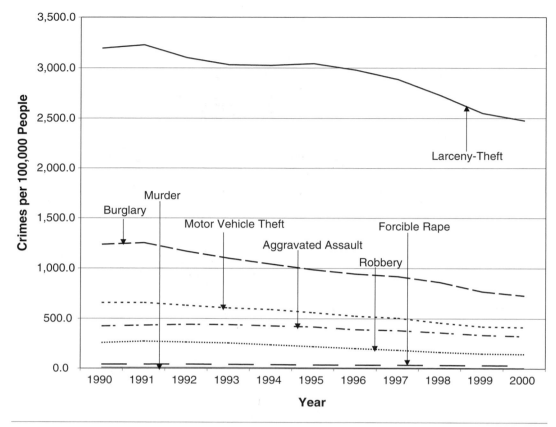

Figure 1.2   UCR Crime Rate, United States, 1990–2000

SOURCE: Bureau of Justice Statistics (1998); PBI (1996, 1997, 1998, 1999, 2000, 2001).
NOTE This graph adapted to alter from color to patterns.

between unreported crime and "crimes known to the police" is as high as 4 to 1. According to the NCVS, the main reasons given by respondent victims for not reporting victimization incidents are general apathy and their belief that there is really nothing the police can do about the crimes.

Although crime in the United States is a serious and prevalent social problem, the good news is that property-related crime and violent victimizations in the year 2000 were the lowest since the NCVS began in 1973. In 1973, the NCVS reported 44 million criminal victimizations; therefore, the estimated 25.9 million victimizations in 2000 show a significant decline in the rates of violent and property crime (BJS, 2001a).

This estimate represents the continued large decline in criminal victimizations from a peak of 52 million in 1994 to 28.8 million in 1999 to a

low point of 25.9 million in 2000. During the year 2000, approximately three fourths of all victimizations were for property crimes such as residential burglary, theft, and motor vehicle theft. Violent crimes such as simple assault, aggravated assault, rape, and robbery against individuals 12 years of age or older constituted the other one fourth of the total victimizations. *In the year 2000, the cumulative violent crime, simple assault, and household theft rates reached the lowest rate ever recorded by the BJS.*

As reported in the NCVS trend data between 1993 and 2000, the overall rate for all violent crimes fell 44%. The most significant decline was the 71% drop for attempted rape, followed by a 52% decline for completed rapes and sexual assaults, a 53% decline for aggravated assault, and a 40% decline for simple assault. With regard to property crimes, the motor vehicle theft rate

Table 1.2 Crime Index Rate, United States, 1990–2000

Year	Crime Index Total	Violent Crime	Property Crime	Murder	Forcible Rape	Robbery	Aggravated Assault	Burglary	Larceny-Theft	Motor Vehicle Theft
1990	5,820.3	731.8	5,088.5	9.4	41.2	257.0	424.1	1,235.9	3,194.8	657.8
1991	5,897.8	758.1	5,139.7	9.8	42.3	272.7	433.3	1,252.0	3,228.8	658.9
1992	5,660.2	757.5	4,902.7	9.3	42.8	263.6	441.8	1,168.2	3,103.0	631.5
1993	5,484.4	746.8	4,737.7	9.5	41.1	255.9	440.3	1,099.2	3,032.4	606.1
1994	5,373.5	713.6	4,660.0	9.0	39.3	237.7	427.6	1,042.0	3,026.7	591.3
1995	5,275.9	684.6	4,591.3	8.2	37.1	220.9	418.3	987.1	3,043.8	560.4
1996	5,086.6	636.5	4,450.1	7.4	36.3	201.9	390.9	944.8	2,979.7	525.5
1997	4,922.7	610.8	4,311.9	6.8	35.9	186.1	382.0	919.6	2,886.6	505.8
1998	4,615.5	566.4	4,049.1	6.3	34.4	165.2	360.5	862.0	2,728.1	459.0
1999	4,266.8	524.7	3,742.1	5.7	32.7	150.2	336.1	770.0	2,551.4	420.7
2000	4,124.0	506.1	3,617.9	5.5	32.0	144.9	323.6	728.4	2,475.3	414.2

NOTE: Rate = number of offenses per 100,000 inhabitants. The numbers presented in this spreadsheet are state-level estimates and therefore may vary from those available from other sources.
SOURCE: Bureau of Justice Statistics (1998); FBI (1996, 1997, 1998, 1999, 2000, 2001).

dropped significantly by 55% from 1993 to 2000. Over the same period, the household burglary rate dropped by 45%, and rates of theft declined 43%.

According to Table 1.3, victimization rates for all crimes decreased by 13.6% from 1999 to 2000, while the decrease for property crimes was 10.1%. The largest drop in the property crime victimization rate from 1999 to 2000 was for completed motor vehicle thefts, a 21.3% drop. The smallest decline in a property crime offense category from 1999 to 2000 was burglary: There was only a 1.7% decrease in the category of burglaries resulting from unlawful entry without force, and burglaries using forcible entry dropped 12.7%. The violent crime category that had the largest percent change from 1999 to 2000 was sexual assault, which dropped by 37.5%, while robberies without injuries dropped 33.3%. (For a discussion and explanation of why the crime rate has dropped, see Chapter 8 of this book.)

COSTS OF CRIMINAL VICTIMIZATIONS

Crime costs billions of dollars each year. There are two major types of costs: monetary and human. Monetary costs can be estimated by identifying criminal justice system operating costs and offender-processing costs. Human costs refer to the lost lives and unused human potential of the thousands of homicide victims each year. Professors James Alan Fox and Jack Levin in Chapter 4 shatter the nine common myths about serial murderers and replace them with facts and empirical knowledge. They also examine what was learned from the criminal investigations and apprehension of Herbert Mullen, Jeffrey Dahmer, Kenneth Biandie, Theodore Bundy, David Berkowitz, John Wayne Gacy, and Danny Rolling.

The total annual cost of crime has been estimated at approximately $625 billion annually. Several component costs of the criminal justice system can be fairly accurately measured. These include tangible police, corrections, and court budgets. Other component costs are more difficult to measure accurately. These include net losses from robbery, burglary and larceny; credit card fraud; computer crime; automated teller machine fraud; federal income tax evasion; and counterfeit notes and currency.

Recent estimates of the total cost of victimization from criminal activity are:

Tangible economic losses—$105 billion

Tangible criminal justice system costs—$70 billion

Intangible psychological/mental health costs—$450 billion

Tangible losses consist of direct losses from victimization, such as the cost of medical and mental health services, victim and witness assistance and concrete services, domestic violence

Injuries experienced by victims of personal violence are not limited to the immediate crime encounter. The crime victims may have been assaulted, robbed, raped, or murdered, or they may have had their automobile stolen, but the perpetrator's actions also have a residual effect—the crime usually causes victim survivors to experience intense psychological pain and emotional torment, medical injuries, trauma, and intense fear that they will be victimized again (Roberts, 1999). To better understand the impact of violent crime victimization, we can put a price tag or monetary value on victimization. A group of economists and criminologists estimated and measured the tangible losses from a rape-related death as $1.03 million, while estimating the intangible cost of a human life at $1.91 million. The tangible value of a crime victim's life was measured in terms of the person's occupation and age (e.g., a 25-year-old homicide victim translates into 40 years of wages and income lost) and the amount of money spent to reduce the risk of death (e.g., emergency medical care and ambulance), among other variables (Miller, Cohen, & Wiersema, 1996).

program and shelter costs, and loss of productivity in terms of wages and salary (e.g., days lost from school and/or work).

Monetary costs to crime victims seem to be astronomical! They include cash and property replacement costs for victims of robbery,

Table 1.3   Rates of Criminal Victimization and Percent Change, 1993–2000

Type of Crime	1993	1994	1998	1999	2000	Percent Change[b]	
						1994–2000	1999–2000
Personal crimes[c]	**52.2[a]**	**54.1**	**37.9**	**33.7**	**29.1**	**−46.2%***	**−13.6%***
Crimes of violence	49.9	51.8	36.6	32.8	27.9	−46.1*	−14.9*
Completed violence	15.0	15.4	11.6	10.1	9.0	−41.6*	−10.9+
Attempted/threatened violence	34.9	36.4	25.0	22.6	18.9	−48.1*	−16.4*
Rape/Sexual assault	2.5	2.1	1.5	1.7	1.2	−42.9*	−29.4*
Rape/attempted rape	1.6	1.4	0.9	0.9	0.6	−57.1*	−33.3*
Rape	1.0	0.7	0.5	0.6	0.4	−42.9	−33.3+
Attempted rape	0.7	0.7	0.4	0.3	0.2	−71.4 *	−33.3
Sexual assault	0.8	0.6	0.6	0.8	0.5	−16.7	−37.5*
Robbery	6.0	6.3	4.0	3.6	3.2	−49.2*	−11.1
Completed/property taken	3.8	4.0	2.7	2.4	2.3	−42.5*	−4.2
With injury	1.3	1.4	0.8	0.8	0.7	−50.0*	−12.5
Without injury	2.5	2.6	2.0	1.5	1.6	−38.5*	6.7
Attempted to take property	2.2	2.3	1.2	1.2	0.9	−60.9*	−25.0
With injury	0.4	0.6	0.3	0.3	0.3	−50.0*	0.0
Without injury	1.8	1.7	0.9	0.9	0.6	−64.7*	−33.3*
Assault	41.4	43.3	31.1	27.4	23.5	−45.7*	−14.2*
Aggravated	12.0	11.9	7.5	6.7	5.7	−52.1*	−14.9+
With injury	3.4	3.3	2.5	2.0	1.5	−54.5*	−25.0*
Threatened with weapon	8.6	8.6	5.1	4.7	4.2	−51.2*	−10.6
Simple	29.4	31.5	23.5	20.8	17.8	−43.5*	−14.4*
With minor injury	6.1	6.8	5.3	4.4	4.4	−35.3*	0.0
Without injury	23.3	24.7	18.2	16.3	13.4	−45.7*	−17.8*
Personal theft[d]	2.3	2.4	1.3	0.9	1.2	−50.0*	33.3

(Continued)

9

Table 1.3   Rates of Criminal Victimization and Percent Change, 1993–2000 (continued)

Type of Crime	1993	1994	1998	1999	2000	Percent Change[b]	
						1994–2000	1999–2000
Property crimes	318.9	310.2	217.4	198.0	178.1	−42.6%*	−10.1%*
Household burglary	58.2	56.3	38.5	34.1	31.8	−43.5*	−6.7
Completed	47.2	46.1	32.1	28.6	26.9	−41.6*	−5.9
Forcible entry	18.1	16.9	12.4	11.0	9.6	−43.2*	−12.7+
Unlawful entry without force	29.1	29.2	19.7	17.6	17.3	−40.8*	−1.7
Attempted forcible entry	10.9	10.2	6.4	5.5	4.9	−52.0*	−10.9
Motor vehicle theft	19.0	18.8	10.8	10.0	8.6	−54.3*	−14.0+
Completed	12.4	12.5	7.8	7.5	5.9	−52.8*	−21.3*
Attempted	6.6	6.3	3.0	2.4	2.7	−57.1*	12.5
Theft	241.7	235.1	168.1	153.9	137.7	−41.4*	−10.5*
Completed[e]	230.1	224.3	162.1	149.0	132.0	−41.2*	−11.4*
Less than $50	98.7	93.5	58.6	53.2	43.4	−53.6*	−18.4*
$50 to $249	76.1	77.0	57.8	54.0	48.9	−36.5*	−9.4*
$250 or more	41.6	41.8	35.1	31.7	29.3	−29.9*	−7.6
Attempted	11.6	10.8	6.0	5.0	5.7	−47.2*	14.0

NOTE: Victimization rates may differ from those reported previously because the estimates are now based on data collected in each calendar year rather than data about events within a calendar year. Completed violent crimes include rape, sexual assault, robbery with or without injury, aggravated assault with injury, and simple assault with minor injury. In 1993 the total population age 12 or older was 211,524,770; in 1994, 213,135,890; in 1998, 221,880,960; in 1999, 224,568,370; and in 2000, 226,804,610. The total number of households in 1993 was 99,927,410; in 1994, 100,568,060.

*The difference between the indicated years is significant at the 95%-confidence level.

+The difference between the indicated years is significant at the 90%-confidence level.

[a]Victimization rates (per 1,000 persons age 12 or older or per 1,000 households)

[b]Differences in annual rates shown in each column do not take into account any changes that may have occurred during interim years.

[c]The NCVS is based on interviews with victims and therefore cannot measure murder.

[d]Includes pocket picking, purse snatching, and attempted purse snatching.

[e]Includes thefts with unknown losses.

SOURCE: Bureau of Justice Statistics (1998).

burglary, personal and household larceny, and motor vehicle theft. Annual monetary cost estimates include $37 billion net loss from robbery, burglary, and larceny of banks; almost $60 billion from drug abuse costs, plus an additional $2 billion for health care related to drug abuse and drug treatment services; $500 million for credit card fraud; $13.2 billion for drunk driving costs resulting from DWI motor vehicle crashes; $31.7 billion for private security costs; and over $81 billion for federal income tax evasion. These cost estimates do not include the billions of dollars lost as a result of white-collar crime.

LAW ENFORCEMENT

Federal, state, and local law enforcement in the United States is a growth industry. Many billions of dollars are allocated annually to protect public safety and prevent crime. Our society's continuing concern for safe streets and neighborhoods and law and order, the availability of federal and state funding, increasing population size, and the possibility of a protracted war on terrorism all make the possibilities for future employment for criminal justice graduates excellent.

Local Law Enforcement

Local police departments were developed to protect public safety by enforcing laws, maintaining order, patrolling neighborhoods, and investigating crimes. Police also provide community and crime prevention services. During the past decade, most police departments throughout the United States implemented a community-oriented policing philosophy. Community policing is a collaborative model that involves community leaders and citizens volunteering for domestic violence crisis response teams, neighborhood crime watch or block programs, auxiliary police activities, police-sponsored athletic programs, neighborhood councils, police bicycle patrols, police Youth Explorer programs, and federally funded "Weed and Seed" programs (i.e., programs that remove violent criminals and drug traffickers from targeted neighborhoods and housing developments and use a number of social service programs to stabilize neighborhoods and prevent crime). Chapter 9 in this book, by

Joseph F. Ryan, provides a detailed examination of community policing programs nationwide and the impact of the COPS federal grants program.

From 1990 to 2000, 62 large-city police departments, in cities with populations of 250,000 or more, saw the number of full-time sworn police personnel increase by approximately 17%, from 130,242 to 152,858. More specifically, the five largest police departments were able to increase the number of full-time sworn police personnel as follows:

- New York City Police Department (NYPD): increase to 40,435 by the end of 2000
- Chicago Police Department: increase to 13,466 by the end of 2000
- Los Angeles Police Department: increase to 9,341 by the end of 2000
- Philadelphia Police Department: increase to 7,024 by the end of 2000
- Houston Police Department: increase to 5,343 by the end of 2000

The total number of police has grown steadily from 1990 to 2000, yet the size of each police department in relation to population and their deployment varies from place to place. Although most of the 15 largest city police departments increased in size during the 1990s, particularly as a result of the COPS federal grant program, two police departments in major cities decreased in size–Detroit's police department decreased from 4,595 in 1990 to 4,154 in 2000, and the police department of Washington, D.C., decreased from 4,506 in 1990 to 3,612 in 2000. For details on the number of sworn police personnel in the 15 largest local police departments, see Table 1.4.

For fiscal year 2000, the operating budgets for the 62 police departments serving large cities with a population of 250,000 or more were about $13.1 billion (Reaves & Hickman, 2002, p. 4). After inflation was factored in, total operating budgets were 20% higher in 2000 than in 1990. The average police department budget in fiscal year 2000 was almost $212 million, compared to approximately $176 million for fiscal year 1990 (see Table 1.5). In 2000, the average minimum police chief's starting salary in cities with populations of over 500,000 residents was $105,500, while the average minimum starting salaries for a sergeant or

Table 1.4 Sixty-Two Largest Local Police Departments Serving Cities With a Population of 250,000 or More, by Number of Full-Time Sworn Personnel and Number of Full-Time Sworn Personnel per 100,000 Residents Served, 1990 and 2000

No. of Full-Time Sworn Personnel				*No. of Full-Time Sworn Personnel per 100,000 Residents*			
City	*1990*	*City*	*2000*	*City*	*1990*	*City*	*2000*
New York (NY)	31,236	New York (NY)	40,435	Washington (DC)	742	Washington (DC)	631
Chicago (IL)	11,837	Chicago (IL)	13,466	Detroit (MI)	447	Newark (NJ)	536
Los Angeles (CA)	8,295	Los Angeles (CA)	9,341	New York (NY)	427	New York (NY)	505
Philadelphia (PA)	6,523	Philadelphia (PA)	7,024	Chicago (IL)	425	Baltimore (MD)	466
Detroit (MI)	4,595	Houston (TX)	5,343	Philadelphia (PA)	411	Chicago (IL)	465
Washington (DC)	4,506	Detroit (MI)	4,154	Atlanta (GA)	396	Philadelphia (PA)	463
Houston (TX)	4,104	Washington (DC)	3,612	Baltimore (MD)	389	Detroit (MI)	437
Baltimore (MD)	2,861	Baltimore (MD)	3,034	St. Louis (MO)	389	St. Louis (MO)	428
Dallas (TX)	2,635	Dallas (TX)	2,862	Newark (NJ)	368	Cleveland (OH)	381
Boston (MA)	2,053	Phoenix (AZ)	2,626	Boston (MA)	357	Boston (MA)	367
Phoenix (AZ)	1,949	San Francisco (CA)	2,227	Cleveland (OH)	348	Atlanta (GA)	354
Milwaukee (WI)	1,866	Las Vegas (NV)	2,168	Buffalo (NY)	315	New Orleans (LA)	343
San Diego (CA)	1,816	Boston (MA)	2,164	Pittsburgh (PA)	312	Milwaukee (WI)	335
Honolulu (HI)	1,781	San Diego (CA)	2,022	Miami (FL)	310	Buffalo (NY)	317
San Francisco (CA)	1,777	Milwaukee (WI)	1,998	Milwaukee (WI)	297	Cincinnati (OH)	311

SOURCE: Reaves and Hickman (2002).

12

equivalent rank was $51,265. Entry-level patrol officers' minimum beginning salary in all small, medium, and large cities was approximately $35,000 in fiscal year 2000.

Local police agencies in large cities have a number of specialized units that focus on specific crime-related areas and the delivery of services to persons and families. In both 1990 and 2000, over three fourths of the police departments had full-time specialized units assigned to child abuse units, domestic violence units, general crime prevention or burglary squads, gang resistance units, missing children bureaus, juvenile aid or juvenile crime units, victim assistance, bias-related crimes, repeat or habitual offenders, and drug education in the schools. For details on the percentage of police agencies that have full-time and/or part-time personnel assigned to special crime units, see Table 1.6.

State Law Enforcement

State police and state highway patrols each have several thousand sworn officers in their respective states. The largest state highway patrol agency is the California Highway Patrol, with approximately 6,000 officers, followed by the New York State Police, Illinois State Police, Indiana State Police, Pennsylvania State Police, Massachusetts State Police, and New Jersey State Police. Each of the latter six statewide police departments has several thousand sworn officers responsible for patrolling the interstate highways and providing police services and criminal investigations in rural areas of each state that do not have a local full-time police department.

Federal Law Enforcement

The FBI is responsible for enforcing over 250 federal laws and therefore has a wider jurisdiction than any of the remaining large federal law enforcement agencies: the Immigration and Naturalization Service (INS), the U.S. Customs Service, or the Federal Bureau of Prisons.

The FBI is the chief investigative agency of the U.S. Department of Justice. In general, it has jurisdiction over all federal crimes that are not the sole responsibility of a different federal enforcement agency. Since its formation in 1908, the FBI's jurisdiction has focused on kidnapping, crimes against banks, organized crime, aircraft piracy, violations of the Civil Rights Act, interstate gambling, and interstate flight to avoid prosecution. After the horrendous terrorist attacks of September 11, 2001, at the World Trade Center and the Pentagon, the FBI has made it a priority to investigate all tips on potential terrorist activities throughout the United States. In view of the numerous active terrorist threats surfacing during the spring of 2002, the FBI announced plans to hire 900 new FBI agents in order to mount a full-scale investigation of terrorism throughout the United States.

Issues in Law Enforcement

The section of this book entitled "Critical Issues in Law Enforcement" focuses on such questions as: Do international counterterrorism strategies seem to be effective in reducing terroristic acts? On the basis of a national evaluation, does community policing seem to work? Does COMPSTAT greatly improve the efficiency of police management and reduce crime? Do external control measures and early warning systems reduce incidents of police use of excessive force? Do mandatory arrest policies reduce domestic violence incidents significantly? And do thorough homicide investigations involving the collection of essential forensic evidence result in apprehending the murderer?

The study of law enforcement policies, counterterrorism policies, management practices, patrol functions, operational procedural guidelines, community policing, problem-solving policing strategies, COMPSTAT police management and crime, and crime-mapping technology is critically important to the efficiency of federal, state, and local criminal justice agencies. In addition to Chapter 9, by Joseph Ryan, on community policing, Part III of this volume includes Chapter 7, by Richard Ward and Sean Hill, dealing with international perspectives on counterterrorism; Chapter 8, by Vincent Henry, on current and futuristic perspectives on COMPSTAT, and its success in significantly reducing crime and improving the quality of life in New York City; Chapter 13, by Norman Cetuk, on homicide investigations and trends; and Chapter 10, by Mark Blumberg and Betsy

Table 1.5   Annual Operating Budget of Police Departments Serving Cities With a Population of 250,000 or More, 1990 and 2000

| Population | Annual Operating Budget | | | | | | | | | |
| | Per Agency | | Per Resident | | Per Employee | | Per Sworn Employee | |
Served	1990	2000	1990	2000	1990	2000	1990	2000
Total	**$176,134,761**	**$211,581,036**	**$242**	**$266**	**$64,493**	**$64,323**	**$83,814**	**$85,786**
1,000,000 or more	$586,494,149	$695,642,921	$272	$290	$64,024	$60,061	$82,450	$79,804
500,000–999,999	138,511,331	165,117,265	228	250	64,667	69,391	84,109	91,127
350,000–499,999	77,063,388	99,818,601	200	237	67,715	73,811	91,352	102,539
250,000–349,999	59,757,009	71,229,731	196	229	62,470	65,788	81,747	89,364

NOTE: All data are presented in 2000 dollars.
SOURCE: Reaves and Hickman (2002).

Table 1.6   Special Units Operated by Police Departments Serving Cities With a Population of 250,000 or More, 1990 and 2000

| | Percent of Agencies With | | | |
| | Full-Time Special Unit | | Full-Time Special Unit or Part-Time Personnel | |
Type of Special Unit	1990	2000	1990	2000
Victim assistance	32	47	45	71
Crime prevention	95	76	100	97
Repeat offenders	68	34	77	57
Prosecutor relations	66	31	76	58
Domestic violence	50	81	61	97
Child abuse	87	77	95	92
Missing children	89	66	95	95
Juvenile crime	81	68	94	84
Gangs	69	84	89	98
Drug education	90	73	98	95
Drunk drivers	56	40	76	81
Bias-related crimes	34	26	58	71

SOURCE: Reaves and Hickman (2002).

Wright Kreisel, on the various mechanisms, including civil litigation, criminal prosecution, and internal police departmental policies, used to control and lessen police officers' excessive use of force, avoid police misconduct, and attain the goal of police accountability. In Chapter 3, Mary Jackson focuses on a much overlooked area of law enforcement response, namely response to the activities of illegal street gangs, especially the Bloods, Crips, and Latin Kings, which continue to expand throughout the United States.

COURTS AND SENTENCING

Sentencing involves a difficult decision on the part of judges. It involves balancing justice for the community at large, the crime victim, and the defendant. Basically, the trial court judge is making a decision as to what sanctions or punishment should be imposed upon a person found guilty of a crime. In general, the criminal laws grant judges extensive discretionary powers in sentencing the offender. In most jurisdictions, judges have the leeway/discretion to sentence the offender to a wide range of options, such

as a long or short period of imprisonment, a suspended prison term and probation, probation and restitution, victim-offender mediation, and other community sanctions customized to the individual needs of the offender and the victim (e.g., home confinement with electronic monitoring, boot camp and shock incarceration, intensive drug treatment, and/or intensive probation supervision). Chapter 19 of this book, by Cassia C. Spohn, provides a detailed discussion of the critical issues related to sentencing reforms, court processing decisions, sentencing disparity, and the responsibilities of criminal court judges.

The overwhelming majority of adult offenders are given a sentence to be served in the community, usually probation. Approximately 4 million adults are currently on probation or parole in the United States. Eighty percent of offenders on probation are men, and a disproportionate segment are minorities. Probationers are better educated than incarcerated felons, with more than half (58%) having a high school diploma and 17.7% having a year or more of college (BJS, 1997). For detailed information on the strengths and weaknesses of probation programs and services, see Chapter 21 of this book.

Table 1.7   Selected Prison Statistics

Dec. 31	Number of Inmates		Sentenced Prisoners per 100,000 Resident Population		Population Housed as % of Highest Capacity	
	Federal	State	Federal	State	Federal	State
1990	65,526	708,379	20	272	—	115%
1995	100,250	1,025,624	32	379	126%	114
1999	135,246	1,228,455	42	434	132	101
2000	145,416	1,236,476	45	432	131	100

—, Data not available

SOURCE: Bureau of Justice Statistics (2001b).

Sentencing decisions weigh heavily on the length of prison time served by offenders as well as parole decisions. Eligibility for parole depends on the requirements established by the criminal code and on the sentence given by a judge. In approximately half the states, the actual release date of each offender is based on the parole board's discretion. In the other states that have determinate sentences or strict parole guidelines, the inmate's release is mandatory once the required length of sentence is served. Most state parole boards make decisions based on statutory criteria, individual sentences, level of participation in inmate rehabilitation programs, and a prison record of no disciplinary infractions.

Although there is wide variation among state prison systems with regard to the average time served by first-time inmates for selected offenses, the average time served prior to release is slightly over 2 years. For example, a burglary offender may be given 36 to 48 months and serve 67% of his minimum term before being released after 24 months. By contrast, a convicted murderer may be sentenced from 8 to 20 years, or a maximum of 240 months. However, the average convicted murderer spends 127 months, or 10 years and 6 months, incarcerated in a state prison. If we examine the most common felony convictions, the average sentence for first-time incarcerated offenders is 17 months for illegal possession of controlled substances, 17 months for larceny, and 21 months for drug trafficking (BJS, 1997).

Some of the other most critical issues relate to probation failures and successes, the death penalty, and capital punishment. Within Part V of this book, "Critical Issues in the Criminal Court System," some of the issues discussed include: Have excessive probation caseloads led to the failure of probation? With the implementation of new technology such as kiosks, are probation agencies documenting successful outcomes? Is it humane to execute a poor, black, mentally ill person? Does the death penalty help family members of murder victims to obtain closure and peace of mind? How likely is it that innocent persons are executed because of mistaken eyewitness testimony at their murder trial? In Chapter 20, Beau Breslin and David Karp discuss and debate the above-mentioned critical issues related to capital punishment. Chapter 21, by Michael Jacobson and Charles Lindner, focuses on the failures of probation and on the currently operating promising programs.

AMERICAN CORRECTIONS

The American correctional system is the most populated and overcrowded prison system in the world. By the end of 2000, the total number of prisoners incarcerated in federal and state correctional facilities and penitentiaries had increased dramatically to approximately 1.3 million–an 80% increase above the number imprisoned during the 1990s. In 1990, the total number of prisoners confined in federal or state correctional facilities had reached a record of 773,905, and by 1991 it had climbed slowly to 823,414 inmates (BJS, 1992, p. 8) (see Table 1.4).

As noted by Michael Welch in the first edition of this book, the majority of inmates in most county and city jails are pretrial detainees. Most of these pretrial detainees are too poor to make bail. Jails are populated by a disproportionate number of black, Hispanic, poor, uneducated, and unemployed men. More than half have drug and alcohol problems.

To assist parolees in the transition from prison life to re-entry into the community, reintegration programs have been established throughout the United States. These include work-release programs, prerelease guidance centers, and halfway houses. One of the most successful halfway house programs is San Francisco's Delancey Street, which prepares ex-offenders for jobs in the moving, restaurant, and florist businesses. The annual cost of incarcerating over 1.3 million individuals in our prisons and penitentiaries is estimated at $20 billion. Prisons are expensive to build and operate. As states and counties have tried to handle the growing prison populations, prison budget allocations have also grown. During the past decade, state legislatures have appropriated over $30 billion in capital spending to build new prisons and additional units. The annual operating and capital expenditures budget for adult and juvenile correctional institutions is $16.06 billion for adult corrections and $1.74 billion for juvenile corrections (American Correctional Association, 1993).

The primary purpose of most prisons for the past 100 years has been to change the offender into a law-abiding citizen. Positive behavior change is difficult for convicted felons who have failed in public schools, failed in work, and failed at crime by being caught, convicted, and incarcerated. In Chapter 25 of this book, Mary Bosworth examines education and work programs and drug treatment programs currently operational in two large state correctional systems, California and New York State, as well as throughout the Federal Bureau of Prisons. As Bosworth points out, over 600,000 men and women inmates return to the community each year. Expanding the academic education, vocational training, and work-release programs currently available only on a limited scale is crucial for preparing inmates to lead productive, law-abiding lives.

THE VICTIMS' MOVEMENT AND VICTIM ASSISTANCE

There are two major types of local programs designed to aid crime victims and witnesses. The first type is under the auspices of a city or county prosecutor and is located in a prosecutor's office, a county office building, or the local courthouse. The major goal of these victim/witness assistance programs is to alleviate the stress and trauma for victims and witnesses who testify in court. For example, prior to the court date, program staff accompany the victim to an empty courtroom to orient the individual to the physical layout and the courtroom procedures. Other services may include transportation to court, child care while the victim or witness is appearing at the court, apprising the victim of the progress of the court case, victim advocacy during a criminal trial, and referral to social service agencies.

The second major type of program operates under the auspices of a nonprofit social service agency, a city or county police department, or a probation department. These types of programs often provide victims with concrete services (e.g., emergency food vouchers) as well as crisis intervention at the crime scene, in the program office, or in the person's home. Victim service programs sometimes provide a range of additional services such as accompaniment to and advocacy in court, repair or replacement of broken locks, and emergency financial aid (Roberts, 1990).

In a growing number of communities, victim services and witness assistance programs have expanded to meet the special needs of child, adult, and elderly crime victims and their families. In some communities throughout the United States, victim services have been expanded to include crisis intervention, support groups, emergency food vouchers and financial aid, shelter and transitional housing for battered women, lock repair and replacement, child care for witnesses' children while they testify in court, victim advocacy in the courtroom, home visits, short-term therapy, relocation assistance to transitional housing, and intervention with witnesses' employers. However, a number of cities and towns still do not have a fully staffed and comprehensive victim assistance program (Roberts, 1997).

The victims' movement has made slow and steady progress since 1975, when a federal agency, the Law Enforcement Assistance Administration (LEAA), allocated funding for just 14 projects geared to helping crime victims and witnesses. With the demise of LEAA in 1981, many cities and counties did not have the funds to continue victim assistance projects. Therefore, the majority of the prosecutor-based witness assistance programs ended in the late 1970s and early 1980s. However, in the mid-1980s and mid-1990s, three important events led to increases in funding to help crime victims:

1. The passage of the Victims of Crime Act (VOCA) of 1984, funded by federal criminal penalties. Federal VOCA funding has been distributed to the states to fund prosecutor-based witness assistance programs, sexual assault programs, and battered women's shelters. By 1992, VOCA was funding close to 2,500 victim-oriented programs.

2. The enactment of legislation in 28 states to fund new programs to aid crime victims and witnesses (Roberts, 1990). The majority of these states fund victim services through penalty assessments and fines.

3. The passage of the Violence Against Women Act (VAWA I) in 1994 with $1.2 billion of federal funding and the passage of VAWA II in 2000 with $3.3 billion in federal funding earmarked for the next 5 years for domestic violence training, programs, and services (Roberts, 2002).

Between 1984 and 2002, the U.S. Department of Justice's Office for Victims of Crime and state attorney generals have allocated several billion dollars to aid crime victims and witnesses. Victim/witness assistance and domestic violence programs are now a major growth industry.

The criminalization of woman battering and the implementation of policies mandating police arrest of batterers have improved markedly in the past decade. The terms *spousal* and *partner abuse* refer to intentional abuse by adult men or women of their intimate partners by methods that cause bruises, scratches, cuts, bleeding, injuries, pain, or suffering. Two significant events led to widespread social and legal reforms in the area of domestic violence. The first was the 1984 Minneapolis Police Experimental Study, which indicated that arresting batterers seemed to deter further incidents of family violence. Five replication studies in different regions of the United States also indicated that arresting batterers who were employed at the time of the police complaint usually led to a decrease in battering incidents at follow-up. The second influential event was the landmark Supreme Court decision in *Thurman v. City of Torrington* (1984), which held the police liable for their negligence in failing to protect Tracy Thurman from severe and repeated injuries from her husband. Tracy Thurman was awarded $1.9 million in damages from the city of Torrington. Other major lawsuits were filed during the 1980s against city and town police departments for failing to protect battered women victims. As a direct result of the above two events, police departments throughout the United States passed mandatory arrest policies as well as probable-cause arrest policies. In addition, both recruit and in-service police training on domestic violence rapidly expanded in county and city police training academies nationwide. For a detailed discussion of the criminal justice system's response to woman battering and the evolution of change to batterers' accountability, see Lisa Frisch's chapter in this book (Chapter 11). Also, for a qualitative pilot study on battered women's perceptions of the usefulness of police arrest policies as well as temporary restraining orders, see Gina Pisano-Robertiello's chapter in this book (Chapter 12).

When compared to the almost complete lack of support services given to victims during the 1960s and 1970s, victim services have certainly improved. However, there is still a way to go. Only about 200 police departments have either 24-hour crisis intervention units or victim service programs. One major stumbling block to creating comprehensive 24-hour victim assistance programs is the shortage of trained forensic mental health specialists, family therapists, and victim advocates willing to work at night or on weekends, when most violent victimizations take place. Finally, whereas most state prisons employ an average of 350 correctional officers, most prosecutor-based witness assistance programs and police-based domestic violence units have only two to four full-time staff. Each victim or witness assistance program has the potential to serve thousands of violent crime victims in its area, but most states and counties refuse to allocate the funds for critically needed new staff positions to treat crime victims.

CHARACTERISTICS OF OFFENDERS

Michael Welch, in his chapter in the first edition of this book, characterized the inmates in our city jails as the growing underclass: a group of people who have the misfortune of populating our jails as pretrial detainees because they do not have the money to post bail. Cole (1992) estimated that about 5 million of the 33 million Americans with incomes below the official poverty line are members of this new group. This group tends to be overrepresented by minorities: African Americans and Hispanics. Being part of this underclass is more a way of life than a social or economic condition. Their behavior patterns typically include antisocial behavior, habitual criminality, chemical dependency, out-of-wedlock births, very erratic work histories and long periods of unemployment, welfare dependency, school failure, and illiteracy. For the most part, this group lacks hope for leading a productive life. Their lives have been filled with pain and despair as they see their friends and relatives incarcerated, addicted to drugs, dealing in drugs, sick or dying from AIDS, or victims of brutal crimes. The inner-city youths that get caught up in the criminal justice system are usually products of decaying urban neighborhoods and families where the primary role model is a parent or older sibling who is either dealing drugs or serving time in a state prison for a felony offense.

Almost 30 years ago, the author identified and discussed the five most prevalent handicaps among prison inmates: (a) character disorder: an antisocial defect resulting from undersocialization and inappropriate acting-out responses to the stresses of daily living; (b) unemployability:

lack of motivation to work as well as marketable vocational skills; (c) relationship hangups: lack of close interpersonal ties with family and friends; (d) social stigma: being labeled a felon; and (e) immaturity: lack of the capability to take responsibility for one's actions and a general inability to make socially acceptable decisions (Roberts, 1974, p. 6). Many of the above handicaps characterize today's juvenile and adult offenders. However, with the phenomenal increase in drug abuse and mental illness during the 1990s, incarcerated offenders have become increasingly prone to violence when they fail to take their neuroleptic medications.

Some of the most critical concerns among criminal justice professionals in the 1990s are what to do with young murderers, the proliferation of youth gangs, the chronic use of drugs by juvenile and adult offenders, and the increased spread of infectious and deadly diseases (i.e., AIDS and tuberculosis). The rates of AIDS and drug abuse in the criminal justice system continue to rise each year. Chapter 15, by Kathleen Heide, examines the etiology and biopsychosocial characteristics of youth homicide, as well as effective treatment approaches with young killers. Chapter 3, by Mary Jackson, reviews the growth in youth gangs and the emerging role for police gang suppression units. As increased numbers of sex offenders are sentenced to probation and other community-based alternatives, the threat of transmission of the HIV virus to victims of sexual assault intensifies. In Chapter 6, Lori Scott reports on the prevalence of reported and investigated sex crimes over the last decade and the mostly inadequate management of these crimes by the criminal justice system to date. Scott also describes several community-based sex offender treatment programs in Arizona, Connecticut, and California.

Most convicted offenders do not possess the academic or vocational skills for legitimate employment. Unfortunately, with only limited opportunities to earn money legally, they turn to illegal activity at an early age. It is extremely difficult to convince youthful offenders that it is better in the long run to stay in school and prepare for a law-abiding career when they see their peers making $1,000 a week trafficking in illegal drugs. Their future is dismal, many die before they reach the age of 30, and many more spend their most productive years in jail or state prison. Chapter 27, by Joan Petersilia, focuses on the political, economic, and social realities of recently released prisoners and their need for reintegrative, transitional, and aftercare services.

The majority of crimes are committed by repeat offenders. Generally, these individuals have no respect for others and are unable to maintain normal relationships of love, respect, and trust with a significant other. Recent surveys indicate that only one third of the inmates in state prisons have completed high school and that 20% dropped out in the eighth grade or earlier.

The profile of the typical offender is that of a young black or Hispanic male who has grown up in poverty and has not received the nurturance of a close family. His values and morals are typically learned from television or the street, and he tends to be a drug abuser or heavy drinker. His career in crime usually begins at an early age. He is intermittently involved in antisocial and delinquent acts and shows no remorse.

REMEDIES

Basically, criminal justice practitioners and policy makers hold one of three philosophies with regard to the criminal law, crime, and offenders: (a) punishment, (b) rehabilitation and treatment, or (c) a combination of punishment and rehabilitation.

As a result of tougher and determinate sentencing laws administered by judges, primarily in response to public outcries, the number of inmates under the jurisdiction of federal or state correctional agencies climbed to a record high of 1.3 million by 2000. Examining the trend in the offense composition of state prison admissions reveals a significant increase in offenders convicted of drug offenses and offenders violating probation or parole conditions. In 1980, drug offenses accounted for only 1 out of every 15 court-committed admissions to state correctional facilities. However, by 1990, approximately 1 in 3 newly sentenced offenders had been convicted of drug offenses (Gilliard, 1993). During the same 12-year period, probation and parole violators committed by the courts to state prisons increased from just under 17% to approximately 30% of the total admissions.

Jails and prisons are human warehouses for the new underclass: the thousands of drug-abusing, unemployed, and repeat offenders. Prisons are overpopulated, and recidivism rates are high. We have learned that building more prisons is not the solution. Many of our nation's prisons have become schools for crime, places where offenders become more incorrigible and violent as a result of the unsanitary, degrading, and inhumane conditions in many prisons.

By 2000 there were approximately 1.3 million inmates in our nation's state and federal correctional facilities. How do they typically spend their time each day? Usually in one or two ways: (a) "hanging out" in the prison yard or recreation area or locked in their cells or (b) working a full day in prison industries that provide skills in dead-end jobs like making license plates for the state or desks and file cabinets for state offices. Productivity standards are comparable to those of manufacturers on the outside. The inmate workers receive token amounts of pay below minimum wage. They are required to turn over their wages to the state or federal prison in order to provide monetary restitution to their victims and to send payments to the inmates' dependents. A small sum of money is put aside each week in a "going home" savings account in preparation for the inmates' release.

Putting inmates to work in labor-intensive manufacturing and production jobs is a realistic alternative to inmate idleness and potential prison riots. Several different organizational models of prison industries had emerged on a selective basis in the 1990s:

1. The state government model, in which prison industries produce products solely for state and local governmental agencies' use, such as desks, chairs, and file cabinets or laundry for the state hospitals.

2. The joint venture model, in which prison industries contract with a private corporation to produce products that the corporation will market and distribute.

3. The corporate model, similar to that of a private sector business, in which inmates are hired,

trained, and transported daily (work-release) to a manufacturing facility near the prison.

In the first edition of this book, Professors McNally and Dwyer provided a review of new and expanded prison industries during the first half of the 1990s. Their chapter described inmates' work assignments at Zephyr Products, Inc., in Kansas and discussed the past, present, and future of prison industry alliances with the private sector. McNally and Dwyer documented the development of prison industry partnership programs with private sector companies on a limited basis in 23 states. They concluded by enumerating the specific benefits of prison industry enhancement programs to taxpayers, corporations, prisons, labor, and the offenders themselves. In sharp contrast to federal correctional institutions, where the overwhelming majority of inmates work in prison industries (Unicor), the majority of inmates in state prisons are locked in their cells and taking naps, "hanging out" in the prison yards, or "pumping iron" in weightlifting rooms. Back in 1994, it was my earnest hope that correctional administrators would follow the recommendations of McNally and Dwyer. Unfortunately, as documented in Bosworth's chapter in this second edition, correctional administrators have done very little in the past 9 years to improve state prison industry partnerships with the private sector.

The remedies that need to be expanded in the 21st century include juvenile structured wilderness programs, juvenile family counseling programs, boot camps for juveniles, inmate transitional and aftercare programs, prison industry-private sector partnership programs, and intermediate sanctions such as intensive probation and parole supervision, electronic monitoring, community residential centers, community service, fines, victim-offender mediation, and restitution. See Chapter 14, by Albert Roberts, on juvenile justice policies and programs, Chapter 18, by Gaylene Armstrong,

Doris Layton MacKenzie, and David Wilson, on the effectiveness of different boot camp models nationwide, and Chapter 17, by Scott K. Okamoto and Meda Chesney-Lind, on the critical issues regarding female juvenile delinquency.

Conclusion

Victimization rates for violent crimes have declined significantly from 1993 to 2001. A growing number of large cities seem to be safer, cleaner, and much less crime ridden.

Several surveys confirm that a geographically and demographically representative sample of citizens in Pennsylvania, Delaware, and Alabama firmly support alternatives to prison sentences. The three preferred sanctions for nonviolent offenders are boot camps, carefully monitored probation plus restitution, and intensive probation supervision (IPS) plus community service. The public is beginning to realize that mandating the offender to a work assignment or to giving money back to the victim is less expensive than prison and more likely to rehabilitate the offender (Farkas, 1993, pp. 1, 15).

The underlying philosophy rooted in all intermediate sanctions is that convicted offenders can be punished fairly, consistently, and humanely in the community. Offenders in IPS are required to maintain employment or attend school, abide by a strict curfew, be available for routine drug testing, and provide community service or restitution to victims. Offenders committing infractions are quickly punished with additional fines, restitution, or incarceration.

The main problem is that without adequate funding to hire needed staff, these programs can become the failures of the 21st century. With adequate planning, program development, and increased monitoring and accounting, intensive probation surveillance programs and restitution programs for certain types of offenders could save states and counties billions of dollars.

REFERENCES

American Correctional Association. (1993). *ACA directory of juvenile and adult correctional departments, institutions, agencies and paroling authorities.* Laurel, MD: Author.

Bureau of Justice Statistics. (1997). *Crime victimization 1996.* Washington, DC: U.S. Department of Justice.

Bureau of Justice Statistics. (1998). *National crime surveys: National sample, 1973–1983* (6th ICPSR ed.) [Electronic data file]. Conducted by U.S. Bureau of the Census. Ann Arbor, MI: Inter-University Consortium for Political and Social Research [Producer and Distributor].

Bureau of Justice Statistics. (2001). *Crime victimization 2000: Changes 1999-2000 with trends 1993-2000.* Washington, DC: U.S. Department of Justice.

Bureau of Justice Statistics. (2001b, August). *Prisoners in 2000* (Bureau of Justice Statistics Bulletin, NCJ 188207). Washington, DC: U.S. Department of Justice, Office of Justice Programs.

Cole, G. (1992). *The American system of criminal justice.* Pacific Grove, CA: Brooks/Cole.

Farkas, S. (1993, April). Pennsylvanians prefer alternatives to prison. *Overcrowded Times, 4*(1), 13-15.

Federal Bureau of Investigation. (1996). *Crime in the United States, 1995* (Uniform Crime Reports). Washington, DC: Government Printing Office.

Federal Bureau of Investigation. (1997). *Crime in the United States, 1996* (Uniform Crime Reports). Washington, DC: Government Printing Office.

Federal Bureau of Investigation. (1998). *Crime in the United States, 1997* (Uniform Crime Reports). Washington, DC: Government Printing Office.

Federal Bureau of Investigation. (1999). *Crime in the United States, 1998* (Uniform Crime Reports). Washington, DC: Government Printing Office.

Federal Bureau of Investigation. (2000). *Crime in the United States, 1999* (Uniform Crime Reports). Washington, DC: Government Printing Office.

Federal Bureau of Investigation. (2001). *Crime in the United States, 2000* (Uniform Crime Reports). Washington, DC: Government Printing Office.

Gilliard, D. K. (1993). *Prisons in 1992* (Bureau of Justice Statistics Bulletin). Washington, DC: U.S. Department of Justice.

Miller, T. R., Cohen, M. A., & Wiersema, B. (1996). *Victim costs and consequences: A new look.* Washington, DC: U.S. Department of Justice.

Police Departments in Large Cities, 1990-2000, Bureau of Justice Statistics, Special Report, U.S. Department of Justice, Office of Justice Programs.

Reaves, B. A., & Hickman, M. J. (2002). *Police departments in large cities, 1990-2000* (BJS Special Report). Washington, DC: U.S. Department of Justice, Office of Justice Programs.

Roberts, A. R. (1974). *Correctional treatment of the offender.* Springfield, IL: Charles C Thomas.

Roberts, A. R. (1990). *Helping crime victims.* Newbury Park, CA: Sage.

Roberts, A. R. (1997). The role of the social worker in victim/witness assistance programs. In A. R. Roberts (Ed.), *Social work in juvenile and criminal justice settings.* Springfield, IL: Charles C Thomas.

Roberts, A. R. (1999). Victims of violence. In R. Gottesman (Ed.), *Violence in America: An encyclopedia* (Vol. 3, pp. 366-372). New York: Charles Scribner's Sons.

Roberts, A. R. (Ed.). (2002). *Handbook of domestic violence intervention strategies.* New York: Oxford University Press.

Thurman v. City of Torrington, 595 F. Supp. 1521 (1984).

Victims of Crime Act, 18 U.S.C. §§ 3013, 3681, 3682. 42 U.S.C. §§ 10601 to 10603, 10603a, 10603b, 10604, 10605 (1984).

Violence Against Women Act, Pub. L. No. 103-322, 1994, 108 Stat. 1902 (1994).

Violence Against Women Act, Pub. L. No. 106-386, 108 Stat. 1905–1995 (2000).

2

Case Exemplar on Criminal Justice Processing: Illegal Possession of a Weapon, Domestic Violence, and Murder

Donald J. Sears

The criminal justice system is grounded in a number of fundamental principles that guarantee to all citizens certain rights that cannot be denied. It is these rights that our criminal justice system seeks to preserve and protect.

Whether these rights are protected is a function of how the criminal justice system operates. To fully understand and appreciate the dynamics of the system, a detailed look at how one case is processed is most appropriate. What follows, therefore, is a step-by-step walk through the criminal justice process, from arrest to parole.

Saturday night, March 23, a report of gunshots is received by the dispatch officer in Precinct #505. She broadcasts the report to officers on patrol in the area, who approach the scene of the call with caution. As they arrive, they see and hear no indication that a crime has been committed. The ranking officer on the scene directs the others as they take up positions around the house.

After several knocks at the door, one of the officers manages a glimpse through a partially opened bedroom window. He sees the body of a black female, lying face down on the floor. No movement is detected. Nor does there appear to be anyone else in the house. Another officer radios that he has found the back door open.

Gathering this information, the ranking officer on the scene determines that probable cause exists to enter the house, and he orders that the officers enter the house to investigate.

Inside, they find signs of an argument. Overturned furniture and broken china are strewn throughout the living room. Attending to the woman on the floor, they find three gunshot wounds to the chest. There is no sign of life. She has apparently been killed with a small-caliber handgun, fired at close range. Inside the drawer of a nightstand they find ammunition for a .38 caliber revolver. This is taken as evidence. The murder weapon, however, is not recovered.

A further search of the house reveals the woman's handbag, found in the hall closet. Police find her driver's license inside and identify her as Joanne Brown. They also find a photograph of her and a black male inside the purse. Detectives are called to the scene, as is the medical examiner. As the coroner's officers photograph the body and document the crime scene, detectives begin to canvass the area, questioning neighbors about Joanne. Many indicate that she was separated from her husband and was seeking a divorce. Her husband, Rodney Brown, had left the home approximately 10 days ago, after Joanne had filed some type of legal papers with the family court. They positively identify Rodney as the man in the photograph with Joanne. Although several of the neighbors heard the gunshots, none saw anyone run from the scene.

At this point, homicide detectives believe that Rodney Brown is the prime suspect in the killing of Joanne. Pay stubs found in a desk drawer of the house indicate that Rodney works as a meatpacker in a downtown meatpacking plant. The following morning, detectives speak to the foreman of the plant. He states that Rodney has not shown up for work for the last several days. The personnel manager tells police that Rodney filled out a change of address card about 1 week ago and that his paychecks are now mailed to an apartment on West Clover Avenue.

An arrest warrant is drawn up for Rodney, charging him with the murder of Joanne. Police also contact a judge in order to secure a search warrant for the apartment on West Clover. In a detailed affidavit, homicide detectives outline the information that has been gathered so far and assert that this leads to the conclusion that there is probable cause to believe that Rodney killed Joanne. In response to the judge's questions, the detectives assert that Rodney may be hiding out in his apartment and may still be armed with the murder weapon. In light of all of the circumstances presented to her, the judge agrees that there is probable cause to believe that Rodney is the murderer. She also agrees

that there is probable cause to believe that evidence of the crime may be found in the West Clover apartment. As a result, she signs the search warrant.

Several hours later, police surround the apartment building on West Clover. Detectives knock on the door of Rodney's apartment, only to find that he is not home. They contact the manager, who unlocks the apartment door after seeing the search warrant. Police scour the apartment, looking for evidence of the crime. Inside a bedroom closet, they find a fully loaded .38 caliber handgun, which is immediately seized as evidence.

Believing that Rodney will eventually return to his apartment, police set up a surveillance of the building. At approximately 11:30 A.M., they see Rodney get out of a taxicab in front of the building. Not wanting a confrontation on the street, which would provide more opportunity for Rodney to escape, they allow him to enter the building and arrive at his apartment door. Just as he unlocks the door, police move in. They seize him and immediately place him under arrest. He is handcuffed and forced to sit at his own dining room table until a patrol car arrives to transport him to the precinct house. One of the detectives asks why he killed Joanne. Rodney replies that he never meant to, the gun just went off while they were struggling.

Rodney is placed into the rear of the patrol car and transported to the station. On the way, he is advised that he has the right to remain silent and the right to an attorney. He has the right to refuse to answer questions, and if he does answer questions, he can stop answering at any time. He is also told that if he does make any statements, they can be used against him. When asked at the station if he wants to make a confession, he refuses.

Rodney is taken to the booking room, where he is told to empty his pockets. The booking officer takes an inventory of Rodney's belongings and places them in a locker for safekeeping. An arrest report is filled out, recording information such as name, address, telephone number, date of birth, height, weight, descriptive markings (e.g., scars, tattoos), nicknames, and social security number. The crime charged, name of the victim, date of arrest, and police file number are also recorded on this form. At first, Rodney does not want to tell police this information. When they tell him that he cannot refuse, he reluctantly furnishes these details. His photograph is taken, first from the front and then from the side, with the police file number recorded on the photographs. He is then fingerprinted.

The officer first takes prints of each individual finger, rolling each one from side to side. He then prints the four fingers of each hand as a group. Next, the thumbs are printed without rolling—each is simply pressed flat. Finally, the officer takes the prints of Rodney's palms. This process is repeated four times because the local department retains one set of prints, the federal government requires one set, and the state government must have two. Rodney is allowed to wash his hands and is then placed in an interview room to await the arrival of a detective.

Before attempting to take a statement from Rodney, the detective in charge of the investigation types out an investigation report detailing all of the events leading up to Rodney's arrest. After approximately 1 hour, he is ready to take a statement from Rodney. He advises Rodney again of his constitutional rights and asks him if he would like to talk about the incident. Rodney indicates that he first wants to speak to a lawyer and asks the detective to contact the public defender's office for him. The officer refuses, saying that he is not required to provide an attorney. Nor will he tell Rodney the phone number for the public defender's office.

Not wanting to speak with the detective without first talking to a lawyer, Rodney refuses to give a statement. He is placed in a holding cell while the desk officer contacts a judge to determine bail. The judge considers the crime charged, Rodney's family support and ties to the community (how long he has lived in the area, employment history, etc.), and any prior convictions that exist. He is trying to determine if Rodney poses a danger to society and if there is a risk that he will flee the jurisdiction if released. After considering all of these factors, the judge sets bail at $100,000. Because Rodney cannot afford this and there are no family members he can turn to for help, he is remanded to the county jail until his initial appearance on Monday.

Police transport him to the county facility that evening, where his clothes are taken away and he is given a uniform and assigned to a cell. The jail is very crowded and noisy. The guards seem oblivious to the noise and the uncomfortable conditions and indifferent to Rodney's well-being. He must share a cell with five other inmates, all of whom are awaiting trial for committing violent crimes. He is told that he has just missed dinner and thus must wait until breakfast for something to eat. He is also told that he will be taken to the county courthouse in the morning for his initial appearance. This will require him to get up at 5:00 A.M. so that he can eat breakfast, be shackled to other prisoners, be placed in a transport van, and arrive at the courthouse by 8:30 A.M.

On Monday, March 25, Cynthia Durham, an attorney with the county public defender's office, picks up her caseload for the day. It consists of 17 new cases, representing just a portion of those persons arrested over the weekend. One of these new cases is *State v. Rodney Brown*. She glances through the police reports as she walks through the hallway to the criminal courtroom. The judge is already on the bench, calling the list for the day. He will take care of the initial appearances first. In the jury box sit all of the prisoners brought from the county jail, including Rodney. When his name is called, he is instructed to stand up.

Cynthia tells the judge that she has been assigned to the case. The judge in turn tells Rodney that Ms. Durham will be his lawyer. He then reads the charges to Rodney. They include murder, unlawful possession of a weapon, and possession of a weapon for an unlawful purpose. Cynthia enters a not-guilty plea on behalf of Rodney and asks the judge to reduce the bail, arguing that Rodney is not likely to flee. The prosecutor opposes her request and advises the judge that this matter will be presented to the grand jury within the next 10 days. The judge denies Cynthia's request and sends Rodney back to the county jail, where he will stay for the next 3 weeks without hearing anything from his attorney or the court.

Shortly after Rodney's initial appearance, the assistant county prosecutor assigned to handle Rodney's case schedules a time to present the case to the grand jury. He subpoenas the homicide detective who had the main responsibility in the investigation and arrest of Rodney. No other witnesses are necessary.

On April 8, the grand jury considers the case of *State v. Rodney Brown.* This group of 23 citizens from the community listens as the homicide detective describes the details of the investigation. He is allowed to tell the jurors what other people said and did because hearsay evidence is allowed at the grand jury hearing. The prosecutor is also allowed to ask leading questions in an effort to make the process move more quickly. The defendant is not present, nor is his attorney. Indeed, no evidence as to the defendant's version is given to the grand jury.

After the detective testifies, the prosecutor tells the members of the grand jury what the law is regarding the crimes that have been charged. He also says that they can indict the defendant for any lesser included offenses or originate new charges that they feel are warranted. He then dismisses them to their deliberation room, where they consider whether there is probable cause to indict the defendant for the crimes charged.

After 2 hours of deliberation, the grand jury determines that there is sufficient evidence upon which to base an indictment, and they return a three-count indictment against Rodney Brown, charging him with murder in the first degree, unlawful possession of a weapon, and possession of a weapon for an unlawful purpose. The indictment is delivered to the prosecutor, where it will remain secret for 2 days. This waiting period is designed to allow the prosecutor time to arrest the defendant if he is free on bail. Because Rodney is still in jail, this is of no consequence. Eventually the indictment is released to the public, and the court schedules an arraignment date.

On April 18, Rodney is brought back to court for his arraignment. Once again, he sits in the jury box with a number of other inmates from the jail. He has not seen or spoken to his attorney at all. When his name is called, he stands up, and his attorney approaches counsel table. The judge states that the grand jury has indicted him and describes the crimes charged and the penalties he is exposed to, which include the death penalty if he is convicted of first-degree murder. Cynthia advises the judge that Rodney is pleading not guilty to all charges and once again asks the judge to lower bail. This request is denied.

The prosecutor provides Cynthia with discovery in the case, consisting of all of the police reports, the autopsy report, and a list of all of the witnesses that the state may call at trial. The prosecutor also advises the judge that he will make a decision within the next 10 days as to whether he will seek the death penalty in this case. The judge then signs an order to that effect, which also provides that any defense motions that Cynthia wishes to make on behalf of her client must be filed within 30 days. Finally, the judge schedules a plea-bargaining session, called a pretrial conference, for June 5. Without speaking to his lawyer at all, Rodney is taken back to the courthouse holding cell, where he remains until transported back to the jail later that day.

One week after the arraignment, Cynthia visits Rodney at the jail. This is the first time the two actually meet to speak about the case. They are placed in an attorney interview room where they can speak in private, although a guard waits right outside the door.

Rodney tells her his version of what occurred between him and Joanne on March 23. Their marital problems had been ongoing for several months, and Joanne filed a domestic violence complaint against him with the family court. Without hearing Rodney's side of the story, the family court judge issued a restraining order against Rodney, forbidding him to stay in their home. He stayed several nights with friends but then needed to get clean clothes and some personal belongings. Although he tried to get the police to accompany him to the house (which was a requirement of the restraining order), the local officers told him they were too busy and could not spare the manpower. He thereafter took it upon himself to go to the house alone.

When he got there, Joanne at first let him in. They began to talk, and he asked why she was doing this to him. An argument erupted, and the two began to throw things at each other. Joanne ran into the bedroom and tried to get the revolver that they kept in the nightstand. He grabbed it, and the two struggled for the weapon. It eventually went off, killing Joanne. Frightened, Rodney fled out the back door with the gun.

Although Cynthia listens intently, she is somewhat skeptical about Rodney's story.

She already knows that the domestic violence complaint was signed by Joanne because Rodney had a history of violent behavior. Three prior domestic violence complaints were signed against him, and he was even convicted of criminal assault on one of those occasions. In addition, the autopsy report showed three bullet wounds to the chest with no powder burns on the clothing. This told Cynthia that the gun had been fired repeatedly from at least 5 feet away and thus could not have gone off accidentally while the two were struggling for it. Although she confronts Rodney with these discrepancies, he maintains his version of the incident.

When she returns to her office, Cynthia drafts up a number of motions for filing with the court. These include various suppression motions (designed to prevent the state from using or referring to any evidence seized as a result of an illegal search), a *Miranda* motion (designed to prevent the state from using or referring to any statements made by Rodney after he was placed in custody), and a motion to prevent the application of the death penalty in this case. The latter is rendered unnecessary because, in her mail, she finds a letter from the prosecutor indicating that he will not seek the death penalty in this case. Finally, she asks her investigator to interview Rodney's neighbors to see if anyone saw or heard anything that might help in his defense.

On June 5, the date of his pretrial conference, Rodney is brought to the courthouse early. Cynthia meets with him in the holding cell of the courthouse to discuss the case. Her investigator has found nothing that will help. Cynthia's plan for his defense is that it was not an intentional act but rather an accidental shooting. In this way, Rodney would be guilty of the lesser included offense of manslaughter rather than murder. Although he would still be sentenced to prison, it would be for much less than the life sentence he is facing for first-degree murder.

She explains to Rodney that there are some problems with pursuing this line of defense. Because no other witnesses have been found to corroborate Rodney's version of a struggle, Rodney is the only witness who can tell the jury of the fight with Joanne. If he testifies, however, the prosecutor can tell the jury about his prior conviction for assault as well as his history of

domestic violence. If he never takes the stand to testify, this information cannot be brought out. Thus there is a danger in testifying in his own behalf. She also emphasizes again that the physical evidence does not support his version of the killing. Nevertheless, it is the only defense that Rodney has.

When court begins, the prosecutor presents the plea bargain that he is willing to offer to Rodney. He will amend the charge from first-degree to second-degree murder in exchange for a guilty plea. The weapons charges will thereafter be "merged" into the murder conviction, and Rodney will be sentenced only on the murder charge. In addition, the prosecutor will agree not to seek an enhanced prison term. Because Rodney has a prior record, the prosecutor could seek to increase the mandatory term of incarceration that Rodney is facing. If he accepts the plea offer, the prosecutor will not ask for this.

Cynthia discusses this with Rodney, who tells her that he cannot agree to any time in prison. If he has to go to prison, he would rather take his chances with a jury. Cynthia rejects the offer on behalf of Rodney, and the judge sets out a scheduling order for the trial. Because there are a number of pretrial motions to be decided, he schedules a pretrial hearing for July 17. The trial will take place on August 21.

On July 17, the prosecutor is ready to proceed with the pretrial motions. Even though the defendant has asserted that the police procedures used in the investigation were improper, it is the prosecutor's burden to prove that everything the police did was correct. Thus the prosecutor must present witnesses to explain to the judge the procedures that were used. In this regard, the prosecutor must show that *every* search was legal and that the questioning of Rodney at the scene was constitutionally proper.

Cynthia points out that there are actually seven searches that must be reviewed by the judge: five at the scene of the crime and two at Rodney's apartment. At the scene of the crime, these include the initial entry into the house; opening of the nightstand drawer; opening of the hall closet; opening of the handbag found inside the closet; and opening of the desk drawer where Rodney's pay stubs were found.

Cynthia is trying to convince the judge that these searches are unconstitutional because the officers did not have probable cause to conduct the search. Therefore any evidence found as a result must be suppressed, including the body of Joanne, the ammunition from the nightstand, the handbag from the closet, the photograph of Rodney found inside the handbag, and Rodney's pay stubs.

As to Rodney's apartment, Cynthia's approach is that there was not enough evidence for the initial judge to issue the search warrant. As a result, the warrant was defective, and therefore the search was improper. The revolver found in Rodney's bedroom would thereafter have to be suppressed.

Finally, Cynthia believes that any statement made by Rodney at the scene of the crime must be suppressed because Rodney was not advised of his constitutional rights prior to making the statement.

After hearing the testimony of the officers involved in the investigation, the judge finds that the police were justified in entering Joanne's house. The information available to the officers at the time clearly indicated that a crime had been committed. The officers knew that gunshots were heard just moments before they arrived, they saw an apparent victim on the floor, and there was evidence that the assailant had escaped out the back door. Taken in the totality of the circumstances, these facts were sufficient to establish probable cause to believe that a crime had occurred within the house. This alone would be enough to justify a search of the house.

Because it appeared that Joanne was injured, it was critical that they enter immediately. Moreover, there was a likelihood that the suspect could be fleeing the jurisdiction. Because time was critical, it was appropriate for the officers to act immediately rather than seek a written search warrant from a judge. Thus, all of the evidence found at the scene of the crime would be admissible at the trial.

Next, the judge addresses Cynthia's motion to suppress the evidence found at Rodney's apartment. He states that there was more than enough evidence to find that Rodney was a likely suspect in the murder, that he lived at the apartment identified in the search warrant, and that he could still be armed with the murder weapon. Once inside the apartment, the police

had the right to conduct a search of the premises, including the closet where the revolver was found. Thus, this evidence would not be suppressed.

Finally, the judge hears the argument on whether Rodney's statement at the scene should be suppressed. The prosecutor tries to convince the judge that Rodney's statement was voluntary. Cynthia argues that because Rodney was in custody at the time, police were required to advise him of his rights before any questioning. Because they did not, any statements given by Rodney should be suppressed. The judge agrees with Cynthia and rules that the prosecutor may not use Rodney's statement against him.

Although both sides are ready to begin the trial on August 21, the judge informs them that he is still in the middle of another trial, which will last for several more days. Thus Rodney's trial will have to be adjourned. After three more adjournments, the trial is ready to begin on October 9.

The actual trial begins with jury selection. Each side attempts to obtain jurors who will be more sympathetic to their side. Fourteen people will ultimately be chosen to sit as jurors on Rodney's case. These 14 are chosen from a panel of almost 100 prospective jurors.

The judge advises all of the prospective jurors that this is a criminal case involving the death of a woman. He introduces the attorneys and Rodney and reads a list of the expected witnesses. If any of the jurors know any of these individuals, they will be excused. The clerk then picks 14 names at random, and these people are seated in the jury box. The judge asks all of them their name, occupation, and marital status, if they have ever been involved in any type of lawsuit (civil or criminal), if they have ever served on a jury before, if they have ever been the victim of a crime, and if they have any affiliation to any law enforcement agencies. Ultimately, the judge is looking for people who can be fair and impartial and decide the case on the basis of the evidence that is presented to them.

If a juror indicates that he or she cannot be fair and impartial, the judge will excuse him or her from service in this case "for cause." Even if the juror indicates that he or she can be fair and impartial, each attorney has a number of peremptory challenges. These challenges can be used to excuse jurors from service when the attorney feels that the juror might not be sympathetic to his or her side. Conversely, when a challenge is used, it is hoped by the attorney that the next prospective juror chosen will be more likely to find in their favor.

Eventually, 14 people are chosen as acceptable jurors to both sides, and the jury is sworn in. Opening statements now take place.

The prosecutor delivers his opening statement first. He tells the jurors what he hopes to prove in order to obtain a conviction of the crimes charged. In this regard, he reads the indictment to the jury so they know exactly what the defendant is facing. He then relates to the jury a summary of what he thinks the testimony will be. He assures them that at the conclusion of the case they will be convinced that the defendant is guilty.

Cynthia emphasizes in her opening statement that the defendant is not required to prove his innocence. On the contrary, the prosecutor has the heavy burden of proving Rodney guilty beyond a reasonable doubt. She urges them to listen carefully to the testimony and to ask themselves a critical question throughout the case: Where are the eyewitnesses who saw Rodney? She advises that at the conclusion of the case they will be left with so many questions that they will have a reasonable doubt as to the guilt of Rodney and will have to find him not guilty.

After a short recess, the prosecutor begins his case. His first witness is a neighbor who heard the gunshots. Although she did not see the killer, she sets the stage for the rest of the witnesses' testimony. The next witness is the chief investigating detective. He testifies about the chronology of events leading up to the arrest of Rodney, including arrival at the scene, discovery of the body, questioning of neighbors, and the search of Rodney's apartment. He is prevented from telling the jury what the neighbors told him, however, because this constitutes impermissible hearsay testimony. It is prohibited because it would be unfair to allow the prosecutor to present evidence that could not be probed for accuracy, by way of cross-examination, when the person who actually made the statement is not present in court. Thus, the

matrimonial problems that neighbors related to police are never revealed to the jury.

The personnel manager from Rodney's job testifies about his unexplained absence from work and his change of address. Finally, the medical examiner testifies. He details the injuries received by Joanne and how they resulted in her death. He is allowed to give his opinion, as a medical expert, that the cause of death was definitely the gunshot wounds. On cross-examination, Cynthia tries to establish that the wounds could have been inflicted accidentally during the course of a struggle. He is adamant, however, that the shots were fired from several feet away.

The prosecutor introduces into evidence the revolver found in Rodney's closet, the picture found in Joanne's handbag, the ammunition found in the nightstand, Rodney's personnel records, and the bullets that were removed from Joanne's body. At that point, the prosecutor advises the judge that he will rest, indicating that he has presented all of the evidence that he feels is relevant to the case.

Cynthia makes a motion, asking the judge to dismiss the charges against Rodney because the prosecutor has failed to meet his burden of proof. The judge denies her request, indicating that, at this stage of the case, the prosecutor is entitled to the benefit of all of the favorable evidence, as well as all of the reasonable inferences that can be drawn from the evidence. In this light, the jury could possibly find Rodney guilty of the crimes charged.

It is now time for Cynthia to present her witnesses. She advises the judge that she has no witnesses other than perhaps Rodney. No one can force Rodney to testify, however, and thus she asks for some time to discuss this with him. A short recess is taken so that Cynthia can discuss with Rodney whether he should testify in his own behalf. They review the evidence presented by the prosecutor so far. He has shown that neighbors heard gunshots and police thereafter found Joanne dead. No one saw who did it. Nor did anyone see the murderer running from the scene. The house showed signs of a struggle prior to the murder, and Rodney was identified. Neighbors told how Rodney had left the house several days prior to the murder, and the jury knew that Rodney was absent from work during this same time. They also knew that Rodney had rented an apartment and that a revolver was found inside. Because Rodney's statement was suppressed, however, the jury does not know about his confession.

Cynthia and Rodney decide that the prosecutor's case seems weak. The prosecutor has not even shown that Rodney was there that night. If Rodney testified, his past history could be exposed, and this could convince the jury to convict him. Indeed, the jury might believe that Rodney has a tendency to violence, and is therefore the type of person who could kill, simply because of his past record. In addition, there is no evidence to corroborate Rodney's story that the shooting was accidental. Thus, they decide that Rodney will not testify in his own behalf. Cynthia will simply argue that the prosecutor has not proven his case beyond a reasonable doubt. She therefore advises the judge that the defense rests as well.

Closing arguments are given, but in reverse order of the opening statements. Because it is the prosecutor's burden to prove the case, he is allowed to go last so that he can rebut the arguments of the defense.

Cynthia argues that there is a lack of evidence to prove that Rodney is guilty. Although she expresses sadness at the death of Joanne, she implores the jury not to convict Rodney for a crime that he did not commit. She reminds the jury that Rodney has a constitutional right to remain silent and that they should not hold it against him simply because he did not testify. Finally, she reiterates that the evidence must show that Rodney is guilty beyond a reasonable doubt and that if they have a "reasonable doubt" as to whether he is guilty, they *must* find him not guilty.

The prosecutor reviews all of the evidence and argues that it all points to Rodney as the guilty party. He infers for the jury that Rodney and Joanne had a fight that culminated in her death. Why else would Rodney leave his home and his job without explanation? He tells the jury that although he must prove Rodney's guilt beyond a reasonable doubt, he is not required to prove the case beyond *all* doubt, or beyond *a shadow of a doubt,* but merely beyond a *reasonable* doubt. He maintains that he has done so.

After closing arguments, the judge instructs the jury on the applicable law that they are to

consider in their deliberations. This includes the elements of the crimes charged, definitions of key terms in criminal law, certain rights of the defendant, and the prosecutor's burden of proof. He advises that the jury's verdict must be unanimous and that they are to deliberate until they reach a verdict or until they are hopelessly deadlocked. The two alternates are then chosen at random by the clerk. These two will not participate in the deliberations but will remain in the courthouse just in case they are needed.

Because it is late in the day, the judge tells the jury that they must return in the morning for their deliberations. He advises them not to speak to anyone about the case and not to start their deliberations until they are all together again. The next morning, they begin deliberations, and after 4 hours they reach a verdict.

The foreman of the jury announces that they have found Rodney not guilty of first-degree murder but guilty of the lesser included offense of murder in the second degree. They have found him not guilty of unlawful possession of a weapon and possession of a weapon for an unlawful purpose. The jury is "polled," and each juror confirms that this is the unanimous verdict. The judge excuses the jurors and thanks them all for their service.

Cynthia requests that the verdict be set aside because it is against the weight of the evidence and because it is inconsistent. On this latter point, she asserts that it is inconsistent for the jury to find Rodney guilty of murder by shooting but find him not guilty of ever possessing a weapon.

The judge denies her request. He explains that it is not his function to substitute his view of the evidence for the jury's. Rather, he must only determine whether there is enough evidence for the jury to believe that Rodney was guilty of second-degree murder. In this regard, the judge feels that although there was no direct testimony that placed Rodney at the scene of the crime, there was enough circumstantial evidence to find him guilty of the murder. He also believes that although the verdict may appear to be inconsistent, the jury is allowed to consider each count of the indictment separately and reach an independent conclusion as to each. He is not allowed to probe into the rationale of the jury as to why they may have convicted on one count while acquitting on another.

The judge advises Rodney that he will have to return to court on November 20 for sentencing. In the interim, he will order the probation department to prepare a presentence report. In this regard, a probation worker reviews the police reports and statements of witnesses, may question those involved, evaluates the impact of the crime on the victim's family as well as society, and interviews the defendant. The worker then makes a recommendation to the judge as to what the appropriate sentence should be. The judge is not bound by the recommendation but quite often finds it helpful in arriving at an appropriate sentence.

On November 20, Rodney returns to court to be sentenced. The judge, the prosecutor, and Cynthia have reviewed the presentence report, which recommends incarceration in state prison for 7½ years, with a parole ineligibility period of 2½ years (one third of the sentence). The prosecutor argues that Rodney should be given the maximum sentence allowable for a second-degree crime and urges the judge to sentence him to 10 years, with a $3^1/_3$-year parole ineligibility period. Cynthia, on the other hand, tries to convince the court that the minimum sentence would be sufficient to punish Rodney and deter him and others like him from committing similar crimes in the future. She requests that the sentence be 5 years, with no parole ineligibility period at all.

Considering the aggravating and mitigating circumstances of both the crime and Rodney's background, including his prior history of violence, the judge imposes the sentence recommended in the presentence report and commits Rodney to state prison for 7½ years, with a parole ineligibility of 2½ years. He is given credit for the time he has already served awaiting trial and is thereafter turned over to the custody of the commissioner of corrections for processing into the state prison system.

Rodney is taken back to the county jail, where he has spent the last 8 months of his life. In early December, a decision is made to transfer him to the state's medium-security prison upstate. Shortly before Christmas, he is shackled and placed on a transport bus with other prisoners for the long ride upstate.

As they approach the prison, he is surprised to find that it does not have the high stone walls, barbed wire, and guard towers he envisioned. Although it does have walls surrounding it, they are not ominous looking but rather smooth and brightly painted. Passing through the gate, he enters a central compound where prison guards are waiting. He files off the bus with the other prisoners, and they are directed to the intake area of the prison.

When it is Rodney's turn, he sits at a desk, and a prison intake official fills out the necessary paperwork for admitting him into the prison. Information about Rodney's background, family, medical history, employment history, and details about the crime are discussed. He is assigned a prisoner number, given a uniform to wear, and led to a two-man cell. His cellmate is already there, having arrived 1 week ago. He is serving time for armed robbery.

As the weeks and months pass, Rodney spends most of his time either in his cell or hanging around the compound. Occasionally he receives word on the progress of the appeal filed by the public defender's office. Fourteen months after his conviction, he is advised that the appellate court has denied his appeal, affirming both the conviction and the sentence imposed by the trial judge.

Boredom is a way of life here because there are few programs to occupy the prisoners' time. What few programs exist are overcrowded and uninteresting. Eventually, he works at the prison laundry. Although it is a menial job, it at least gives him something to do during the day.

A social worker from the probation department visits him about once a month. She seems genuinely interested in his welfare and how he is getting along in prison. She tries to arrange stimulating outlets for Rodney that will pass the time and also provide some type of vocational as well as social retraining for when Rodney eventually leaves prison. At first he resists her efforts, but eventually he begins to participate in a variety of programs.

After 20 months at state prison, he is advised that his case is being reviewed by the parole board for release in another 2 months, which coincides with the expiration of his parole ineligibility period. The process includes a review of the facts of the crime, any opposition to release filed by the victim's family, and an interview with Rodney. The latter is sometimes the most critical because the parole board is very interested in seeing whether Rodney has made an adequate readjustment of his life to the point where he will not be a danger to himself or others if he is released. They are also interested in seeing that Rodney is remorseful for what he did and has been reformed by the process.

On the day of the hearing, the parole board seems somewhat sympathetic to Rodney, noting that he has made a good adjustment to prison life and has been a model prisoner. No response has been received from any of Joanne's family members regarding his release. Nevertheless, the board denies his parole, reasoning that he has not served enough time in prison. The crime, they state, was a very serious one, and if they let him out now, the deterrent effect of the justice system will be undermined. He will be eligible for another parole review in 6 to 8 months.

Rodney passes the next 6 months in much the same manner as before. He merely "exists," trying to find ways to pass the endless hours of idleness. When he again comes up for review, the parole board grants his application for release. In total, he has served almost 3 years.

Upon his release, he is given the name of his parole officer and advised to contact him within 2 weeks. He eventually makes an appointment to see the parole officer, and the two discuss some of the requirements of parole. Rodney must obtain a steady job or attend school full time. He must avoid any opportunity for criminal behavior and cannot be arrested for anything. He must even avoid associating with "criminal types." He is not allowed to leave the state. If he finds that he must leave, he must first seek permission from the parole officer. The parole officer will require him to report once a month so that the officer can monitor Rodney's progress in society. If Rodney fails to comply with these requirements, his parole can be revoked, and he can be sent back to prison to serve out the remainder of his term.

Finding a job is difficult for Rodney, but eventually he locates a small grocery store that is willing to hire an ex-convict. He works for minimum wage, sweeping floors and doing general cleanup work. He abides by his parole requirements and faithfully reports once a month. After being under the control of his parole officer for 2½ years, he is released completely. This finally concludes his involvement with the criminal justice system, a process that has taken almost 5½ years to complete.

NOTE

1. All of the names in this case study are fictitious, and any resemblance to real persons, either living or deceased, is purely coincidental.

PART II

UNDERSTANDING CRIME AND CRIMINALS

3

Law Enforcement Officers' Response to Illegal Street Gang Activity

Mary S. Jackson

Almost since its inception, American society has experienced gang activity in one form or another. Further, both gang activities themselves and strategies to deal with gangs and their activities have often played an important role in boosting local economies. Consider for example, the poor Missouri dirt farmers who are said to have received a share in the loot from the James Gang's bank and train robberies. These ordinary people held the James Gang in high regard. The railroad justified the need to increase the protection of its valuables from the James Gang by increasing the number of Pinkerton men (private security officers) they employed and assigned to that task. Thus, the James Gang gave a boost to the income of the poor farmers and, inadvertently, created railroad law enforcement jobs for individuals who would not otherwise have been employed.

Today, gangs similarly help to strengthen some local economies both by providing money to gang members' own families and neighborhoods from their drug trafficking and other illegal activities and by attracting federal, state, and local dollars for antigang programs to their town or city. Government funding supports not only the hiring of more police officers but also the hiring of more civilians for prevention programs. Even former gang members may be employed in some antigang community programs. For example, in one case the federal government funded the Black P-Stone Nation, a Chicago-based street gang, to assist in reducing gang and drug violence. Jeff Fort, one of the leaders of the Black P-Stone Nation, was convicted and sentenced to prison for misappropriation of government funds ("Black P-Stone Nation," 2002).

Unfortunately, the response of law enforcement officers (LEOs) to gang activity has not been carefully planned. Rather, it has been for the most part dictated by economic needs, media coverage, and the prevailing political atmosphere.

Thrasher's classic book *The Gang* (1927/2000) was first published in 1927, when the economy was weak and unemployment and illegal "gangster" activity was rampant throughout the United States. However, gang literature did not become prolific until the 1980s and 1990s, when there were more movies, books, articles, videos, musical lyrics, and newspaper stories about gangs and gang activity than at any other time in American history.

During this period, the Great Depression was but a fading memory; the United States had a growing economy, and prosperity seemed to be assured. The civil rights and Vietnam War protests that had swept the nation were over and the Cold War had ended, so military and LEO roles had to be reevaluated and reassigned. Increasingly, military and LEO roles came to focus on the war on drugs. Thus, although initially law enforcement's attitude toward gangs seemed to be complacent—as if to say, "They are out there, but they pose no problem"—that attitude would change as the involvement of gang members with drug trafficking became more common and more publicized.

The widespread increase in gang activity was an asset to adult criminals with their drug enterprises. Drug traffickers often recruited young gang members to move their drugs. They viewed youth as excellent employees because juvenile laws were not as stringent: Young drug runners would face less harsh punishment than adults and would be off the streets for less time if arrested, thereby cutting the dealer's downtime to a minimum.

As drug traffickers increased their profits in drug money by using gang members, politicians, regardless of their political affiliation, began to see the political value of capitalizing on this issue and campaigning for a get-tough policy on gang members that would rid the neighborhoods of gangs. And as TV and newspaper stories played on the public's fears and movies glamorized violence to youths, the public perception of a need for more antigang personnel and programs increased. As a result, harsher laws were enacted to specifically target gang activity.

Since September 11, 2001, however, the United States has been so preoccupied with defending itself against terrorist attacks on its own soil, fighting the Taliban in Afghanistan, and engaging in the search for Osama bin Laden that there has been limited media coverage of gang activities. The diversion of media attention away from gangs has led to diversion of funding from antigang activity, and this in turn has led to cutbacks in antigang programs.

In the 1960s and 1970s, media coverage was similarly limited: Information about gangs was seldom broadcast on television or depicted in the movies, and back then no recording artists portrayed themselves as gang members or described gang life in their videos or lyrics. Yet gangs did exist and were just as real then as they are now (Jackson & Springer, 1997). Therefore, it appears that the length of the war on terrorism will determine when the spotlight will shift once again to gangs and gang activity. However, even when the media are ignoring the issue of gang activity, LEOs must continue to assert their presence and address illegal activities perpetuated by gangs in their jurisdictions.

Gang members cause more frustration for LEOs than any other segment of the population. One major reason for this is that they are primarily youthful offenders. The laws governing them and the procedures for handling them are different from those for adult criminal offenders. Many LEOs forget (or have not been exposed to) juvenile law when interacting with gang members and rely on their knowledge of adult criminal law. This places them at a disadvantage and makes their gang intervention techniques ineffective. Effective gang intervention skills are extremely important, yet LEOs receive little or no ongoing training to understand juvenile law or to work with youthful offenders. Thus, even the LEOs assigned to youth gang units become so dismayed as to question their own effectiveness. According to Roberts and Waters (1998), one of the major issues associated with youth in the juvenile justice system is gang involvement (p. 47).

One tool that has been used as a response to street gangs has been community policing. In general, community policing approaches, which were common in the 19th century, have been overwhelmingly restored across the nation as a response to increasing violent crime rates in the 1980s and 1990s. The Violent Crime Control and Law Enforcement Act of 1994, a.k.a. the Crime Act, one response to the increasing share of violent crimes perpetuated by younger offenders, provided an added impetus to revive community policing with its appropriation of millions of dollars for law enforcement organizations to recruit, hire, and train more officers who would engage in community policing. In this model of law enforcement, LEOs are supposed to work in partnership with the community in reducing crime, with "the community" being defined to include not only residents living in the area but also other professionals working in community agencies.

Policy changes in law enforcement agencies can be very abrupt, so that programs are implemented before their details are thoroughly planned. When this happens with a new community policing policy, the officers are told that community policing is a top priority but are not armed with the essential skills to make it work. For example, some local communities may encourage a multidisciplinary approach to reducing gangs and gang activity using a team that may consist of social workers, guardians *ad litem,* probation officers, youths, and parents. Although LEOs are an integral part of the team, they are often excluded from discussions and viewed by the rest of the team only as the "enforcers." In addition, LEOs are not provided with a definitive understanding of the role of each team member. Though the team approach appears to be one of the most effective responses to gangs and gang activity, LEOs are simply told in Basic Law Enforcement Training (BLET) that the team approach is effective and that they should be working in teams whenever possible (see, e.g., North Carolina Justice Academy, 2002). Logically, how can LEOs be expected to become an effective part of a team if they have not interacted with different agencies proactively and if they have only limited knowledge of the expectations, roles, and responsibilities of the other team members? In such circumstances, LEOs are likely to feel that their authority is being threatened, leading to a less than positive outcome for gang reduction, collaborative teamwork, and attitudes toward community policing as an effective response to gang reduction. Further, although community policing requires officers to interact with community residents and work with them to identify and reduce crime in the neighborhood, many law enforcement organizations do not take a strong position to ensure that officers develop a sensitivity to and an understanding of the different cultures with which they are expected to interact. Consequently LEOs often struggle with their assigned task and feel torn by conflicting demands: projecting the image of an authority figure while at the same time acting as a

peacemaker or mediator in the community and maintaining solidarity with their fellow officers, who are supposed to stand together no matter what.

It is not surprising that under these pressures many LEOs may "cross the line" and become involved in the very illegal activities that they are assigned to work against. When the goal is higher arrest rates, other officers as well as administrators can turn a blind eye to the methods a unit is using to reach this goal. In Los Angeles, New York, and Manatee and St. Petersburg, Florida, special antigang units had to be eliminated because the officers' behaviors turned out to be no different than that of the gang members they were fighting. In Los Angeles, for example, gang units were deactivated in March 2000 because the officers had been using "rogue cop tactics" (McCarthy, 1997): They were "operating like a gang themselves, beating, framing and even shooting suspects" ("LAPD Disbands CRASH Anti-Gang Units," 2000). In Florida, the *St. Petersburg Times* described St. Petersburg's antigang "Delta Squad" as overly aggressive blatant criminals who "embraced a solidarity within the unit which is similar to street gang values" ("Tarnished Police Trust," 2000, p. 1).

In fact, LEOs and illegal street gang members share more than a few similarities. Both are vested with powers of discretion over their group and in some instances in the neighborhood. Both use their powers to achieve their goals. Illegal gang members are involved in illegal activities, yet, as stated earlier, some of them provide money to their poor families, and in some instances gangs provide money to political campaigns (Cuthbertson, 2001). LEOs are sworn to protect and serve but may engage in "rogue" tactics to suppress gang activity.

To combat illegal gang activities, it is necessary to understand the types of gangs, their roles, and how they function. According to Emile Durkheim (1897/1997), it is "normal" for societies to have crime, in the sense that crime is found in every society and can never be entirely eradicated. Similarly, it is normal for American cities to have gangs: According to CNN (1997), gangs exist in 94 percent of all U.S. cities. A look at socially acceptable "legal gangs" such as police and fraternities can shed light on how illegal street gangs work as well. Therefore, this chapter will describe the dynamics of both.

All of the aforementioned are critical issues for the LEOs who are given the task of responding to criminal gang activity. This chapter will focus specifically on the impact that developments such as the enactment of laws and policies have had on LEO response to gang activity. It will suggest that even today we are continuing to do what Ryan (1976) called "blaming the victims"—that is, blaming LEOs for using ineffective and criminal tactics to suppress illegal gang activity when they are pressured for results by the public, the media, and police administration but are given only limited options to obtain those results, limited funds with which to work and acquire the needed personnel, and limited training, if any, on juvenile law or how to intervene when interacting with juvenile offenders. The intent of this chapter is not to pass judgment on LEOs and their methods of gang suppression but simply to suggest the need to consider how circumstances beyond officers' control, such as get-tough policies, or officers' lack of options because of limited training or education, affect LEOs themselves, the communities they serve, and youths who are gang members. These circumstances are often not understood simply because the literature has a very narrow perspective on gangs and gang activity in American society.

The public's usual image of gangs is that of youths holding 9 mm weapons and dressed in baggy clothes and bandanas, scowling and assuming a "beat down pose." This scary image has funneled numerous federal, state, and local dollars into law enforcement organizations and other social agencies for the purpose of gang suppression. In many cities, gang suppression— the aggressive arrest, prosecution, and incarceration of gang members (Fritsch, Caeti, & Taylor, 1999)—is the primary strategy used against street gangs (Decker & Curry, 2000). But rarely do discussions consider issues that LEOs have to deal with to make gang suppression happen and the effects of legislative policies on their options.

LEOs' situation can be likened to that of African Americans in American society, whom W. E. B. DuBois (2000) described as experiencing a "warring of souls." LEOs have to identify with society's expectations of them to suppress illegal gang activity, yet they are covertly persecuted and denigrated by society, they are provided improper training, they receive insufficient pay and personnel to accomplish assigned tasks, and they must deal with the passage of laws that make it more difficult for them to perform their jobs. Officers can feel frustrations similar to those of illegal street gang members. They may feel that they are torn between two worlds—that of their close-knit group, the police, and that of the larger society—and trying to survive in both. The LEO is, in a sense, a member of a legal gang that is forced to suppress an illegal gang, using rules that constantly change. How the LEO responds when faced with the issue of gang suppression is a critical matter.

DEFINING A GANG

When referring to youth gangs, one needs to be as specific as possible because the term *gang* has many conflicting definitions and because definitions can vary depending on locality and depending on what funds are being sought. LEOs' response to gang activity varies from state to state, and variations are largely based on state funding needs (Katz, 2001). As Grennan, Britz, Rush, and Barker (2000), summarizing the argument of Miller (1980), have pointed out, "Law enforcement and the media (you could add corrections, juvenile justice, etc.) broaden or narrow their definition [of *gang*] as it suits their own particular cause" (p. 9).

The popular perception of gangs appears to be that gangs primarily break laws and are involved in criminal behavior. The father of gang theory, Frederic Thrasher (1927/2000), had a different notion. He defined a gang as "an interstitial group originally formed spontaneously and then integrated through conflict" (p. 18). He further characterized gangs in terms of "milling, movement through space as a unit, conflict, and planning" (p. 19). For Thrasher, gangs' collective behavior was "the development of tradition, unreflective internal structure, esprit de corps, solidarity, morale, group awareness, and attachment to local territory"

(p. 19). Thus, Thrasher's perception of a gang was not predicated on criminality; "planning" did not necessarily involve planning for crimes, and the "conflict" with external forces that he described was not necessarily violent by his definition. Further, Thrasher's definition of gangs was not age specific or class specific. Only later did researchers define gangs specifically in terms of violence and criminality and focus specifically on adolescents from lower-class neighborhoods.

From the 1960s to the present, most researchers' definitions, though widely differing, have agreed on the characteristic of illegal activity. Thus, in their classic book *Delinquency and Opportunity: A Theory of Delinquent Gangs,* Cloward and Ohlin (1960) defined gangs in terms of their patterns of behavior in a criminal subculture. They divided gangs into three categories: criminal gangs, conflict or violent gangs, and retreatist gangs (p. 1). According to the authors, the criminal gang functioning in the criminal subculture is viewed as a stable entity in the neighborhood. It is composed of older individuals who serve as role models for the young adolescents and are often referred to by gang members as "Original Gangstas" (OGs). They proudly tell the newer members the history of the gang and teach their skills and values to the younger ones in the neighborhood, who they hope will follow in their criminal footsteps. They are well integrated into the community, and though they are an integral part of the illegal subculture, some of them function well in the legitimate business world on a daily basis. The second type of gang, the conflict or violent gang, is less stable than the criminal gang. It is more fragmented and less certain with regard to goals and objectives. Its primary goal is to establish and live up to a reputation as "bad." Finally, the third type of gang, the retreatist gang, does not possess the needed organizational skills to become a cohesive unit and is not tough enough to recruit many followers and build a reputation as enforcers. Thus, members retreat into a world of their own that is filled with sex and drugs (Regoli & Hewitt, 2000).

More recent researchers have similarly emphasized illegality in their definitions: Huff (1993) stated that gangs are primarily involved in "illegal activity" (p. 4), and Klein and Maxson (1998) emphasized "illegal incidents" (p. 205). Klein (1995), by suggesting that a gang is "any group of youth the police call a gang" (quoted in Hagedorn, 1988, p. 83), implicitly defined gangs as groups targeted by law enforcement. Further, researchers have defined gangs as age and class specific, describing them, for example, as "quasi-institutionalized structures within poorer minority communities" (Hagedorn, 1988, p. 6) or as being characterized by a "focus . . . on adolescents and young adults" (Huff, 1993, p. 4).

For our purposes, rather than using such definitions, which often have highly negative associations, we will rely on the more neutral definition that Thrasher originally supplied. This definition allows us to look at similarities between what we will call "legal gangs" and illegal street gangs. Legal gangs are gangs that are socially acceptable—for example, police and the nationally sanctioned fraternities and sororities that exist on almost every college campus in America. Occasionally legal gangs may engage in unsanctioned or unacceptable activities. One prime example is the unit within the Los Angeles Police Department, Community Resources Against Street Hoodlums (CRASH), that was formed in the early 1980s with the primary goal of illegal street gang suppression (OJJDP, 2000a, p. 22). In 1999, a CRASH member who was caught stealing $1 million worth of cocaine from the evidence room turned state's evidence against the other members, describing numerous instances of beating offenders, dealing drugs with offenders, intimidating witnesses, planting evidence, and committing perjury. As a result, in 2000 the unit was disbanded, over 20 officers resigned, were prosecuted, or received suspensions, and over 200 gang convictions in the state were overturned (Knowland & Nebbia, 2002). Knowland and Nebbia's description of CRASH as "a secret fraternity" suggests a resemblance between rogue police units and college fraternities that, in their own ways, can "cross the line" and engage in illegal activities. The extreme cruelty that some fraternity chapters exhibit in hazing during rush week has placed many on notice or caused them to be suspended from campuses. Many fraternity parties are places where illegal drugs are abundant, and unfortunately some of the drug use, especially in the case of GHB (gamma hydroxybutyrate), may lead to date rape, coma, or brain damage. Yet the public appears to be more tolerant of these legally sanctioned gangs than of street gangs, even though the victims who die as a result of their activities are as dead as those who are shot by a gang member in a drive-by.

Illegal street gangs are not sanctioned by society. Their members are viewed negatively whether they move on to become millionaires or remain in their neighborhood gang-banging. Many street gang members have become prominent in the entertainment industry and have become successful entrepreneurs. For example, Ice T was a gangbanger who became a successful recording artist. He topped law enforcement's list of gang members to watch after he released his album *Kop Killer* in the 1980s. He is currently seen by millions weekly on television playing the role of a New York police detective on *Law and Order: Special Victims Unit*. In addition, many gang members are attending colleges and universities and are able to move from the illegal gang membership to membership in the legal gangs called fraternities. LEOs focus on the young street gang members who become involved in the drive-bys and the drug subculture. These are the gangs that America seems to glamorize through movies, magazines, and music videos and lyrics, and as a result, many children and adolescents join gangs.

Why Join a Gang?

Whether a gang is legal or illegal, members join it for the same reasons. The reasons are rooted in the five basic human needs that Abraham Maslow (1998) described: physiological needs, the need for safety, the need for love/belongingness, the need for self-esteem, and the need to feel that one has the potential to reach specific goals (self-actualization). Any reason given for joining a gang, whether a fraternity, a street gang, or a police organization, can be categorized under one or more of Maslow's basic human needs. LEOs and college fraternity brothers all want the security of being able to identify with a group and find solidarity through the group membership. As a result of the pledge of solidarity, they receive the love, protection, and security of the group, which then allows them an opportunity to perform at a higher level and reach their full potential (self-actualization).

The imperative of maintaining solidarity in a gang, whether a law enforcement agency or a street gang, is so overwhelming that it can create confusion and frustration for members, especially for members of legal gangs, such as LEOs, who may be faced with the problem of engaging or covering up for an illegal activity conducted by their fellow officers. They may not be sure whether to do something illegal, from a wish to fulfill individual needs or from feelings of loyalty to their group, or whether to uphold the law and even turn evidence against their group.

Individuals can have two types of families: natural families (families that are biologically formed) and social families (produced by a general agreement among individuals that they will function as a family unit). If basic needs are not being met in the natural family, there is a tendency to spend more time and exert more effort in the social family. Many LEOs, because of their work schedules, tend to spend so much time with their social family (other members of the legal gang of LEOs) that it creates stress for their natural family. Similarly, street gang members recruit youths from their natural family, and the social family (the gang) takes over the functions of their primary family. Because the organization (the gang) takes precedence and functions as the primary family, the LEO and the street gang member will go to great lengths to protect its members and their activities.

Another similarity between legal and illegal gangs is that they all take an oath of loyalty that is usually followed by an orientation period. Within the orientation period comes an initial stage of testing, ranging from minor pranks to violent physical actions, by which the membership assesses the recruit for full acceptance. As acceptability for membership is determined, so are decisions about roles and responsibilities. The criteria for which one officer is teamed with another are based on what the individual brings into the organization and what his or her abilities are. For example, is the LEO an intergenerational officer—in other words, from a family of LEOs? If so, were they detectives working

Narcotics or Homicide? The same criteria are applied to illegal gang members. What role did their parents or relatives play in the gang? Were they Original Gangstas? Did they hold a rank in the organization? Whether the gang is legal or illegal, intergenerational connections can hinder or enhance members' role and status in the organization as they undergo the initial assessment stage.

The second stage allows members an opportunity to demonstrate their capabilities to each other. Recruits can rise in rank on the basis of their capabilities, be placed at a lower rank and left out of action, be pushed out of the organization, be sacrificed (caught doing something not sanctioned by the group and receive no support), or be allowed to maintain membership but to perform only minor tasks.

The third stage is full acceptance: The member is accepted fully into the organization as a loyal member with all membership rights. This implies that the member can count on backup from others in the group when needed.

Other similarities are that all gang members wear colors (gang identifiers). These are visual signs that immediately state who the gang members are and distinguish them from others. When gang members want to be identified quickly and easily, they wear their colors proudly so that others may be intimidated or feel safe in their presence, depending on what message they are attempting to send. LEOs' uniforms serve this function. Further, when one sees LEOs in helmets and protective shields, one can expect that they are ready to engage in physical altercations. Similar rules apply to illegal gangs. When they are "decked out" and wearing their identifiers, get out of the way because there is likely to be a violent physical altercation (Jackson, 1994).

Although there are many similarities between legal and illegal gangs, it is necessary for societies, and particularly for legal gangs like LEOs, to focus and capitalize on the differences between the two groups so as to maintain some sense of order and social control. Only when a member or a group of members in a legal gang are discovered to be engaging in illegal activity do similarities between legal and illegal gangs become especially visible.

NOTORIOUS ILLEGAL STREET GANGS

The Bloods and the Crips

There may be as many different street gangs as there are law enforcement organizations, and many of them are prominent and notorious. The Bloods and the Crips are singled out for discussion here because their members not only have risen to levels of power and prestige and influence among young people (via movies and music) but also are making powerful political allies through large financial contributions to both Republican and Democratic political parties. Historically, Bloods and Crips have been bitter foes. They have been the source of much heated debate over the last 20 years, most of which has focused on legislation to suppress them. Yet it appears that in light of all the legislation, the two organizations continue to recruit and expand their territory throughout the country. It has been the primary task of LEOs to suppress the recruitment and expansion of the illegal activities of these organizations.

Bloods

The Bloods (Black [or Blood] Lords Of Our Destiny) evolved in Los Angeles in the late 1960s in response to the need for neighborhood unity, mutual support, and protection. Many of the original members (Sylvester Scott, Jordan Downs) lived on Piru Street; thus, they espouse what they called "Piru love," or brotherly love, and will often say to another Blood, "Show me some love," a greeting that can also mean "I'm here for you; show me that you're here for me." Some OGs (Original Gangstas) say the gang was started by two Jamaicans who were "tired of being hassled by the cops in the area" and decided it was time to stick together like other groups in the community. The Bloods' most distinctive identifying characteristic is their colors, with their primary color being red, symbolizing the "blood shed by their brothers." However, a Blood will wear blue as a sign of disrespect toward Crips. Their war threat or motto is Crip Killers (CK), as Crips are their primary enemies. Their goal of showing complete disrespect for the Crips involves communication patterns that highlight that goal. This disrespect is best denoted in their total disregard of the letter C, which is primarily replaced with the letter B or in some instances other letters. The Bloods have extended their sets (segments of their organization) to all parts of the country, including Native American reservations.

Crips

The stories of the origin of the Crips vary a bit, but the two main versions are credible enough to deserve mention. One is that the original Crips were formed by Larry Hoover (and patterned after the Black Panthers) in the late 1960s in Los Angeles. Another version points to Ray Washington, who attended Washington High School, as the founder. Gang members originally referred to themselves as "Cribs," and the news media labeled them Crips (Avendorph, 2001). Their primary color is blue and their main opponent is the Bloods. Sometimes, to show their disrespect for the Bloods, they disregard the letter B and replace it with the letter C. Crips is an acronym that is variously spelled out depending on the geographical location. For example, some refer to the Crips as the California Revolutionary Independent Pistol Slingers. Law enforcement intelligence suggests that the Crips outnumber the Bloods in membership. The recruitment and expansion of the organization has been enormous, and it is believed that there are Crips in every part of society.

Hispanic Gangs

Hispanics are one of the fastest-growing minority groups in American society. The American Hispanic population is composed primarily of Mexicans, Puerto Ricans, and Cubans. Gang activity among Hispanics poses the need for LEOs to develop an efficient and effective response to their activities.

Latino gangs have a long history beginning in Cuba (Grennan et al., 2000). However, Latino street gang activity reached the forefront of media coverage in the United States in the 1950s with the Mexican Mafia in Los Angeles, which started in the East Los Angeles barrios in the 1920s (p. 308). The Mexican Mafia membership increased so much that another group

was formed in 1958, La Nuestra Familia (Our Family), which has formed a partnership with the Wah Chin and Chung Ching Yea Chinese gangs (p. 310).

All Hispanic gang members, whether Mexican, Puerto Rican, or Cuban, view life as a street gang member as a very serious undertaking. They believe that they must instill stark fear in their opponent while maintaining a fearless macho demeanor (*vato loco*) (National Alliance of Gang Investigators Associations [NAGIA], 2002). Although similar in many respects to other street gangs, Hispanic gangs are unique in some ways. Many current members' gang affiliations can be traced back for generations through family members who were also active participants in gang activity (Grennan et al., 2000).

A distinctive feature of Hispanic gangs is their strategic use of two languages. Even when they are second- or third-generation Americans and are not fluent in Spanish, they can still pretend that they have no knowledge of the English language when approached by LEOs. (Of course, some gang members are not fluent in English and truly do not understand; in this case, the language barrier adds to the problems that can occur during interactions between them and LEOs.) Some Hispanic gang members will use Calo (a dialect that mixes Spanish and English) to confuse LEOs (NAGIA, 2002).

Like other street gang members, the Hispanics have their particular monikers (street names) that are based on their attitudes, behaviors, and appearance (e.g., *Maton,* which means "killer," *Flaco,* which means "skinny," or *Pata,* which means "feet"). The monikers single them out and simultaneously describe them and their particular skills to the community (Grennan et al., 2000). Also, like members of other street gangs, Hispanic gang members engage in a particular style of dress. Overall, their dress usually appears to be neat and carefully ironed, though some do wear the baggy pants. Their shirts are worn either open completely or buttoned completely to the top button.

Latin Kings

One of the oldest and largest Hispanic gangs is the Latin Kings (a.k.a. Almighty Latin King Nation), which originated in the Chicago area

(Daily, 2002). The Kings are considered to be one of the most violent street gangs in the country (NAGIA, 2002). They have affiliates throughout the United States and recruit members aged 8 years and up from both urban and rural areas (Daily, 2002). Their colors are black and gold, and their emblem is a crown represented by two crossed pitchforks with the tines pointed down (Daily, 2002). Their gangs are composed primarily of Mexicans and Puerto Ricans, with each nationality using some symbols of their own particular culture to distinguish them from other Kings. For example, most Mexican Kings exhibit the five-pointed crown, and Puerto Rican Kings will exhibit the three-pointed crown (Daily, 2002).

Police intelligence suggests that major factions of the Kings are in Chicago, Connecticut, Massachusetts, and New York (NAGIA, 2002). They have their own constitution, and LEOs consider them to be very well organized. Their motto is, "Once in, one is always a King [or, if female, a Queen]." Yet LEO intelligence suggests that there may be some rivalry within the Kings. For example, the Chicago Kings are rivals to the Connecticut Kings, who are known as the Almighty Latin Charter Nation (ALCN) (Security Threat Intelligence Unit [STIU], 2002). However, the Kings (regardless of where they are located in the country) have a primary nemesis in the Folk Nation (a multiracial Chicago-based alliance of several gangs) and anyone who associates with the Folk Nation.

The Kings are grounded in tradition and maintain a religious focus rooted in Catholic practices. In the United States, they have adopted observance of January 6 as the Kings Holy Day and the first week in March as Kings Week (STIU, 2002). According to STIU (2002), the Kings have established strongholds in California, Illinois (especially Chicago), Connecticut, Florida, Ohio, Massachusetts, Michigan, Minnesota, New York, Texas, Wisconsin, and Puerto Rico and are prevalent in other regions as well.

GIRLS AND WOMEN IN GANGS

More often than not, girls and women are ignored in discussions of gangs, and because of the continued paternalistic nature of American society (Chesney-Lind & Shelden, 1997) they are not deemed to be "as dangerous" as their male counterparts. Yet they play pivotal roles in the many gangs (Chesney-Lind & Hagedorn, 1999) and may participate both directly and indirectly in gang activities. Female members often provide key strategic intelligence information to the gang leaders. When gangs decide to expand and migrate to other areas, female members can be essential to the gang members' gaining a foothold in the new community. They know who resides in the community, who the leaders in the community are, who is "using" in the community, who is receiving assistance from the government (in the form of welfare, counseling, or drug treatment services), who has been incarcerated and who is currently on probation, which children and adolescents are "ripe" for recruitment, and which LEOs are assigned to the community as well as the LEOs' routine. With this type of intelligence, the gang recruiter observes the patterns in the community and then, at the appropriate time, moves in and begins to act on the information by recruiting from the community. Recruitment targets are primarily younger children and adolescents who have spent time incarcerated, are experiencing difficulties in their natural families, or are considered social isolates because they do not "fit" the society's norms.

The direct involvement of girls and women in gangs consists of their taking part in gang-banging (fights, distribution and trafficking of drugs, etc.). Law enforcement has learned that female gang members should be taken very seriously. At one time, girls and women who associated with these two notorious gangs were considered as no more than window dressing. That is, many LEOs thought that female gang members were just girlfriends of male gang members who provided sexual favors to keep the male gang members happy or that the male gang members were forcing them to be there and using them as prostitutes. However, LEOs have discovered that many female gang members are not simply girlfriends or forced into the gang but are voluntarily involved in the gang activity and in many instances have formed their own subgroup within the gang. The female gangsters are as lethal as, and in many instances more lethal than, their male counterparts. They can conceal contraband (not only drugs but weapons) on their person better than males. They can solicit information better than their male counterparts because they are still considered "the weaker sex" by many and are not as readily suspected of illegal criminal activity. As a result, they can camouflage their activities very well, and many LEOs are either reluctant to take the necessary action or are deceived by their perceived innocence.

In Hispanic gangs in particular, male gang members demonstrate prominent paternalistic attitudes and behavior towards female gangs and gang members. Hispanic male gang members proudly display their machismo (exaggerated masculinity) and feel that even though girls and women may form or join gangs and participate in criminal and violent activities, males still have the dominant and most important role. The female, regardless of rank in the gang, is still expected to be subservient to the male. Paternalism is still prevalent within the traditional Latino culture and thus persists in the Hispanic gang subculture as well.

Characteristically, the female gang member is quite similar to her male counterpart. She has probably been involved in numerous deviant behaviors (runaway, truancy, curfew violations, etc.) and has established a reputation as someone who does not conform to traditional expectations. Although her level of confidence within the gang milieu is usually very high, she is very insecure when she is away from the gang. Like the LEO, she is struggling in an attempt to find her niche. On the outside, she has to show that she is as tough as the male members of the gang, yet her psyche is fragile. She is constantly trying to prove her worth not only as a gang member but as a woman. This manifests itself in many ways. For example, she is very attentive to her dress, is involved in numerous sexual relationships, and is very meticulous about makeup and hair style (being particular to include the gang colors). Unlike her male counterpart, she will use tattoos sparingly and infrequently; when she does have one, it will be strategically located so that it serves to highlight what she considers a beautiful part of the body.

A major challenge for LEOs as they respond to gang activity is to understand the role of the female and female gangs. Often the focus is on male gang members with very little attention given to the female. The trifling amount of attention given to female members is usually rooted in the old paternalistic thinking that there is a need to protect and care for young girls. That response can be appropriate if, for example, officers apply it equally to 13-year-old male gang members, and if officers remember the importance of self-protection and are aware that even a young girl or a young boy may be armed and may use a weapon against them.

LAWS TARGETING GANG ACTIVITIES

Some argue that tougher laws and specific gang initiatives reduce gang activity. Therefore, laws have been enacted to specifically focus on illegal street gang activity.

The Violent Crime Control and Law Enforcement Act of 1994 (the Crime Act) is one of the largest commitments of federal dollars (over $15 billion) in the history of the United States to target gang activities. The bill authorized over $3 billion for the hiring and training of LEOs primarily for purposes of community-oriented policing and over $100 million for programs aimed specifically at reducing gang activities and the use of illegal drugs by adolescents. The antigang provision also stipulated new and stiffer penalties for crimes committed by gang members and for adults' instigation of youths who were gang members to engage in illegal activities. Roberts and Fisher (1997), in their discussion of the Crime Act, suggested that the "future looks promising" (p. 141) and argued strongly for interdisciplinary training for professionals working with children, adolescents, and their families (p. 141). However, if LEOs are expected to work on interdisciplinary teams, there is a need to include LEOs in such interdisciplinary training as well. Although in many instances LEOs are not included, they are involved in training among themselves that will improve their response efforts to violent situations.

A number of other laws have been implemented to increase LEOs' response efforts to reduce gang activity. Knox (2000) has provided a comprehensive discussion of the methods employed by LEOs, ranging from gang loitering laws in Chicago, which allow LEOs to arrest gang members who create a menace on public property (p. 378), to the Gang Resistance Education and Training (G.R.E.A.T.) programs sponsored by the Bureau of Alcohol, Tobacco, and Firearms (BATF). The G.R.E.A.T. programs, popularized first in Phoenix, Arizona, are similar to D.A.R.E. programs (p. 381): Just as in D.A.R.E. youths are taught strategies to use in response to pressures to use or traffic in drugs, so in G.R.E.A.T. youths are taught strategies to use when faced with gang members' recruitment techniques. Prosecutorial methods are also used by LEOs to combat gang activities. For example, California's Street Terrorism Enforcement and Prevention (STEP) Act allows for the prosecution of gang members on conspiracy charges (p. 381) and provides for increased prison sentences (p. 382). In addition, the coordination of antigang efforts has been facilitated through the use of coordinated Internet resources such as the Statewide Organized Gang Database (SWORD), which is computer software used primarily for the purpose of tracking, monitoring, and reporting gang activities (p. 376).

As a result of the Crime Act and other state and local legislation, a great deal of money has been spent in the name of zero tolerance and the suppression of gang activity. Yet gang activities continue to flourish. Thus, LEOs continue to experience frustration, and the numerous ways that their frustration is vented create critical issues for citizens as well as the LEOs and their families.

COMMUNITY POLICING APPROACHES TO GANG ACTIVITY

Community-oriented policing is not a new idea: It is described in the "Guiding Principles" set forth by Sir Robert Peel in the London Metropolitan Police Act of 1829. Peel's thinking encompassed the notion that LEOs are public servants who are sworn to protect and serve all citizens. Thus, one of their primary duties is to work collaboratively with community members to reduce crime in their neighborhoods.

Peel suggested that they walk around in the communities and talk with the citizens. This proactive approach was first brought into American culture in the cities that were the trailblazers in policing practice: New York in 1844, Boston in 1832, and Philadelphia in 1834. LEOs in the United States during the 1800s functioned much like social workers in that they provided whatever services were needed to citizens, from collecting coal and ice to lecturing the corner bully. They did not carry guns, only nightsticks and badges. But over time, although Peelian principles continued to be cited in law enforcement organizations, the actual practice of policing underwent change as the country experienced changes brought on by the Great Depression, war, changing demographics, and social disorder within American society.

As LEOs became less visible in communities and more reactive, community policing became community public relations, where LEOs attended community town hall meetings to defend and explain their actions after the fact. Now, however, the pendulum has swung in the other direction, and law enforcement agencies are emphasizing community-oriented policing as a response to gang activities.

Community-oriented policing methods are a high priority in many law enforcement municipalities. Law enforcement agencies are some of the main sources of information about street gangs (National Youth Gang Center, 2000), and millions of dollars are given to researchers to design and implement community models to combat gang activity. But neighborhood citizens are still reporting that gang activity is more visible than law enforcement activity. According to Knox (2000), in the 1990s "FBI spy chasers"—agents who had worked on counterintelligence in the Cold War and who had to be reassigned to domestic duties when the Cold War ended—were frequently directed to assignments that focused on gang convictions (p. 381). Now the speculation among gang members in light of September 11, 2001, terrorist attack is that federal convictions against street gang members will decrease due to the new focus on homeland security and international terrorism.

Working as an LEO is a major responsibility in American society. It is the only profession in

American society that allows the individual to possess a weapon and use it with almost total discretion. It is also the only profession holding such life-and-death powers that requires only a high school education for admission. In small jurisdictions, the entry-level salaries average around $17,000, and even in larger urban areas the beginning salaries are equally low, so police departments do not attract individuals with degrees in higher education or inspire LEOs to seek higher education. Although many law enforcement agencies are moving in the direction of seeking applicants with 4-year college degrees, there is still a lack of thorough initial training and ongoing training updates for LEOs.

If community policing is to be the prevailing philosophy of law enforcement agencies, the organizational administrators must commit to providing academic stipends and release time to officers so that they can continue their education, either by obtaining a 4-year degree or by going on to achieve an advanced degree, and improve their skills by acquiring ongoing training in areas other than weapons certification and physical qualifications. An all-out effort should be made to assist officers with the enhancement of their mental qualifications because many are placed in roles as educators.

LEOs are placed in a catch-22 position when dealing with gangs and gang members. They are supposed to be all things to all persons. They must try to protect and serve everyone, including gangs and gang members, while meeting the community's expectation that their tasks will include proactive community policing practices. One proactive strategy is that of educating youth. This is an area in which the officers receive minimal, if any training, yet those who depend on them maintain the highest expectations of them.

LEOs in many instances are the primary educators for G.R.E.A.T. programs. They are expected to go into the middle schools and spend approximately 9 weeks educating middle school students about the importance of remaining gang free. They are expected to cover an array of topics with the students according to a curriculum that covers crime, victims, and citizens' rights; cultural sensitivity and prejudice; conflict resolution; meeting basic needs; drugs and neighborhoods; responsibility; and goal setting (Esbensen, 2000, p. 7). Although the lesson plans are specifically designed for police officers to use (Esbensen, 2000), the majority of the officers do not receive ongoing evaluations on teaching techniques or style that might enhance their teaching effectiveness.

Even when evaluations of the effectiveness of the G.R.E.A.T. projects are undertaken, the focus is on self-reports from the student participants. Although, as a result of participation in G.R.E.A.T., students have reported the ability to resist joining gangs (Palumbo & Ferguson, 1995, cited in Esbensen, 2000, p. 7), the issue here is whether the officer's ability to deliver the G.R.E.A.T. curriculum was also a major factor in helping student participants resist joining gangs. Like D.A.R.E. officers, G.R.E.A.T. officers are thrust into a situation where they are expected to perform well but receive very limited initial training, no ongoing training or supervision, and few, if any, incentives such as pay increases. Many of their supervisors do not have the resources to offer constructive criticism. This impedes officer-teachers from improving in areas that may benefit not only the student participants but the officer as well. It inhibits their teaching techniques, preventing them from becoming more effective G.R.E.A.T. instructors able to deter more children from becoming involved in gangs.

Like D.A.R.E. officers, officers teaching the G.R.E.A.T. curriculum are assigned specifically and entirely to teaching that curriculum. Often they volunteer for this duty, but in some cases it is viewed as punishment because the assignment comes from the supervisor with limited or no prior discussion or input from the officer. The officer moves from the primary responsibility of suppression of gang activity to one of prevention. Although many officers are able to make the transition successfully, such a dramatic transition warrants initial monitoring and feedback. Even when an officer is newly hired to deliver the G.R.E.A.T. curriculum, he or she is still in need of continuing occasional monitoring, feedback, and supervision.

The G.R.E.A.T. curriculum can be a very effective gang deterrent, but as the programs are implemented, more support in the form of readily accessible resources should be made available to the officer delivering the curriculum.

A multidimensional community approach would serve to improve the quality of the program. The G.R.E.A.T. officer could team up with the schoolteacher, social worker, guidance counselor, and parents to deliver the curriculum. This approach would prove very useful because all these people interact at some point with the youth who becomes a gang member.

Jackson and McBride (1996) suggested that the LEO is "the key authority figure with whom the gang member has dealt" (p. 109). However, it is the school classroom teacher who is the primary figure in the gang member's life because decisions to join are rarely if at all made by those youths who are experiencing success in school. The majority of the youths who join street gangs do so because they are seeking to gain successes and the gang appears to them to be their most viable means to experience success.

The LEO teaching the G.R.E.A.T. program plays a pivotal role in prevention because this officer may in fact be the only adult who talks with the youths about gangs and gang prevention strategies before they are actually approached by a gang member to join a gang. This same officer may also be the officer who must arrest them if they do become gang members and violate the law.

One prominent result of the increased concern about street gang activity has been the creation of specialized gang units (Katz, 2001). Many officers are placed in gang units, some as punishment, though many will willingly volunteer for the assignment. They seldom receive initial training and after placement in the unit, they are on their own. They have few written standard operating procedures for dealing with gangs, limited understanding about how to work effectively with other jurisdictions and departments, and a lack of clarity about the roles and responsibilities of social workers and other professionals they must interact with. As a result, they may revert to the territorial attitudes that they have as members of a legal gang and may feel no obligation to work with social workers or other professionals who can provide them with relevant information about gang members and their families.

Some states have not implemented gang units per se but have placed school resource officers, who are either sheriff's deputies or police officers, in the public school system as a means of deterring gang activities. These officers are suppose to interact with the students in an effort to reduce the incidence of school violence and gang activity and to help create a safe learning environment for students (Johnson, 1999). Their role is to be proactive and be in a position to prevent a situation before it starts or defuse it before it escalates. This sort of community policing has been considered a viable response to gang activity in school systems.

IMPLICATIONS FOR PRACTICE

Community-oriented policing as a response to gang activities is great in theory, but in practice it can be effective only if LEOs from the top administrators down to the line officers on patrol are totally sold on the idea. Many law enforcement administrators view community policing simply as a way to obtain more funds to purchase more equipment and to hire more officers without placing them in communities that are gang infested. Many officers who are placed in gang-infested areas or who are members of gang units are seldom monitored by supervisors, and very few practice a multidisciplinary team approach. Yet as numerous researchers have concluded, illegal street gang activity is a community problem that requires the collaborative response of a team approach involving mobilization of the whole community and good interagency communication and coordination of services (Sandoval, 2001; Spergel, 1995; Spergel, cited in Mays, 1997; Spergel & Grossman, 1997; Vogel & Torres, 1998). The team approach should extend to other disciplines and not just law enforcement personnel. Social workers, schoolteachers, guardians *ad litem,* probation officers, judges, adolescents, and parents should all play important roles in the reduction and elimination of illegal street gang activity. The fragmentation of services on the part of LEOs leads to inefficient and ineffective methods of responding to gang activity.

Gang members, especially leaders or high-ranking members, appear to recognize the poor coordination and communication among agencies that deal with gang activity and indeed to rely on it as a factor that allows them to continue their illegal activities with less interference. Some gang members have suggested that there would be no illegal street gangs if professionals were better organized and if their responses were more coordinated (Jackson, 1994).

If multidisciplinary teams were created to address prevention of gang activity and were maintained regardless of political fluctuations, antigang efforts would be more successful. Instead, we have a situation in which the gang issue is politically manipulated and its facts are distorted. McCorkle and Miethe (1998) argued that the severity of the gang problem can be overstated to justify political action. For example, although from 1996 to 1998 there was an estimated 7% decrease in the number of youth street gangs and an 8% decrease in the number of actual gang members (OJJDP, 2000b, Table 9), many politicians and law enforcement administrators continued to promote a "get tough" philosophy and to assert that gang activity was increasing. The authors even suggested that some jurisdictions "manufacture" their gang crisis to expand their authority and obtain more funding (see also Katz, 2001). Meehan (2000) similarly has described the "gang problem" as a catchall category for many other social problems and has asserted that "for the officer . . . gangs are a political problem, one that the police socially construct in response to political pressure" (p. 338).

Law enforcement organizations have been given large amounts of federal money to reduce and eliminate gang activity. Yet the response has focused on zero tolerance and suppression ("lock 'em up"). How effective has this response been? Gangs are in every major city in the country. They have become computer literate and are in some instances able to tap into law enforcement Web-based communication systems to monitor the information collected on them. In some instances, they have even suggested that the data be corrected to reflect their activities correctly.

They have become politically connected, and youths emulate them via dress and dance as they watch videos and listen to their music during their daily routines. Ask almost any youth about Snoop Dog, Ice Cube, Lil Bow Wow, or Death Row Records, and they can provide, with complete details, information on the conflicts and the gang affiliations. Yet the federal government continues to allocate billions of dollars in the name of gang suppression. This response can only create more confusion within the ranks of LEOs as they continue to grapple with the contradictions. It will be interesting to see where the pendulum swings now that the federal dollars are needed to fight a war on terrorism.

CONCLUSION

Gangs have been defined in many different ways. Here Thrasher's original, broad definition, which does not highlight illegal activity, and can describe both legal gangs and illegal gangs, is used. Legal gangs are organizations that receive social acceptance (such as fraternities and law enforcement organizations); street gangs (such as Bloods and Crips) fall into the category of illegal gangs. Since its inception, American society has experienced both types of gangs, but large amounts of federal dollars were not allocated to suppress illegal gang activity until the passage of the Crime Act in 1994. This event marked the first time that the federal government had ever earmarked such a large amount of money as a response to crime reduction and prevention of illegal gang activity.

Suppression of gang activity and harsher penalties for gang members and adults who participate themselves or inspire youths to participate are major provisions of the bill's focus on gang activity. In response to federal and state funding initiatives, law enforcement organizations have undertaken a variety of antigang activities. One such activity is community policing. This is not a new concept but simply an old response to a contemporary issue. Community policing has had many drawbacks, but a major one has been that although police departments have placed great emphasis on ensuring that officers meet regular weaponry and physical qualifications, they have not placed the same emphasis on mental qualifications such as improved education of officers.

Thus, LEOs are still allowed to enter and rise through the ranks with only a high school education. They can shoot well but may not be able to outthink the suspect and weigh and develop more options than the suspect because inmates are still able to study for and receive degrees in higher education while incarcerated. Officers with little education can become more easily frustrated and narrowly focused. Thus, their options become more and more limited as they try in their ill-equipped manner to consider alternatives for reducing gang activity in their jurisdictions. Initially they have a tendency to deny the existence of gang activity, but by the time they observe the overt signs (such as gang graffiti) the gang already has a strong foothold in the neighborhood. Not only would officers with more education have more flexible and innovative responses, but they would be less likely to feel intimidated by the use of a multi-disciplinary team approach, as they would feel that they had more ideas to offer the team than the simple "beat 'em down" approach. They would be more receptive to working with others because their worldview would be broader. It is imperative that law enforcement organizations begin to invest in their officers by providing them with educational stipends, education release time, and ongoing continuing education.

Many LEOs feel uncomfortable with community policing because they have a limited understanding of the different cultures that have emerged in American society. As the demographics continue to change, LEOs will be faced with many new additional challenges. Because they are the only legally sanctioned group authorized to carry a weapon and badge and use them with almost absolute discretion, citizens have the right to expect that an enlightened, educated progressive LEO will be policing them. This is within the realm of probability, and the impact on illegal gang activity can be greatly reduced by more educated responses from LEOs.

REFERENCES

Avendorph. T. (2001, March). *Chicago gangs*. Paper presented at the conference "Gangs in Our Midst," North Carolina Gang Intelligence Association, Greensboro, NC.

Black P-Stone Nation. (2002). Retrieved October 12, 2002, from Know Gangs Web site: http://knowgangs.com.

Cable News Network, Inc. (CNN). (1997, April 23). Youth gangs no longer a big city problem: Gangs spring up when families relocate. Retrieved February 8, 2002, from www.cnn.com/US/9704/23/gangs/index.html

Chesney-Lind, M., & Hagedorn, J. (1999). *Female gangs in America: Essays on girls, gangs and gender.* Los Angeles: Lakeview.

Chesney-Lind, M., & Shelden, R. G. (1997). *Girls, delinquency, and juvenile justice.* Belmont, CA: Wadsworth.

Cloward, R., & Ohlin, L. (1960). *Delinquency and opportunity: A theory of delinquent gangs.* New York: Collier-Macmillan.

Cuthbertson, E. (2001, March). *Rap music and gangs.* Unpublished paper presented at the conference "Gangs in our Midst," North Carolina Gang Intelligence Association, Greensboro, NC.

Daily, W. (2002, January 30). "Almighty Latin Kings." Retrieved March 7, 2002, from National Alliance of Gang Investigators Associations Web site: www.nagia.org./almightylatinkingsnation.htm

Decker, S. H., & Curry, G. D. (2000). Responding to gangs: Comparing gang member, police and task force perspectives. *Journal of Criminal Justice, 28*(2), 129–137.

DuBois, W. E. B. (2000). *The souls of black folk.* Bensenville, IL: Lushena. (Original work published 1903)

Durkheim, E. (1997). *Suicide: A study in sociology.* New York: Free Press. (Original work published 1897)

Esbensen, F. (2000, Spring). *Preventing adolescent gang involvement* (Juvenile Justice Bulletin). Washington, DC: U.S. Dept. of Justice.

Fritsch, E. J., Caeti, T. J., & Taylor, R. W. (1999). Gang suppression through saturation patrol, aggressive curfew, and truancy enforcement: A quasi-experimental test of the Dallas anti-gang initiative. *Crime and Delinquency, 45*(1), 122–139.

Grennan, S., Britz, M. T., Rush, J., & Barker, T. (2000). *Gangs: An international approach.* Upper Saddle River, NJ: Prentice Hall.

Hagedorn, J. M. (1988). *People and folks: Gangs, crime and the underclass in a Rustbelt city.* Chicago: Lakeview.

Huff, C. R. (1993). Gangs in the U.S. In A. Goldstein & R. Huff (Eds.), *The gang intervention handbook.* Champaign, IL: Research Press.

Jackson, M. S. (Director). (1994, March 9). *A gang member speaks* [Video]. Personal video taped for training purposes.

Jackson, M. S., & Springer, D. W. (1997). Social work practice with African American juvenile gangs: Professional challenge. In C. A. McNeece & A. R. Roberts (Eds.), *Policy and practice in the justice system* (pp. 231–248). Chicago: Nelson-Hall.

Jackson, R., & McBride, W. (1996). *Understanding street gangs.* Incline Village, NV: Copperhouse.

Johnson, I. M. (1999). The effectiveness of a school resource officer program in a southern city. *Journal of Criminal Justice, 27,* 173–192.

Katz, C. M. (2001). The establishment of a police gang unit: An examination of organizational and environmental factors. *Criminology, 39*(1), 37–73.

Klein, M. (1995). *The American street gang: Its nature, prevalence, and control.* New York: Oxford University Press.

Klein, M., Maxon, C., & Miller, J. (1995). *The modern gang reader.* Los Angeles: Roxbury.

Knowland, D., & Nebbia, G. (2002, March 13). The L.A. police scandal and its social roots. Retrieved October 7, 2002, from World Socialist Web site: www.wsws.org.

Knox, G. (2000). *An introduction to gangs.* Peotone, IL: New Chicago School Press.

LAPD disbands CRASH anti-gang units. (2000, March 3). Retrieved October 15, 2002, from CrimeLynx Web site: www.crimelynx.com/crash.html.

Maslow, A. (1998). *Toward a psychology of being* (3rd ed.). Hoboken, NJ: John Wiley.

Mays, G. L. (1997). *Gangs and gang behavior.* Chicago: Nelson-Hall.

McCarthy, T. (1997). The return of L.A. gangs. Retrieved February 8, 2002, from Time Magazine Web site: www.time.com/time/photoessays/gangs/story.html.

McCorkle, R. C., & Miethe, T. D. (1998). The political and organizational response to gangs: An examination of a "moral panic" in Nevada. *Justice Quarterly, 15*(1), 41–64.

Meehan, A. (2000). The organizational career of gang statistics: The politics of policing gangs. *Sociological Quarterly, 41,* 337–368.

National Alliance of Gang Investigators Associations. (2002). *Executive summary.* Retrieved May 23, 2002, from www.nagia.org/executive_summary.htm

National Youth Gang Center. (2000). *1998 National Youth Gang Survey summary* (NCJ 183109). Washington, DC: U.S. Department of Justice, Office of Justice Programs, Office of Juvenile Justice and Delinquency Prevention.

North Carolina Justice Academy. (2002). *BLET (Basic Law Enforcement Training) manual: Instructor's training.* Raleigh: North Carolina Department of Justice.

Office of Juvenile Justice and Delinquency Prevention. (2000a). *Executive summary* (NCJ 171154). Washington, DC: U.S. Dept. of Justice, Office of Juvenile Justice Programs.

Office of Juvenile Justice and Delinquency Prevention. (2000b, November). *Survey results: Number of youth gangs and gang members* (OJJDP Summary, 1998 Youth Gang Survey). Washington, DC: U.S. Dept. of Justice.

Regoli, R., & Hewitt, J. (2000). *Delinquency in society.* New York: McGraw-Hill.

Roberts, A. R., & Fisher, P. (1997). Policy, administration, and direct service roles in victim/witness assistance programs. In C. A. McNeece & A. R. Roberts (Eds.), *Policy and practice in the juvenile justice system* (pp. 127–142). Chicago: Nelson Hall.

Roberts, A. R., & Waters, J. A. (1998). The coming storm: Juvenile violence and juvenile justice responses. In A. R. Roberts (Ed.), *Juvenile justice: Policies, programs, and services* (pp. 40–70). Chicago: Nelson-Hall.

Ryan, W. (1976). *Blaming the victim.* New York: Vintage.

Sandoval, G. (2001). *Protecting our future: The children of Los Angeles.* Los Angeles: Los Angeles School Police Department, Office of the Chief of Police. Retrieved February 8, 2002, from the Los Angeles School Police Department Web site: www.laspd.com

Security Threat Intelligence Unit (STIU), Tallahassee, Florida. (2002). Gang and Security Threat Group Awareness. Retrieved September 30, 2002, from Florida Dept. of Corrections Web site: www.dc.state.fl.us/pub/gangs/

Spergel, I. A. (1995). *The youth gang problem: A community approach.* New York: Oxford University Press.

Spergel, I. A., & Grossman, S. F. (1997). The Little Village Project: A community approach to the gang problem. *Social Work, 42,* 456–470.

Tarnished police trust [Editorial]. (2000, June 23). *St. Petersburg Times,* p. 1.

Thrasher, F. (2000). *The gang: A study of 1,313 gangs in Chicago.* Chicago: New Chicago School Press. (Original work published 1927)

Violent Crime Control and Law Enforcement Act, Pub. L. No. 103-322, 108 Stat. 1796 (1994).

Vogel, R. R., & Torres, S. (1998). An evaluation of Operation Roundup: An experiment in the control of gangs to reduce crime, fear of crime and improve police community relations. *International Journal of Police Strategies and Management, 21*(1), 38–53.

4

SERIAL MURDER

Popular Myths and Empirical Realities

JAMES ALAN FOX AND JACK LEVIN

Since the early 1980s, Americans have become more aware of and concerned about a particularly dangerous class of murderers, known as serial killers. Characterized by the tendency to kill repeatedly (at least three or four victims) and often with increasing brutality, serial killers stalk their victims, one at a time, for weeks, months, or years, generally not stopping until they are caught.

The term *serial killer* was first used in the early 1980s (see Jenkins, 1994), although the phenomenon of repeat killing existed, of course, throughout recorded history. In the late 1800s, for example, Hermann Webster Mudgett (aka H. H. Holmes) murdered dozens of attractive young women in his Chicago "house of death," and the infamous Jack the Ripper stalked the streets of London, killing five prostitutes. Prior to the 1980s, repeat killers such as Mudgett and Jack the Ripper were generally described as mass murderers. The need for a special classification for repeat killers was later recognized because of the important differences between multiple murderers who kill simultaneously and those

who kill serially (Levin & Fox, 1985). *Mass killers*—those who slaughter their victims in one event—tend to target people they know (e.g., family members or coworkers), often for the sake of revenge, using an efficient weapon of mass destruction (e.g., a high-powered firearm). As we shall describe below, serial murderers are different in all these respects, typically killing total strangers with their hands to achieve a sense of power and control over others.

A rising concern with serial killing has spawned a number of media presentations, resulting in the perpetrators of this type of murder becoming a regular staple of U.S. popular culture. A steady diet of television and movie productions could lead viewers to believe that serial killing is a common type of homicide. An increasing interest in serial homicide, however, has not been limited solely to the lay public. During the past two decades, the number, as well as the mix, of scholars devoting their attention to this crime has dramatically changed. Until the early 1980s, the literature exploring aspects of multiple homicide consisted almost

Fox, J. A., & Levin, J. (1998). Serial Murder: Popular Myths and Empirical Realities. In M. D. Smith and M. A. Zahn (Eds.), *Homicide: A Sourcebook of Social Research* (pp. 165-175). Thousand Oaks, CA: Sage. Reprinted with permission.

exclusively of bizarre and atypical case studies contributed by forensic psychiatrists pertaining to their court-assigned clients. More recently, there has been a significant shift toward social scientists examining the cultural and social forces underlying the late 20th-century rise in serial murder as well as law enforcement operatives developing research-based investigative tools for tracking and apprehending serial offenders.

Despite the shift in disciplinary focus, some basic assumptions of psychiatry appear to remain strong in the public mind. In particular, it is widely believed that the serial killer acts as a result of some individual pathology produced by traumatic childhood experiences. At the same time, a developing law enforcement perspective holds that the serial killer is a nomadic, sexual sadist who operates with a strict pattern to victim selection and crime scene behavior; this model has also contributed to myopic thinking in responding to serial murder. Unfortunately, these assumptions from both psychiatry and law enforcement may have retarded the development of new and more effective approaches to understanding this phenomenon. In an attempt to present a more balanced view, this chapter examines (serially, of course) several myths about serial killing/killers, some long-standing and others of recent origin, that have been embraced more on faith than on hard evidence.

MYTH 1: THERE IS AN EPIDEMIC OF SERIAL MURDER IN THE UNITED STATES

Although interest in serial murder has unquestionably grown, the same may not necessarily be true for the incidence of this crime itself. Curiously enough, there may actually be more scholars studying serial murder than there are offenders committing it. Regrettably, it is virtually impossible to measure with any degree of precision the prevalence of serial murder today, or even less so to trace its long-term trends (see Egger, 1990, 1998; Kiger, 1990). One thing for certain, however, is that the problem is nowhere near epidemic proportions (Jenkins, 1994).

It is true that some serial killers completely avoid detection. Unlike other forms of homicide, such as spousal murder, many of the crimes committed by serial killers may be unknown to authorities. Because serial murderers usually target strangers and often take great care in covering up their crimes by disposing of their victims' bodies, many of the homicides may remain as open missing persons reports. Moreover, because the victims frequently come from marginal groups, such as persons who are homeless, prostitutes, and drug users, disappearances may never result in any official reports of suspicious activity.

Even more problematic than the issue of missing data in measuring the extent of serial murder, law enforcement authorities are often unable to identify connections between unsolved homicides separated through time or space (Egger, 1984, 1998). Even if communication between law enforcement authorities were to improve (as it has in recent years), the tendency for some serial killers to alter their *modus operandi* frustrates attempts to link seemingly isolated killings to the same individual.

The lack of any hard evidence concerning the prevalence of serial homicide has not prevented speculation within both academic and law enforcement fields. The "serial killer panic of 1983-85," as it has been described by Jenkins (1988), was fueled by some outrageous and unsupportable statistics promulgated by the U.S. Department of Justice to buttress its claim that the extent of serial murder was on the rise. Apparently, some government officials reasoned that because the number of unsolved homicides had surged from several hundred per year in the early 1960s to several thousand per year in the 1980s, the aggregate body count produced by serial killers could be as high as 5,000 annually (Fox & Levin, 1985; for commentary on homicide clearance rates, see Chapter 6 by Marc Riedel). Unfortunately, this gross exaggeration was endorsed in some academic publications as well (see Egger, 1984; Holmes & DeBurger, 1988).

More sober thinking on the prevalence issue has occurred in recent years (Egger, 1990, 1998; Holmes & Holmes, 1998). Although still subject to the methodological limitations noted above in the identification of serial crimes, Hickey (1997) has attempted the most exhaustive measurement of the prevalence and trends in serial murder. In contrast to the Justice Department's estimate of thousands of victims annually, Hickey (1997) enumerated only 2,526 to 3,860 victims slain by 399 serial killers between 1800 and 1995. Moreover, between 1975 and 1995, the highest levels in the two centuries, Hickey identified only 153 perpetrators and as many as 1,400 victims, for an average annual tally of far less than 100 victims. Although Hickey's data collection strategy obviously ignored undetected cases, the extent of the problem is likely less than 1% of homicides in the country. Of course, that as much as 1% of the nation's murder problem can potentially be traced to but a few dozen individuals reminds us of the extreme deadliness of their predatory behavior.

MYTH 2: SERIAL KILLERS ARE
UNUSUAL IN APPEARANCE AND LIFESTYLE

As typically portrayed, television and cinematic versions of serial killers are either sinister-appearing creatures of the night or brilliant-but-evil master criminals. In reality, however, most tend to fit neither of these descriptions. Serial killers are generally White males in their late 20s or 30s who span a broad range of human qualities including appearance and intelligence.

Some serial killers are high school dropouts, and others might indeed be regarded as unappealing by conventional standards. At the same time, a few actually possess brilliance, charm, and attractiveness. Most serial killers, however, are fairly average, at least to the casual observer. In short, they are "extraordinarily ordinary"; ironically, part of the secret of their success is that they do not stand out in a crowd or attract negative attention to themselves. Instead, many of them look and act much like "the boy next door"; they hold full-time jobs, are married or involved in some other stable relationship, and are members of various local community groups. The one trait that tends to separate prolific serial killers from the norm is that they are exceptionally skillful in their presentation of self so that they appear beyond suspicion. This is part of the reason why they are so difficult to apprehend (Levin & Fox, 1985).

A related misconception is that serial killers, lacking stable employment or family responsibilities, are full-time predators who roam far and wide, often crossing state and regional boundaries in their quest for victims. Evidence to the contrary notwithstanding, serial killers have frequently been characterized as nomads whose compulsion to kill carries them hundreds of thousands of miles a year as they drift from state to state and region to region leaving scores of victims in their wake. This may be true of a few well-known and well-traveled individuals, but not for the vast majority of serial killers (Levin & Fox, 1985). According to Hickey (1997), only about a third of the serial killers in his database crossed state lines in their murder sprees. John Wayne Gacy, for example, killed all of his 33 young male victims at his Des Plaines, Illinois, home, conveniently burying most of them there as well. Gacy had a job, friends, and family but secretly killed on a part-time, opportunistic basis.

MYTH 3: SERIAL KILLERS *ARE ALL INSANE*

What makes serial killers so enigmatic—so irrational to many casual observers—is that they generally kill not for love, money, or revenge but for the fun of it. That is, they delight in the thrill, the sexual satisfaction, or the dominance that they achieve as they squeeze the last breath of life from their victims. At a purely superficial level, killing for the sake of pleasure seems nothing less than "crazy."

The basis for the serial killer's pursuit of pleasure is found in a strong tendency toward sexual sadism (Hazelwood, Dietz, & Warren, 1992) and an interest reflected in detailed fantasies of domination (Prentky, Burgess, & Rokous, 1989). Serial killers tie up their victims to watch them squirm and torture their victims to hear them scream. They rape, mutilate, sodomize, and degrade their victims to feel powerful, dominant, and superior.

Many individuals may have fantasies about torture and murder but are able to restrain themselves from ever translating their sadistic dreams into reality. Those who do not contain their urges to kill repeatedly for no apparent motive are assumed to suffer from some extreme form of mental illness. Indeed, some serial killers have clearly been driven by psychosis, such as Herbert Mullen of Santa Cruz, California, who killed 13 people during a

4-month period to avert an earthquake—at least that is what the voices commanded him to do (the voices also ordered him to burn his penis with a cigarette).

In either a legal or a medical sense, however, most serial killers are not insane or psychotic (see Levin & Fox, 1985; Leyton, 1986). They know right from wrong, know exactly what they are doing, and can control their desire to kill—but choose not to. They are more cruel than crazy. Their crimes may be sickening, but their minds are not necessarily sick. Most apparently do not suffer from hallucinations, a profound thought disorder, or major depression. Indeed, those assailants who are deeply confused or disoriented are generally not capable of the level of planning and organization necessary to conceal their identity from the authorities and, therefore, do not amass a large victim count.

Many serial killers seem instead to possess a personality disorder known as sociopathy (or antisocial personality). They lack a conscience, are remorseless, and care exclusively for their own needs and desires. Other people are regarded merely as tools to be manipulated for the purpose of maximizing their personal pleasure (see Harrington, 1972; Magid & McKelvey, 1988). Thus, if given to perverse sexual fantasy, sociopaths simply feel uninhibited by societal rules or by conscience from literally chasing their dreams in any way necessary for their fulfillment (see Fox, 1989; Levin & Fox, 1985; Vetter, 1990).

Serial killers are not alone in their sociopathic tendencies. The American Psychiatric Association estimates that 3% of all males in our society could be considered sociopathic (for a discussion of the prevalence of antisocial personality disorder, see American Psychiatric Association, 1994). Of course, most sociopaths do not commit acts of violence; they may lie, cheat, or steal, but rape and murder are not necessarily appealing to them—unless they are threatened or they regard killing as a necessary means to some important end.

MYTH 4: ALL SERIAL KILLERS *ARE SOCIOPATHS*

Although many serial killers tend to be sociopaths, totally lacking in concern for their victims, some actually do have a conscience but are able to neutralize or negate their feelings of remorse by rationalizing their behavior. They feel as though they are doing something good for society, or at least nothing that bad.

Milwaukee's cannibalistic killer, Jeffrey Dahmer, for example, actually viewed his crimes as a sign of love and affection. He told Tracy Edwards, a victim who managed to escape, that if he played his cards right, he too could give his heart to Jeff. Dahmer meant it quite literally, of course, but according to Edwards, he said it affectionately, not threateningly.

The powerful psychological process of *dehumanization* allows many serial killers to slaughter scores of innocent people by viewing them as worthless and therefore expendable. To the dehumanizer, prostitutes are seen as mere sex machines, gays are AIDS carriers, nursing home patients are vegetables, and homeless alcoholics are nothing more than human trash.

In a process related to this concept of dehumanization, many serial killers compartmentalize the world into two groups—those whom they care about versus everyone else. "Hillside Strangler" Kenneth Bianchi, for example, could be kind and loving to his wife and child as well as to his mother and friends yet be vicious and cruel to those he considered meaningless. He and his cousin started with prostitutes, but later, when they grew comfortable with killing, branched out to middle-class targets.

MYTH 5: SERIAL KILLERS *ARE INSPIRED BY PORNOGRAPHY*

Could Theodore Bundy have been right in his death row claim that pornography turned him into a vicious killer, or was he just making excuses to deflect blame? It should be no surprise that the vast majority of serial killers do have a keen interest in pornography, particularly sadistic magazines and films (Ressler, Burgess, & Douglas, 1988). Sadism is the source of their greatest pleasure, and so, of course, they experience it vicariously in their spare time, when not on the prowl themselves. That is, a preoccupation with pornography is a reflection, not the cause, of their own sexual desires. At most, pornography may reinforce sadistic impulses, but it cannot create them.

There is experimental evidence that frequent and prolonged exposure to violent pornography tends to desensitize "normal" men to the plight of victims of sexual abuse (Malamuth & Donnerstein, 1984). In the case of serial killers, however, it takes much more than pornography to create such an extreme and vicious personality.

MYTH 6: SERIAL KILLERS ARE PRODUCTS OF *BAD CHILDHOODS*

Whenever the case of an infamous serial killer is uncovered, journalists and behavioral scientists alike tend to search for clues in the killer's childhood that might explain the seemingly senseless or excessively brutal murders. Many writers have emphasized, for example, Theodore Bundy's concerns about being illegitimate, and biographers of Hillside Strangler Kenneth Bianchi capitalized on his having been adopted.

There is a long tradition of research on the childhood correlates of homicidal proneness. For example, several decades ago, Macdonald (1963) hypothesized a triad of symptoms— enuresis, fire setting, and cruelty to animals— that were seen as reactions to parental rejection, neglect, or brutality. Although the so-called Macdonald's Triad was later refuted in controlled studies (see Macdonald, 1968), the connection between parental physical/sexual abuse or abandonment and subsequent violent behavior has remained a continuing focus of research (Sears, 1991). It is often suggested that because of such deep-rooted problems, serial killers suffer from a profound sense of powerlessness that they compensate for through extreme forms of aggression in which they exert control over others.

It is true that the biographies of most serial killers reveal significant physical and psychological trauma at an early age. For example, based on in-depth interviews with 36 incarcerated murderers, Ressler et al. (1988) found evidence of psychological abuse (e.g., public humiliation) in 23 cases and physical trauma in 13 cases. Hickey (1997) reported that among a group of 62 male serial killers, 48% had been rejected as children by a parent or some other important person in their lives. Of course, these same types of experiences can be found in the biographies of many "normal" people as well. More specifically, although useful for characterizing the backgrounds of serial killers, the findings presented by Ressler et al. and Hickey lack a comparison group drawn from nonoffending populations for which the same operational definitions of trauma have been applied. Therefore, it is impossible to conclude that serial killers have suffered as children to any greater extent than others.

As a related matter, more than a few serial killers—from New York City's David Berkowitz to Long Island's Joel Rifkin—were raised by adoptive parents. In the adopted child syndrome, an individual displaces anger for birth parents onto adoptive parents as well as other authority figures. The syndrome is often expressed early in life in "provocative antisocial behavior" including fire setting, truancy, promiscuity, pathological lying, and stealing. Deeply troubled adopted children may, in fantasy, create imaginary playmates who represent their antisocial impulses. Later, they may experience a dissociative disorder or even the development of an alter personality in which their murderous tendencies become situated (Kirschner, 1990, 1992).

The apparent overrepresentation of adoption in the biographies of serial killers has been exploited by those who are looking for simple explanations for heinous crimes, without fully recognizing the mechanisms behind or value of the link between adoption and criminal behavior. Even if adoption plays a role in the making of a serial murderer, the independent variable remains to be specified—that is, for example, rejection by birth parents, poor health and prenatal care of birth mother, or inadequate bonding to adoptive parents.

Some neurologists and a growing number of psychiatrists suggest that serial killers have incurred serious injury to the limbic region of the brain resulting from severe or repeated head trauma, generally during childhood. As an example, psychiatrist Dorothy Lewis and neurologist Jonathan Pincus, along with other colleagues, examined 15 murderers on Florida's death row and found that all showed signs of neurological irregularities (Lewis, Pincus, Feldman, Jackson, & Bard, 1986). In addition, psychologist Joel Norris (1988) reported excessive spinal fluid found in the brain scan of serial killer Henry Lee Lucas. Norris argued that this abnormality reflected the possible damage caused by an earlier blow or a series of blows to Lucas's head.

It is critical that we place in some perspective the many case studies that have been used in an attempt to connect extreme violence to neurological impairment. Absent from the case study approach is any indication of the prevalence of individuals who did not act violently despite a history of trauma. Indeed, if head trauma were as strong a contributor to serial murder as some suggest, then we would have many times more of these killers than we actually do.

It is also important to recognize that neurological impairment must occur in combination with a host of environmental conditions to place an individual at risk for extreme acts of brutality. Dorothy Lewis cautions, "The neuropsychiatric problems alone don't make you violent. Probably the environmental factors in and of themselves don't make you a violent person. But when you put them together, you create a very dangerous character" ("Serial Killers," 1992). Similarly, Ressler asserts that no single childhood problem indicates future criminality: "There are a whole pot of conditions that have to be met" for violence to be predictable (quoted in Meddis, 1987, p. 3A). Head trauma and abuse, therefore, may be important risk factors, but they are neither necessary nor sufficient to make someone a serial killer. Rather, they are part of a long list of circumstances—including adoption, shyness, disfigurement, speech impediments, learning and physical disabilities, abandonment, death of a parent, academic and athletic inadequacies—that may make a child feel frustrated and rejected enough to predispose, but not predestine, him or her toward extreme violence.

Because so much emphasis has been placed on early childhood, developmental factors in making the transition into adulthood and middle age are often overlooked. Serial killers tend to be in their late 20s and 30s, if not older, when they first show outward signs of murderous behavior. If only early childhood and biological predisposition were involved, why do they not begin killing as adolescents or young adults? Many individuals suffer as children, but only some of them continue to experience profound disappointment and detachment regarding family, friends, and work. For example, Danny Rolling, who murdered several college students in Gainesville, Florida, may have had a childhood filled with frustration and abuse, but his eight-victim murder spree did not commence until he was 36 years old. After experiencing a painful divorce, he drifted from job to job, from state to state, from prison to prison, and finally from murder to murder (Fox & Levin, 1996).

MYTH 7: SERIAL KILLERS CAN *BE IDENTIFIED IN ADVANCE*

Predicting dangerousness, particularly in an extreme form such as serial homicide, has been an elusive goal for those investigators who have attempted it. For example, Lewis, Lovely, Yeager, and Femina (1989) suggest that the interaction of neurological/psychiatric impairment and a history of abuse predicts violent crime, better even than previous violence itself. Unfortunately, this conclusion was based on retrospective "postdiction" with a sample of serious offenders, rather than a prospective attempt to predict violence within a general cross section.

It is often said that "hindsight is 20/20." This is especially true for serial murder. Following the apprehension of a serial killer, we often hear mixed reports that "he seemed like a nice guy, but there was something about him that wasn't quite right." Of course, there is often something about most people that may not seem "quite right." When such a person is exposed to be a serial murderer, however, we tend to focus on those warning signs in character and biography that were previously ignored. Even the stench emanating from Jeffrey Dahmer's apartment, which he had convincingly explained to the neighbors as the odor of spoiled meat from his broken freezer, was unexceptional until after the fact.

The methodological problems in predicting violence in advance are well known (Chaiken, Chaiken, & Rhodes, 1994). For a category of violence as rare as serial murder, however, the low base rate and consequent false-positive dilemma are overwhelming. Simply put, there are thousands of White males in their late 20s or 30s who thirst for power, are sadistic, and lack strong internal controls; most emphatically, however, the vast majority of them will never kill anyone.

MYTH 8: ALL SERIAL KILLERS ARE SEXUAL SADISTS

Serial killers who rape, torture, sodomize, and mutilate their victims attract an inordinate amount of attention from the press, the public, and professionals as well. Although they may be the most fascinating type of serial killer, they are hardly the only type.

Expanding their analysis beyond the sexual sadist, Holmes and DeBurger (1988) were among the first to assemble a motivational typology of serial killing, classifying serial murderers into four broad categories: visionary (e.g., voices from God), mission-oriented (e.g., ridding the world of evil), hedonistic (e.g., killing for pleasure), and power/control-oriented (e.g., killing for dominance). Holmes and DeBurger further divided the hedonistic type into three subtypes: lust, thrill, and comfort (see also Holmes & Holmes, 1998).

Although we applaud Holmes and DeBurger for their attempt to provide some conceptual structure, we must also note a troubling degree of overlap among their types. For example, Herbert Mullen, believing that he was obeying God's commandment, "sacrificed" (in his mind) more than a dozen people to avert catastrophic earthquakes; his motivation was both "visionary" and "mission-oriented." Furthermore, the typology is somewhat misaligned: Both the "lust" and "thrill" subtypes are expressive motivations, whereas "comfort" (e.g., murder for profit or to eliminate witnesses) is instrumental or a means toward an end.

Modifying the Holmes-DeBurger framework, we suggest that serial murders can be reclassified into three categories, each with two subtypes:

1. Thrill
   a. Sexual sadism
   b. Dominance

2. Mission
   a. Reformist
   b. Visionary

3. Expedience
   a. Profit
   b. Protection

Most serial killings can be classified as thrill motivated, and the *sexual sadist* is the most common of all. In addition, a growing number of murders committed by hospital caretakers have been exposed in recent years; although not sexual in motivation, these acts of murder are perpetrated for the sake of *dominance* nevertheless.

A less common form of serial killing consists of mission-oriented killers who murder to further a cause. Through killing, the *reformist* attempts to rid the world of filth and evil, such as by killing prostitutes, gays, or homeless persons. Most self-proclaimed reformists are also motivated by thrill seeking but try to rationalize their murderous behavior. For example, Donald Harvey, who worked as an orderly in Cincinnati-area hospitals, confessed to killing 80 or more patients through the years. Although he was termed a mercy killer, Harvey actually enjoyed the dominance he achieved by playing God with the lives of other people.

In contrast to pseudoreformists, *visionary* killers, as rare as they may be, genuinely believe in their missions. They hear the voice of the devil or God instructing them to kill. Driven by these delusions, visionary killers tend to be psychotic, confused, and disorganized. Because their killings are impulsive and even frenzied, visionaries rarely remain on the street long enough to become prolific serial killers.

The final category of serial murder includes those who are motivated by the expedience of either profit or protection. The *profit-oriented* serial killer systematically murders as a critical element of the overall plan to dispose of victims to make money (e.g., Sacramento landlady Dorothea Puente murdered 9 elderly tenants to cash their social security checks). By contrast, the *protection-oriented* killer uses murder to cover up criminal activity (e.g., the Lewington brothers systematically robbed and murdered 10 people throughout Central Ohio).

MYTH 9: SERIAL KILLERS SELECT VICTIMS WHO SOMEHOW *RESEMBLE THEIR MOTHERS*

Shortly after the capture of Hillside Strangler Kenneth Bianchi, psychiatrists speculated that he tortured and murdered young women as an expression of hatred toward his mother, who had allegedly brutalized him as a youngster (Fox & Levin, 1994). Similarly, the execution of Theodore Bundy gave psychiatrists occasion to suggest that his victims served as surrogates for the real target he sought, his mother.

Although unresolved family conflicts may in some cases be a significant source of frustration, most serial killers have a more opportunistic or pragmatic basis for selecting their victims. Quite simply, they tend to prey on the most vulnerable targets—prostitutes, drug users, hitchhikers, and runaways, as well as older hospital patients (Levin & Fox, 1985). Part of the vulnerability concerns the ease with which these groups can be abducted or overtaken. Children and older persons are defenseless because of physical stature or disability; hitchhikers and prostitutes become vulnerable as soon as they enter the killer's vehicle; hospital patients are vulnerable in their total dependency on their caretakers.

Vulnerability is most acute in the case of prostitutes, which explains their relatively high rate of victimization by serial killers. A sexual sadist can cruise a red-light district, seeking out the woman who best fits his deadly sexual fantasies. When he finds her, she willingly complies with his wishes—until it is too late.

Another aspect of vulnerability is the ease with which the killers can avoid being detected following a murder. Serial killers of our time are often sly and crafty, fully realizing the ease with which they can prey on streetwalkers and escape detection, much less arrest. Because the disappearance of a prostitute is more likely to be considered by the police, at least initially, as a missing person rather than a victim of homicide, the search for the body can be delayed weeks or months. Also, potential witnesses to abductions in red-light districts tend to be unreliable sources of information or distrustful of the police.

Frail older persons, particularly those in hospitals and nursing homes, represent a class of victims that is at the mercy of a different type of serial killer, called "angels of death." Revelations by a Long Island nurse who poisoned his patients in a failed attempt to be a hero by resuscitating them and of two Grand Rapids nurses aides who murdered older patients to form a lovers' pact have horrified even the most jaded observers of crime.

Not only are persons who are old and infirm vulnerable to the misdeeds of their caretakers who may have a particularly warped sense of mercy, but hospital homicides are particularly difficult to detect and solve. Death among older patients is not uncommon, and suspicions are rarely aroused. Furthermore, should a curiously large volume of deaths occur within a short time on a particular nurse's shift, hospital administrators feel in a quandary. Not only are they reluctant to bring scandal and perhaps lawsuits to their own facility without sufficient proof, but most of the potentially incriminating evidence against a suspected employee is long buried with the victim.

MYTH 10: SERIAL KILLERS *REALLY WANT TO GET CAUGHT*

Despite the notion that serial killers are typically lacking in empathy and remorse, some observers insist that deeply repressed feelings of guilt may subconsciously motivate them to leave telltale clues for the police. Although this premise may be popular in media portrayals, most serial killers go to great lengths to avoid detection, such as carefully destroying crime scene evidence or disposing of their victims' bodies in hard-to-find dump sites.

There is an element of self-selection in defining serial killing. Only those offenders who have sufficient cunning and guile are able to avoid capture long enough to accumulate the number of victims necessary to be classified as serial killers. Most serial killers are careful, clever, and, to use the FBI's typology (see Ressler et al., 1988), organized. Of course, disorganized killers, because of their carelessness, tend to be caught quickly, often before they surpass the serial killer threshold of victim count.

Murders committed by a serial killer are typically difficult to solve because of lack of both motive and physical evidence. Unlike the usual homicide that involves an offender and a victim who know one another, serial murders are almost exclusively committed by strangers. Thus, the usual police strategy of identifying suspects by considering their possible motive, be it jealousy, revenge, or greed, is typically fruitless.

Another conventional approach to investigating homicides involves gathering forensic evidence—fibers, hairs, blood, and prints—from the scene of the crime. In the case of many serial murders, however, this can be rather difficult, if not impossible. The bodies of the victims are often found at desolate roadsides or in makeshift graves, exposed to rain, wind, and snow. Most of the potentially revealing crime scene evidence remains in the unknown killer's house or car.

Another part of the problem is that unlike those shown in the media, many serial killers do not leave unmistakable and unique "signatures" at their crime scenes. As a result, the police may not recognize multiple homicides as the work of the same perpetrator. Moreover, some serial killings, even if consistent in style, traverse jurisdictional boundaries. Thus, "linkage blindness" is a significant barrier to solving many cases of serial murder (Egger, 1984).

To aid in the detection of serial murder cases, the FBI operationalized in 1985 the Violent Criminal Apprehension Program (VICAP), a computerized database for the collection and collation of information pertaining to unsolved homicides and missing persons around the country. It is designed to flag similarities in unsolved crimes that might otherwise be obscure (Howlett, Haufland, & Ressler, 1986).

Although an excellent idea in theory, VICAP has encountered significant practical limitations. Complexities in the data collection forms have limited the extent of participation of local law enforcement agencies in completing VICAP questionnaires. More important, pattern recognition is far from a simple or straightforward task, regardless of how powerful the computer or sophisticated the software. Furthermore, even the emergence of a pattern among a set of crime records in the VICAP database does not ensure that the offender will be identified.

In addition to the VICAP clearinghouse, the FBI, on request, assembles criminal profiles of the unknown offenders, based on behavioral clues left at crime scenes, autopsy reports, and police incident reports. Typically, these profiles speculate on the killer's age, race, sex, marital status, employment status, sexual maturity, possible criminal record, relationship to the victim, and likelihood of committing future crimes. At the core of its profiling strategy, the FBI distinguishes between *organized nonsocial* and *disorganized asocial* killers. According to Hazelwood and Douglas (1980), organized killers typically are intelligent, are socially and sexually competent, are of high birth order, are skilled workers, live with a partner, are mobile, drive late model cars, and follow their crimes in the media. In contrast, disorganized killers generally are unintelligent, are socially and sexually inadequate, are of low birth order, are unskilled workers, live alone, are nonmobile, drive old cars or no car at all, and have minimal interest in the news reports of their crimes.

According to the FBI analysis, these types tend to differ also in crime scene characteristics (Ressler et al., 1988). Specifically, organized killers use restraints on the victim, hide or transport the body, remove the weapon from the scene, molest the victim prior to death, and are methodical in their style of killing. Operating differently, disorganized killers tend not to use restraints, leave the body in full view, leave a weapon at the scene, molest the victim after death, and are spontaneous in their manner of killing. The task of profiling involves, therefore, drawing inferences from the crime scene to the behavioral characteristics of the killer.

Despite the Hollywood hype that exaggerates the usefulness of criminal profiling, it is an investigative tool of some, albeit limited, value. Even when constructed by the most experienced and skillful profilers, such as those at the FBI, profiles are not expected to solve a case; rather, they provide an additional set of clues in cases found by local police to be unsolvable. Simply put, a criminal profile cannot identify a suspect for investigation, nor can it eliminate a suspect who does not fit the mold. An overreliance on the contents of a profile can misdirect a serial murder investigation, sometimes quite seriously (see, for example, Fox & Levin, 1996). Clearly, a criminal profile can assist in assigning subjective probabilities to suspects whose names surface through more usual investigative strategies (e.g., interviews of witnesses, canvassing of neighborhoods, and "tip" phone lines). There is, however, no substitute for old-fashioned detective work and, for that matter, a healthy and helpful dose of luck.

From Myth to Reality

The study of serial homicide is in its infancy, less than two decades old (O'Reilly-Fleming, 1996). The pioneering scholars noted the pervasiveness and inaccuracy of long-standing psychiatric misconceptions regarding the state of mind of the serial killer (see Levin & Fox, 1985; Leyton, 1986; Ressler et al., 1988). More recently, these unfounded images have been supplanted by newer myths, including those concerning the prevalence and apprehension of serial killers.

The mythology of serial killing has developed from a pervasive fascination with a crime about which so little is known. Most of the scholarly literature is based on conjecture, anecdote, and small samples, rather than rigorous and controlled research. The future credibility of this area of study will depend on the ability of criminologists to upgrade the standards of research on serial homicide. Only then will myths about serial murder give way to a reliable foundation of knowledge.

References

American Psychiatric Association. (1994). *Diagnostic and statistical manual of mental disorders* (4th ed.). Washington, DC: American Psychiatric Association.

Chaiken, J., Chaiken, M., & Rhodes, W. (1994). Predicting violent behavior and classifying violent offenders. In A. J. Reiss, Jr. & J. A. Roth (Eds.), *Understanding and preventing violence* (Vol. 4, pp. 217-295). Washington, DC: National Academy Press.

Egger, S. A. (1984). A working definition of serial murder and the reduction of linkage blindness. *Journal of Police Science and Administration, 12,* 348-357.

Egger, S. A. (1990). *Serial murder: An elusive phenomenon.* Westport, CT: Praeger.

Egger, S. A. (1998). *The killers among us: An examination of serial murder and its investigation.* Upper Saddle River, NJ: Prentice Hall.

Fox, J. A. (1989, January 29). The mind of a murderer. *Palm Beach Post,* p. 1E.

Fox, J. A., & Levin, J. (1985, December 1). Serial killers: How statistics mislead us. *Boston Herald,* p. 45.

Fox, J. A., & Levin, J. (1994). *Overkill: Mass murder and serial killing exposed.* New York: Plenum.

Fox, J. A., & Levin, J. (1996). *Killer on campus.* New York: Avon Books.

Harrington, A. (1972). *Psychopaths.* New York: Simon & Schuster.

Hazelwood, R. R., Dietz, P. E., & Warren, J. (1992). The criminal sexual sadist. *FBI Law Enforcement Bulletin, 61,* 12-20.

Hazelwood, R. R., & Douglas, J. E. (1980). The lust murderer. *FBI Law Enforcement Bulletin, 49,* 1-5.

Hickey, E. W. (1997). *Serial murderers and their victims* (2nd ed.). Belmont, CA: Wadsworth.

Holmes, R. M., & DeBurger, J. (1988). *Serial murder.* Newbury Park, CA: Sage.

Holmes, R. M., & Holmes, S. T. (1998). *Serial murder* (2nd ed.). Thousand Oaks, CA: Sage.

Howlett, J. B., Haufland, K. A., & Ressler, R. J. (1986). The violent criminal apprehension program—VICAP: A progress report. *FBI Law Enforcement Bulletin, 55,* 14-22.

Jenkins, P. (1988). Myth and murder: The serial killer panic of 1983-85. *Criminal Justice Research Bulletin* (No. 3). Huntsville, TX: Sam Houston State University.

Jenkins, P. (1994). *Using murder: The social construction of serial homicide.* New York: Walter de Gruyter.

Kiger, K. (1990). The darker figure of crime: The serial murder enigma. In S. A. Egger (Ed.), *Serial murder: An elusive phenomenon* (pp. 35-52). New York: Praeger.

Kirschner, D. (1990). The adopted child syndrome: Considerations for psychotherapy. *Psychotherapy in Private Practice, 8,* 93-100.

Kirschner, D. (1992). Understanding adoptees who kill: Dissociation, patricide, and the psychodynamics of adoption. *International Journal of Offender Therapy & Comparative Criminology, 36,* 323-333.

Levin, J., & Fox, J. A. (1985). *Mass murder: America's growing menace.* New York: Plenum.

Lewis, D. O., Lovely, R., Yeager, C., & Femina, D. D. (1989). Toward a theory of the genesis of violence: A follow-up study of delinquents. *Journal of the American Academy of Child and Adolescent Psychiatry, 28,* 431-436.

Lewis, D. O., Pincus, J. H., Feldman, M., Jackson, L., & Bard, B. (1986). Psychiatric, neurological, and psychoeducational characteristics of 15 death row inmates in the United States. *American Journal of Psychiatry, 143,* 838-845.

Leyton, E. (1986). *Compulsive killers: The story of modern multiple murderers.* New York: New York University Press.

Macdonald, J. M. (1963). The threat to kill. *American Journal of Psychiatry, 120,* 125-130.

Macdonald, J. M. (1968). *Homicidal threats.* Springfield, IL: Charles C Thomas.

Magid, K., & McKelvey, C. A. (1988). *High risk: Children without a conscience.* New York: Bantam.

Malamuth, N. M., & Donnerstein, E. (1984). *Pornography and sexual aggression.* Orlando, FL: Academic Press.

Meddis, S. (1987, March 31). FBI: Possible to spot, help serial killers early. *USA Today,* p. 3A.

Norris, J. (1988). *Serial killers: The growing menace.* New York: Doubleday.

O'Reilly-Fleming, T. (1996). *Serial and mass murder: Theory, research and policy.* Toronto, Ontario: Canadian Scholars' Press.

Prentky, R. A., Burgess, A. W., & Rokous, F. (1989). The presumptive role of fantasy in serial sexual homicide.

American Journal of Psychiatry, 146, 887-891.

Ressler, R. K., Burgess, A. W., & Douglas, J. E. (1988). *Sexual homicide: Patterns and motives.* Lexington, MA: Lexington Books.

Sears, D. J. (1991). *To kill again.* Wilmington, DE: Scholarly Resources Books.

Serial killers. (1992, October 18). NOVA. Boston: WGBH-TV.

Vetter, H. (1990). Dissociation, psychopathy, and the serial murderer. In S. A. Egger (Ed.), *Serial murder: An elusive phenomenon* (pp. 73-92). New York: Praeger.

5

RELIGION-RELATED CRIME

Documentation of Murder,
Fraud, and Sexual Abuse

KAREL KURST-SWANGER AND MARGARET RYNIKER

Religion is an integral part of American life, providing social connections, spiritual guidance, and a reinforcement for law conformity. Faith communities have been active players in the fight against drug abuse, delinquency, and other crime problems. There is no denying the powerful role that religion can play in healing communities ravaged by crime and other social problems. Religious leaders can play a unique role in this endeavor.

However, this chapter will explore a different dimension of the relationship between religion and crime. We will examine specific instances in which crime is committed within the realm of religious ideology. Crime and religion intersect in a number of circumstances in which religious beliefs are used to legitimate criminal behavior, religious ideology is used to instigate and justify violence, and religious institutions cover up criminal behavior to avoid scandal. Murder, suicide, financial exploitation, neglect, and sexual abuse are but a few of the crimes that may have religious overtones. Clergy have committed child sexual abuse, parents have been prosecuted for manslaughter when they refuse to access traditional medical treatments for their children, and religious

movements have been held responsible for a number of killing sprees. Left in the wake are innocent children and adult victims whose intentions were only to believe in something bigger than themselves. Religious institutions are consequently confronted with the challenge of protecting their image and defending their beliefs in light of unrelenting media attention and public scrutiny.

Lines are blurred between religious freedom and the commission of abuse, neglect, and criminal behavior. The U.S. government has historically maintained a "hands-off" stance in dealing with all organized religions. American political doctrine celebrates the separation of church and state and the free exercise of religion. Only when a complaint is brought that is so clear cut that it cannot be ignored will the state intervene. When a group has practices that lead to crime, those crimes are to be prosecuted as crimes, not as religious crimes. The courts are left with the formidable task of balancing the right to practice religion against the rights of states to intervene when those practices place others in harm's way. Religious institutions must reexamine their boundaries between spiritual obligations and civil responsibility, recognizing the right to pursue religious freedom *only* within the context of civil and criminal law.

Our intention here is to explore specific examples of religion-related crime within the context of American culture. This chapter is divided into five main sections. The first section addresses the lack of a scholarly framework to discuss religion-related crime in the United States. We suggest a three-pronged typology as a potential conceptual framework to discuss the unique, dynamic relationship between religion and crime. We also present some critical issues that are likely to be faced by scholars, theologians, practitioners, and policy makers in the future. The next three sections provide illustrations of each type of religion-related crime and some of the social, political, and legal issues raised by each. We conclude the chapter with some recommendations for future discourse and inquiry.

RELIGION-RELATED CRIME: A TYPOLOGY

We struggle to find a place for such crimes in our criminological vocabulary, but to date little research has focused broadly on crimes within religious environments. We will refer to such crimes as *religion related* to better address the often disparate relationship between religious and secular law in the United States and the variety of factors and motivations that may be involved in the commission of crime within a particular religious context. Because theology is not necessarily the motivational factor behind every criminal incident, *religion related* serves as an all-inclusive term. It allows us to describe crimes that have been committed within a variety of religious contexts. We define religion-related crime as any illegal or socially injurious act that is committed within the auspices of religious practice or as a result of a particular religious belief. We distinguish three types of religion-related crimes: reactive crimes, theological crimes, and abuses of religious authority.

Reactive crimes are those in which a religious leader or group commits a crime in reaction to some external force, such as a social or political force. For example, a reactive crime might be committed to facilitate change in the social order on the basis of religious ideology. For example, individuals who have bombed

abortion clinics or murdered physicians who perform abortions commit such murder as a reaction or response to the current sociopolitical order, which goes against their particular religious ideology. Cult groups, such as the Branch Davidians of Waco, Texas, or MOVE of Philadelphia, Pennsylvania, reacted to law enforcement intervention with tragic results. Reactive forms of religion-related crime reflect the inherent conflicts between church, state, and the broader secular society. Conflicts are often resolved with violence. In this sense, the reactive type of religion-related crime is somewhat similar to political crime aimed at maintaining or altering the existing political structure, regime, or ideology. This chapter presents several crimes committed within new religious movements, generally referred to as cults, as examples of reactive religion-related crime.

Theological crimes are committed as a result of particular religious practice. Violating the law is not the motive but a consequence of the dissonance between theology and secular law. Examples are Native American tribes' use of peyote as a religious rite, parents' beating of their children to rid them of evil, and the practice of polygamy. This chapter presents the refusal to access medical treatment for children, based on religious belief, as an example of theological religion-related crime.

Abuses of religious authority are situations in which a religious leader or institution abuses charismatic or legitimate religious authority for the benefit of the individual or the institution. Examples include clergy who embezzle from church coffers, evangelists who commit fraud, and religious leaders who physically, sexually, or emotionally abuse church members. This type of religion-related crime may also involve religious institutions that participate in a criminal scheme as an organization or, through their neglect, place members of their organization in danger of victimization. In this sense, the abuse of religious authority as a type of religion-related crime is most closely associated with occupational crime. This chapter presents the recent disclosure of sexually abusive priests and the role of the Catholic Church in perpetuating the problem as an example of abuse of religious authority.

This chapter's three specific examples of religion-related crime—crimes committed within new religious movements, the medical neglect of children, and the abuse of religious authority as evidenced by the recent disclosures of child sexual abuse involving Catholic priests— were chosen because of their current salience for scholars and professional practitioners from a variety of disciplines, including legal, medical, social work, psychological, criminal justice, sociological, and religious studies. We focus on recent debates between the religious and secular communities, the legal issues raised by such debates, and some of the dilemmas of public policy making. We raise the following critical issues:

- Should the Catholic Church face civil and/or criminal liability charges for their protection of known child molesters?
- Should new religious movements be monitored by governmental authorities?
- Should religious communities participate in scientific inquiries into faith healing?
- Is there a connection between traditional forms of family violence and the types of abuses commonly found in religious communal living?
- What types of rational policies should be developed to further protect victims of religion-related crime?

REACTIVE RELIGION-RELATED CRIME: DANGER IN NEW RELIGIOUS MOVEMENTS?

The last several decades have been marked by the proliferation of numerous new religious movements (NRMs), often referred to as "cults" or "fringe religious groups." Some have estimated that 3,000 new alternative religious groups have emerged in recent years worldwide (Dawson, 1998). These groups represent a wide range of beliefs, practices, and customs and therefore cannot be generalized into one category (Mayer, 2001). For the most part, new religious movements have enjoyed a peaceful existence with little interference from the wider social order, and the majority have not participated in malevolent acts.

There are some groups, however, whose operations involve deviant behavior, including murder, suicide, and sexual abuse (among other crimes). These groups have often drawn law enforcement officials into no-win situations in

which death and destruction were the only result. Examples of such horrific situations include the following (Rush, 2000):

- In 1985, an armed conflict between MOVE and the police left 11 people killed and an additional 270 homeless after a firebomb (set by police) destroyed housing in Philadelphia, Pennsylvania.
- On April 19, 1993, a 51-day siege was concluded in the Branch Davidian compound in Waco, Texas, in which 86 Branch Davidians and four law enforcement agents were killed and an additional 16 other officers were injured. Only 9 Branch Davidians survived the ordeal.
- On March 26, 1997, 39 members of the Heaven's Gate religious group committed mass suicide by ingesting food laced with phenobarbitol and washed down with vodka.

Here we will explore the unique dynamics of NRMs specifically referred to as cults and their potential for the commission of criminal acts when external forces provide a sense of urgency or emergency, moving them to react or respond.

The Term *Cult*

The term *cult* is derived from the Latin word *cultus,* meaning "care" or "adoration," which is in turn derived from *colere,* meaning "to cultivate." The term was originally used by scholars studying the sociology of religion simply to refer to new religious organizations (Richardson, 1993). However, as its use has proliferated in the media and the anticult movement, it has taken on sinister implications. The term has evolved to serve as a weapon against groups that are not well understood (Richardson, 1993). It is associated with three elements: peculiar practices, crazed leadership, and total brainwashing of members (Flinn, 1987). Its negative connotations are often amplified by the addition of such descriptors as *extremist, fanatical, fatalistic,* and *pathological.*

Keeping in mind that all major religions of the world, including Christianity, Islam, and Buddhism, began as religious movements considered deviant by the larger society of the time (Zellner & Petrowsky, 1998), it is important that we recognize the effect that a term with such negative connotations may have on theoretical and empirical analysis. Richardson (1993) has eloquently argued that we should abandon the term *cult* in scholarly endeavors because it is difficult to objectively study phenomena with such social baggage.

Besides the problem of terminology, another problem in studying cults is the conflicting and competing discourses regarding them that have come from stakeholders including cult members, the anticult movement, psychotherapists, the media, scholars, and various professional disciplines (Barker, 1995). As Barker pointed out, these groups have such different agendas and constructs regarding NRMs that documenting objective accounts of the evolution and behavior of such organizations is difficult. Scholarly and popular writings provide numerous examples in which various segments of the NRM debate refer to each other as the anticult movement (ACM), the countercult movement, cult defenders, or cult apologists. Sociological criteria, theological standards, and accounts of actual events are mixed with the rhetoric of concerned parents, law enforcement officials, and the media.

Those in search of information regarding NRMs are left with the arduous task of separating the rhetoric from scholarly discourse. We are inevitably struck with the dilemma of how to best refer to the religious organizations under analysis in this chapter. The truth is, some NRMs have acted in such ways as to have earned the negative title of *cult*. Some groups have committed heinous acts of physical and sexual violence against members, moved large numbers of people to commit suicide, or engaged in acts of terrorism. We have chosen to use the term *cult* because our purpose is to describe the connection between such groups and crime. However, we fully acknowledge the danger in painting all NRMs with the same broad brush, given that most groups are peaceful. Therefore, our intention is to describe some of the organizational elements that appear to create a climate in which abuse can occur within the organization and to set in motion the commission of crimes against the external community. Obviously, it is difficult to clearly identify which NRMs are a potential threat to members or the public at large and which organizations

are truly seeking alternative spiritual living. Social scientists have begun studying the unique features of such religious organizations in an attempt to understand their development and evolution (see Dawson, 1998; Lewis, 1996; Zellner & Petrowsky, 1998). The tragic consequences of organizations such as the People's Temple, MOVE, the Branch Davidians, the Order of the Solar Temple, Heaven's Gate, and Aum Supreme Truth during the 1990s have prompted criminal justice professionals to seek to understand how to best predict which organizations will engage in criminal behavior and which ones will not.

Organizational Features Enhancing the Potential for Harm

Certain NRMs, by design or default, appear to be positioned to commit a wide range of criminal behaviors, including fraud, sexual and physical abuse, murder, suicide, coercion, and neglect. Therefore, understanding the complex phenomenon of such group behavior—the relationship between the group and its surrounding environment, the division of power within the group's organization, and the group's unique religious doctrine—is likely to facilitate understanding "the dynamics that sometimes transform idealistic truth-seekers into ruthless murderers or terrorists" (Mayer, 2001, p. 373).

NRMs, particularly those posing a threat, can be distinguished from other religious groups on the basis of organizational factors such as the presence of a single charismatic authority, development of a new social system to accommodate the belief system of the leader, and religious beliefs that tend to be apocalyptic or fatalistic. Each of these organizing elements, when considered alone, is not likely to result in criminal behavior. However, when these three organizational features come together, the potential for abuse and violence increases.

Charismatic Leadership

Most scholars agree that charismatic leadership and the authority vested in the leader by the group are a central organizational element in all cult groups. Cult groups are totally reliant on a single charismatic leader, whereas other organizations, such as churches, denominations, and sects, are highly structured organizations diffusing leadership and power among many members within the organization. Members of the latter type of organizations, although observant of the belief system, move freely within the broader society and have lives outside the organization. Cults, on the other hand, are characterized by a single dynamic leader who is central to the tenets and practices of the group and without whom the group would not survive (Zellner & Petrowsky, 1998). Operating in seclusion, cult members are encouraged to reject aspects of the broader culture and are focused around a particular belief system as touted solely by the leader. Membership in the group, therefore, is dependent upon individual members' shedding their individual identity, viewpoints, and perspectives to become true followers of the leader. Thus, the concepts of charismatic leadership and conversion are closely linked in that "the authority exercised by charismatic leaders has its roots in their followers' matrix of assumptions about reality" (Balch, 1998, p. 21).

Therefore, the eventual outcome of violence or other criminal behavior is likely to be influenced by the behavior patterns of the leader. It is unclear whether some cult leaders are truly guided by some higher spiritual vision, consciously manipulate a leadership role to further their own personal agenda, or struggle with a form of mental illness. Regardless, what separates problematic cults from other religious organizations is the self-proclaimed supreme power of the leader over the membership and the inherent power differential that is created by singular leadership.

Additionally, leadership grounded solely in charisma is likely to be an unstable form of authority in the long run, creating an environment for volatility (Robbins & Anthony, 1995). Because charismatic authority is not reinforced by institutional supports or contained by institutional restraints, charismatic leaders are confronted with the challenge of maintaining their authority in the face of dissent and discord within the group. Such religious leaders are generally distrusted by the broader culture, so they face external pressures as well. Problems surface when the authority of the leader is challenged, creating a circumstance where the leader must continuously find new ways to assert authority. This may require the leader to root out dissenters, reorganize the membership to destabilize the development of institutional structures within the group, or demand that members demonstrate their faith by performing a variety of tasks (Robbins & Anthony, 1995).

Creation of a New Social System

Charismatic leaders often demand the creation of new social systems by which the group defines its activities and interacts with its external environment. NRMs that require members to abandon all family ties and to live in seclusion from the broader social community are more likely to have the potential for harm because of their isolation and overreliance on the authority of one leader. Such groups "establish a radical separation between themselves and the established social world, which they regard as hopelessly evil" (Hall, 1989, p. 78). Such separation from the world and one's previous life makes maintaining family ties difficult. This, of course, is the most frightening aspect for parents and other family members and a critical piece that fuels the anticult movement. However, it is difficult to ascertain whether seclusion from the outside world is mandatory or a natural and voluntary surrender on the part of group members. Most scholars dismiss the notion that members are "brainwashed" into shedding their previous social connections; in fact, after 20 years of research there is little evidence to support such claims (Dawson, 1998). Richardson (1994) argued that individuals who join such groups are actively seeking such personal change and that consequently conversion is not coercive. As Levine (1984/1998) noted, radical departure from a lifestyle is more likely to occur for individuals in late adolescence and early adulthood because individuals of this age group are generally not fully committed to careers or families of their own. Like the elderly, they are in a position to freely move into or to "try out" an alternative living environment.

Another perspective compares cult social systems to other social systems. George Knox (1999) argued that cults have much in common with groups such as "gangs" in that they restrict the freedom of thought by members, demand total allegiance to the self-proclaimed leader, and encourage dependence on group identity and estrangement from the larger establishment. In addition, both gangs and cults look to recruit people in need of social and personal acceptance and often have internal sanctions for members who are disobedient. Yet in this sense cults and gangs have much in common with many social organizations, including sports teams, fraternities, sororities, clubs, and occupational groups such as the military, police, and firefighters. These social organizations have some of the same properties as cults, but they generally do not operate in seclusion, nor do the members rely solely on the words and wisdom of one charismatic leader. Additionally, these types of organizations tend not to reject the core principles of the broader society. Therefore, cults present a unique social organization in which members are ready and willing to shed their lifestyle, interpersonal relationships, and autonomy for the sake of a particular spiritual belief under the charismatic authority of the leader.

The social system provided by the cult organization is one that appears to meet the needs of its members. Some may ask, Why would someone want to join a group that represents such a radical departure from the broader culture? Individuals seeking membership in a cult organization probably do so for the very same reasons that individuals seek membership in other types of religious organizations. Some find peace and solace in the answers that the religious organization offers to their core spiritual questions, others are attracted to the rigidity of rules and boundaries set forth by religious authorities, and others seek the fulfillment that comes with being part of a group. Research suggests that conversion is driven by economic, social, organismic, psychic, and ethical deprivation (Glock, 1964) or by the belief that the enduring acutely felt tensions of living are best resolved within a religious problem-solving perspective (Lofland & Stark, 1965). Levine (1984/1998) acknowledged the conversion of seemingly healthy young people into cult organizations in search of self-identity as a transition from childhood to adulthood. Wright and Piper (1986) noted that the semblance of a family system appears attractive to young recruits who lack meaningful family relationships. Therefore, individuals, often

young people, join religious organizations for a variety of reasons, serving perhaps a variety of purposes. Research suggests that most members leave such groups within 2 years of their conversion, a finding that runs "flatly counter to the assertion of the anti-cult movement that converts are the naive victims of 'brainwashing'" (Dawson, 1998, p. 145).

However, this is not to dismiss the reality that cults need to continuously recruit new converts for future survival. Groups cannot function without members or financial support. The long-term survival of any group is dependent upon the procurement of financial and people resources, and cults are certainly no exception. Therefore, it is incumbent upon the leader to ensure that his or her flock maintains an appropriate balance of members, and perhaps it is in the best interest of the cult leader that members come and go at regular intervals. As we have discussed previously, charismatic authority is difficult to maintain over time, so the rotation of members in and out of a group may in fact serve as a stabilizing factor. Members who appear discontented are likely to be encouraged to leave the group so as not to cause any disruption of the leader's authority.

Members appear to join NRMs voluntarily, but it is also true that cult groups engage in conscious, deliberate recruitment practices. Cult leaders appear to understand the "profile" of likely candidates and how to seek them out for potential membership. Wright and Piper (1986) noted that young people between the ages of 18 and 25 are more likely to be influenced by the answers offered by cult membership and therefore are likely recruits. However, some evidence suggests that elderly members are targeted for their financial assets (Collins & Frantz, 1994).

If membership or conversion into the cult type organization is voluntary, what occurs to predispose a cult toward violence? Isolation from the broader community and the power differential created by the leader create circumstances in which abuse can occur internally within the organization. Members live in communal living environments, creating, in effect, a new family system. As in traditional family structures, isolation and power differentials fuel physical, emotional, sexual, and financial abuse. The same factors that predispose traditional families to abusive behavior—intensity of involvement over time in a wide range of activities, ascribed roles based on age or gender, engagement in activities likely to cause internal conflict and strife, guarding of privacy from the outside world, and stress (Gelles, 1997)—are likely to be present in any communal living environment. Thus, NRMs that fit the title of *cult* are also likely to share a risk for internal abuse similar to that of many traditional families. The group members, particularly women and children, will be vulnerable to physical, sexual, and emotional abuse. Women are at particular risk of sexual exploitation (Lalich, 1997), and children are at risk of medical and educational neglect as well as physical, sexual, and emotional abuse (Schwartz & Kaslow, 2001). Schwartz and Kaslow (2001) noted that the deliberate recruitment of senior citizens into cult organizations for the procurement of their financial assets is a relatively recent phenomenon. As in the broader culture, senior citizens are particularly vulnerable to financial exploitation.

Keeping in mind that the social system created in cult organizations is completely controlled by the leader, we must look toward the leader for clues regarding internal abuse toward members. It is not clear if religious tradition guides abusive behavior or if abusive leaders rely on religious tenets to rationalize or justify their behavior. For example, Sharon Ringe (1992) noted that acceptance of the patriarchal text of the Bible has been used by many to support oppressive rule over women as justification for spousal abuse. Some cult leaders, therefore, may use such interpretations of the Bible as a justification for the use of physical punishment to keep members of their flock in line. In fact, some may even perceive it as their duty to "educate" members through emotional, physical, or sexual violence. As stated earlier, cult leaders use coercive power to maintain their authority and may do so in line with what they believe is the word of God. Cult leaders may also engage in abusive acts as a result of their own history of victimization, mental illness, or substance abuse. Little research has been conducted from this perspective.

Belief Systems

The third organizing principle that appears to be linked to the criminal behavior of some NRMs is what has been described as an apocalyptic, millennial, or fatalistic belief system, in which the end of the world appears to be the central concern of group members.

Focused on imminent catastrophe, atonement, and transformation and the establishment of a new world order, this worldview offers a perspective on the coming of God and the end of history as we know it. The terms *apocalypse* and *millennium* are drawn from the New Testament Book of Revelations. Within the context of such a worldview, themes of conspiracy and paranoia appear to be centrally linked to technological and scientific advancement, government regulation, and fascination regarding the possibility of extraterrestrial life (Stewart & Harding, 1999).

Such notions are not new; in fact, there is evidence to suggest they have been popularized and expressed in many ways throughout American history, including through popular culture, politics, and the media (Stewart & Harding, 1999). Therefore, apocalyptic and millennial views are not, in and of themselves, explanations for violence. In fact, many such groups actually view themselves as being on the other side of the apocalypse, saved by the grace of God, and therefore as having no need to prepare for violent conflicts (Hall, 1989). Yet marginal or fringe religious organizations that have experienced violent episodes have been characterized as having such worldviews. Groups such as the Branch Davidians, the People's Temple, early Mormons, and Radical Reformation Protestants such as the early Anabaptists have been characterized in this manner (Robbins & Anthony, 1995). Others engaged in violence, such as the Ruby Ridge group, the Oklahoma City bombers, the Unabomber, and Heaven's Gate, have also embraced such notions.

Although apocalyptic views do not ignite violence, they may provide a sufficient level of urgency to motivate group members to action (Mayer, 2001). Contemporary disasters, social problems, and examples of corporate and government collusion dominate the global media, invoking a sense of crisis or emergency (Stewart & Harding, 1999). The tragic events of September 11, 2001, and the resulting actions and discourse from government, military, and media officials are likely to fuel further the apocalyptic vision of impending doom. Worries about terrorism, weapons of mass destruction, and bioterrorism dominate contemporary American political and social dialogue. Outbreaks of anthrax and other violent outbreaks shortly after September 11, coupled with continued warnings of the inevitability of future terrorist attacks and ongoing conflicts in the Middle East, heighten the anxiety of most Americans and are likely to have an even greater impact on those with apocalyptic worldviews. Contemporary fear and anxiety are likely to resonate with paranoia, giving rise to the type of urgency that is often exhibited by apocalyptic groups prior to their violent episodes. Consequently, we predict that new tragedies involving cult groups are likely to arise in the near future.

Violence may also be related to the belief that the group or the leader is being persecuted by the government, the broader society, a particular group of people, or individuals who are believed to represent evil. Real or imaginary assaults against the group promote a defensive stance. Persecution may result in survivalistic behaviors, such as stockpiling weapons and supplies and developing fortress barriers for protection. Again, the literature is unclear on whether leaders' felt persecution is real, a manipulative response to internal or external challenges of authority, or the result of an untreated mental illness. In any case, such behaviors should signal authorities that some sort of escalation is likely to follow.

Professional Challenges

The challenges that lie ahead for law enforcement, criminal justice, and mental health professionals in reference to NRMs are great. Unraveling the unique relationship of psychology and religion and the evaluation of appropriate intervention techniques will be necessary. A delicate balance must be achieved whereby the right of religious freedom is secured, yet the lives of cult members and the broader community are protected. Families remain concerned about the safety and well-being of their loved ones perceived to be "lost" to the cult way of life.

Learning to predict which groups will ultimately resolve their issues through crime commission is an arduous task, yet what we have learned from the past is instructive for future patterns. Understanding how individuals guided by particular belief systems interact within their own communities is as important as having a full understanding of the thought and behavior patterns of charismatic leaders. Mayer (2001) presented the various internal and external factors associated with violence as including organizational elements, environmental influences, and emotional responses to various stimuli. Mayer restated what other scholars have argued, that no one factor can predict violent episodes and that violence within NRMs appears to be the result of a complex set of internal and external factors.

However, the red flags as presented here are likely to warrant the official attention of authorities. Monitoring such groups will require law enforcement officials to undertake such investigations with great care. They must understand and respect the ideology of the group, and this will require them to listen to the concerns of the group. Intervention strategies should be chosen to deescalate fear, anxiety, and potential violence and to preempt millennial enthusiasm (Whitsel, 2000). This will require law enforcement to institute different techniques to fit the unique psychological dynamics of different cult groups so that they can avoid becoming an "instrument of the prophecy" (Baker, 2001).

THEOLOGICAL RELIGION-RELATED CRIME: DOES FAITH HEALING ENDANGER THE WELFARE OF CHILDREN?

The nexus of religion and crime is also revealed in the area of medical neglect. Within this context, parents who, on the basis of religious belief, withhold necessary medical care from their children may find themselves subject to civil or criminal litigation for endangering the welfare of their children. The deaths and illnesses of many children are at the center of a heated controversy between religious leaders and child welfare advocates and public health professionals. Some argue that it is the parents' right to determine whether their child should receive medical care consistent with religious doctrine; others vehemently argue such practices are blatantly irresponsible and endanger the welfare of children and the community.

The courts have generally acknowledged the rights of adults to reject their own medical care consistent with religious tradition. However, the dispute over the right to withhold medical treatment for a child on the basis of religious beliefs is putting courts in the precarious position of balancing the free exercise of religion against the responsibility to protect children. Thus, a number of public health, medical, and legal issues have surfaced.

A number of religious groups refuse to use medical and/or preventative health treatments or diagnostic measures on religious grounds. Some faiths refuse any and all medical interventions, whereas others object only to specific types of procedures or treatments. For example, Jehovah's Witnesses do not believe in blood transfusions, and Christian Scientists choose faith healing over traditional medical treatment. Faith healing has been touted as superior to traditional medical intervention by many different denominations, at great cost of human life. For example, Asser and Swan's (1998) study associated child fatalities with over 23 different denominations from 34 different states, although the majority of child fatality cases (83%) were associated with the Church of the Firstborn (23 deaths), the End of Time Ministries (12 deaths), the Faith Assembly (64 deaths), the Faith Tabernacle (16 deaths), and Christian Science (28 deaths).

Medical Consequences

The debate is of life-and-death concern, for the consequences of medical neglect are often tragic. Asser and Swan (1998) conducted a study to determine if standard medical treatment would have made the difference in the lives of 172 children who died between 1975 and 1995. Cases of child fatality were chosen on the basis of evidence that parents withheld medical care because of religious beliefs. Their findings warrant public concern. Of the 172 deaths, 140 were attributed to medical conditions in which the typical survival rate would have exceeded 90% had the standard medical treatment been employed. Eighteen other cases had an expected survival rate greater than 50%. In all but 3 of the remaining cases, medical treatment would have likely had some benefit. Asser and Swan concluded that "when faith healing is used to the exclusion of medical treatment, the number of preventable child fatalities and the associated suffering are substantial and warrant public concern" (p. 625).

Outbreaks of communicable diseases, normally controlled through vaccination, are of special concern to the public health community. According to CHILD (Children's Healthcare Is a Legal Duty), outbreaks of polio, measles, whooping cough, and diphtheria can be traced to sects claiming exemptions from state immunization laws. For example, in 1991 there were 492 cases of measles in Philadelphia among children associated with Faith Tabernacle and the First Century Gospel Church, and 6 of the children died (CHILD, 2001). Two studies demonstrate the medical and legal dilemma when a communicable disease is involved. Feikin et al. (2000) found that exemptors of immunizations for measles and pertussis placed their children at a 22.2 times greater risk of acquiring measles and a 5.9 times greater risk of acquiring pertussis than vaccinated children. The study also indicated that at least 11% of the vaccinated children who contracted measles did so by contact with a child who had not been vaccinated due to religious exemption. In a similar study, Salmon et al. (1999) found that exemptors were 35 times more likely to contract measles than children who were vaccinated.

Kaunitz, Spence, Danielson, Rochat, and Grimes (1984) found that members of religious groups in Indiana who refused prenatal care and gave birth at home without trained assistants substantially increased the risk of death for themselves and their babies. Their findings indicated that the perinatal mortality rate was 3 times higher and the maternal mortality rate was 100 times higher than statewide rates. In addition, Simpson (1989, 1991) found evidence that Christian Science adults had overall higher mortality rates than the general public.

In addition to physical consequences, cases of medical neglect may be associated with psychological distress. Bottoms, Shaver, Goodman, and Qin (1995) conducted a survey of 19,272 clinical psychologists, psychiatrists, and clinical social workers and found substantial evidence of religion-related child abuse among their caseloads, particularly victimization by medical neglect. Interestingly, many of the parents perpetrating religion-related child abuse and neglect were described by the mental health providers as mentally ill. In addition, some of those victimized by medical neglect were also physically, sexually, and emotionally abused by their parents or associates.

Compounding the debate is the reluctance of the faith healing community to submit to medical scrutiny through scientific study. Faith healing is encouraged through the use of testimonials, with little verification that individuals have been appropriately diagnosed by trained, licensed medical professionals or that their specific ailments would not normally heal without medical intervention. Although we do not want to dispute the existence of medical miracles or the important role that prayer can play, there is little scientific evidence to suggest that faith healing is anywhere near as successful as traditional medical practices. Until religious leaders allow faith healing to be scientifically tested, with the same vigorous methods as medicine, it will be difficult for many to support the use of faith healing as the only source of medical intervention in life-threatening situations involving children.

Legal Issues

Religious groups stand behind the First Amendment's prohibition of government interference with religion to maintain the right to reject medical treatment for their children. The U.S. Supreme Court grappled with this very issue in 1944 in *Prince v. Massachusetts,* stating, "The right to practice religion freely does not include the liberty to expose the community or child to communicable disease, or the latter to ill health or death." Despite this ruling from the Supreme Court, several specific legal issues remain in controversy, spanning public health, educational, civil, and criminal statutes. Legislators and the courts have struggled to find compromises regarding this issue in an attempt to balance the right of free exercise of religion with the responsibility to protect children from harm. For example, 39 states have religious exemptions within their civil codes and 31 states have a religious defense available for criminal charges.

States continue to embed exceptions for a variety of medical treatments and/or preventive and diagnostic measures in legislative mandates as a result of political pressure from religious organizations. Exemptions include items such as immunizations, metabolic and hearing testing and prophylactic eyedrops for newborns, lead screening, bicycle helmets, vitamin K, and physical medical exams. Six states allow exemptions from studying about disease in school (CHILD, 2001). All states, with the exception of Mississippi and West Virginia, allow religious exemptions from immunizations. Rota et al. (2001) found that the process of claiming exemptions from immunizations requires less effort than fulfilling the immunization requirement.

The Child Abuse Prevention and Treatment Act of 1996 (CAPTA) currently mandates that the states require parents to provide proper medical care for their children, but the act also allows states to give parents the right to file an exemption to withhold treatment on the basis of religious doctrine. Thirty-nine states allow religious exemptions within their child abuse statutes.

Child advocacy groups, medical associations, and other groups have contested such state exemptions on behalf of children by bringing forth lawsuits and amicus briefs. Several organizations, including the United Methodist Church, the National Association of Medical Examiners, Justice for Children, the National Child Abuse Council, and Child Healthcare Is a Legal Duty (CHILD), are arguing that Congress must readdress the exemption issue. Central to such legal battles is CHILD, a nonprofit organization whose mission is to protect children from abusive religious practices, especially religion-based medical neglect (CHILD, 2001).

Such exemptions, however, do not preclude the possibility of civil and/or criminal sanctions being applied to such cases. For example, in the case of *McKown v. Lundman,* a $1.5 million (reduced from an original verdict of $14.2 million) judgment was issued in 1995 against a mother and three other Christian Scientists who chose not to seek secular medical treatment for her 11-year-old son who died from untreated diabetes. The boy's father brought suit against the mother, the stepfather, a hired prayer practitioner, and a Christian Science nurse for not accessing traditional medical treatment for the boy, even after he slipped into a diabetic coma. After 3 days of Christian Science prayer, the boy died. The father also brought suit against the church, and a jury awarded $9 million in punitive damages, but a state appellate court overturned the award, finding the church not liable for individual actions (Mayer, 1996a, 1996b). Both sides appealed their cases to the U.S. Supreme Court, but the court denied certiorari in January 1996 (Mayer, 1996b).

The case of Dennis and Lorie Nixon provides an example of the type of criminal charges brought forth regarding medical neglect. The Nixons are members of the Faith Tabernacle Church in Pennsylvania, which believes that illness should be healed by prayer. The defendant's daughter, Shannon (age 16), became ill and died of complications related to diabetes acidosis. At no time during her illness did she request medical treatment, nor did her parents make such a request. Her parents were arrested on charges of involuntary manslaughter and endangering the welfare of children, in which they were convicted and sentenced to 2½ to 5 years in prison and a $1,000 fine. On appeal, the Nixons argued that their daughter was a mature minor and could therefore refuse medical treatment and that consequently their responsibility as parents was dissolved. The Superior Court judges disagreed consistent with past decisions (*Commonwealth v. Cottam,* 1992, and *Commonwealth v. Barnhart,* 1985), which ruled that a child's religious beliefs do not relieve the parents of their responsibilities, especially in cases in which a life is in immediate danger. The Superior Court upheld their convictions. In addition, the Nixons attempted to cite the state's child abuse and neglect statutes, which allow exemptions from medical treatment on the basis of religious tenets. However, Judge Del Sole responded by stating that an action can qualify as involuntary manslaughter even if it does not constitute child abuse and that, in effect, CAPTA protects the Nixons against being labeled child abusers but does not preclude them from criminal

responsibility (Rodier, 1999). The court affirmed their convictions.

ABUSES OF RELIGIOUS AUTHORITY: THE CATHOLIC CHURCH ON TRIAL

The abuse of religious authority involves religious leaders using their unique position in the community to commit a wide range of offenses, most notably fraud, embezzlement, and sexual abuse. Religious leaders from all faiths have been charged with such offenses. For example, the Associated Press reported on May 22, 2002, that a Roman Catholic priest had pled guilty to conspiring with five other men to manufacture and distribute GHB (gamma hydroxybutyric acid), commonly known as the "date rape drug." Court documents indicated that 25 gallons of a precursor chemical used in the making of GHB had been sold to St. Patrick's Church in Sheffield, Illinois, where Rev. Jeffrey Windy worked. He was scheduled for sentencing in August 2002 (Leitsinger, 2002). Another case is that of the former pastor from Rivers of Life Fellowship in Steelville, Minnesota, who pled guilty of felony charges for stealing at least $10,000 in church funds in December 2001. He was also implicated on arson charges after his church was destroyed by fire, but the arson charges were dismissed for lack of evidence during a preliminary hearing. David Cartee (age 33) was sentenced to 7 years, the maximum, for his crime (O'Neil, 2001).

The crimes described here fit the description of occupational crime as defined by Gary Green (1990): that is, "any act punishable by law that is committed through opportunity created in the course of an occupation which is legal" (p. 12). One could argue that members of the clergy who commit such acts, and the religious institutions that protect them, are not substantially different from other professional groups who may abuse their authority, such as police officers, corrections officers, physicians, educators, attorneys, and psychologists, and thus that they ought to be held to the same set of legal standards and self-policing mechanisms that other professional groups have been required to maintain. These professions have also learned the hard way that no one wins when an organization goes to great lengths to protect one of its own without first safeguarding the safety of the consumers it serves.

Victimization by clergy, regardless of the type of crime, is painful for primary victims, their families, and the congregation. A clergy member who steals from the church coffer violates the unique bond and trust that parishioners have come to rely on. Victims of sexual abuse endure an especially painful experience, often with lifelong consequences. For example, Bottoms et al. (1995) found significant evidence of child sexual abuse in their survey of mental health professionals. Although many mental health professionals did not consider the perpetration of sexual abuse by a member of a religious organization differently from abuse by any other perpetrator, Bottoms et al. argued that victimization by religious authorities presents unique circumstances in that cases are less likely to be reported and more likely to "promote painful confusion in young victims that make its long-term psychological consequences difficult to bear" (p. 94). They found that cases involving the abuse of religious authority tend to be overwhelmingly sexual in nature and that "religious professionals' role as unquestioned moral leaders apparently gave them special access to children" (p. 95).

Sexual Abuse and the Catholic Church

Although the abuse of religious authority has gained media attention in recent years, nothing has rocked the religious community like the recent charges of sexual abuse of children by Roman Catholic priests, placing the sacred institution of the church on public trial. Some of the charges involve multiple victims, over a number of years, with the additional scandal of institutional cover-ups. As this book goes to print, the story continues to unfold, with the Catholic community struggling to come to grips with the reality of such abuses. Many argue that the church should be held liable for their actions. Arthur Austin, an alleged victim of sexual abuse by Rev. Paul Shanley, stated, "If the Catholic Church in America does not fit the description of organized crime, then Americans seriously need to examine their concept of justice" (Robinson & Farragher, 2002, p. A1).

Some cases have impressive documentation, but we will focus on one case in particular because of the obvious criminal and civil legal issues it raises. The Rev. Paul R. Shanley was arrested in California in May 2002 and was extradited back to Massachusetts to face charges of rape. The Boston Archdiocese, under court order, released over 800 pages of documentation establishing the church's knowledge of Rev. Shanley's misconduct over the years. Paul Shanley was a street priest in the Roxbury section of Boston in the 1970s. He had long hair, wore jeans, and ministered to the "street people." The church was in possession of Shanley's diaries, his correspondence with his superiors, and his public statements. In his diaries, he wrote of helping young people to learn to use drugs: "My God, I've even taught kids how to shoot up properly." Shanley wrote also of his struggles with venereal disease and how he showed young drug users how to avoid contracting the disease (Rezendes & Farragher, 2002, p. A1). These entries alone should have been enough to cause alarm.

As far back as 1967, there were complaints of molestation of boys by Shanley. Further, his attendance at a North American Man-Boy Love Association (NAMBLA) conference in 1978, and repeated reports from Catholics in Rochester, New York, that Shanley had spoken openly about sexual relationships between adults and children (Robinson & Farragher, 2002), should have signaled the Church to respond. Yet he was promoted from associate pastor to pastor of his church following the NAMBLA conference (Rezendes & Farragher, 2002). Even after the Boston Archdiocese had paid monetary settlements to several of Shanley's victims, Cardinal Bernard F. Law "did not object to Shanley's request to be director of a church-run New York City guest house frequented by student travelers" (Robinson & Farragher, 2002, p. A1).

The church is still unwilling to take responsibility for its complicity in a child molester's career of abuse. The church was notified of complaints of sexual assaults for four decades. It took a number of temporary steps, moving Rev. Shanley from one parish to another. A similar story is that of convicted pedophile John Geoghan within the same archdiocese.

Rev. John Geoghan is now serving 9 to 10 years in a Massachusetts state prison for sexual assault. Cardinal Law of the Boston Archdiocese was subpoenaed to testify in May 2002 regarding his lack of response to repeated and credible complaints of criminal behavior in a lawsuit brought by 86 plaintiffs accusing him and others in the Boston Archdiocese of negligent behavior. Historically, the church has chosen to police its own, and now it will endure civil, and perhaps even criminal, scrutiny for its past and present institutional responses to crimes of child sexual abuse.

The Catholic Church is not eager to concede power and authority to laypersons in the civil government. The church has always seen itself as above, or at least separate from, the government of the United States. It does not wish to be governed by elected officials and politicians. Separation of church and state, a founding plank in U.S. philosophy, makes these types of cases more complicated. So where does this leave religious leaders, ethically, spiritually, and legally? No group is effective at self-policing. The Catholic Church is a prime example. One cannot fairly discipline one's colleagues and friends, particularly within the context of a religious ideology that stresses God's forgiveness. The Catholic Church has struggled to do so and failed miserably and publicly. The National Conference of Bishops in Washington, D.C., in 1992, reiterated *voluntary* guidelines for dealing with allegations of sexual misconduct. At that time, some bishops who had had heartbreaking experiences with pedophilia within their ranks argued to strengthen the guidelines to require bishops to report such allegations to the legal authorities. But many bishops, like Cardinal Law, stated that they would report to civil authorities only when compelled to by law. Cardinal Law, among others, wanted to treat clergy sexual abuse as a family matter (Franklin, 1992).

The Vatican issued new rules for the churches on May 18, 2001, with regard to such matters, delegating the authority to rule on cases of sexual abuse to the Congregation for the Doctrine of the Faith, which would be responsible for deciding if a tribunal would hear the case or the case should be referred to Rome. These rules, referred to as May 18 *motu proprio* (i.e., issued under his personal authority),

dictate that cases should be handled by priest-staffed courts, within a 10-year statute of limitations from the "accuser's" 18th birthday, with absolute confidentiality. Ecclesiastic courts would have the final authority to defrock a priest, while maintaining principles of judicial review (Allen, 2001). These rules contain no mention as to when or if civil authorities should be notified. Tod M. Tamberg, a spokesman for the Archdiocese of Los Angeles, stated that secret proceedings "protect the rights of both the accusers and the accused . . . from the trauma of being put on public display" (Lofland & Stark, 2002, p. A4).

Dioceses around the United States have been struggling with their sexual abuse policies for a number of years. The Denver Archdiocese, operating under a 1991 policy, "calls on all employees of the 24-county archdiocese to notify law enforcement whenever they have a credible reason to believe child prostitution, molestation, assault or abuse has or is occurring" (Culver, 2002, p. A1).

The Chicago Diocese has the most highly touted policy on sexual misconduct of minors. Its 1997 policy cites its primary purposes of policies and procedures to be "the safety of children, the well being of the community, and the integrity of the Church." This statement alone places the parties of interest in the proper order. The policy demands facilitation of cooperation with and avoidance of interference with the civil authorities who are responsible for investigating allegations of sexual misconduct. The Chicago plan calls for a separate Victim Assistance Minister to be designated to minister to the victim, the victim's family, or other persons affected. It further requires a review process for continuation of ministry. Although the intent of this review is healing and reconciliation, the safety and well-being of the community is the primary concern. Chicago policy requires that the church comply with all civil reporting requirements related to sexual misconduct with a minor and to cooperate with official investigations (Archdiocese Chicago, 1997).

The strongest language found in various Catholic diocesan policies comes from New Jersey, which has taken on strict adherence to state statutes regarding child abuse. For example, the Roman Catholic Diocese of Paterson in Clifton, New Jersey, states:

Regarding our own personnel, the Diocese of Paterson will remove an alleged abuser from the environment—school, parish, etc.—and place that individual on leave pending investigation by the appropriate authorities. We do not investigate ourselves. We cooperate and comply with all investigating agencies. We defer comment to those agencies. No one will be permitted to return to duty until written evidence has been received by the Division of Youth and Family Services and civil authorities indicating that the individual poses no risk and may return to his/her position. (Roman Catholic Diocese of Paterson, 2002)

Civil and Criminal Legal Dimensions

Historically, the Catholic Church, as well as leaders from other faiths, has hidden behind state legislation that has not routinely considered religious leaders mandated reporters of child abuse and neglect. Because most child protective systems intend to respond specifically to cases of abuse and neglect involving parents and/or guardians of children, cases in which a member of the clergy is the perpetrator are likely to be considered an issue for law enforcement rather than child protective services unless the member of the clergy is in a caretaking role. Thus, even though religious leaders are not considered mandated reporters, this does not preclude their legal responsibility to report potential crimes. Interestingly, religious leaders appear to have less difficulty involving law enforcement in cases in which a church leader is suspected of financial exploitation.

Although the issue of institutional protection of child molestation is a significant issue, we must also look to victims and their families and their reluctance to involve the authorities. Religious institutions have been successful in keeping such allegations private, in part because families have not chosen to report such incidents to the authorities.

Although privacy of the family is a legitimate concern, the fact that the church has disregarded such instances and continued to reassign priests in positions in which they continue to have access to children is very problematic. Nathaniel Pallone (in press) described "the endemic difficulty in distinguishing sin from crime that peculiarly and differentially affects members of the Roman Catholic hierarchy and the exemption from the laws of the nation governing crime that members of the hierarchy appeared at least implicitly to have granted both to their subordinates and to themselves. Arrogance and betrayal sounded its principal themes." If deemed negligent, the church is likely to be sanctioned by litigation. Until citizens and church leaders come to understand the serious nature of child sexual abuse and the long-term physical and emotional consequences of victimization (Bottoms et al., 1995), it is likely that such abuses of religious authority will continue. Child sexual abuse must be viewed as a crime problem, worthy of criminal justice intervention.

CONCLUSION

This chapter focused on the nexus of crime and religion. We suggested a three-pronged typology to categorize and further define religion-related crime. Three specific examples were chosen to illustrate the dimensions where legal and religious principles conflict. As we discussed, it is important to embrace and protect the principles of religious freedom, but it is clear that legal boundaries must be set to protect others from injurious religious practices. The Constitution protects religious freedom, but it does not give license to religious individuals or institutions to be above the law.

A number of questions remain unanswered. How can religious principles ultimately result in criminal or neglectful behavior? How can those who give their lives to religious ideology commit such heinous acts? Are cult leaders who engage in abusive practices that lead a group to violent outcomes really living out a spiritual calling, or are some of them suffering from a mental illness? Do clergy who engage in pedophilia do so because of the unique social loneliness created by celibacy policies, or are some drawn to take clergy positions because of the unfettered access to children such a career will provide? Do mothers beat their children to rid them of evil or because of a psychotic episode? Do women and men engage in polygamous marriages because they truly believe God prescribes such arrangements, or is there another psychological explanation?

In addition to understanding individual behavior, we must also further investigate the dynamics of group and institutional behavior. Groups tend to behave in ways that will protect members against unduly external influences, meaning that group members often bond when external threats are made to the group. This is true of sports teams, families, gangs, corporations, and governments. Therefore, is the Catholic Church's protection of sexually abusing priests inherently any different from cult members' protection of their leader from external persecution? In this sense, are there any similarities between abuses found within families and the abuses described in this chapter? If a religious community is a pseudofamily, could we expect similar positive and negative behavior to result? Ultimately, we may find that human behavior is such a complex, dynamic phenomenon that religious meaning can be only one dimension of it, so that attempts to understand criminal and/or neglectful behavior within a religious framework may not be fruitful. Or we may find that religious and spiritual understanding is the core psychological principle upon which such behavior is based.

In either case, it is important that both religious leaders and secular practitioners engage in an open dialogue, among themselves and with each other, regarding these social, political, and legal issues. Religious institutions must reexamine their practices and should be held accountable for any negligent behavior. Religious leaders have a responsibility to educate their followers as to the legality of any particular religious practice and must be held liable when they abuse such authority. Leaders of religious organizations and law enforcement officials should make a conscientious effort to meet and discuss ways in which religious organizations can prosper within the confines of the law. Meaningful problem solving and boundary setting before an incident occurs would provide a preventative mechanism by which both religious and secular institutions could benefit. At the same time, public policy must be refined to assist clergy in clarifying the boundaries in which they must operate. The vagueness in current state statutes regarding medical exemptions and the reporting of child abuse must be revised. Much can be accomplished by understanding the root philosophical positions and motivations of all perspectives. The goal should be to protect religious freedom while maintaining secular law as the fundamental organizing principle for the larger society. Until such reforms are instituted, civil and criminal remedies will continue to test the limits of religious freedom and impose consequences on those who harm by neglect or purposeful actions.

REFERENCES

Allen, J. L. (2001, December 14). Doctrinal congregation given authority over sex abuse cases: Vatican move seen as attempt to balance priests' right, victims' need for rapid action. *National Catholic Reporter, 38*(7), 10–11.

Archdiocese Chicago. (1997). Policies of the Archdiocese of Chicago. Book 2, Title III, Chapter 1, Section 1100, Sexual Misconduct With Minors. Retrieved September 23, 2002, from http://policy.archchicago.org.

Baker, T. E. (2001). Hostage negotiations: Paranoid and suicidal cults. *Journal of Police Crisis Negotiations, 1,* 99–111.

Balch, R. W. (1998). The evolution of a new age cult: From Total Overcomers Anonymous to death at Heaven's Gate. In W. W. Zellner & M. Petrowsky (Eds.), *Sects, cults, and spiritual communities: A sociological analysis.* Westport, CT: Praeger.

Barker, E. (1995). The scientific study of religion? You must be joking. *Journal for the Scientific Study of Religion, 34,* 287-310.

Bottoms, B. L., Shaver, P. R., Goodman, G. S., & Qin, J. (1995). In the name of God: A profile of religion-related child abuse. *Journal of Social Issues, 51,* 85–111.

Child Abuse Prevention and Treatment Act. 42 USC 5101 et seq.; 42 USC 5116 et seq. (1974, amended 1996).

Children's Healthcare Is a Legal Duty. (2001). Religious exemptions from health care for children. Retrieved September 23, 2002, from www.childrenshealthcare.org.

Collins, C., & Frantz, D. (1994, June). Let us prey. *Modern Maturity, 37,* 22–32.

Commonwealth v. Barnhart, 497 A.2d 616 (Pa. Super. 1985).

Commonwealth v. Cottam, 616 A.2d 988 (Pa. Super. 1992).

Culver, V. (2002, March 20). DA will review abuse policy; Ritter plans to meet with Catholic officials. *Denver Post*, p. A1.

Dawson, L. L. (1998). *Cults in context: Readings in the study of new religious movements*. New Brunswick, NJ: Transaction.

Feikin, D. R., Lezotte, D. C., Hamman, R. F., Salmon, D. A., Chen, R. T., & Hoffman, R. E. (2000). Individual and community risks of measles and pertussis associated with personal exemptions to immunization. *Journal of the American Medical Association, 284*, 3145–3150.

Flinn, F. K. (1987). Criminalizing conversion: The legislative assault on new religions et al. In J. M. Day & W. S. Laufer (Eds.), *Crime, values and religion*. Norwood, NJ: Ablex.

Franklin, J. L. (1992, November 22). Catholics struggle with delay. Breaking the silence: The Church and sexual abuse. *Boston Globe*, p. 1.

Gelles, R. J. (1997). *Intimate violence in families*. Thousand Oaks, CA: Sage.

Glock, C. Y. (1964). The role of deprivation in the origin and evolution of religious groups. In R. Lee & M. Marty (Eds.), *Religion and social conflict*. New York: Oxford University Press.

Green, G. S. (1990). *Occupational crime*. Chicago: Nelson-Hall.

Hall, J. (1989). Jonestown and Bishop Hill. In R. Moore & F. McGehee III (Eds.), *New religious movements, mass suicide and the People's Temple* (pp. 77-92). Lewiston, NY: Edwin Mellon.

Kaunitz, A. M., Spence, C., Danielson, T. S., Rochat, R. W., & Grimes, D. A. (1984). Perinatal and maternal mortality in a religious group avoiding obstetric care. *American Journal Obstetrics and Gynecology, 150*, 826–831.

Knox, G. W. (1999). Comparison of cults and gangs: Dimensions of coercive power and malevolent authority. *Journal of Gang Research, 6*(4), 1–39.

Lalich, J. (1997). Dominance and submission: The psychosexual exploitation of women in cults. *Cultic Studies Journal, 14*, 4–21.

Leitsinger, M. (2002, May 22). Priest, 5 others plead guilty in date-rape drug case. *Press and Sun Bulletin*, p. 22.

Levine, S.V. (1998). The joiners. In L. L. Dawson (Ed.), *Cults in context: Readings in the study of new religious movements*. New Brunswick, NJ: Transaction. (Reprinted from S. V. Levine, *Radical departures: Desperate detours to growing up*, 1984, New York: Harcourt Brace Jovanovich)

Lofland, J., & Stark, R. (1965). On becoming a world-saver: A theory of conversion to a deviant perspective. *American Sociological Review, 30*, 863–874.

Lofland, J., & Stark, R. (2002, January 9). Vatican sets new rules for priests. *Los Angeles Times*, p. A4.

Lundman v. First Church of Christ, Scientist, No. 95–534 (1996).

Mayer, J. (1996a, February 5). Faith healer held liable for death. *National Law Journal*, p. A8.

Mayer, J. (1996b, February 5). Reductions; verdicts reduced after trial. *National Law Journal*, p. C18.

Mayer, J. (2001). Cults, violence and religious terrorism: An international perspective. *Studies in Conflict and Terrorism, 24*, 361–376.

McKown v. Lundman, No. 91–8197, Dist. Ct., Hennepin Co., Minn. (1995).

O'Neil, T. (2001, December 21). Pastor will serve 7-year prison term for thefts. *St. Louis Post-Dispatch*, p. C11.

Pallone, N. J. (in press). Sin, crime, arrogance, betrayal: A psychodynamic perspective on the crisis in American Catholicism. *Journal of Brief Treatment and Crisis Intervention, 2*(4).

Prince v. Massachusetts, 321 U.S. 158 (1944).

Rezendes, M., & Farragher, T. (2002, April 26). Files show Shanley tried blackmail. *Boston Globe*, p. A1.

Richardson, J. T. (1993). Definitions of cult: From sociological-technical to popular-negative. *Review of Religious Research, 34*, 348–356.

Richardson, J. T. (1994). The ethics of "brainwashing" claims about new religious movements. *Australian Religious Studies Review, 7*, 48–56.

Ringe, S. H. (1992). The word of God may be hazardous to your health. *Theology Today, 49*, 367–376.

Robbins, T., & Anthony, D. (1995). Sects and violence: Factors enhancing the volatility of marginal religious movements. In S. Wright (Ed.), *Armageddon in Waco*. Chicago: University of Chicago Press.

Robinson, W. V., & Farragher, T. (2002, April 9). Files show Law, others backed priest. *Boston Globe*, p. A1.

Rodier, D. N. (1999, November 1). State court rulings. *Pennsylvania Law Weekly*, p. 5.

Roman Catholic Diocese of Paterson. (2002). Protecting our children. Retrieved October 28, 2002, from www.patersondiocese.org/moreinfo.cfm?Web_ID=671.

Rota, J. S., Salmon, D. A., Rodewald, L. E., Chen, R. T., Hibbs, B. F., & Gangarosa, E. J. (2001). Processes for obtaining nonmedical exemptions to state immunization laws. *American Journal of Public Health, 91*, 645–653.

Rush, G. E. (2000). *The dictionary of criminal justice*. Guilford, CT: Dushkin/McGraw-Hill.

Salmon, D. A., Haber, M., Gangarosa, E. J., Phillips, L., Smith, N. J., & Chen, R. T. (1999). Health consequences of religious and philosophical exemptions from immunization laws: Individual and societal risk of measles. *Journal of the American Medical Association, 282,* 47–53.

Schwartz, L. L., & Kaslow, F. W. (2001). The cult phenomenon: A turn of the century update. *American Journal of Family Therapy, 29,* 13–22.

Simpson, W. (1989). Comparative longevity in a college cohort of Christian Scientists. *Journal of the American Medical Association, 262,* 1657–1658.

Simpson, W. (1991). Comparative mortality in two college groups. *Mortality and Morbidity Weekly Report, 40,* 579–582.

Stewart, K., & Harding, S. (1999). Bad endings: American apocalypsis. *Annual Review of Anthropology, 28,* 285–310.

Whitsel, B. C. (2000). Catastrophic new age groups and public order. *Studies in Conflict and Terrorism, 23,* 21–36.

Wright, S. A., & Piper, E. S. (1986). Families and cults: Familial factors related to youth leaving or remaining in deviant religious groups. *Journal of Marriage and the Family, 48,* 15–25.

Zellner, W., & Petrowsky, M. (1998). Freedom Park. In W. W. Zellner & M. Petrowsky (Eds.), *Sects, cults, and spiritual communities: A sociological analysis.* Westport, CT: Praeger.

6

THE SEX OFFENDER

Prevalence, Trends, Model Programs, and Costs

LORI KOESTER SCOTT

Jack was a 53-year-old businessman whose 13-year-old stepdaughter accused him of molesting her. She told her best friend that he had been groping her since she was 8 years old, forcing her to lie on top of him, and, most recently, coming in to stare at her while she was bathing. When the police showed up at his house to interview him, he denied everything and threatened to kill himself.

By the time of his court date, Jack admitted that he had touched his stepdaughter but only "because she was an overly affectionate child, and she was such a pest that he had to pinch her breasts to get rid of her." He was allowed to plead to a charge of attempted child molestation, was ordered to have no contact with children, and was sentenced to 5 years' probation with a 6-month jail term. After a year of treatment with a traditional therapist, he was still minimizing his participation and blaming the victim for being seductive.

He was then placed on a specialized caseload and sent to a cognitive-behavioral therapist. Once he began working specifically on his sexual deviancy, he began taking total responsibility for his actions. He eventually explained to the victim in detail how he had fantasized every touch and set up every situation in which she was molested.

His periodic polygraphs show no additional contact or sexual fantasies of children. All other family members have been in counseling, and the victim is now in college.

Richard, another offender, was arrested at age 25 on suspicion of rape. He was charged with breaking into 23 homes or apartments over the course of 2 years, threatening his victims, all in their 20s, with a knife, tying them up, and raping them repeatedly. Evidence found in his apartment, cheap jewelry and half-used bottles of perfume that he had taken from the victims and brought home to his wife, linked him conclusively to the crimes. A serial rapist, he was convicted of eight of the charges and was sentenced to 75 years in prison.

Above are two cases seen by the courts these days. The first example is much more common than the second, but both men are typical sex offenders. Chances are that Jack, having been diagnosed as not primarily attracted to children, and responding well to treatment and supervision, will continue to be in control of a problem he knows he must always manage. Richard, seen as very dangerous, out of control, and responsible for damaging numerous lives, will remain in prison. Should he ever be eligible for release, it will probably be to another institution, a secured facility for civilly committed sex offenders.

In the 1980s and 1990s, the attention of the public and the criminal justice system was increasingly focused on sex offenders and their victims. This crime had been an issue for many hundreds of years, but the women's movement and a new courage to speak out on the part of victims brought more offenders into the court system. More numerous cases here and there gradually led to a huge increase in the number of investigations, prosecutions, and educational programs; expanded media attention; increased treatment for victims; and the inception of experimental treatment programs for offenders. Most recently, the issues of civil commitment and community notification have presented new and costly dilemmas for state criminal justice systems. However, as this review will reflect, sexual abuse in all its manifestations is a crime that is still not well understood and is often inadequately managed by many jurisdictions.

Some of the most recently published data on sex offenses in the United States show that 1 in 6 women and 1 in 33 men reported experiencing a completed or attempted rape at some time in their life (Tjaden & Thoennes, 1998). In this confidential survey, 22% of the women were under 12 when victimized and 32% were aged 12 to 17. It is estimated that only 16% to 33% of sex crimes are reported (Ringel, 1997, p. 3). Despite this low reporting rate, in recent years an increasing number of sex offenders have been prosecuted and are under some form of incarceration or supervision by the criminal justice system. Since 1980, the number of imprisoned sex offenders has grown by more than 7% per year. In 1994, nearly 1 in 10 state prisoners were incarcerated for committing a sex offense. Additionally, of the 234,000 sex offenders under control in 1997, about 134,000 were under some form of community supervision (usually probation or parole; Greenfeld, 1997).

Adult rape and sexual assault cases have been legitimized by the courts for some time, although the struggle to reduce sexual aggression in what many label our "rape-prone" society has not yet been successful (Herman, 1988; Koss, Gidycz, & Wisniewski, 1987; Scully,

1990). Incest and child molestation cases were infrequently acknowledged up to 1980 and were not aggressively prosecuted. As recently as 1953, Kinsey and his associates stated that there was no logical reason why children should be disturbed about sexual abuse (cited in Salter, 1988). Not until 1963 did the state of Minnesota criminalize incest, a behavior that was thought to be extremely rare (Patton, 1991). Freud's attempt to explain the tormented sexual memories of his female patients as unconscious desires for their fathers was generally discredited in the 1980s, yet as recently as 1991 a Phoenix psychiatrist evaluated an incest case by stating that his 13-year-old female patient had unconsciously manipulated her father and that the offender's problems had evolved in part from "unresolved Oedipal feelings for his mother." He went on to describe the offense as "immature hysterical sexual behavior on both parts" (author's personal files).

Zorza (1999) stated that "courts continue to take a skeptical approach to actively prosecuting this crime" (p. 33-2). She documented the lack of training about family violence in medical schools and the lack of training about abuse, sexual abuse, and family violence in schools of psychology and counseling. The 1996 American Psychological Association Presidential Task Force on Violence and the Family reported that "many child protective service and mental health workers were not adequately trained or equipped to treat maltreated children" and "have not been trained to look for or assess for signs of abuse" (p. 69). This report also stated that many judges, therapists, and other professionals harbor stereotypes that impede their ability to look for or find signs of child abuse (American Psychological Association, 1996).

Still, the growing commitment to prosecute sex crimes of all types has greatly affected the makeup of police, courts, prisons, and community corrections. Efforts are underway to expand the psychological knowledge and effective treatment and supervision of this most difficult, highly secretive, and consistently manipulative offender. However, in spite of the increased national attention paid to this issue in the past two decades, there still remains a great deal of ignorance and denial on the part of the general public. At a distance they know it exists, but they most frequently see the offender as the anonymous and frightful creature of headlines, books, and movies.

What has been the professional history of evaluating and dealing with sex offenders and their victims? When the psychological literature finally began acknowledging and discussing incest, a careful analysis shows that at first the victim was blamed and then the spouse. Today, and still with reluctance by some professionals, there is a commitment to holding the adult perpetrator responsible for his actions. As the statistics reveal, the sad fact is that the riskiest place for a child may be his or her own home. In 90% of the rapes of children less than 12 years old, the child knew the offender (Greenfeld, 1997). At present, laws vary widely from state to state, with wide discrepancies in the very definitions of forms of sexual abuse and the sentencing guidelines worked out by legislators.

ETIOLOGY

In the past 20 years, much more attention has been paid to the issue of understanding sexual abuse than at any previous time in history. A search of the literature up through the 1970s would have yielded little research or documentation, but with the increased focus on the frequency of victimization by the women's movement and by researchers, a growing body of work has been produced by professionals in the psychological and criminal justice fields. It is generally believed necessary to define and classify sex offenders in order to (a) assess risk for sentencing purposes; (b) determine amenability to treatment; and (c) understand the etiology of such disorders so that prevention and/or early intervention may reduce the frequency of victimization.

Among the first contributions and incentives for further research were studies by Groth (1979) and Burgess, Groth, Holmstrom, and Sgroi (1982). Groth (1979), who had worked for many years with convicted sex offenders in Connecticut and Massachusetts, attempted to categorize different types of rapists and child molesters. He classified rapists as motivated by either "power" or "anger" and child molesters as "fixated" or "regressed," with the fixated

molesters more commonly identified as true pedophiles, those adult individuals who are interested exclusively in children for satisfying their sexual needs. The "regressed" molester was considered to be primarily attracted to adult partners but to have temporarily forsaken his normal adult desires for sex with an age-appropriate partner.

During the same time period, Dr. Gene Abel, a behavioral psychiatrist, and his associates, including Dr. Judith Becker, were conducting a landmark study of sex offenders at the New York State Psychiatric Institute. Four hundred and eleven offenders who were guaranteed confidentiality admitted that they had attempted 238,711 offenses and had completed 218,900 of them, for an average of 533 acts and 336 victims each. Offenders who had been identified with only one paraphilia (sexual disorder), such as rape, were found to have engaged in an average of 3.5 additional paraphilias, such as exhibitionism, voyeurism, obscene phone calls, child molestation, and bestiality. Many so-called incest offenders were in reality pedophiles; a large number who had avowed no attraction to same-sex children disclosed a history of victimization of children of both sexes; and 44% of incest offenders revealed an astonishing number of nonrelated child victims in addition to numerous other paraphilias (Abel, Mittelman, & Becker, 1985).

Clinicians and corrections personnel who have used physiological testing, including the polygraph, in recent years to counteract the inadequate self-reporting of sex offenders have validated that many offenders have a wide range of victims, age levels, and sexual deviancies. For example, a recent Colorado study of a sample of imprisoned sex offenders found them to have extensive criminal histories, committing sex crimes for an average of 16 years before being caught (Ahlmeyer, English, & Simons, 1999). The authors also concurred with the riskier data on incest offenders.

In a 1992 study of 118 incestuous fathers, 16% of the men were classified as "sexually preoccupied," possessing a "clear and conscious (often obsessive) sexual interest in their daughters." Some of the men in this category were further classified as "early sexualizers" who admitted that they had regarded their daughters as sex objects from birth. "One father reported that he had been stimulated by the sight of his daughter nursing and that he could never remember when he did not have sexual feelings for her. He began sexually abusing her when she was four weeks old" (Williams & Finkelhor, 1992, p. 8).

On the basis of increasing information and extensive clinical experience, it has become more and more apparent to professionals that trying to fit sex offenders neatly into categories is not always possible. A continuum of behaviors seems to be more accurate.

An obsession with pornography, voyeurism, obscene phone calls, and acts of indecent exposure are all examples of "hands-off" behaviors that sometimes lead to more serious "hands-on" invasive attacks of child molestation and rape. Some offenders do not progress beyond certain stages; others, left unchecked and unnoticed by the system, may go on to cause irreparable damage to countless lives.

PSYCHOLOGICAL THEORIES

Not all aberrant sexual behavior involves victims. The continuum of sexual behavior or "addiction" was explored by Patrick Carnes in the 1980s as he attempted to identify and treat sex offenders using the 12-step model. Carnes (1983) had been working with addicts and their families in Minnesota and had become increasingly aware that some patients who came in addicted to alcohol or drugs revealed an addiction to sexual behavior. These individuals gradually developed a pattern of needing to perform certain sex acts in a compulsive manner in an attempt to satisfy other needs. Carnes's definition of addiction also included a person's willingness to sacrifice to the point of self-destructiveness for the taking of a drug or a specific experience, a phenomenon he noted in many sex offenders.

This model implies that sexual offending is an illness, which it is not. However, many offenders like to use it as an excuse. Family theories that view sexual abuse as a problem that exists in the family network lead to partial blaming of the victim instead of the offender's acceptance of responsibility. Psychodynamic

theories have been dismissed, at least in North America, as being irrelevant and ineffective as a treatment strategy. To date, the majority of the practitioners and researchers in the field follow what can best be described as a cognitive-behavioral model based on social learning theory, which posits that sexually deviant behavior is the result of gradual conditioning to a powerful reinforcer (Knopp, 1984; Knopp, Freeman-Longo, & Stevenson, 1992). A large professional organization, the Association for the Treatment of Sexual Abusers (ATSA), was formed in the late 1980s and has grown to an international organization of therapists, psychiatrists, and criminal justice professionals who work with sex offenders in a variety of settings. The latest in research and clinical procedures is presented annually at ATSA conferences and in journals. In their most recent practice and standards guidelines, they state:

> Contemporary treatment programs for sexual abusers employ cognitive-behavioral techniques to target clinical needs that are associated with risk to re-offend. . . . Structured, skills-oriented, and pragmatic treatment programs appear to be the most effective in reducing rates of re-offending. Unstructured, insight-oriented, and abstract treatment programs are much less likely to be effective in reducing rates of re-offending, and some may even increase the risk of re-offending. (ATSA, 2001, p. 18)

Offenders frequently present with a significantly disturbed developmental history. Early feelings of emotional and social isolation, often combined with physical or sexual abuse, have led to cognitive distortions or systems of mistaken beliefs about themselves, others, and the way in which the world operates. They often present in a "victim posture." Their belief systems form the beginning components of the "offense cycle" or chain of behaviors, as the offender develops a habit of using fantasies to manage emotional needs not met through healthy connectedness with others, sometimes along the lines of wealth, power, control, or revenge. At some point, deviant sexual fantasies become part of this repertoire.

The offender also begins to incorporate into his fantasy a pattern of rationalization and justification that further disinhibits him and desensitizes him to the taboo nature of the behavior (Wolf, 1984). Escaping to his fantasies places him in control.

> This phase of the cycle objectifies sexual partners, increases the frequency of thoughts related to sex, reduces discrimination or selectivity of sex partners, rehearses sexual misbehavior and reinforces the individual's belief that the primary goal of sexual relations is to feel better about oneself. Behaviorally, the individual is likely to engage in compulsive masturbation, experience deficits with respect to touch discrimination, and make sexual decisions based upon opportunity. (Emerick, 1991, p. 58)

At some point, the offender begins cruising and grooming behavior around victims who match his deviant sexual interest.

> Once the offender acts out at least part of his deviant interest, he feels very guilty and develops ineffective strategies to discontinue his deviant behavior. These strategies are partially contained in the final step, reconstitution, which is the use of socially appropriate behavior to an extreme to disguise the offender's deviant behavior and interest and to manage his guilt associated with his sexual misconduct. Because nothing has truly changed in the offender's life, he is again at step one, which is victim posture thinking, and begins the whole process anew. (Gray, 1991, p. 54)

The "relapse prevention" model of sex offender treatment, developed by Marlatt and Pithers (Laws, 1989), demonstrates that there are a number of common factors or behaviors preceding a relapse into the offense cycle. Offenders often make apparently irrelevant decisions that bring them closer and closer to high-risk situations, lapse, and possible relapse, also supported by rationalization and denial (Laws, 1989).

> John was arrested for molesting a niece and was given a year in jail plus lifetime probation on a specialized sex offender caseload. He began treatment with a specialist and was doing moderately well. He was given permission to visit with his childless brother on Saturdays. However, as soon

> as he arrived at his brother's he would walk down to the tavern on the corner, where he would try to make friends with various women he saw there. He eventually was invited to the home of one of the women who had a 10-year-old daughter. When he was asked to go to the store for dinner supplies, he offered to take the girl with him. He asked her if she wanted to "steer the car." While she was sitting on his lap, he molested her.

The above cycle of relapse demonstrates how John gave himself permission at every step to go further and further into a situation leading to reoffense.

The primary emotion reported by child molesters preceding their offense is depression; the primary emotion reported by rapists is anger. Many offenders also present with a profile consistent with a diagnosis of character disorder, which includes qualities of narcissism and distorted perceptions of the world about them; they tend to be ruminative, chronically depressed, and sexually preoccupied (Wolf, 1984). Only a small percentage are diagnosed as psychotic or suffering from a chronic mental illness. A study of sex offender prisoners at a federal institution showed that offenders diagnosed as psychopaths were not likely to complete the treatment and could be screened out before using valuable time and space (Norris, 1991). Dr. Robert Hare and his colleagues studied psychopathy in a large sample of forensic populations and found that 26% of rapists were diagnosed as psychopathic but only 5.4% of child molesters (cited in Salter, 1998).

A great many studies examine the role of intimacy deficits and lack of attachment in the development of sex offenders (Marshall, 1989; Marshall, Barbaree, & Fernandez, 1995; Ward, McCormack, & Hudson, 1997). It is hypothesized that one consequence of a lack of intimacy skills and the subsequent experience of emotional loneliness is that sexual offenders may indirectly seek intimacy through sex, even if they have to force a partner to participate. A "relationship" with a child may seem less risky, more comfortable, safe, and less likely to lead to rejection. Offenders' childhood relationships may be characterized by physical and/or sexual abuse or by emotional detachment or abuse, or an unemotional, rigid, distant, or absent parent.

According to Marshall, Serran, and Cortoni (2000), the role of paternal attachment may be particularly important and deserves further theoretical and empirical examination.

> Ken was a strong and handsome 19-year-old youth who worked in construction. Late one night, high on cocaine, he went to a fast food restaurant, and when two teenage girls left and walked to their car, he forced himself in the car, said he had a gun, and made them drive to an isolated part of town. For the next few hours, he repeatedly raped both of them. Finally one of the girls ran down the road and signaled a passing car.
>
> After his arrest and conviction, his mother came to see the probation officer who prepared the presentence report. She said that Ken's father had been an alcoholic who had beaten all the children when young, particularly Ken. By the time he was a teenager, his mother was drinking one night and, in a state of despair, shot herself in front of the children. Miraculously, she survived. When she was in the hospital recovering, a psychiatrist told her and her husband that the children needed to come in for some counseling because of what had happened. Ken's father said that he "wasn't going to let any shrink tell him what to do," leaving the family as chaotic as before.

ASSESSING RISK

To make appropriate judgments about the disposition of sex offenders, many professionals in the system usually look at the severity of the offense, together with perceived risk to the community. This begins with the child protective services worker or police officer who initially investigates the case. Prosecuting attorneys often use criteria of degrees of intrusion, force, or frequency to determine a basis for prosecution leading to trial or acceptable plea bargains. Presentence writers include the impact on the victim in summarizing the case for the sentencing judge. All these individuals bring to the case their own personal beliefs and biases regarding this emotionally charged and still controversial crime.

At present, in spite of several years in which many studies have been conducted and analyzed, there is little uniformity in any of these assessment processes, which vary widely from state to state. For example, prosecutors or judges may decide that if there is actual intercourse in a case involving a child victim, they will accept no less than prison. Yet Hindman (1989), a nationally respected authority on the effect of sexual abuse on victims, suggests that many other factors may make the offense more traumatic and long lasting to the victim, such as the reaction of family and others to disclosure of the crime. In children, especially, the violation of trust by the person who should have been protecting them has been shown to dramatically increase the level of trauma suffered by the victim (Center for Sex Offender Management [CSOM], 2002b). Unfortunately, past sentencing procedures often gave incest offenders the most lenient sentences and frequently allowed and still do allow many of these offenders to move back into the family home without treatment or restrictions on their contact with children.

Much has been written and discussed in the past 15 years on the likelihood of reoffending by sex offenders. The media sometimes describe sex offenders as an all-encompassing group with recidivism rates as high as 70% to 80%. A search of the literature shows no studies with recidivism rates even approaching that number. First, it is necessary to carefully define the terms used. Many studies mean very little because of methodological variability and ambiguity. The criteria may include rearrests, convictions, or possibly an act that was processed as a technical parole or probation violation instead of a sex offense. Many systems still allow sex offenders to plead to nonsex crimes, such as aggravated assault or criminal trespass. Exhibitionists, supposedly the highest recidivists, may be more easily caught because of their visibility and easy victimization of strangers. Children who are molested by a father or other relative or trusted adult, such as a teacher or minister, are often very reluctant to report such a crime, especially if the person has already been through the system and has been allowed recontact with children.

In a newly published survey of recidivism studies, CSOM (2002a) stated that

> in many instances, policies and procedures for the management of sex offenders have been driven by public outcry over highly publicized sex offenses. However, criminal justice practitioners must

avoid reactionary responses that are based on public fear of this population. Instead, they must strive to make management decisions that are based on the careful assessment of the likelihood of recidivism. The identification of risk factors that may be associated with recidivism of sex offenders can aid practitioners in devising management strategies that best protect the community and reduce the likelihood of further victimization. (p. 5)

A widely recognized study of sex offender recidivism was published by Hanson and Bussiere in 1998. They gathered 61 research studies on 28,972 adult sex offenders that met the criteria of using a longitudinal design and a comparison group. The results of this meta-analysis showed a sex-related rearrest or reconviction rate of 18.9% for rapists and 9.9% for child molesters. The recidivism rate for *any* reoffense was 46.2% for rapists and 36.9% for child molesters over a 4- to 5-year period. In isolating various factors consistent across all studies, the authors found that those most likely to recidivate were those who were younger and who had a prior sex offense, male victims, or stranger or extrafamilial victims. Failure to complete treatment was a moderate factor, but being sexually assaulted as a child was not related to repeat sex offending.

In another study of 400 sex offenders under community supervision, half of whom had committed a new sex offense during that time, Hanson and Harris (1998) found a number of significant "dynamic" risk factors accounting for the differences between the two groups. Those included unemployment (especially for rapists), substance abuse problems, and attitudinal differences such as little remorse, seeing themselves at no risk to reoffend, and being less likely to avoid high-risk situations. They concluded that sex offenders are most likely to reoffend when they become preoccupied with sex, have access to victims, do not acknowledge their recidivism risk, and show rapid mood changes, particularly increases in anger.

To date, meta-analysis findings suggest that characteristics generally found in recidivists include

- Multiple victims
- Diverse victims
- Stranger victims
- Juvenile sexual offenses
- Multiple paraphilias
- History of abuse and neglect
- Long-term separations from parents
- Negative relationships with their mothers
- Diagnosed antisocial personality disorder
- Unemployed
- Substance abuse problems
- Chaotic, antisocial lifestyles

Experienced probation and parole officers know that alcohol and drug use are often involved both in the original offense and as a part of the cycle toward reoffending. Although the use of alcohol does not cause such behavior (even though many offenders tend to use it as an excuse), it and other drugs can serve as disinhibitors to deviant activities, particularly rape (Valliere, 1997).

In the same way, pornography or any sexually stimulating material is often found to be a precursor to offending. Sex offenders are frequently discovered to have large collections of pornography, which can include hard-core or child pornography. This fits with Hanson's analysis of "sexual preoccupation" as a dynamic risk factor. In treatment, offenders admit to using their collections to fantasize and masturbate, although some may use innocent photographs and pictures from magazines and catalogs. Child molesters sometimes force their young victims to watch and then imitate sexually explicit photos or videos.

Hard-core pedophiles (those who are primarily sexually attracted to children) often have secret networks linking them to sources of illegal child pornography, the production of which escalated rapidly in the late 1970s (Finkelhor, 1984). The FBI and U.S. Postal Inspector's Office have tried to maintain surveillance over such connections and arrest when possible. Unfortunately in the new age of the Internet, pornography, and particularly child pornography, is more accessible than ever. Pedophiles can easily find, share, and trade; they can congregate on the Web with others of their persuasion and reinforce their deviant fantasies. In addition, many easily contact children via chat rooms, sometimes sending them pornography or arranging to meet a potential victim in person.

For those who would dismiss Internet crimes as "victimless," a study conducted in the Sex Offender Program of the Butner, N.C., Federal Prison revealed that out of 62 child pornographers convicted on one offense, admissions were made to an additional 1,379 contact sexual crimes that were never detected (Hernandez, 2000). FBI and other law enforcement personnel have been forced to dedicate more and more staff to technologically tracing and apprehending these individuals.

Most risk assessments in use today have traditionally been used on the general prison population and often do not apply to some sex offenders, who, except for their hidden deviance, have led much more stable and prosocial lifestyles than the average inmate (Bonta & Hanson, 1995). A further sampling of sex offender recidivism studies shows that McGrath (1991) found that untreated exhibitionists had the highest reoffense rates. Recidivism rates of rapists varied widely (8–36%). And among untreated extrafamilial child molesters studied by Marshall and Barbaree (1990), 43% reoffended within a 4-year period. Prentky, Lee, Knight, and Cerce (1997), in a 25-year follow-up of rapists and child molesters, showed a cumulative failure rate of 41%.

Max was a music school teacher who taught in several areas of the United States, Mexico, Africa, and the Middle East. When arrested at the age of 51 for soliciting the sexual favors of two 12-year-old boys, he disclosed a long history of pedophilia. He had moved frequently because even though parents would find out what had happened and force him to leave, no one wanted to prosecute. He kept a diary of "well over 1,000 partners," which was found by police upon his arrest.

Max related a childhood of being raised in an orphanage, in which he claimed that sexual experimentation was common. He defended his subsequent sexual behavior, stating that wherever he went, "young men flocked to him and wanted to have sex with him." Unfortunately, even in this arrest, the charge was pled to a minor felony because the children's parents did not want them to testify. He was originally placed on probation and ordered into intensive treatment but refused to attend. His probation was revoked and he was sent to prison. After a short term of incarceration, unfortunately, he was free at age 53 to go wherever he wished.

A mistake made by many professionals supervising sex offenders was to treat the offender as though he would reoffend only with children of the same age as the original victim and that he would not cross age or gender lines. Judges, for example, would often allow an offender who had molested a female child to live with young boys or allow a rapist to live with children. The use of the polygraph in many jurisdictions has helped convince judges of the need for extreme caution.

Don, age 35, had been placed on probation after a year in jail for the molestation of his 5-year-old stepdaughter. As part of the plea bargain, charges had been dropped involving his 13-year-old niece. He immediately became involved in therapy, working hard on all assignments. Three months after meeting a much younger woman at his church, he married her. The probation officer suspected that it was a way to obtain supervised contact with his two sons, but he was doing so well in therapy that he was allowed to go through with the marriage.

Before his therapist discharged Don from treatment, his probation officer ordered a polygraph. Don admitted that he was still occasionally fantasizing about young girls he saw in public places, such as the grocery store. Communication with police was increased as a precaution. Two months later Don approached a 12-year-old girl at a convenience market and tried to get her into his car, to the point of offering her money. Even though he had not committed a new hands-on offense, the girl called police, who alerted the probation officer. Don was jailed for 30 days, placed on formal intensive probation, and moved up to high-risk status.

DISPOSITION: PROBATION OR PRISON?

Because the determination of risk and/or offender typology is not definitive, and because many difficulties are involved in prosecuting and sentencing these offenders, it is not surprising that some type of community corrections is the final outcome of over half the cases. Some states use a combination of prison, jail, or mental hospital confinement followed by community release. Arizona, for example, routinely sentences certain sex offenders convicted of "dangerous crimes against children" to lifetime probation, with or without prison, with the rationale that such offenders cannot ever be cured but can learn to control their behavior. Studies such as Prentky et al.'s (1997) long-term survival analysis data on released child molesters confirm such a sentencing philosophy. Many other states have expressed interest in lifetime follow-up for certain offenders, and some have begun to pass similar legislation.

In 1996 the American Probation and Parole Association (APPA) published a comprehensive book on managing adult sex offenders in which they surveyed jurisdictions all over the United States regarding this issue, paying on-site visits to several innovative programs. They found that the most promising programs were those in which teamwork and collaboration between several parties formed a system that "contained" the offender, interfering with his habits of secrecy and manipulation. The team is most likely to consist of the probation or parole officer, the offender's therapist, a polygrapher, a police representative regarding notification, and possibly, a victim advocate, representative, or therapist (English, Pullen, & Jones, 1996).

One of the highlighted programs that has been functioning for over 15 years is based in Maricopa County (Phoenix), Arizona. Offenders placed on probation tend to be considered less dangerous and presenting somewhat less risk to the community. Although a large percentage are incest offenders with female victims, the remainder are male-targeted pedophiles, rapists, and high-risk exhibitionists who need surveillance. In the past few years, caseloads have grown to include a number of serious offenders who were sentenced to prison for 5- to 15-year sentences on one or two counts and are now on lifetime probation, rather than parole, following their release from incarceration.

Operating on the principle that such offenders need external controls while they are learning to develop internal control, the program uses the services of surveillance officers as part of a probation team to monitor the evening and weekend activities of offenders. A list of 16 specialized terms for sex offenders is added to regular terms of probation, calling attention to the known behavior patterns of offenders and imposing prohibitions against situations that might lead to a reoffense. Using the concepts of relapse prevention, cognitive-behavioral therapists work closely with probation officers to break through the offenders' entrenched denial, rationalizing, minimizing, and manipulating.

Maricopa County Probation Conditions for Sex Offenders:

1. Do not initiate, establish or maintain contact with any male or female child under the age of 18, including relatives, or attempt to do so, without the prior written approval of the probation officer. Sign and abide by the probation department definition of "no contact."

2. No contact with victim(s) without prior written approval of probation officer.

3. Do not go to or loiter near schools, school yards, parks, playgrounds, arcades, swimming pools or other places primarily used by children under the age of 18, or as deemed inappropriate by probation officer, and without the prior written approval of probation officer.

4. Do not date, socialize, or enter into a sexual relationship with any person who has children under the age of 18 without prior written approval of probation officer.

5. At the direction of the probation officer, attend, actively participate, and remain in sex offender treatment. Authorize therapists to disclose to the Court and the probation department information about your progress in treatment.

6. Submit to any program of psychological or physiological assessment at the direction of probation officer, including, but not limited to, the penile plethysmograph and/or the polygraph, to assist in treatment, planning and case monitoring.

7. Abide by all current Arizona law requiring the registration and DNA testing of sex offenders.

8. Residence, employment, and education, including any temporary changes, must have the advanced written approval of the probation officer.

9. Do not travel outside Maricopa County without the advanced written approval of the probation officer.

10. Abide by any curfew imposed by probation officer.

11. Do not possess any sexually stimulating or sexually oriented material, in any form, without the prior written approval of the probation officer, or patronize any adults-only establishment where such material or entertainment is available.

12. Do not possess children's clothing, toys, games, videos, etc. without prior written approval of probation staff.

13. Be responsible for your appearance, including the wearing of undergarments and clothing in locations where another person might see you.

14. Do not hitchhike or pick up hitchhikers.

15. Do not operate a motor vehicle without prior written approval of probation officer.

16. Prior written approval must be granted before the utilization of any computer equipment or internet access; if granted, Maricopa County Adult Probation computer usage guidelines must be followed.

Probation departments throughout the country are increasingly adopting terms similar to the above restrictions. It is important that judges impose these specialized terms at sentencing and that prosecutors collaborate with probation officers in asking for specialized terms. In some jurisdictions, prosecutors have the specialized terms already printed as part of plea agreements. It is almost impossible for community supervision agents to enforce treatment, polygraphs, nonconfidentiality, and noncontact orders without the backing of the court.

Treatment and Teamwork

In the Arizona program, all sex offenders are initially required to attend 45 hours of classes on human sexuality, the development of sexual deviancy, understanding the offense cycle, societal attitudes on male-female roles and stereotypes, and the impact of victimization. Partners are encouraged to attend. Following the classes, the offender is placed into group treatment for as long as all involved believe that intense treatment is necessary. Because sexual deviance is often ingrained, with lifelong habits of fantasizing and/or acting out behaviors in total secrecy, offenders find that the group works as a supportive network of people who share the same problems, which could never be discussed previously. At the same time, group members are not hesitant to confront each other about rationalizing, minimizing, and other thinking errors. Once the offender learns his offense cycle, he and the other group members share ways to "unlearn" it and interfere with the beginning stages of the cycle, following the concepts of relapse prevention.

Supervising officers are encouraged to attend these groups on a random basis, not only to learn more about how treatment works, but to make informed office and field contacts. In turn, therapists can receive valuable information from staff about the offender's home and work situation, family dynamics, hobbies, and other activities that occupy the time when he is not in treatment. In New Haven, Connecticut, which has recently developed an intensive and effective sex offender program for high-risk offenders, a victim advocate is an integral part of the team. He or she offers valuable feedback on the victim's progress, providing input on restrictions and privileges granted the offender. While working with victims has not historically been a field officer's role, agencies are increasingly finding that officers can be powerful catalysts in a victim's recovery and in using the offender's admissions in treatment to relieve victims of much of their anxiety and confusion.

APPA and CSOM recommend that, if at all possible, those working in this field should be trained to work with sex offenders and should be part of a specialized unit, depending, of course, on the size of the jurisdiction. Even if there are only a few sex offenders, the person supervising them should have as much specialized training as possible. Agencies can collaborate initially with the prosecutor, writing a well-informed presentence report that explains to judges the rationale for certain specialized conditions. Because of the increasing body of knowledge surrounding these offenders, use of that knowledge can make the officer's job more efficient and the community safe.

Specialized terms of probation also ensure that incest offenders, for example, will not be allowed to return to the family home without appropriate treatment and safeguards in place for all family members. Well-designed guidelines for such contact or reunification should be developed, as it makes little sense to allow a convicted, untreated child molester back into the intimacy of family life, where children might be unclothed, bedroom doors unlocked, a spouse sleeping through the night, and bathroom doors "accidentally" left open. If offenders are allowed to return home prematurely, this gives a message to the victim that the system is willing to risk her vulnerability once again and that reporting the abuse did very little to help her situation. Many children do indeed still love their fathers, and just want the abuse to end. Professionals know that without separation, intensive treatment, and good supervision, it will not.

A major new component of the containment approach in an increasing number of jurisdictions is the polygraph. This is being used in community supervision programs as well as treatment programs in many prisons. Sex offender polygraphy is a specialization that should not be undertaken without specific training on sex offender dynamics. The American Polygraph Association has endorsed a special course of instruction and offers several qualified persons as instructors. Most of the agencies that now use the polygraph as a component of treatment and supervision say that they could not do without it (Ahlmeyer et al., 2000).

Use of the polygraph serves two purposes. Initially, soon after sentencing or release from jail, the offender participates in a disclosure polygraph, designed to cover the defendant's sexual history, including first sexual experiences, any sexual relations with relatives, masturbation practices, exhibitionism, voyeurism, bestiality, other paraphilias, use of force, and sexual fantasies. The offender is given an opportunity to discuss these issues before the exam, often resulting in significant disclosure. He or she is also given a chance to discuss and perhaps clarify any questions on the actual polygraph test that may have indicated deception. The results are shared and discussed with the therapist, probation officer, and the offender's therapy group. The goal is to begin treatment with a "clean" sexual history, giving the offender the opportunity to learn to control all components of his deviant behavior, not just those he has chosen to disclose of his own accord.

Dr. H., a local chiropractor, had been accused of molesting a 6-year-old neighbor girl. The original and lengthy court proceedings resulted in a mistrial. He had already lost his license for allegedly groping a female patient inappropriately. After numerous legal delays, during which Dr. H. continued to deny his guilt, he signed a "no contest" plea to sexual contact with a minor and was sentenced to 15 years' probation.

A local psychologist had stated that there was nothing in Dr. H.'s testing to suggest that he was "anything but a conscientious individual who wants to be helpful—particularly through his own specific fields of expertise" and that he "does not have to go outside of his own marital relationship to have his [sexual] needs satisfied."

While on a general probation caseload, Dr. H. saw a therapist for individual sessions, who reported that the client was cooperative, was "doing well," and, after 15 sessions, did not need any more therapy. (Dr. H. was also continuing to deny any deviancy.) He was placed on a specialized sex offender caseload because of his continuing denial and manipulative and domineering behavior. When the female officer attempted her first field visit, he hid behind the door, stating that he was nude. He highly objected to her subsequent decision to place him in a new therapy group for sex offenders and resisted for months when he was asked to take a polygraph. He continually demanded hearings before the judge, who continued to order him to take the tests. Dr. H. also began writing or calling legislators and public officials to complain about the probation department's "harassment."

After finally submitting to the polygraph, after years of fighting the "system," he not only admitted to molesting the victim in the present case but stated that he had engaged in sexual relations with at least 30 other children. He also disclosed that he had become involved in a wide range of sexual activities with female patients, sexually touching an "unquantifiable" (his term) number of these patients.

He reported possessing a large variety of pornographic materials, including child pornography. He related that his current marriage was struggling due to his impotency and herpes. Finally, he revealed that he was still fantasizing about sexual relations with children. He was placed in group treatment, where he made significant progress through the rest of his probation.

Salter (1998) discussed these types of offenders who lead a double life. They are often churchgoers, sometimes clergy, and are kind and generous. They have a likeability that misleads the average person. They may be teachers or pillars of the community. They are good at bonding with and manipulating therapists. Salter maintained that people who are not malevolent have trouble seeing it in others. Churchgoers, for example, can be gullible and trusting; they want to believe in goodness and are easy to convince. Therapists can be deceived as much as anyone.

The disclosure polygraph is used as a component of treatment and is extensively reviewed by ATSA (2001) in its most recently published guidelines for treatment. It can save valuable treatment time by overcoming the offender's denial, a common and frustrating issue for therapists. It makes therapists and supervision staff aware of the seriousness of many offenders who may have presented with a very limited history.

The maintenance polygraph is used as the offender progresses through treatment and is a useful tool for long-term supervision. Both disclosure and maintenance polygraphs are invaluable for planning, for family reunification procedures, and for help in determining risk. Probation officers report discovering behaviors or behavior patterns that are lapses that could lead to relapse. Examples are driving to work past a school at a certain time; continuing to sexually fantasize about the victim in a possible reunification; leaving town without permission; joining a club or church group that might lead to contact with children; and accessing pornography by computer or other means.

The Jackson County, Oregon, Probation Department was one of the first jurisdictions in the country to use the polygraph on a regular basis with sex offenders. In a recidivism study involving 173 offenders, only 5% were convicted of committing new sex crimes under supervision, although 16% absconded and 13% were revoked to prison for various reasons, including noncompliance with treatment or prohibited contact with children (Aytes, Olsen, Zakrajsek, Murray, & Ireson, 2001). Legal challenges to the use of the polygraph have been upheld in California (Abrams, 1989). Tennessee, Colorado, and Wisconsin are among the states that have recently removed statutory barriers to its use with sex offenders (CSOM, 2001). As of 2000, 33% of states were using the polygraph with sex offenders. Some states, such as Colorado, have produced an extensive and detailed statewide protocol governing the treatment, assessment, and supervision of both adult and juvenile sex offenders.

COST

Treatment and surveillance as described in the above programs can certainly increase the cost of supervising high-risk sex offenders. However, a commitment to prevention, increased community safety, and concern for victims can ultimately decrease the cost to society. A study by Prentky and Burgess (1990) of sex offenders in Massachusetts showed that the cost of prosecuting and incarcerating one sex offender who offends against a new victim greatly exceeds the relatively low cost to the state of appropriate supervision. Using Marshall and Barbaree's (1990) well-documented and conservative study of recidivism in treated (25%) versus untreated (40%) sex offenders, Prentky and Burgess (1990) estimated the costs of investigating, arresting, prosecuting, and incarcerating for 7 years a sex offender who had one victim at $183,333. In examining the costs of recidivism in untreated offenders, they devised a rather complicated formula using the above information to show a differential of $68,000 per case, concluding hypothetically that releasing 1,000 untreated sex offenders from prison would cost society about $68 million.

There is no good way to estimate the ultimate cost to our country from the millions of victims who have suffered some form of sexual abuse. A recent report stated that rape has the highest annual victim costs of all crimes at $127 billion per year (National Institute of Justice, 1996). This does not include child victims. After murder, child sexual abuse is the most serious and expensive crime to the victim and society. Nearly two thirds (62%) of pregnant and parenting adolescents have experienced contact molestation, attempted rape, or rape prior to their first pregnancy. Rape victims are 9 times more likely than nonvictims to have attempted suicide (CSOM, 2002a). Other effects of sexual abuse include physical injuries, sexually transmitted diseases, depression, subsequent substance abuse, and posttraumatic stress disorder (McMahon & Puett, 1999). Finally, effects include the emotional trauma suffered by families of victims and the financial difficulties experienced by families in incest cases.

Specialized and intensive supervision of sex offenders in the community, while more costly than that of the average probationer, can be cost effective in its prevention of further victimization. In 1997, Maricopa County, Arizona, estimated its average cost per year per offender to be $2,750. This included supervision by both a probation officer and a part-time surveillance officer. Offenders pay the cost of treatment, although indigent offenders can be subsidized by the state. The same is true for polygraphs. With probationers also paying probation fees every month, the cost is kept down. By the time an offender has moved to maintenance (which could take anywhere from 3 years to never), the cost can decrease significantly. Approximately 73% of the sex offenders under supervision are granted lifetime probation. This is considered a worthwhile expense, considering the integrity and success of the program and the ability to monitor certain offenders without expensive prison beds (Scott, 1997).

JUVENILE OFFENDERS

The sex offender issue cannot be examined fully without a discussion of the juvenile offender population. Historically, juveniles were, for the most part, not held accountable for their actions, which were often considered "adolescent experimentation" or were met with a "boys will be boys" attitude. Only in the late 1970s and 1980s were the extent and significance of the problem finally acknowledged. Programs specializing in the treatment of juvenile offenders exist in the majority of states. An excellent example is the Utah model, which includes various levels from incarceration and lengthy treatment to community supervision on probation, depending on risk.

Groth reported that 60% to 80% of adult offenders admitted that they had begun their deviant sexual behaviors as adolescents (cited in Sgroi, 1982). Fifty-seven percent of Abel's group described their deviant arousal beginning in adolescence, many when they were 12 years old or younger (Abel et al., 1985). Treatment centers for victims of sexual abuse report that up to 56% of the assailants were under age 18. Groth (cited in Sgroi, 1982) described molesters and rapists as young as age 9, cases that are not uncommon today. A third of Abel's subjects progressed from compulsive masturbatory activity, repetitive exhibitionism, and/or persistent voyeurism to more serious "hands-on" offenses.

Why does a child act out sexually? Most experts agree that aggressive social behaviors are learned primarily through observation and by direct experience. Of a sample of Canadian prisoners convicted of a variety of crimes, 68.4% reported early childhood abuse (Weeks & Widom, 1998). CSOM (1999) listed child maltreatment as a very significant factor, along with exposure to pornography, poor impulse control, exposure to violence, substance abuse, social competency deficits, empathy deficits, emotional regulation difficulties, and sexual victimization. Sexual victimization tends to be a more important factor when it occurs at an early age, when it lasts a long time, and when reporting is considerably delayed. Because of the pervasiveness of pornography and child pornography on the Internet, children are exposed to a variety of images as never before. According to Marshall, Hudson, and Hodkinson (1993):

> Circulating levels of the sexually activating hormone testosterone [increase] fourfold among boys during the brief 2-year period of pubescence. At this time, therefore, insecure adolescents will be exquisitely responsive to the images we have

described. . . . Forcing a woman to have sex or having sex with a child requires none of the social skills that these boys have failed to acquire; it provides a rare opportunity in the lives of these young males to experience power and control, and to be relatively unconcerned with rejection. . . . Of all the various messages that our society, especially through the media, provides to growing boys, these, in particular, will be most appealing to insecurely attached young males. (p. 175)

Knopp (1984) reported that sex offenders do not acknowledge a close, nurturing relationship with their fathers, who seem to be either abusive or physically or emotionally absent from the child's life. Problems can occur if the family is extremely strict or rigid and often repressive about dealing with sexuality. In many young offenders without fathers present, there can be a succession of mother's partners, some of whom may sexually abuse the child or be consistently sexual in front of the child. These children do not develop a sense of boundaries, have difficulty with attachments, have little sense of empathy or remorse, and can be abusive and take advantage of others.

There is considerable agreement among professionals who treat juveniles charged with sex offenses that the majority of them can be treated very successfully (ATSA, 2000). This is, in part, because they are treated at the beginning and formative stages of their psychosexual development. They are often removed (at least temporarily) from their very dysfunctional families; they become involved with counselors and others who can be positive role models; they can be very responsive to learning the offense cycle and relapse prevention and applying it to themselves; and they can benefit from learning positive methods of conflict resolution and developing a sense of healthy identity and mutual respect in male-female relationships.

Unfortunately, progress in this area has been impeded in part by political and policy issues. Many states have passed laws mandating the prosecution and sentencing of juveniles as adults, some as young as 13. Adult systems are not set up to deal with juvenile issues such as school attendance, family involvement, financial support, and alternative housing. Probation officers know that the juvenile's environment is critical to success. The labeling of a juvenile through community notification can unnecessarily mark him for life and make his recovery more difficult. Suggestions have been made that juveniles in these systems could be reassessed at ages 21 or 25 to determine if they need to be supervised longer.

Becker (1988) reported that the average juvenile offender seen in her clinical studies had listed a total of 7 victims, as compared to the average adult offender's total number of 380 victims. That statistic alone, even if it were one fourth as dramatic, should emphasize the importance of juvenile sex offender treatment as a preventive measure.

PRISON PROGRAMS

Because most early legislation regarding sex offenders centered on the concept of the "sexual psychopath," the uncontrollable rapist who was mentally ill or even criminally insane, most attempts at treating sex offenders took place in state mental institutions. In some states, programs were operated under both jurisdictions, under dual and sometimes conflicting goals of treatment versus supervision and custody. In Alabama, offenders were treated in a comprehensive program located in a maximum-security prison, but little community supervision was available. In Arizona, extensive community treatment and surveillance in Phoenix and Tucson contrast with a prison program that so far has little more than an educational component.

The Adult Diagnostic and Treatment Center at Avenel, New Jersey, is a semiautonomous, medium-security prison exclusively for the evaluation and treatment of adult sex offenders, offering a wide range of programs and incorporating a prerelease component. Minnesota maintains an intensive sex offender treatment program using cognitive-behavioral therapy. Offenders are released to the supervision of specialized parole and probation agents. Vermont, because of its small population, has been able to develop a program of treatment within the prison setting, good transitioning into the community, and a continuum of treatment that sometimes involves the same therapists who work in the prison. Specialized parole and probation officers and trained community volunteers assist in "containing" the offender after release. In a Vermont study of 191 offenders with a follow-up period of 28 to 126 months, 3.85% of the treatment group were rearrested for a new sex offense versus 24% to 26.1% of those who had just some or no treatment (McGrath, Cumming, Livingston, & Hoke, 2000).

Unfortunately, budget cutbacks often force a reduction in staff that will keep many offenders from getting the treatment they need. Budget problems and changes in administrators affect prison programs much more than probation or parole, probably because offenders in the community are able to pay for their treatment. By the end of the 1990s, many good prison programs had been closed. Legislators faced with budget crises see "feel good" programs for sex offenders as very expendable, but the fact is that any effort to treat and rehabilitate a sex offender who will be back out on the street helps prevent more victims. More and more, the victim community is joining with offender therapists and criminal justice personnel to lobby for treatment and community supervision for offenders (D'Amora & Smith, 1999).

Most experts agree that if a program is located within a prison, it should at least be in a separate facility or wing of the institution. All staff should be dedicated exclusively to that facility and the treatment of sex offenders, and all administrative and security decisions should be made by the director of the unit. Programs that are integrated into the general prison population constitute the majority, and least worthwhile, of treatment models. "The great rolling momentum of the imprisonment experience overwhelms and overshadows the relatively brief therapeutic contact and nullifies their emotional impact" (Knopp, 1984).

For many offenders, a 2- or 3-year residential program followed by gradual release into the community under specialized supervision would be practical. Yet only a small percentage of the 100,000 sex offender inmates nationwide will receive such treatment. The remainder, having spent 4 or 6 or 10 years obsessing on their fantasies, will eventually be back on the streets.

CIVIL COMMITMENT

The picture of sex offender incarceration has changed in many states since new civil commitment laws have recently been passed. These laws are an attempt to identify the most serious, highest-risk sex offenders as they are ready for prison release, keep them incarcerated in a "mental health" facility after they have served their prison sentence, and treat them until such time as the institution believes they can be released. Such laws have been the target of legal challenges concerning double jeopardy, ex post facto laws, and right to due process. Also challenged is the very definition of "sexually violent predator." The Kansas statute, upheld by the U.S. Supreme Court, includes "any person who has been convicted of, or charged with, a sexually violent offense and who suffers from a mental abnormality or personality disorder which makes the person likely to engage in predatory acts of sexual violence" (Walsh & Flaherty, 1999, p. 34). Further challenges are pending.

Psychologists, prison personnel, prosecutors, and defense attorneys have been busy ever since in the 17 states with these laws, examining actuarial risk studies and treatment protocols, litigating in court, filing appeals, and remodeling and/or building institutions. It has proven to be an expensive undertaking, with per-bed costs estimated at $90,000 to $180,000 per year (Abel, 1999). Many states, in fact, have decided not to pass such a law because they cannot afford it.

COMMUNITY NOTIFICATION LAWS

The second recent development in the management of sex offenders has been the passage of notification laws in every state. Although some states used to have sex offender registration laws on the books, they were sporadically enforced. The occurrence of two high-profile cases, the murders of Polly Klaus in California and Megan Kanka in New Jersey, resulted in the passage of "Megan's Law" in many states and a federal law soon after, forcing compliance by the remaining states. Community notification laws, however, remain extremely varied and inconsistent among states regarding method of notification, type of offense, age of offender, agencies in charge, and possible appeal process. Some states notify only on certain risk levels, which are determined differently by state; some have Internet postings; some states notify on juveniles; others notify only on offenders being released from prison.

Again, as in the civil commitment laws, ex post facto challenges have arisen, and legal challenges continue to occur. The expense to agencies has been high in terms of both technology and manpower. At this point, there are no data on how effective this legislation has been in preventing sexual abuse. A Wisconsin study reviewed the effects of the law on attitudes of agency personnel, citizens who attended meetings, and perceptions of offenders. The authors concluded that the dilemma of community notification is balancing the public's right to know with the need to successfully reintegrate offenders within the community (Zevitz & Farkas, 2000). The majority of citizens surveyed found the information useful, but 18% were disappointed that they could do nothing about removing the offender from their community. Labor expenditures were cited as the major issue concerning agencies involved, who considered notification to be an unfunded mandate by the state. Additional problems encountered were media sensationalism and overreaction by the public. Agencies did comment positively on the increased sharing of information. Offenders and their probation or parole agents commented on the increased difficulty of finding stable residences and employment. Ironically, those are two factors associated with decreased recidivism.

PHARMACOLOGIC TREATMENT

Use of antiandrogenic drugs in working with sex offenders has been practiced for many years, but it remains controversial and relatively expensive. Sapp and Vaughn (1991) reported that of the 73 prison programs they studied at the time, 9 used organic treatment to some degree; however, 45 of the program directors stated that they would use biological techniques if they could. Knopp et al. (1992) reported that Depo-Provera was used in 11% of community-based programs.

Depo-Provera, or medroxyprogesterone acetate (MPA), is a hormone that reduces testosterone levels and thus the sex drive. Most notable for their work in this area are Dr. John Money and Dr. Fred Berlin of Johns Hopkins Medical School in Baltimore. At the Hopkins clinic, patients, such as highly compulsive exhibitionists who request help in controlling their urges, are evaluated to determine their degree of compulsivity. Offenders must come to the clinic every week for an injection that costs them approximately $50. Marshall, Jones, Ward, Johnston, and Barbaree (1991) cautioned that therapists who use these procedures do not expect these medications to eliminate sex offending:

> Rather they are principally used as a way of reducing sexual activity to controllable levels in those offenders whose sex drive seems so excessively high as to put them at serious immediate risk to reoffend and to render them unresponsive to psychological interventions. (p. 471)

Berlin and Meinecke (1981) reported that only 3 of 20 patients in one study showed recurrence of sexually deviant behavior while taking medication. However, 11 patients discontinued taking MPA against medical advice, and 10 or those 11 reoffended. Five were homosexual pedophiles. The authors concluded that the men seemed to do well in response to MPA as long as they continued taking it.

A newer drug, Depo-Lupron (leuprolide acetate), has fewer side effects than Depo-Provera but is significantly more expensive. A 1997 ATSA policy statement cautioned that antiandrogen drugs should never be used as exclusive treatment for aggressive sexual behavior but "should be coupled with appropriate monitoring and counseling within a comprehensive treatment plan" (p. 1). In addition to the antiandrogen drugs, numerous therapists report that the use of the serotonin reuptake inhibitors (SRIs), the new antidepressants, can be extremely helpful as part of the treatment of some sex offenders. According to Prentky (1997), "The literature on the use of SRI's is, by now, substantial and quite promising" (p. 342).

FEMALE OFFENDERS

The number of convicted female offenders is increasing, although official reports can be misleading if crimes of prostitution are included. Finkelhor (1984) reported that approximately 34% of sexually abused males and 13% of sexually abused females had been victimized by females. In approximately half the cases, the female offender was acting in the company of others. Viewing females as perpetrators of sexual abuse, perhaps parallel to viewing males as victims, challenges society's stereotypes.

Although there is a paucity of research and data on female offenders, some clinicians have attempted to focus on specialized groups and compare them to male offenders. In a study of 25 female offenders in a Minnesota program, the women expressed many of the same needs and emotions as males in events leading up to their offenses. Two factors seem to differ: Half the women offended in collusion with male partners, and all but one related a history of sexual abuse as children, either in or out of the family or both.

Many women carry their childhood baggage with them:

Gwen, age 24, was placed on probation after she orally raped her 6-year-old son. She was originally uncooperative and resistant to therapy, even though she took responsibility for her actions. She spent much of her time defending her relationship with her abusive boyfriend, who had watched her molest her child. After he was gone, she was placed in a women offenders' group and began examining her behavior in depth.

She had been sexually used as a child by her father and uncle, as had her sisters. By the time she was 14, she had left the house and spent a year as a prostitute. When she came home, her father began having sex with her again. She and the father of her child were heavy drug users. Her next relationship was equally dysfunctional. However, after 2 years in therapy, and after meeting a man totally different from the others in her life, she was recovered enough to again gain custody of her son from the state and try to live a normal life.

As previously described, both male and female offenders in therapy frequently describe instances of childhood sexual exploitation by mothers, baby-sitters, older sisters, and other caretakers, suggesting that instances of female abuse have long gone unreported and unresolved. Cases that attract media attention are often female teachers who have sexually assaulted male, and sometimes female, students.

FUTURE DIRECTIONS

While theoretical debates and research continue, courts and communities are faced with sex offenders in record numbers. Unfortunately, as corrections agencies and therapists have been learning how to assess and control sex offenders more effectively, the press and state legislatures have taken this emotional issue and passed laws quickly, without much input or discussion. When budgets grow smaller, treatment always suffers.

But much has been gained in the past 20 years. ATSA reports a membership of 2,010, from a few hundred in 1988. CSOM has become a national resource center for up-to-date information on the issues discussed here and has provided invaluable training to numerous probation and parole agencies throughout the country. Both the criminal justice and the treatment communities are beginning to realize that the combined efforts of both will be necessary to manage the problem. Few offenders remain in treatment on their own.

Courts must demand thorough risk evaluations from professionals who have developed the special skills and knowledge to effectively treat the sex offender. Prosecutors and defense attorneys alike should demand that legislatures implement rational and enlightened sentencing and punishment policies, taking a close look at the monetary implications of those policies. We can reduce sexual violence cost-effectively in community corrections by using the services of treatment experts, physiological checkups, specialized officers, and a continuum of supervision. Most important, greater focus should be placed in the area that will have the greatest impact on victims—the juvenile offender. If we can intervene early and effectively with a teenage abuser, we may save 300 or 400 child victims from harm.

Increasingly, enlightened professionals are asking that we look at sex crimes not only as a criminal issue but as a public health crisis. When there is such an epidemic, public health policy not only tries to heal those who have been hurt but looks deeper into the causes of a problem and will ultimately focus on prevention. Many victim advocates are realizing that supporting intensive and specialized supervision and treatment for many offenders is in their best interests. We know from their histories what contributes to the making of a sex offender, yet very few parents will ask what they can do to keep their child from becoming one.

The survivors of sexual trauma affect our society in a hundred ways: teenage girls who become self-destructive drug abusers, premature mothers, prostitutes, school dropouts, wives of abusive men, untreated mothers of another generation of abused children. Conflicted young boys can grow up to repeat their own victimization, some as prostitutes, some as offenders, exponentially and predictably. It is an issue that we cannot, in any sense, afford to ignore.

REFERENCES

Abel, G. G. (1999, September). *Clinical consultations regarding adult and juvenile sex offenders.* Paper presented at the annual meeting of the Association for the Treatment of Sexual Abusers, Buena Vista, FL.

Abel, G. G., Mittelman, M. S., & Becker, J. V. (1985). Sexual offenders: Results of assessment and recommendations for treatment. In M. H. Ben-Aron, S. J. Hucker, & C. D. Webster (Eds.), *Clinical criminology: The assessment and treatment of criminal behavior* (pp. 191–205). Toronto: University of Toronto Press.

Abrams, S. (1989). Probation polygraph surveillance of child abusers. *Prosecutor, 22,* 29–36.

Ahlmeyer, S., Heil, P., McKee, B., & English, K. (2000). The impact of polygraphy on admissions of victims and offenses in adult sex offenders. *Sexual Abuse: A Journal of Research and Treatment, 12,* 123–138.

American Psychological Association. (1996). *Violence and the family: Report of the American Psychological Association Presidential Task Force on Violence and the Family.* Washington, DC: Author.

Association for the Treatment of Sexual Abusers. (1997). *Policy statement on anti-androgen therapy and surgical castration.* Beaverton, OR: Author.

Association for the Treatment of Sexual Abusers. (2000). *Policy statement on the effective legal management of juvenile sexual offenders.* Beaverton, OR: Author.

Association for the Treatment of Sexual Abusers. (2001). *Practice standards and guidelines for members.* Beaverton, OR: Author.

Aytes, K., Olsen, S., Zakrajsek, T., Murray, P., & Ireson, R. (2001). Cognitive/behavioral treatment for sexual offenders: An examination of recidivism. *Sexual Abuse: A Journal of Research and Treatment, 13,* 223–231.

Becker, J. V. (1988). Adolescent sex offenders. *Behavior Therapist, 11,* 185–187.

Berlin, F. S., & Meinecke, C. F. (1981). Treatment of sex offenders with antiandrogenic medication: Conceptualization, review of treatment modalities, and preliminary findings. *American Journal of Psychiatry, 138,* 601–607.

Bonta, J., & Hanson, R. K. (1995). *Violent recidivism of men released from prison.* Paper presented at the annual meeting of the American Psychological Association, New York.

Burgess, A. W., Groth, N., Holmstrom, L. L., & Sgroi, S. M. (1982). *Sexual assault of children and adolescents.* Lexington, MA: Lexington.

Carnes, P. (1983). *Out of the shadows.* Minneapolis, MN: CompCare.

Center for Sex Offender Management. (1999). *Understanding juvenile sexual offending behavior: Emerging approaches and management practices.* Washington, DC: U.S. Department of Justice, Office of Justice Programs.

Center for Sex Offender Management. (2001). *Recidivism of sex offenders.* Washington, DC: U.S. Department of Justice, Office of Justice Programs.

Center for Sex Offender Management. (2002a). *Additional sexual assault statistics and links.* Washington, DC: U.S. Department of Justice, Office of Justice Programs.

Center for Sex Offender Management. (2002b). *Characteristics of sexual assault*. Washington, DC: U.S. Department of Justice, Office of Justice Programs.

D'Amora, D., & Smith, G. (1999). Partnering in response to sexual violence: How offender treatment and victim advocacy can work together in response to sexual violence. *Sexual Abuse: A Journal of Research and Treatment, 11,* 293–304.

Emerick, R. L. (1991). The paraphiliac. In R. L. Emerick (Ed.), *Continuing education manual for providers of management and treatment of convicted sexual abusers and their victims*. Phoenix, AZ: Western Correctional Association.

English, K., Pullen, S., & Jones, L. (Eds.). (1996). *Managing adult sex offenders: A containment approach*. Lexington, KY: American Probation and Parole Association.

Finkelhor, D. (1984). *Child sexual abuse: New theory and research*. New York: Free Press.

Gray, S. (1991). Therapeutic models for the treatment of sex offenders. In R. Emerick (Ed.), *Continuing education manual for providers of management and treatment of convicted sexual abusers and their victims*. Phoenix, AZ: Western Correctional Association.

Greenfeld, L. A. (1997). *Sex offenses and offenders: An analysis of data on rape and sexual assault*. Washington, DC: U.S. Department of Justice, Bureau of Justice Statistics.

Groth, N. (1979). *Men who rape: The psychology of the offender*. New York: Plenum.

Hanson, R. K., & Bussiere, M. (1998). Predicting relapse: A meta-analysis of sexual offender recidivism studies. *Journal of Consulting and Clinical Psychology, 66,* 348–362.

Hanson, R. K., & Harris, A. (1998). *Dynamic predictors of sexual recidivism*. Ottawa: Solicitor General of Canada.

Herman, J. (1988). Considering sex offenders. *Signs, 13,* 695–724.

Hernandez, A. E. (2000, November). *Self-reported contact sexual offenses by participants in the Federal Bureau of Prisons' Sex Offender Treatment Program: Implications for Internet sex offenders*. Paper presented at the annual meeting of the Association for the Treatment of Sexual Abusers, San Diego, CA.

Hindman, J. (1989). *Just before dawn*. Ontario, OR: Alexandria.

Knopp, F. H. (1984). *Retraining adult sex offenders: Methods and models*. Orwell, VT: Safer Society.

Knopp, F. H., Freeman-Longo, R., & Stevenson, W. (1992). *Nationwide survey of juvenile and adult sex-offender treatment programs and models*. Orwell, VT: Safer Society.

Koss, M. P., Gidycz, C. A., & Wisniewski, N. (1987). The scope of rape: Incidence and prevalence of sexual aggression and victimization in a national sample of students in higher education. *Journal of Consulting and Clinical Psychology, 55,* 162–170.

Laws, D. R. (1989). *Relapse prevention with sex offenders*. New York: Guilford.

Marshall, W. L. (1989). Intimacy, loneliness, and sexual offenders. *Behavior Research and Therapy, 27,* 491–503.

Marshall, W. L., & Barbaree, H. E. (1990). Outcome of comprehensive cognitive-behavioral treatment programs. In W. L. Marshall, D. R. Laws, & H. E. Barbaree (Eds.), *Handbook of sexual assault: Issues, theories and treatment of the offender* (pp. 363–385). New York: Plenum.

Marshall, W. L., Barbaree, H. E., & Fernandez, Y. M. (1995). Some aspects of social competence in sexual offenders. *Sexual Abuse: A Journal of Research and Treatment, 7,* 113–127.

Marshall, W. L., Hudson, S. M., & Hodkinson, S. (1993). The importance of attachment bonds in the development of juvenile sex offending. In H. E. Barbaree, W. L. Marshall, & S. M. Hudson (Eds.), *The juvenile sex offender*. New York: Guilford.

Marshall, W. L., Jones, R., Ward, T., Johnston, P., & Barbaree, H. E. (1991). Treatment outcome with sex offenders. *Clinical Psychology Review, 11,* 465–485.

Marshall, W. L., Serran, G. A., & Cortoni, F. A. (2000). Childhood attachments, sexual abuse, and their relationship to adult coping in child molesters. *Sexual Abuse: A Journal of Research and Treatment, 12,* 17–26.

McGrath, R. (1991). Sex offender risk assessment and disposition planning: A review of empirical and clinical findings. *International Journal of Offender Therapy and Comparative Criminology, 35,* 329–351.

McGrath, R., Cumming, G., Livingston, J., & Hoke, S. (2000). *The Vermont treatment program for sexual aggressors: An evaluation of a prison-based treatment program*. Paper presented at the annual meeting of the Association for the Treatment of Sexual Abusers, San Diego, CA.

McMahon, P., & Puett, R. (1999). Child sexual abuse as a public health issue: Recommendations of an expert panel. *Sexual Abuse: A Journal of Research and Treatment, 11,* 255–266.

National Institute of Justice. (1996). *Victim costs and consequences: A new look* (NIJ Research Report). Washington, DC: Author.

Norris, C. (1991, November). *The feasibility of treating the psychopath in a residential sex offender program*. Paper presented at the annual meeting of the Association for the Treatment of Sexual Abusers, Fort Worth, TX.

Patton, M. Q. (1991). *Family sexual abuse: Frontline research and evaluation.* Newbury Park, CA: Sage.

Prentky, R., & Burgess, A. (1990). Rehabilitation of child molesters: A cost-benefit analysis. *American Journal of Orthopsychiatry, 60,* 108–117.

Prentky, R., Lee, A., Knight, R., & Cerce, D. (1997). Recidivism rates among child molesters and rapists: A methodological analysis. *Law and Human Behavior, 21,* 635–659.

Prentky, R. A. (1997). Arousal reduction in sexual offenders: A review of antiandrogen interventions. *Sexual Abuse: A Journal of Research and Treatment, 9,* 335–347.

Ringel, C. (1997, November). *Criminal victimization in 1996; Changes 1995–1996 with trends 1993–1996* (NCJ-165812). Washington, DC: Bureau of Justice Statistics, U.S. Department of Justice.

Salter, A. (1988). *Treating child sex offenders and victims: A practical guide.* Newbury Park, CA: Sage.

Salter, A. (1998). *Risk assessment of sexual offenders.* Seminar presented at Specialized Training Services, Phoenix, AZ.

Sapp, A. D., & Vaughn, M. S. (1991). Sex offender rehabilitation programs in state prisons: A nationwide survey. *Journal of Offender Rehabilitation, 17,* 55–75.

Scott, L. K. (1997). Community management of sex offenders. In B. K. Schwartz & H. R. Cellini (Eds.), *The sex offender* (Vol. 2). Kingston, NJ: Civic Research Institute.

Scully, D. (1990). *Understanding sexual violence.* Cambridge, MA: Unwin Hyman.

Sgroi, S. (1982). *Handbook of clinical intervention in child sexual abuse.* Lexington, MA: D. C. Heath.

Tjaden, P., & Thoennes, N. (1998). *Prevalence, incidence, and consequences of violence against women: Findings from the National Violence Against Women Survey.* Washington, DC: National Institute of Justice.

Valliere, V. N. (1997). Relationships between alcohol use, alcohol expectancies, and sexual offenses in convicted offenders. In B. K. Schwartz & H. R. Cellini (Eds.), *The sex offender* (Vol. 2). Kingston, NJ: Civic Research Institute.

Walsh, E., & Flaherty, B. (1999). Civil commitment of sexually violent predators. In B. K. Schwartz (Ed.), *The sex offender:* Vol. 3. *Theoretical advances, treating special populations, and legal developments.* Kingston, NJ: Civic Research Institute.

Ward, T., McCormack, J., & Hudson, S. M. (1997). Sexual offenders' perceptions of their intimate relationships. *Sexual Abuse: A Journal of Research and Treatment, 9,* 57–74.

Weeks, R., & Widom, C. S. (1998). Self-reports of early childhood victimization among incarcerated adult male felons. *Journal of Interpersonal Violence, 13,* 346–361.

Williams, L., & Finkelhor, D. (1992). *The characteristics of incestuous fathers.* Unpublished manuscript, University of New Hampshire Research Lab.

Wolf, S. C. (1984, November). *A multi-factor model of deviant sexuality.* Paper presented at the Third International Conference on Victimology, Lisbon, Portugal.

Zevitz, R. G., & Farkas, M. A. (2000). *Sex offender community notification: Assessing the impact in Wisconsin* (National Institute of Justice Research Brief). Washington, DC: U.S. Department of Justice.

Zorza, J. (1999). Why courts are reluctant to believe and respond to allegations of incest. In B. K. Schwartz (Ed.), *The sex offender:* Vol. 3. *Theoretical advances, treating special populations, and legal developments.* Kingston, NJ: Civic Research Institute.

Part III

CRITICAL ISSUES IN LAW ENFORCEMENT

7

THE TERRORISM PHENOMENON: A WORLD ENGAGED

RICHARD H. WARD
SEAN HILL

Undoubtedly, the events of September 11, 2001, marked a turning point in the American public's perception of terrorism. For most people in the United States, terrorism had been viewed as something that happened somewhere else, and despite the initial warnings related to the first bombing of the World Trade Center in 1993 and the Alfred P. Murrah Federal Building in Oklahoma City in 1995, most people had considered the United States insulated from anything other than random and sporadic actions of a few radicals.

Within the criminal justice community—including practitioners, researchers, and academics—terrorism has generally not been a high-profile concern. Approximately 100 colleges and universities offered courses on terrorism in 2000.[1] Additionally, most of the research conducted has been in a few research centers and think tanks, such as RAND, the Center for Strategic International Studies, and a relatively small number of academic programs. The most predominant aspect of academic and professional study of terrorism over the last three decades has been the definitional argument over what exactly constitutes terrorism.

From a definitional standpoint, terrorism is generally defined as "the unlawful use of force or violence against persons or property to intimidate or coerce a government, civilian population or any segment thereof, in furtherance of political or social objectives."[2] Some policy makers advocate the development of a universal definition, whereas others insist on using ambiguous and nonconsequential terms such as *transnational* (Brady, 1998). The effect of this lack of agreement is that one country may categorize an act as criminal terrorism, whereas another country may classify it as a liberation movement deserving support rather than punishment (Meyer, 1989). Thus, one man's terrorist is another man's freedom fighter. Additionally, some researchers believe that the state is the main perpetrator of terrorism today, whereas several others contend that terrorism must be defined as a type of unwarranted rebellion that arises at the grassroots level. As a result of the spread of international terrorism that started with Yasser Arafat's al Fatah movement in the late 1960s, the term *terror* has taken on a much broader meaning.

In many ways, definitional issues have proven difficult from both a research and a practical standpoint. Undoubtedly, U.S. citizens view the bombing of the World Trade Center as international terrorism and the Oklahoma City bombing as domestic terrorism, but we do not view the Unabomber as a terrorist or the murder of citizens by drug dealers to create fear in a community as acts of terror. From another perspective, the media have come to use *terror* in a much broader context. It is in this context that we must look at terrorism of the future in a new light. It has long been held that terrorism and the media are bound together in a symbiotic relationship; however, the heart of the issue remains whether the effect of the media on public opinion and government decision making is favorable or negative for the continuity of extremist movements (Hoffman, 1998).

The Office of International Criminal Justice (OICJ), when located at the University of Illinois at Chicago, sponsored annual conferences on issues related to terrorism and international crime from 1986 through 1998.[3] OICJ also maintains files on more than 400 active terrorist groups in more than 100 countries. A number of professional organizations have been offering courses and meetings related to terrorism for more than a decade. These include the American Society for Industrial Security (ASIS) and the International Association of Chiefs of Police (IACP). Following the major terrorist attacks in 1993 and 1995, greater emphasis was placed on so-called "first responder training." Since 1983, the State Department Bureau of Diplomatic Security has sponsored the Antiterrorism Assistance Program (ATAP), which has trained over 23,000 law enforcement officers from 112 countries.

The evolution of joint terrorist task forces (JTTFs), sponsored by the Federal Bureau of Investigation, has a longer history, with the first two task forces being established in New York and Chicago. Today there are 16 JTTFs located throughout the country that combine the resources and personnel of federal, state, and local law enforcement agencies to increase communication and coordination when countering the threat of terrorism. The Central Intelligence Agency and other federal organizations, such as the Defense Intelligence Agency (DIA), the Naval Criminal Investigative Service

(NCIS), and the Bureau of Alcohol, Tobacco and Firearms (ATF), have also maintained counterterrorism units for more than a decade.

The literature on terrorism has grown considerably over the past decade. Much of it has been related to defining, classifying, and describing terrorist activities and groups, both globally and domestically. In many ways, the research being conducted during this period could largely be classified as "operational" rather than "theoretical." Very little research was conducted on issues related to the information/intelligence function and the evaluation of strategic approaches to combating terrorism. When studying terrorism, one must critically analyze the reciprocal relationship between terrorism, intelligence, the political climate, socioeconomic conditions, and international legal issues (see Figure 7.1). Only through this multidimensional approach will the etiology and future of terrorism be clarified and a strategy to fight terrorism developed.

Whereas research for the so-called "war on drugs," first implemented during the Nixon administration, has been heavily funded, terrorism research, education, and training have largely been ignored. One exception has been the amount of federal funding devoted to research on weapons of mass destruction (WMD), which became something of a low-profile national priority during the Clinton administration with the development and implementation of the National Domestic Preparedness Consortium. This was due in no small part to the aftermath of the Gulf War and recognition that Iraq was developing a WMD capability. As a result, the United States and the global community erected such measures as the Nuclear Nonproliferation Treaty (NPT), the Chemical Weapons Convention (CWC), and the Biological Weapons Convention (BWC), which put WMD "off limits" and provided an international basis for nonproliferation efforts (Nash, 1996).

Despite media and political criticism of the country's lack of preparation or planning for the threat of a terrorist attack, decision and policy makers largely ignored experts' warnings of the dangers inherent in the country's lack of legal and intelligence capabilities. Included in these warnings was the proposition that terrorists would use the simplest means available to carry

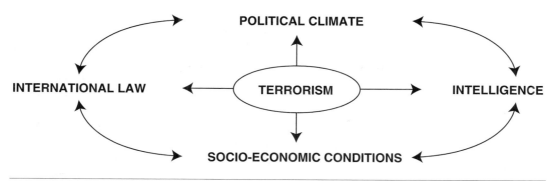

Figure 7.1   Components of Critically Analyzing Trends in Terrorism

out an attack, a notion that became all too real on September 11. John O'Neill, an FBI official who had retired just 1 week before the September 11 attacks and had been in charge of National Security in the New York region, noted in a speech in 1995:

> The fact is that terrorists generally use the simplest technology available for their attacks. Why? It's easier to obtain; it's less expensive; it's often more reliable; it requires less training to use; and, bottom line, it works. We've seen the Semtex in Pan AM 103; the Urea Nitrate in the World Trade Center; the ANFO in the New York conspiracy plot and the Oklahoma City bombing; and the shotgun in the abortion clinic murders. (Hill, 2001, p. 16)

O'Neill, who had taken a job as Director of Security for the World Trade Center, was one of the thousands who died in the collapse of the two towers.

The Role of the Criminal Justice Community

Five key factors affect the level of potential terrorism threats to America: (a) U.S. predominance in world economic, political, and military affairs; (b) the proliferation of WMD; (c) the rise of Islamic extremism and violent nationalism aimed at U.S. citizens abroad; (d) the increasing terrorist use of global mobility and sophisticated communication systems; and (e) the worldwide expansion of uncontrolled criminality and organized crime (Badolato, 2001). The Bush administration's immediate response to the attacks on September 11, in declaring war on the Taliban in Afghanistan and the global al-Qaeda network, was a call for action to combat the threat of terrorism. The appointment of Governor Thomas Ridge as director of the newly formed Office of Homeland Security has served as the mobilization point for a domestic response that, in no small measure, will engage virtually all aspects of the criminal justice system. Additionally, the need for a global response in the areas of international cooperation and a vastly expanded intelligence effort means a need for individuals who are prepared to fill the thousands of new positions devoted to counterterrorism efforts. In addition to an operational agenda is the need for much improved research, both applied and theoretical. Greater preparation must be devoted to language skills, technological capabilities, computer skills, legal resources, tactical models, forensic psychology, strategic security (public and private), cultural awareness, forensic science, victim assistance, and criminal investigation. With regard to WMD, in the aftermath of the anthrax incident that resulted in the deaths of five people, greater attention must be paid to coping with the threat of chemical, biological, and nuclear weapons, from both a prevention and an intelligence perspective, as well as the impact that such threats have on the public and on public security personnel.

The task within the criminal justice community is onerous, involving a level of planning and preparation that may be likened to the overhaul of the U.S. criminal justice system that took place in the 1960s following the report of the President's Commission on Crime and Criminal Justice. In coming years, several hundred billion dollars will be devoted to combating terrorism. Approximately half will no doubt go to national defense and the military, with at least as much going to homeland security.

It is obvious that much of the funding devoted to other efforts in the criminal justice system is being diverted to the terrorist threat, not the least of which is in the areas of crime control, illegal drugs, juvenile delinquency, and gang-related activity. Furthermore, given the international nature of the present threat, issues related to cooperation between countries, extradition, illegal immigration, human rights, and differing legal systems and laws will further tax the justice system's capabilities.

ASSESSING THE THREAT

If America was not prepared for the attacks of September 11, it is reasonable to ask just how critical the issue of terrorism was from a criminal justice perspective. The rapid expansion of security efforts is but one indication that the way we live has been changed forever. Although the most significant changes have focused on air travel and the security precautions being implemented at airports, it is also obvious that virtually every aspect of our lives in a democratic society will come under some form of scrutiny—from food and water to communications and technology, infrastructure vulnerability, and health care. The loss of jobs as a result of terrorism also affects virtually every aspect of the economy, and the cost of waging a war against terrorism has a psychological impact that is immeasurable.

Terrorism is not a new phenomenon and countries throughout the world have been living with this threat in one way or another for decades (Marks, 1997). The resilience of Americans is perhaps the best hope for the future, for despite the broad nature of terrorism, most people will never directly experience the consequences of a terrorist attack. And although

the indirect consequences can be debilitating, the September 11 attack led to a resurgence of feelings of patriotism and American values that many older generations had felt were all but gone in the youth of today. With books such as *The Greatest Generation* and *The Greatest Generation Speaks* by Tom Brokaw that detailed the experiences of World War II veterans, there has been much comparison between American spirit in the 1940s and the fortitude of "Generation Xers" and beyond. Perhaps the character of Americans never diminished but only lay dormant until another great challenge.

Although the current focus is on the al-Qaeda terrorist network and will be for some time to come, one of the most serious outcomes of September 11 is the fact that a new script for terrorism has been written in the ashes of the World Trade Center and the Pentagon. In the aftermath of this tragedy is the recognition that

- The weapons of terrorism are not just bombs and guns but things we use every day—planes, trucks, mail, food, and infrastructure, which may be referred to as ad hoc weapons.
- Large-scale loss of life creates widespread psychological stress and to some degree overreaction by authorities.
- Major attacks cause economic crises in a world engaged in the global market system.
- Fear causes people to change the way they live, affecting not only the quality of life but the career choices of youth.
- Issues of victim services and compensation have gained national attention.

Prior to the September 11 attacks, the major terrorist threat in the United States was considered to be from so-called "single-issue" terrorist groups, such as animal rights groups, ecological extremists, antiabortionists, right-wing hate groups, and militant antigovernment militia organizations. In the 1990s, right-wing extremism overtook left-wing terrorism as the most dangerous domestic terrorist threat to the country, a change that may be attributed to the fall of the Soviet Union and the movement of China away from communism toward a market economy. During the past several years, special-interest extremism, as characterized by the

Animal Liberation Front (ALF) and the Earth Liberation Front (ELF), has emerged as a serious terrorist threat. The FBI estimates that the ALF/ELF have committed approximately 600 criminal acts in the United States since 1996, resulting in damages in excess of $42 million (White, 2002).

The combination of the decentralized nature of the ALF and ELF movements with the notion that issues such as animal rights and the environment are favored by the general public has been noted by many analysts as the hallmark of the future of terrorism. In fact, within the United States, single-issue groups, such as animal rights and ecological movements, have carried out terrorist attacks more frequently, and often these attacks have been carried out by individuals or "cells" loosely associated with the movement. According to the State Department, there were no international terrorist attacks in North America in 2000; however, the FBI maintains that there were seven attacks in North America that could be attributed to either the ELF or the ALF, with other incidents still under review (U.S. Department of State, 2000; Watson, 2002).

Disenchanted fanatics have discovered that acts of extreme violence do have an impact on our way of life, even if they do not necessarily achieve the ultimate goal. In this regard, we can view terrorism as the use of violence to make a statement and affect an audience. This is particularly true with regard to issue-oriented terrorism, which has been the most significant change over the past decade. Do such typologies as "right wing/left wing" have any relevance from a research or investigative standpoint? The answer is that terrorism is no longer simple enough to be sorted into two categories. There is strong evidence to support the hypothesis that the actors and their political philosophies cover a much broader spectrum than those of the 1960s and 1970s. For example, the Southern Poverty Law Center's Klanwatch identifies 858 so-called patriot organizations, of which 380 are militia groups. The single most prevalent characteristic of these groups is a distrust of government; however, the ideology of these groups ranges from religious to antitax and antiabortion, with many groups having multiple ideologies. Therefore, it

may be accurately stated that the line between right- and left-wing terrorism has eroded over the last decade.

Before September 11, global terrorism against American interests, with the exception of the 1993 attack on the World Trade Center, was largely aimed at targets abroad. Here again, although the number of radical fundamentalist attacks on U.S. interests abroad received a great amount of publicity, the number of attacks on Americans abroad have been more frequent in South America over the years than in the Middle East (see Figure 7.2).

A recent U.S. Department of Defense (1998) report on international and domestic terrorism concluded that

- The nation's borders are easily penetrated.
- Terrorist cells could infiltrate easily with nuclear, chemical, and biological weapons.
- The United States should reduce the number of its military bases.
- The military should place greater emphasis on terrorism.
- Computerized information systems and satellites are vulnerable.
- WMD are a real threat.
- The Coast Guard should prepare to defend against cruise missiles.
- Local authorities should be trained in chemical and weapons detection, defense, and decontamination.

The external or international threat against American citizens and interests abroad is expected to be a growing problem in the years ahead. In large measure, such attacks have been and continue to be tied to a variety of issues, both political and economic. Americans remain a better target when traveling outside the United States, for there is less security in most countries for terrorists to penetrate and it is easier for terrorists to maintain operational security (keep their operation a secret) in countries that do not have a sophisticated law enforcement and intelligence capability.

Internationally, there were 423 terrorist incidents in 2000, 392 in 1999, and 274 in 1998 (U.S. Department of State, 2000). Of the 423 incidents in 2000, the majority were committed in Latin America (193); 98 in Asia; 55 in Africa; 31 in Eurasia; 30 in western Europe; and 16 in the Middle East.[4] In terms of casualties, the majority occurred in Asia, where 898 people were injured or killed. In Eurasia, the number injured or killed was 103, and in Africa the number was 102. The majority of anti-U.S. attacks were against business (178); 6 were against military targets, 3 against diplomats, and 2 against government. Over time the number of incidents has ranged from a high of 666 in 1987 to a low of 274 in 1998. In 2000, bombs were used in 179 incidents, kidnapping in 11, and armed attacks in 4.

A Department of Defense report noted that "terrorist groups that acquire NBC (nuclear, biological, chemical) weapons and stridently oppose U.S. polices could pose significant potential dangers to U.S. interests. Terrorists armed with these weapons can gain leverage for their demands because of the weapons' nature" (Office of the Secretary of Defense, 1996, p. 43).

Finally, one cannot rule out the economic impact of terrorism on the quality of life enjoyed by Americans. The effects of the events of September 11 on the global economy have yet to be precisely determined, but it is estimated that the attack on the World Trade Center alone will cost in excess of several billion dollars. The costs to airlines throughout the world has been staggering. In Canada, 3,000 airline employees lost their jobs, and airlines in Canada, Switzerland, and Belgium were forced into bankruptcy (Robertson, 2001). An airport executive estimated that airports were expected to lose more than $2 billion in revenues and to incur more than $1 billion in additional security costs in the 12 months following the September 11 attacks (Marchini, 2001). The Manitoba government has had to establish an advisory council to coordinate marketing between the public and private sectors in a variety of tourism interests to stop the industry's losses from spiraling out of control (Saccoccio, 2001). In the United States, the unemployment rate has reached its highest point in 20 years. According to another report, an estimated 9 million hotel and tourism workers will lose their jobs (Hamid, 2001). The International Labour Union has estimated that "24-million people worldwide could be fired" ("South Africa," 2001, p. 1). A nationwide study conducted by the Milken Institute tied the September 11 attacks to 1.8 million job cuts by the end of 2002 ("Attacks," 2002, p. B-2).

World Bank President James Wolfensohn has estimated that tens of thousands of children will die and that "some ten million people are likely to be living below the poverty line of one (US) dollar per day because of the attacks" ("PanAfrica," 2001, p. 3). Closer to the Middle East, the Organization of Petroleum Exporting Countries (OPEC) cut its oil output significantly and was forced to reduce the price per barrel, largely because of the reduction in fuel use by airline cutbacks (Lewis, 2001). And OPEC members have increasingly warned of the threat of a "price war" among the world's oil exporters if they cannot agree on production caps (Stanley, 2001).

THE ACTORS

Briefly, on an international level, those carrying out attacks on American interests are likely to be individuals operating on behalf of a group, organization, or government. Although we continue to be faced with martyrs who are willing to strap explosives on their bodies and commit suicide, most terrorist attacks abroad involve bombing, kidnapping, assassination, and destruction of property. There continue to be a number of active terrorist groups that target American interests, but most terrorism abroad involves acts against an existing power structure.

A number of countries, including Iraq, Libya, Syria, North Korea, and Iran, continue to be listed by the State Department as being involved in state-sponsored terrorism. President Bush, in his 2002 State of the Union message, referred to Iraq, Iran, and North Korea as constituting "an axis of evil, arming to threaten the peace of the world" (Bush, 2002). However, what keeps North Korea on the list and makes a close watch on the country politically and militarily necessary is not so much its endorsement of terrorism (most of which is targeted at South

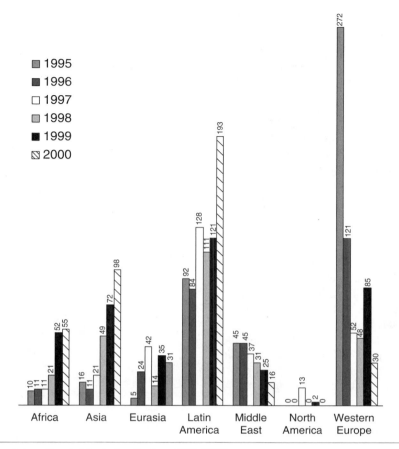

Figure 7.2 International Attacks by Region, 1995–2000.

Korea) as the rapid expansion of its long-range missile development program (which has drained the economy and kept the nation in a perpetual state of poverty). Many of the missiles and platforms developed in North Korea are sold to other state sponsors of terrorism, such as Iran, Syria, and Libya.

Within the United States, we are witnessing what appears to be an increase in attacks by individuals who display some real or imagined grievance. In the past, most of these so-called issue-oriented extremists have been nothing more than a small minority of people who decided that they would split off from the vast majority of individuals who were involved in legitimate protest activities (Dyson, 2001). We

have seen this in the actions of individuals associated with militias, right-wing groups, particular political movements, and local protest groups.

Historically, the American terrorist was most likely to be young, well educated, and from a middle or upper-income family. This has changed. Currently, no demographic profile best defines who is likely to become a terrorist. A pattern is beginning to emerge that may be helpful in helping to identify potential terrorists, but such identifiable characteristics are far from conclusive, and there is a danger of labeling someone a threat on the basis of mere supposition. The terrorist of today is probably not much different from one of our neighbors. The arrest of a citizen in Portland is but one example. An electronics engineer who owed more than $30,000 in back taxes and who described himself as a "chemical hobbyist" was arrested for a number of charges, including the placement of a "stink bomb" at an IRS facility. He pled guilty and reportedly said that an assassination scheme he had put on the Internet was merely an abstract proposal. According to reports, the suspect claimed afterwards that the government had not found his most dangerous chemical weapons, which may have included a stockpile of Sarin nerve gas (Kaplan, 1998).

FUTURE THREATS

From a global perspective, the greatest threat to the United States will continue to be attacks on American businesses and government. In some countries, the threat to tourists will continue to be higher, largely because terrorist groups have found that attacks on tourists can severely damage a country's economy. Future attacks may be less against nation-states and more against global and multinational corporations (e.g., Ford and AOL-Time Warner). Over the last few years, the pattern of targeted attacks has been predominantly against businesspersons. For example, imagine a situation in which Ford builds a factory in North Vietnam because the labor is cheap and North Vietnam has just been granted "most favored nation" status by the State Department. After a few years, Ford has not turned a profit from the factory and pulls out. Who are the jobless, disenfranchised Vietnamese extremists going to target for attacks?

TYPOLOGIES OF TERRORISM

Typologies of terrorism can be useful, but there is much disagreement in the literature as to what constitutes terrorism. Also, as White (2002) noted, classifying a phenomenon or putting it into a taxonomy will result in a certain loss of detail or specifics to make the group fit into a category.

Simply put, and from an American perspective, there are four types of terrorism:

1. International terrorism committed against American interests abroad

2. Domestic terrorism committed by foreigners

3. Domestic terrorism committed by U.S. citizens

4. International terrorism not directly related to U.S. interests

Algeria, Israel, and Colombia are examples of countries where violence is rife and where terrorism affects both the economy and quality of life. Attacks on American businesses have had a detrimental impact on the ability of companies to operate abroad. Many U.S. companies are forced to pay "protection" money, and kidnapping of executives continues to be a major problem. In many countries, our embassies and consulates have become fortresses, contributing in no small measure to strained relationships and the inability of diplomats to maintain open communication with the population.

Within the immigrant community in the United States are a very few individuals, some of whom are sponsored by rogue countries, who are prepared to carry out attacks on American interests. Usually referred to as "sleepers," they continue to represent a threat. Small cells of them hide in pools or pockets of ethnically similar neighborhoods in large cities. The 1993 attack on the World Trade Center that killed six people with a fertilizer-type car bomb is perhaps the best example of a sleeper group's capability. Emerson (2002) has identified the methods used by Islamic radical movements to infiltrate the United States.

Homegrown right-wing and militia groups that preach hate include any number of individuals who are deranged enough to carry out terrorist attacks. Of particular interest and concern is the capability of the Internet to bring people of like interests together and to foster the use of violence. Though the Internet is one of the most important inventions of this century and though it offers tremendous advantages, like most inventions of modern society, it has a downside. The question we must ask is how we cope with the disadvantages. Additionally, it is more difficult for law enforcement and intelligence agencies to infiltrate and prevent attacks by domestic extremists when only two or three people are aware of the operation (as in the case of the Oklahoma City bombing). They can best do so by tracking the materials and the money.

We have also seen the use of violence by cults (usually against their own members), the use of violence by organized criminal groups, and a growing number of individual attacks. These, though not officially designated as being terrorism or investigated as such, create a climate of fear and terror in communities. Keep in mind that the Unabomber, a single individual, was able to successfully threaten the airline industry as well as the academic community.

International terrorism not directly related to U.S. interests, such as that connected to conflict in the Middle East, sectarian violence in India, and separatist movements in Sri Lanka, Chechnya, and Northern Ireland, though not of great interest to most Americans, does have indirect consequences in a number of ways. Trade and economic implications are obvious, but in addition, the racial, religious, and ethnic diversity of the United States means that many Americans have kinship ties to one side or another in these conflicts.

THE WEAPONS

It has been argued that the culture of violence in the United States, coupled with the availability of weapons, contributes to an increasing threat of internal terrorism by a broad but relatively small group of individuals who are willing to kill or destroy property because of real or imagined grievances. It is safe to assume that conventional weapons, firearms and bombs, will be the most common choice of domestic terrorists in the immediate future. Nevertheless, we must be concerned about the threat that WMD pose for the future. The greatest threat externally, and one that concerns U.S. law enforcement, is the use of WMD. Former Secretary of Defense William Cohen has said that at least 25 countries are working on or have the capacity to use WMD. "The front lines are no longer overseas," he said, noting that the threat is "neither far fetched nor far off" ("25 Countries," 1997, p. 13).

Generally, WMD fall into four categories— nuclear weapons, biological weapons, chemical weapons, and, increasingly, means of disabling the nation's communications, computer, and infrastructure components, including those that supply power, food, and water.

Nuclear Weapons

The use of nuclear weapons in unconventional warfare involves a number of logistical and technological problems. Thus, nuclear weapons are generally more likely to be used in state-supported terrorist incidents. However, the fall of the Soviet Union and the sale of nuclear materials on the black market have increased the probability that a terrorist group could obtain a nuclear device. There are an estimated 27,000 nuclear warheads and 1,300 tons of fission materials in Russia, many of which are loosely guarded. According to one report, Russian General Alexander Lebed said that 84 tactical nukes (special atomic demolition munitions) had been found to be missing during a routine inventory (Leifer, 1996). An additional concern is the insufficient security measures being taken to protect the operation of nuclear reactors and the transport of fission material. Due to a lack of resources, Russia is currently using many of its nuclear submarines to power cities on its eastern seaboard, and little is being done to ensure the protection of these ad hoc power plants.

In recent years, there have been a number of arrests in Europe of individuals attempting to transport nuclear materials. In November 1997, for example, four people in Romania were arrested while trying to sell 100 grams of uranium-235 pellets. According to police, the asking price was $30,000 per gram, or $11.4 million for the lot ("Four Uranium Traffickers," 1997). Former nuclear weapons designer Theodore B. Taylor stated that "a small crudely fabricated nuclear device with an unpredictable yield could have toppled the twin towers. . . . Even if the bomb fizzled, gamma ray and neutron effects would still result in thousands of casualties" (Hughes, 1996, p. 6). As Taylor noted, the design for a nuclear weapon is readily available and can be found on the Internet. Many experts believe that it is only a matter of time before a nuclear device is used somewhere in the world by an extremist movement. Think of the devastation to Israel if a low-yield nuclear device were detonated in the heart of Tel Aviv.

Biological Agents

Biological agents are defined as organisms, or toxins developed from living organisms, that can be used against people, animals, or crops. A report by the National Defense University stated that although terrorists do not have a history of interest in biological agents, the potential for their use, internationally and domestically, has become much greater. In addition, the report noted that "law enforcement officials have arrested individuals associated with white supremacist and militia groups who possessed biological agents" (Carus, 1997, p. 2).

Michael Osterholm (1998), chief epidemiologist of the Minnesota Department of Public Health, stated that the ideal bioterrorist weapons

- Would be inexpensive and easy to use
- Could be aerosolized
- Would survive sunlight and heat
- Would cause lethal or disabling diseases
- Could be transmitted from person to person
- Would have no effective medical treatment

The diseases matching these characteristics include

- Bacterial: anthrax, Q-fever, brucellosis, tularemia, plague
- Viral: smallpox, viral encephalitis, viral hemorrhagic fever
- Toxin: botulism, Ricin, staphylococcal enterotoxin B

It is estimated that the greatest threat from biological weapons is the contamination of food or water. There have been attempts to contaminate water systems, but these have failed because municipal water systems are designed to eliminate impurities. To successfully infect a water supply system, a terrorist would have to have large amounts of agent and some knowledge of the water supply network and access to critical locations within the network. Thus, this scenario, though possible, is highly unlikely (Dick, 2001). There have also been terrorist attempts to contaminate food. Because heating kills pathogens, the most likely target would be food that is consumed without being cooked. Agricultural targets are susceptible to biological agents, and the most serious threat is the introduction of anthrax into animals. The World Health Organization has estimated that "50 kilograms of dry anthrax used against a city of one million people would kill 36,000 people and incapacitate another 50,000" (Carus, 1997, p. 2).

The FBI has detected only one bioterrorism incident in the United States. In 1984, a religious cult (the Rajneeshee) in Wasco County (near Portland), Oregon, attempted to infect the community with *Salmonella typhimurium,* an agent that would cause diarrhea and would in turn keep people from voting in a general election. Tests of the agent successfully infected more than 150 people. Two people were arrested in connection with the incident and were subsequently convicted (Carus, 1997). In February 1998, the arrest of two individuals, one allegedly tied to right-wing hate groups, who were suspected of having anthrax bacteria caused a nationwide stir. Although the anthrax proved to be nonlethal and the individuals were released, the threat underscored the dangers associated with biological weapons (Purdom, 1998).

A former deputy chief of research in the Soviet Union who now lives in the United States has said that most of the Russian biologists working in Russia's biowarfare program left the country. Although some went to the United States and Britain, "One can guess that they've ended up in Iraq, Syria, Libya, China, Iran, perhaps Israel, perhaps India—but no one really knows, probably not even the Russian government" (Preston, 1998, p. 53). At one time, Russia had at least 40 research and production facilities.

Although there have been fewer than 15 known incidents involving bioterrorism since the 1950s, the attention being paid to the threat of bioterrorism is perhaps the best evidence of growing concern throughout the world. Of interest are the techniques for disseminating biological agents, given that problems in this regard have hampered their use in the past. A future scenario may include the use of humans as disseminators of a biological agent. Consider the possibility of placing a biological agent in the food of a businessman who does a lot of routine international travel and predictably eats at many of the same restaurants every trip. The agent could be designed to incubate until he returned to the United States and then to release itself and infect many others. How far away is this technology? Ten years ago, experts stated that anthrax could never be developed to be effective in powder form; however, that notion has certainly been dispelled.

Chemical Weapons

A chemical weapon is a poison that kills or incapacitates on contact with the skin. Although chemical warfare has a relatively long history, it is only in the past decade that the threat has been linked to terrorism. Internationally, Iraq's use of nerve gas against both civilian and military targets in its war with Iran and conflict with the Kurdish minority is the most frightening example of the modern use of chemical agents. Thousands of people, including many Kurdish people located near the Turkish border of northern Iraq, died horrible deaths as the world watched.

In what may be described as a more traditional terrorist act, Japan's Aum Shinrikyo cult, led by Shoko Asahara, proved that terrorist groups have the capacity to use lethal chemicals. Their March 1995 attack in the Tokyo subway killed 12 and injured hundreds. According to the FBI, the number of threats to use chemical or biological weapons is on the increase. There were at least 20 incidents in 1997, most of which proved to be no more than threats or hoaxes; however, a number of incidents have been all too real. "Investigators have found biochemical agents in the hands of political extremists, extortionists, murderers and the mentally ill" (Kaplan, 1998, p. 27).

Infrastructure Threats and Cyberterrorism

The increasingly complex nature of our society and increasing globalization have also increased the threat of terrorist attacks against the country's infrastructure and technological framework. It is likely that in the next decade we will see an increase in the number of attacks on electrical, telephone, transportation, water, food, and computer facilities. The term *cyberterror,* coined to describe attacks largely on computer systems, is much broader than it might appear. Virtually all of our "life support" systems today depend in some way on computer technology.

A statement delivered to the National Research Council in 1991 shows that the future dangerousness of cyberwarfare has been a consideration since the dawn of the telecommunications age:

> We are at risk. America depends on computers. They control the power delivery, communications, aviation, and financial services. They are used to store vital information, from medical records to business plans, to criminal records. Although we trust them, they are vulnerable–to the effects of poor design and insufficient quality control, to accident, and perhaps most alarmingly, to deliberate attack. The modern thief can steal more with a computer than with a gun. Tomorrow's terrorist may be able to do more damage with a keyboard than with a bomb. (National Research Council, 1991, p. 2)

Some experts use the term *information warfare* to describe this phenomenon, for virtually every aspect of our lives also depends on our ability to communicate and to provide information. For instance, if you cannot inform a utility company that you have a problem, it is not likely to be fixed. Thus, destroying telephone communication has a major impact on emergency services. As Andrew Riddile (1996), a public policy expert, has pointed out:

> Interconnected networks may be subject to attack and disruption not just by states but also by non-state actors, including dispersed groups and even individuals. Potential adversaries could also possess a wide range of capabilities. Thus, the threat to U.S. interests could be multiplied substantially and will continue to change as ever more complex systems are developed and the requisite expertise is ever more widely diffused. (p. 3)

The American military is the most networked force in the world, so they are extremely vulnerable to information attacks. In 1997, a special team put together by the intelligence community pretended to be North Korea, and some 35 computer specialists, using hacking tools readily available on 1,900 Web sites, managed to shut down large segments of America's power grid and silence the command and control system of the Pacific Command in Honolulu (Center for Strategic and International Studies, 2001). There is a growing literature on this form of terrorism, and it is likely that it will become of greater concern in the new millennium as countries become even more dependent on technology and technological cooperation.

THE WAR ON TERRORISM

Following the events of September 11, America's strategic response to the threat of terrorism has ranged from brilliant to foolish, and at times there have been accusations of overreaction. But when a building is burning, you do not necessarily measure the minimal amount of water it will take to douse the flames. It is quite amazing to see how fast the government has responded with both operational and legislative approaches.

The war on terrorism—a war without frontlines or traditional military responses—is unlike any national conflict of the past. Just how well the country is responding to this new threat is perhaps best measured in terms of our enemy's inability to mount a sustained attack on American interests here and abroad. This is due in no small part to several important initiatives, including the military response, the domestic response, diplomatic efforts, and unprecedented levels of cooperation among government agencies and between the public and the private sector. Of particular importance has been the level of cooperation with other countries, especially Pakistan.

Despite a decade of warnings about the al-Qaeda network from the intelligence community, it was not until after September 11 that the world community reacted strongly, with more than a thousand suspected terrorists being apprehended and a number of plots against American interests foiled. A small glimpse of just a few of the individuals who are either fugitives or in custody in many different countries demonstrates the global reach of current terrorist networks:

1. Ramzi bin al-Shibh (Fugitive—Germany): a Yemeni cleric believed to be responsible for the planning and logistical support in the September 11 attack

2. Imad Yarkas (In custody—Spain): led eight men charged with helping the hijackers prepare for the attack

3. Fawaz Yahya Al-Rabeei (Fugitive—Yemen): suspected of planning future attacks on Americans

4. Ahmed Omar Saeed Sheikh (In custody—Pakistan): top suspect in the kidnapping and murder of *Wall Street Journal* reporter Daniel Pearl

5. Aftab Ansari (In custody—India): suspected of directing money to al-Qaeda terrorists (Johnson, 2002)

Individuals suspected of involvement with al-Qaeda are also in custody in Singapore, Indonesia, the Philippines, Italy, and Malaysia.

The establishment of the Office of Homeland Security in the United States has brought together a broad array of groups at the national level, with corresponding groups at the state and local levels. This has entailed some problems and confusion, but overall the cooperation has been unprecedented. A Law Enforcement Working Group in Washington, D.C., brings together government, military, and justice agencies as part of the domestic planning effort. The FBI, as the lead agency in terrorism investigations, is undergoing a major reorganization, and a new military structure for domestic protection is being implemented. The Federal Aviation Administration has moved rapidly to hire and train sky marshals and airport security personnel. Medical responses to cope with chemical and biological threats have included stepped-up "first responder" and medical personnel training, and antidotes for such threats as anthrax and smallpox are being stockpiled.

Ultimately, the war on terrorism will depend largely on an improved intelligence effort, the use of technology, and the endurance of the American public to deal with a prolonged military operation. The military response in Afghanistan has been swift, resulting in a number of important intelligence successes, and for the moment has apparently crippled efforts by al-Qaeda operatives to communicate. But the global response to terrorism will require a sustained and lengthy effort, which may take years to be successful.

Undoubtedly, the conflict in the Middle East adds yet another dimension that, from a U.S. perspective, increases Arab hostility toward Americans and makes Arab leaders reluctant to be cooperative. Fears of an expanded war in this part of the world are fueled by reports that other countries, such as Iran and Iraq, are actively engaged in the conflict.

To some degree, the Arab-Israeli conflict and U.S. responses in investigating al-Qaeda suspects have raised the concerns of Moslems in this country. Despite numerous public policy statements by public officials that this is not a religious war, an increasing number of attacks, both verbal and violent, on Arabs and Arab interests by hate groups and misguided individuals have heightened tensions. There is no doubt that the war on terrorism has had and will continue to have a profound impact on our way of life, as well as the lives of people in other countries. Most experts agree that there will be future attacks within the country and on American interests abroad.

On a broader scale, the attacks on the World Trade Center and the Pentagon have provided a new script for terrorists in illustrating the devastating impact that a well-planned strategy can have on the U.S. and international economies. It has been alleged that al-Qaeda members sold a bunch of American Airlines stock on September 10 in anticipation of the market crash after the attacks and thereby obtained millions of dollars. Perhaps the most significant outcome of the war on terrorism has been the resilience of the American people and recognition that the country cannot be brought to its knees by those who would threaten our way of life.

On the basis of figures of the past and the current trends in terrorist activities, it is safe to assume that this problem will be with us well into the next millennium. Terrorism can never be fully eradicated. There will always be a disenfranchised portion of the population who believe that they cannot make their grievances known through legitimate channels. However, as noted earlier, we are seeing some significant changes in both who the terrorists are and their tactics. Domestically, the greatest threat is from within, at least in terms of the frequency of acts. There will also continue to be disagreement as to what constitutes a terrorist act. In this regard, the current statistical approaches to measuring terrorism will continue to be suspect, as it is difficult to quantify a human behavior that has no consistent definition. Terrorism may be thought of as the "great ghost": Just when you think you know what it is and can put your hands on it, it changes shape and form and disappears.

REFERENCES

Attacks tied to 1.8 million job cuts; '03 rebound seen. (2002, January 11). *Chicago Tribune,* p. B-2.

Badolato, E. (2001). Terrorism 2001: Where do we go from here? *Journal of Counterterrorism and Security International, 7*(2), 8–9.

Brady, T. J. (1998). Defining international terrorism: A pragmatic approach. *Terrorism and Political Violence, 10*(1), 90–107.

Bush, G. W. (January, 2002). *State of the Union Address.* Presented at the U.S. Capitol, Washington, DC.

Carus, S. (1997). The threat of bioterrorism. *Strategic Forum.* National Strategic Studies of the National Defense University.

Center for Strategic and International Studies. (2001). *Cybercrime, cyberterrorism, cyberwarfare.* Washington, DC: Author.

Dick, R. (2001). *Terrorism: Are America's water resources and environment at risk?* Washington, DC: National Infrastructure Protection Center, Federal Bureau of Investigation.

Dyson, W. E. (2001). *Terrorism: An investigator's handbook.* Cincinnati, Ohio: Anderson.

Emerson, S. (2002). *American jihad: The terrorists living among us.* New York: Free Press.

Four uranium traffickers arrested in Bucharest. (1997, November 21). *ClariNews,* p. 3.

Hamid, H. (2001, November 23). Jobs threat for 9M in world's hotels, tourism industry.

New Straits Times Press (Malaysia), p. 2.

Hill, S. (2001). *Crime and Justice International*, pp. 1-16.

Hoffman, B. (1998). *Inside terrorism*. New York: Columbia University Press.

Hughes, D. (1996). When terrorists go nuclear: The ingredients and information have never been more available. *Popular Mechanics, 56.*

Johnson, K. (2002, March 8). Array of unknowns still troubling U.S. *USA Today*, p. 4A.

Kaplan, D. (1998, November 17). Terrorism's next wave: Nerve gas and germs are the new weapons of choice. *U.S. News and World Report*, pp. 27–28.

Leifer, J. (1996, December 4). Apocalypse ahead. *Washington Monthly*, p. 30.

Lewis, B. (2001, November 13). Oil prices dive on plane crash news. *Toronto Star*, p. C03.

Marchini, D. (2001, November 21). Airports struggle along with airline industry. *Cable News Network*, Transcript No. 112104cb.102.

Marks, K. (1997, December). Right-wing terrorism's renewed threat in the U.S. *Police*, pp. 16–17.

Meyer, J. (1989). German criminal law relating to international terrorism. *University of Colorado Law Review, 60*, 571-594.

Nash, M. (1996). U.S. practice: Contemporary practice of the United States relating to international law. *American Journal of International Law.*

Office of the Secretary of Defense. (1996). *Proliferation: Threat and response*, p. 43. Washington, DC: Author.

Osterholm, M. (1998, March 19). Remarks made at Bio-Terrorism Conference, University of Illinois, Chicago.

PanAfrica, Africa, America and the terrorist menace. (2001, November 8). *Ghanian Chronicle.*

Preston, R. (1998, March 9). The bioweaponeers. *New Yorker*, pp. 52–65.

Purdom, T. (1998, February 22). Tests indicate seized material is non-lethal form of anthrax. *New York Times*, p. 1.

Riddile, A. (1996, August). *The changing face of conflict in the New World Disorder.* Unpublished paper presented at the Bioterrorism Conference of the Office of International Criminal Justice, Chicago.

Robertson, G. (2001, November 13). Black day for airline industry "imperiled" by New York plane crash. *Calgary Herald*, p. C1.

Saccoccio, S. (2001, October). Sept. 11th aftermath: The tourism industry. *CBC News.*

South Africa: Attacks lead to massive job losses. (2001, November 9). *Business Day*, pp. 1-2.

Stanley, B. (2001, November 14). OPEC members admonish non-OPEC producers to share burden of cutting oil output to steady prices. *Associated Press.*

Terrorism: The indefinable threat. (1997, September). *Aviation Security International*, pp. 18–21.

25 countries work on weapons of mass destruction, says study. (1997, November 26). *Baltimore Sun*, p. 13.

U.S. Department of Defense. (1998). *Transforming defense: National security in the 21st century.* Washington, DC: Author.

U.S. Department of State, Office of the Coordinator for Counterterrorism. (2000). *Patterns of global terrorism.* Washington, DC: Author.

U.S. Department of State. (2001). *Patterns of global terrorism: 2000.* Washington, DC: Author.

White, J. (2002). *Terrorism: An introduction* (3d ed.). Belmont, CA: Wadsworth.

APPENDIX: TERRORIST WEAPONS
Explosives

Ammonium nitrate: a common fertilizer that, when mixed with diesel fuel, has an explosive velocity of 3,600 feet per second.

Semtex: a yellowish plastics explosive, about one third more powerful than a similar amount of TNT. It has a texture like putty or clay and can be molded. It is easy to transport because it will not explode without a detonator. Semtex was manufactured in Czechoslovakia with equal parts of RDX and PETN.

C-4: similar to Semtex, a plastic explosive made in the United States and used by the U.S. Army and many allies, as well as by mining companies. It explodes at 26,400 feet per second.

Dynamite: explosive made with nitroglycerine combined with an absorbent material, which is available commercially.

HME (homemade explosives): usually involve use of fertilizer (See **Ammonium nitrate**).

Pipe bomb: device built with length of pipe that is stuffed with explosive (usually black powder) and shrapnel, such as nails or BBs, sealed at both ends, and fitted with detonator.

Plutonium-239: radioactive material; when of weapons-grade quality, can be used in the making of nuclear bombs (See 940817).

Pentaerythritol tetranitrate: an explosive component commonly found in many bombs and surface-to-air missiles.

Rocket-propelled grenade (RPG): a "grenade" launcher with a maximum range of 500 yards.

TNT (trinitrotoluene): approximately twice as powerful as common dynamite but has a lower explosive velocity than plastic explosives. It is made of nitric and sulfuric acid and toluene. It is readily available in the United States and is usually produced in half-pound and 1-pound sticks.

Detonator: used to introduce an electrical charge, which causes the explosive to ignite. Detonation may be achieved in a number of ways, including a timing mechanism, electronically, or by a fuse, all of which complete an electric circuit in some way.

Timing mechanism: Any mechanism that has the capacity for delaying or setting off an explosion at a predetermined time.

Biological Weapons

Bubonic plague: freeze dried
Bacterial: anthrax, Q-fever, brucellosis, tularemia, plague
Viral: smallpox, viral encephalitis, viral hemorrhagic fever
Toxin: botulism, Ricin, Staphylococcal enterotoxin B

Chemical Weapons

Butyric acid: gives off a noxious odor that is difficult to clean. Often used in attacks on abortion clinics.

Sarin and Tabun: nerve agents that kill by short-circuiting the nervous system. They are odorless and colorless and enter the body by inhaling or through the skin. Symptoms include intense sweating, lung congestion, dimming of vision, vomiting, diarrhea, and convulsions. Death comes in minutes or hours (used in Tokyo attack in March 1995).

Mustard gas: causes sores on the skin and sears damp surfaces, eyes, lungs, and open sores. It can kill in a short period at very high doses or can maim a victim.

Atropine sulfate: injections can counter the respiratory paralysis of nerve gas.

VX: code name for deadly U.S. nerve gas. VX is absorbed through the skin or lungs and spreads throughout the body, resulting in paralysis of the respiratory system. Victims go through sweating and itching to panic, disorientation, convulsions, coma, and death, in about an hour. In 1968, the army was testing VX in Utah, where 6,000 sheep were killed accidentally, some as far as 27 miles away.

Radiological Weapons

Nuclear materials: from warheads, laboratories, and power plants, mostly in Russia and the former Soviet states.

NOTES

1. A survey of approximately 1,000 colleges and universities conducted by Sean Hill in 2000 found that about 10% offered a course related to terrorism (Sean Hill, doctoral dissertation, Huntsville, Texas, Sam Houston, State University). RAND was one of the earlier "think tanks" that conducted empirical research on terrorism. Following the Oklahoma City bombing, there was an increase in funding, primarily government funding, for research related to terrorism. Among the academics conducting research worthy of note in this area are William Dyson, Steve Emerson, Bruce Hoffman, Alex Schmid, Brent Smith, and Jonathan White.

2. This is the FBI definition. The State Department and the Pentagon define terrorism as "premeditated, politically motivated violence, perpetrated against a non-combatant target by subnational groups or clandestine state agents, usually intended to influence an audience" ("Terrorism," 1997).

3. The Office of International Criminal Justice was relocated to Sam Houston State University in Huntsville, Texas, in 1999.

4. The State Department does not include Palestinian intrastate violence in the *Global Terrorism* report.

8

COMPSTAT

The Emerging Model of Police Management

VINCENT E. HENRY

One of the most remarkable stories in law enforcement and in criminal justice today is the tremendous decline in crime achieved in New York City since 1993. According to New York Police Department (NYPD) figures, the total number of reported crimes for the seven major crime categories declined an unprecedented 62.45% in 2001 from the levels reported in 1993 (see Chapter 1 of this book for a description of the FBI's seven major offense categories). Only 161,619 of these major crimes occurred in 2001, as compared to 430,460 in 1993, and the 2001 figures represent the lowest annual number of total complaints for the seven major crimes in well over three decades. One of the most remarkable declines occurred in the murder category, which fell 66.68% between 1993 and 2001. The 642 murders recorded in New York City in 2001 represented more than a 71% decline from 1990, the year homicides peaked in New York City with 2,245 murders. Robberies fell 67.56% between 1993 and 2001, felony assaults declined 44.13%, grand larcenies declined 46.41%, burglaries declined 67.70%, and grand larceny autos and forcible rapes respectively declined 73.41% and 40.37% (City of New York, Mayor's Office of Operations, 2001).

According to the FBI's Uniform Crime Report (UCR) data for 1999,[1] New York City's rate of Index Crimes per 100,000 population ranked 165th of the 217 American cities with a population over 100,000 that report their crime statistics to the FBI. This showed a great improvement over the first 6 months of 1996, when New York City ranked 144th, and a vast improvement over the comparable 1993 period, when it ranked 87th of 181 large cities reporting their crime statistics to the FBI (FBI, 1996, 1999). By way of comparison, St. Louis's crime rate in 1999 was 240% higher than New York's; Orlando's was 238% higher; Atlanta's was 229% higher; Flint, Michigan's was 164% higher; Salt Lake City's was 151% higher; Washington, D.C.'s was almost 94% higher; and Denver's was almost 30% higher. New York City is the safest American city with a population over 1 million (City of New York, Mayor's Office, 2000).

The UCR statistics show that crime has been falling in large cities across the nation over the past few years, but New York City's decline in reported crime has been significantly greater—in all crime categories—than the national averages. New York City's crime decline not only has surpassed the national average reduction but has actually pulled the national averages down. Between 1993 and 1999, for example, the UCR data show that the number of murders and non-negligent manslaughters occurring in U.S. cities with a population over 100,000 (excluding New York City) fell 37%, whereas these crimes fell 66% in New York City. The 36% drop in New York City's aggravated assaults was nearly twice the national average decline (19%). Robberies in these cities fell 35% between 1993 and 2000 but fell 58% in New York City. New York City's decline in the forcible rape category (40%) was more than double the decline in other cities (17%), and New York City's 59% decline in burglary was also more than twice the national average decline (26%). New York City's 65% reduction in motor vehicle thefts over this period was more than double the 24% national decline, and its 40% drop in larceny theft was almost quadruple the national big-city decline of 11%. While the overall Total Index Crime in cities with a population over 100,000 (excluding New York City) fell 17% between 1993 and 1999, New York City's Total Index Crime reduction was an astounding 50.1% (City of New York, Mayor's Office, 2000).

The quality of life enjoyed by those who visit and live in New York City has also improved tremendously, and there is a palpable positive change in the sense of safety and civility throughout the city. The vastly improved quality of life, in conjunction with tremendous decline in serious crime, has dramatically improved the city's public image. A nationwide Harris Poll released in July 1997 and a USA Today-Gallup Poll released in September of the same year each listed New York City as the most desirable place to live in the United States (Marks, Egan, & Vogelstein, 1997). A nationwide *New York Times* poll of 782 adults—98% of whom were *not* New Yorkers—conducted in March 1998 showed that more than 60% of those surveyed had a good image of the city, compared to 40% only 2 years earlier. The percentage of poll respondents mentioning "crime" as New York City's most salient attribute was less than half of what it had been in 1996 (Johnson & Connelly, 1988).

Although quality-of-life indicators are much more difficult to quantify than reported crimes, it seems clear that New Yorkers see less graffiti,

encounter fewer hooligans with loud "boombox" radios, and are far less frequently accosted by aggressive panhandlers and "squeegee pests" than they were just a few years ago. Not only do New Yorkers have a statistically much lower likelihood of becoming a crime victim, but they feel safer as well. The tremendous difference in perceptions of safety in New York City became evident in a poll, conducted by the Mayor's Office in late December 1996 and early January 1997, that revealed that 70% of New Yorkers felt safer than they did in 1993 (City of New York, Mayor's Office of Operations, 1998).

These and other data illustrate the remarkable changes that have taken place in New York City over the past several years, and the changes are in large measure the result of a revolution in the way the NYPD conducts its business. The NYPD has been transformed in this relatively brief period from a rather passive and reactive agency that lacked energy and focus to an agency that responds quickly and strategically to crime and quality-of-life trends with an unprecedented vigor. Emerging patterns of crime and quality-of-life problems are identified virtually as they occur, and once they are identified the NYPD reacts immediately and aggressively to address them and does not diminish its efforts until the problem is solved. The NYPD uses timely and accurate intelligence to identify emerging problems, swiftly deploys personnel and other resources to bring a comprehensive array of effective tactics to bear on the problem, and relentlessly follows up and assesses results to ensure that the problem is truly solved. This revolution in the way the NYPD conducts its business is the result of a radically new and thoroughly dynamic police management process known as Compstat.

Because the Compstat process is an intrinsic part of the NYPD's revolution, it has attracted a great deal of attention in the local, national, and international media as well as the attention of police practitioners and academics in the criminal justice field. Compstat is one of the most talked-about issues in the field of policing today, despite many misconceptions about it.

Many prominent criminal justice academicians and police leaders have become convinced that the innovative and strategic problem-solving processes developed and refined in the

NYPD over the past several years are primarily responsible for New York City's falling crime rates (Kelling, 1995; Kelling & Coles, 1996; Silverman, 1998, 1999). Further, this attention and optimism has not been limited to police and academic criminology circles. The NYPD's revolutionary management control and problem-solving processes have been described in feature articles in *Business Week, Forbes,* the *Economist,* the *Wall Street Journal, Newsweek,* and a host of other electronic and print media outlets that do not typically cover issues related to police management. The attention received by Compstat and the new style of results-oriented police management attests not only to their effectiveness but to their applicability in organizations and industries beyond policing. Compstat's influence is also evidenced by the tremendous number of police executives and academicians who have visited the NYPD to study its innovative management methods and problem-solving activities.

Because it is such an effective and successful management tool, Compstat was named one of five recipients of the prestigious Innovations in American Government Awards in 1996. This prestigious award, conferred jointly by the Ford Foundation and Harvard University's John F. Kennedy School of Government, selected Compstat from among 1,500 applicant programs nationwide as one of the five most innovative and successful initiatives at any level of American government. The Innovations in American Government program's Web site describes how Compstat involves an interplay of technology, communication, and organizational change, noting that Compstat is

> a system that allows police to track crime incidents almost as soon as they occur. Included are information on the crime, the victim, the time of day the crime took place, and other details that enable officials to spot emerging crime patterns. The result is a computer-generated map illustrating where and when crime is occurring citywide. With this high-tech "pin-mapping" approach, the police can quickly identify trouble spots and then target resources to fight crime strategically. (Innovations in American Government, 1996)

Although other police departments nationwide are using computers to map crime and improve crime-fighting methods, the NYPD took one other essential step in its anticrime drive. While it was developing Compstat, the department was also undergoing a major management overhaul aimed at bringing the city's 76 precinct commanders and top departmental management closer together. In the process, it knocked down traditional walls between patrol officers, detectives, and narcotics investigators. Where isolation and even turf protection previously reigned, the department now holds weekly meetings that bring together a broad spectrum of police officials to review the computer data and discuss ways to cut crime in specific places. At these meetings, local commanders are held accountable by being required to report on steps they have taken and their plans to correct specific conditions. Also essential to the Compstat process are continual follow-up and assessment of results. Finally, building on its community policing program, the department often invites to the meetings a variety of interested parties, from school officials to neighborhood groups to local business leaders, to help fashion a comprehensive response in crime-ridden areas.

Despite the accolades and attention it has received, Compstat has also been greatly misunderstood. It has been variously portrayed as a high-pressure meeting between executives and middle managers, as a technology system, as a computer program, and as a system for sharing important management information. The fact that the Compstat management style involves all of these things (and a great deal more) may account for some of the misconceptions that surround it.

It should be clearly understood that Compstat, per se, is a management process through which the NYPD identifies problems and measures the results of its problem-solving activities. Compstat involves meetings between executives and managers and uses computer-based technology and other technology systems, but these elements are simply components in a much larger system or paradigm of management that has taken hold in the NYPD and, more recently, in other law enforcement and criminal justice agencies. Compstat meetings have been a key element in crime reduction, but they are only the tip of the iceberg—a great deal more has gone on behind the scenes to bring these unprecedented crime reductions and improvements in New York's quality of life to fruition. Without these other efforts and without a fundamental transformation in the agency's organizational structure, culture, and mind-set, the crime reductions and quality of life improvements could never have been achieved.

Further, despite Compstat's effectiveness as a management tool, the dramatic declines in crime would probably never have been so dramatic without political support and coordination among other agencies in New York City's criminal justice system. It would be misleading to suggest that these changes were solely the work of the NYPD, just as it would be wrong to suggest that they resulted solely from the new Compstat meetings and Compstat technology. To be effective and to achieve such dramatic results, Compstat meetings and Compstat technology had to be two facets of a comprehensive and carefully orchestrated array of management strategies and practices that were implemented throughout the NYPD and other criminal justice agencies.

One should never conceive of an agency of government—perhaps especially a criminal justice agency—as operating independently of other agencies. Within the criminal justice enterprise, for example, police agencies regularly and flexibly interact with prosecutors, courts, corrections, and probation and parole agencies, and to some extent all of these agencies and all of their personnel are interdependent. If a serious breakdown of communications occurred, or if necessary resources and activities in any sphere of the criminal justice enterprise were not forthcoming, the entire system of justice administration could grind to a halt. These interactions are far more complicated than one might first imagine, for they deal not merely with passing along resources and information but with tailoring each agency's protocols and policies in such a way that they do not conflict or interfere with the policies and protocols guiding the activities of other agencies. In this way, criminal justice should be viewed as an enterprise of government involving the coordinated interaction of numerous spheres of interest, function, and responsibility, rather than as a complex of separate agencies each independently pursuing its own goals and agendas.

A schematic depiction of all the lines of communication and interaction among these agencies would resemble a web, with multiple interconnecting lines extending from each agency to every other agency. Although these connections have always existed—that is, there has always been communication and some degree of coordination between the spheres of the criminal justice system—many more of them have been created since 1994 than at any time in the past. A great deal of the increased efficiency and effectiveness of the criminal justice enterprise in New York City over the past several years can be credited to the coordination and direction provided by the mayoral administration, which used its influence over these agencies to facilitate enhanced interaction. What was once simply a web of interconnecting lines has come to resemble a network of complementary policies, practices, and strategies that combine to make the criminal justice enterprise in New York City reach a new level of effectiveness.

A simple example of the need for cooperation and coordination among agencies might be when a police department plans to conduct a major crackdown on those driving while intoxicated (DWI) over a holiday weekend. If the police agency arrests a large number of violators but the agencies responsible for detaining them (often a municipal corrections agency or sheriff's department), prosecuting them, arraigning them, and making arrangements for pretrial release do not have sufficient staff on hand, the police department will encounter serious problems that may backlog the entire system for an extended period. Prior coordination, cooperation, and communication ensure that the system operates more smoothly and with greater efficiency and effectiveness.

One of the most important reasons why Compstat has functioned so well to reduce crime and improve the quality of life in New York City—that is, to make the criminal justice enterprise operate as it should—is that it has political support. From the beginning, Rudolph Giuliani's administration accepted as its mandate the public's demand for a reduction in crime and a restoration of order in a city that seemed to be out of control, and current mayor Michael Bloomberg has largely continued those policies. Prior to 1994, many believed (erroneously, as it turned out) that the city of New York was intrinsically unmanageable. Crime and disorder were major campaign issues in New York City's 1993 mayoral campaign, as was the seeming incapacity of incumbent David Dinkins's administration to effect substantive positive change in this area. Shortly after his inauguration in January 1990, Dinkins and Police Commissioner Lee Brown introduced a vision of community policing to the NYPD and set about making it the agency's dominant philosophy. Following the record-breaking levels of murder and other violent crime in New York in 1990, Dinkins expended considerable political capital to achieve passage of the Safe Streets, Safe City Act in the State Legislature. The Safe Streets Act, passed by the Legislature in February 1991, raised taxes throughout the state of New York to provide a $1.8 billion revenue stream over a 5-year period to increase the size of the NYPD and the Transit and Housing Police Departments (which were separate entities until their 1995 mergers with the NYPD). The Safe Streets legislation mandated that a minimum of 19,747 officers would be assigned to patrol duties and also funded a host of social programs, most of them community based, to facilitate community policing.

On the basis of the NYPD's Staffing Needs Report of October 1990 (NYPD, 1990), Commissioner Lee Brown determined that to implement community policing the NYPD's total uniformed headcount should be 31,351 sworn personnel. Including the Transit and Housing Police, the Safe Streets Act provided a revenue stream to raise the total headcount to 38,310 officers among the three agencies (McKinley, 1994; Strong & Queen, 1994). Given the fairly high rate of attrition and retirement at the time and a number of admitted "fiscal gimmicks" used by the Dinkins administration to delay hiring in order to defer the city's share of the hiring costs, the agency's actual growth was fairly slow. The 38,310 Safe Streets benchmark was not reached until 1996. Because it increased the number of new hires and reduced the minimum hiring age to 20, the agency found that the average age and level of experience among its uniformed patrol force declined significantly.

THE NYPD's MANDATE FOR CHANGE

Three significant events—the Crown Heights riots of August and September 1991 (resulting from a vehicle accident in which a young black child was struck and killed by an auto driven in the motorcade of a prominent Hasidic rabbi), the Washington Heights riots of July 1992 (resulting from the shooting death of an armed drug dealer in a gun battle with a plainclothes officer), and the Mollen Commission's report on police corruption—did little to enhance public confidence in the police or in the capacity of the Dinkins administration to manage the city. Crime, disorder, and the declining quality of life in New York City became the bellwether issues of the 1993 mayoral election. Rudolph Giuliani defeated David Dinkins by a slim 2% margin, but he took these issues as his mandate and made good on his election pledge to vigorously attack them.

Giuliani appointed William Bratton, the highly regarded former chief of the New York City Transit Police[2] and several police agencies in his native Boston, as police commissioner. Bratton immediately set about rousing the department's executive corps from their bureaucratic malaise, and after requesting resignations from the entire upper level of the executive corps (all but one of the agency's top five chiefs were replaced in the first few weeks of the new administration), he assembled a top-notch staff of fairly young but well-seasoned executives who were aggressive risk takers (Bratton, 1998; Krauss, 1994; McQuillan, 1994). At his swearing-in speech, Bratton took the opportunity to begin setting the tone for the agency's new direction and mandate. Bratton invoked John Paul Jones's request to the Continental Congress: "Give me a fast ship, for I intend to sail in harm's way." The words and the sentiment behind them rang true with police officers, who saw themselves as "sailing in harm's way" at work each day. They welcomed leadership that sought to join them in this struggle.

It was clear from the beginning of Bratton's administration that middle managers—particularly precinct commanders—would be given greater authority, discretion, and organizational power at the same time that they would be held highly accountable for these and other resources. Empowering middle managers and an emphasis on quality-of-life enforcement would be essential factors in transforming the NYPD. Just 2 weeks after Bratton took office, a senior police planning officer commented in the *New York Times* that the new administration would give precinct commanders "direct control over resources to carry out enforcement operations, to address chronic crime locations and suppress the low-level irritants to their communities" (Krauss, 1994, p. B3). Once the Compstat meetings began to take shape, Bratton used them in conjunction with other key performance indicators to identify which midlevel managers should be replaced or transferred and which managers should be promoted to positions with additional responsibilities. Within the new administration's first year, more than two thirds of the department's 76 precinct commanders were replaced—either by moving them to positions more suited to their less assertive management style or promoting them to more challenging positions (Bratton, 1998; Silverman, 1996). The strategy here was to match up particular positions with the commanders who had the requisite skills, experience, expertise, and personality to manage them proficiently. Compstat meetings proved to be an essential tool in identifying individual managers' strengths and weaknesses.

The shake-up was calculated to shake off the lethargy, passivity, and drift that had characterized the NYPD's executive corps. Just as important, Bratton's team began immediately to articulate and to demonstrate their belief in the idea that the NYPD could achieve unprecedented levels of performance (Buntin, 1999; Chetkovitch, 2000a, 2000b, 2000c).

One of the first orders of business was the development of Compstat. Until the advent of Compstat in the early days of the Bratton administration, the NYPD had no functional system in place to rapidly and accurately capture crime statistics or use them for strategic planning. Although the department collected crime statistics, they were often 3 to 6 months old by the time they were compiled, and the methods used to analyze them were rudimentary at best. Six-month-old crime data are of little use to any police executive because they say nothing about when and where crimes are

occurring today. The fact that NYPD executives in previous administrations never bothered or never saw a compelling need to get accurate and timely crime intelligence is emblematic of the overall lassitude and lack of concern that characterized many of the agency's managers (Henry, 2002).

This is not to say that every member of the NYPD's management cadre was timid, indecisive, or unconcerned with effectively addressing the kind of crime and quality-of-life issues that plagued the city. Indeed, the agency had many fine and highly skilled managers, but it was only when a sufficient number of less effective managers were weeded out or marginalized that an important shift took place in the agency's management culture. Once indecisive, unimaginative, and ineffective managers were removed or neutralized, the number and percentage of the strong managers who were most capable of leading the department reached a critical mass, and the inept managers no longer impeded the agency's progress.

Bratton recognized that a new management coalition was absolutely essential if substantive change was to take place within the NYPD, and he also recognized that change would require substantial empowerment of middle managers. Whereas earlier attempts at implementing community policing sought to empower beat officers—the individuals at the very bottom of the organizational hierarchy who had the lowest rank and the least legitimate power in the organization—Bratton recognized that a more pragmatic (and ultimately more effective) approach was to expand the power of middle managers. Power, discretion, and authority were decentralized and pushed down the organizational pyramid from headquarters executives to precinct and operational commanders in the field. Bratton (1996) explained his rationale for devolving power from top executives to those at the middle of the organization and rank structures:

> I gave away many of my powers not—as my predecessors wanted—to the cop on the beat, but rather to the precinct commander. I did not want to give more power to the cops on the beat. They were, on the average, only 22 years of age. Most of them never held a job before becoming

New York City police officers, and had only high school or GED qualification. These kids, after six months of training, were not prepared to solve the problems of New York City; sorry, but it just was not going to work that way. However, my precinct commanders typically had an average of 15 years of service, and they were some of the best and the brightest on the police force. All of them were college educated; all were very sophisticated; and they were at the appropriate level in the organization to which power should be decentralized.

My form of community policing, therefore . . . put less emphasis on the cop on the beat and much more emphasis on the precinct commanders, the same precinct commanders who met with community councils and with neighborhood groups. They were empowered to decide how many plain clothes officers to assign, how many to put in community policing, on bicycle patrols, and in robbery squads. They were empowered to assign officers as they saw fit—in uniform or in plain clothes—to focus on the priorities of that neighborhood. . . . Whatever was generating the fear in their precinct, they were empowered to address it by prioritizing their responses. We decentralized the organization, and I eliminated a few levels in the organization of the force and in the hierarchy as well.

Achieving a critical mass of dedicated, decisive, and innovative managers was an important element in changing the agency's management culture, and its impact was akin to what Malcolm Gladwell (1995, 2000) has called a "tipping point." This concept of "tipping points," a term Gladwell borrowed from epidemiology, also helps explain why the NYPD's strategic and highly focused use of quality-of-life enforcement led so quickly to such dramatic crime declines. The tipping point concept involves the idea that some social phenomena (including, according to Gladwell, some forms of crime and social disorder) behave like infectious agents: the frequency of these phenomena increases in a gradual and linear fashion until they reach a certain critical mass or threshold (a "tipping point"), when they explode in an epidemic. Gladwell noted (Lester, 2000) that James Q. Wilson and George Kelling's (1982) "broken windows" theory is fundamentally a tipping point argument, and he pointed out that

the key to controlling crime is to reduce the frequency of quality-of-life offenses to within manageable limits—below the tipping point that made crime explode (Gladwell, 1995, 2000). This is essentially what the NYPD did in its strategic and highly focused enforcement efforts—efforts that were largely based in the Compstat process and in its capacity to develop timely and accurate crime intelligence as well as to direct the rapid deployment of personnel to address emerging crime and quality-of-life issues.

Perhaps few police chief executives saw the necessity for accurate and timely crime statistics in earlier NYPD administrations because few actually believed that they could use them to fight crime and public disorder more effectively. They lacked the vision and the focus and were more concerned with avoiding the kind of scandals that could cost them their jobs than with fighting crime. The new administration's mandate to assertively address crime and disorder was the impetus for the revolutionary Compstat process, and within a few weeks the first affirmative steps were taken to develop appropriate technology systems, policies, and practices that would ultimately and permanently transform the way the NYPD looked at and responded to crime and disorder problems (Buntin, 1999; Chetkovitch, 2000a, 2000b, 2000c).

Eli Silverman (1996) succinctly summarized and described the Compstat process, its emphasis on rapidly compiling, analyzing, and using crime statistics to manage crime problems, and its impact on the NYPD's operational mind-set. He noted that

the most significant aspect of the department's organizational changes within the past few years has been the process known as Compstat. . . . Compstat was originally a document, referred to as the Compstat book, which included current year-to-date statistics for criminal complaints, arrests for major felony categories and gun violations, compiled on a citywide, patrol borough and precinct basis. The initial versions of the Compstat book, which improved steadily over time with regard to overall sophistication and degree of detail, developed from a computer file called "Compare Stats," hence Compstat. . . .

Compstat, through the weekly headquarters meetings, provides the dynamics for precinct and borough accountability, and an arena for testing the mettle of field commanders. As a management tool, Compstat melds upgraded quantitative information on crime locations and times with police deployment arrangements and qualitative quality-of-life information. Precinct problem-solving can be weighed against available resources, and the responsibilities, information-sharing and interaction of different department units can be gauged.

According to Silverman, the establishment of Compstat meant that "for the first time in the agency's long history, key members of the organization began gathering each week to examine various sources of crime information at a meeting devoted solely to the issue of reducing crime and improving the quality of life enjoyed by New York City's residents."

COMPSTAT: A NEW MANAGEMENT PARADIGM

Compstat meetings and Compstat technology are management tools—nothing more, nothing less—that are employed to great effect within a radically new and potentially revolutionary management paradigm. *Revolutionary* is a term that is frequently bandied about in management circles and frequently misused in common discourse. In reality, true revolutions in management (perhaps especially police management) or in any other area of human endeavor are few and far between, so this term is used advisedly. Nevertheless, Compstat *is* a revolutionary method of police management because it involves a fundamental shift in the NYPD's management paradigm and because after its origination in the NYPD and implementation in several other municipal police agencies it began to spread rapidly across the landscape of American police management (Henry, 2002).

The terms *revolution* and *paradigm* are used here in the same sense that scientific historian Thomas Kuhn (1970) applied them when he addressed the idea of revolutions in science and scientific thought. Paradigms are a sort of mindset or a collection of organizing principles and fundamental viewpoints around which we organize our basic understanding of the world. Paradigms can be compared to ideologies, belief systems, philosophical principles, or cognitive models that shape our understanding of something, and because they determine the kind of problems and issues we consider important as well as the way we approach the problems, they influence our behavior as well. In terms of management, a paradigm is a sort of general point of view about human nature and human behavior and about how human organizations operate. It also prescribes the management issues we deem most important and the way we approach their resolution. Our paradigm or outlook on management determines the kind of results we seek to achieve as well as the methods and tools we use to achieve them.

Kuhn (1970) pointed out that under ordinary conditions (what he called "normal science") paradigms guide the development and direction of new knowledge, and knowledge increases incrementally within the paradigm's boundaries. Scientists are guided by the paradigms they follow, and they seldom venture far from them to consider intellectually or experiment with dramatically new ideas. Scientific revolutions occur, Kuhn explained, when paradigms shift radically or when a new paradigm emerges and proves to explain some scientific phenomenon better. As other scientists begin to operate within the new paradigm, they stretch the boundaries of knowledge and develop new theories and new technology based on the paradigm. The paradigm of Newtonian physics, for example, prevailed in science until Einstein's theory of relativity successfully challenged its basic assumptions about the physical world. The theory of relativity eventually supplanted Newton's ideas about physics because it more accurately reflected our knowledge of the world, explained the physical universe better, and opened up new and more productive avenues for scientific research.

Paradigms also guide police managers and executives and shape the way they develop new and innovative approaches to the problems they face. For the most part, the approach many police executives and managers continue to take today is not very different from what they and others have done in the past. Because they continue to operate within a narrow management paradigm that generally does not encourage innovation and generally does not strive to stretch the potential boundaries of performance, most police executives are satisfied with incremental improvements. The paradigm in which many police executives operate, for example, does not truly embrace one of the Compstat paradigm's most important underlying principles and beliefs—that police officers and police agencies can really have a substantial positive impact on the crime and quality-of-life problems facing the communities they serve.

Scientific (and management) revolutions occur when a radical paradigm shift takes place—when there emerges a new set of ideas, ideologies, or controlling principles around which we organize our understanding of a phenomenon. The new understanding of the phenomenon, in turn, points up new insights and better practices. The new paradigm takes hold and gains acceptance when it proves effective— when it does a better job at explaining a phenomenon and when its methods achieve more positive results than the paradigm preceding it.

In similar fashion, the Compstat paradigm presents police managers and executives with a new way of looking at police organizations and police activities. It is radically different from the ideologies and practices that have guided police management through most of this century, and it points to new methods and strategies police can use to pursue their goals. As illustrated in the examples and statistics cited at the beginning of this chapter, the Compstat paradigm's effectiveness in achieving results can scarcely be denied.

THE COMPSTAT MEETING: TECHNOLOGY, INFORMATION, AND COMMUNICATION

To many casual observers, Compstat appears to be simply a meeting at which executives and managers discuss the latest information about emerging crime trends, develop specific tactical plans to address them, and monitor the results of the actions previously undertaken. In one sense, Compstat *is* a meeting. Each week, the commanders of all precincts and operational units in a given geographic area of New York City gather in the NYPD's Command and Control Center at Police Headquarters to give an

accounting of themselves and their officers' activities for the past month (Bratton, 1998; Kelling, 1995; Maple & Mitchell, 1999; Silverman, 1996).[3] The Command and Control Center, informally dubbed the "war room" for its resemblance to the Pentagon's nerve center, is a high-tech conference facility equipped with numerous computer systems, video monitors, video projection screens, and communications equipment.

This level of technology is not absolutely essential for Compstat's effectiveness—the Compstat meetings were initially run in a small room equipped only with easels and flip charts, and they nevertheless produced startling results. Police agencies need not assume that Compstat requires a tremendous investment in hardware or software to achieve its results, for even a complete Compstat technology package that will suit the needs of 99% of American police agencies can be assembled for just a few thousand dollars.

Each precinct commander takes his or her turn at the podium to present his or her activities and accomplishments and to be closely questioned by the police commissioner, several deputy commissioners, the chief of the department, the chief of detectives, the chief of patrol, and other top executives. Precinct commanders are accompanied by detective squad supervisors, narcotics and vice squad commanders, and ranking personnel from just about every operational and investigative unit within their geographic area of responsibility. Because of the intensity of the questioning, the quantity of statistical performance data, and the nature of the technology involved (including computerized pin mapping, comprehensive crime trend analyses, and other graphic presentations of data), Compstat meetings permit the agency's executives to have an unprecedented level of in-depth knowledge about the specific crime and quality-of-life problems occurring at specific locations in each of the NYPD's 76 precincts. With this wealth of highly specific knowledge, executives can ask commanders and managers probing questions about the particular activities and tactics they are using to address these problems at specific locations. Crime and quality-of-life trends and patterns can be more easily discerned through the discussions, and connections between seemingly disparate events and issues are more easily identified. Commanders are expected to have answers and to demonstrate results, and specialized squad commanders must show how they cooperate and coordinate their activities with other operational entities. The focus is on performance and results at Compstat meetings (Bratton, 1998; Henry, 2002; Kelling, 1995; Kelling & Coles, 1996; Silverman, 1999; Witkin, 1998).

But Compstat is much more than a meeting, and Compstat meetings provide executives with an in-depth knowledge of other management performance data beyond just enforcement and productivity information related to crime and quality-of-life conditions. Executives focus on each commander's efforts to ensure that officers in his or her command interact with citizens and with other members of the department in a courteous, professional, and respectful manner. Executives can effectively gauge the morale in each command by examining sick rates, the number and type of disciplinary actions taken, the number of civilian complaints made against officers, and a host of other data. Each commander's performance in managing such important functions as overtime expense, traffic safety, and even the number of automobile accidents involving department vehicles can be evaluated at Compstat meetings. Executives can and do focus on virtually any area of management responsibility, comparing each commander's performance to that of his or her peers. Performance comparisons can easily be made to other similar commands or to patrol borough and citywide averages as well. Changes over time can be calculated and charted for graphical presentation on practically any crime or quality-of-life performance criteria within the commander's scope of responsibility (Bratton, 1998; Kelling, 1995; Silverman, 1999). In short, the Compstat meetings amount to intensive monthly performance evaluations for every commander of practically every operational unit in the agency.

Commanders who fail to achieve results or who otherwise do not make the grade may find that they are no longer commanders, but those who excel and achieve results find promotion and advancement. Compstat meetings have introduced a unique element of competition among the department's management cadre, and they are a stimulus to achieve results that has never before existed in the NYPD. Because the Compstat meetings are also attended by commanders of support units, local prosecutors, and representatives of other criminal justice agencies with whom commanders interact, they allow information to be widely disseminated to appropriate parties. Although the support unit commanders and other attendees may not be asked to make presentations, their presence at the meeting allows for the immediate development of integrated plans and strategies. A precinct commander who intends to commence a major enforcement effort, for example, can get on-the-spot commitments for the resources and assistance he or she needs from ancillary and operational units (Kelling, 1995; Witkin, 1998). Details of the plan can be worked out without crossing bureaucratic lines or scheduling a prolonged series of meetings. Compstat meetings, then, are also about supporting and empowering commanders, sharing information and crime intelligence, communicating management's values and objectives, and ensuring accountability. But again, Compstat meetings are just a part of the story behind the NYPD's transformation and its performance, and the meetings are simply a tool used within the larger Compstat paradigm.

SUPPORTING COMPSTAT THROUGH SYSTEMIC AND PHILOSOPHICAL CHANGES

A host of other organizational, structural, and philosophical changes have been implemented in the NYPD to support its crime reduction and quality-of-life improvement mission, making the crime reductions and quality-of-life improvements possible by enhancing discretion and changing the dimensions of power, responsibility, authority, and accountability throughout the agency. Without these fundamental systemic and philosophical changes in the department and without the support and direction provided by the administration to the police and other criminal justice agencies, the transformation would not have been possible and the dramatic results would not have been achieved.

One can scarcely overstate the importance of these structural and philosophical transformations to the agency's overall success. By themselves and without adoption of other elements of the overall Compstat paradigm, Compstat meetings not only are unlikely to achieve more than temporary results but also could potentially incur long-term damage and possibly undermine the organization's viability as an effective law enforcement agency. The organizational, structural, and philosophical changes are as much a part of the Compstat paradigm and the NYPD's transformation as the Compstat meetings themselves.

Compstat meetings permit executives and managers to monitor virtually every aspect of the agency's activities—from fulfilling the primary mission of reducing crime and making the city's streets safer to closely observing and controlling virtually every systemic change instituted in the agency's systems, practices, structures, and culture. Compstat meetings are, in a sense, a window through which the department's executives and managers can glimpse every aspect of its operations as well as the progress and directions of every change taking place. They are also a mechanism by which the agency's operations and practices can be continually assessed and fine-tuned to ensure their continued success, and through which important messages can be subtly or overtly transmitted and reinforced.

The Compstat meetings also present opportunities for the agency to temporarily (and, in a sense, artificially) break free of the constraints typically imposed by bureaucracy and by rigid hierarchical organizational structures. Instead of operating within a hierarchical framework where lines of communication, authority, and responsibility are precisely defined by straight horizontal and vertical lines on an organizational chart, for the duration of the Compstat meeting the organizational structure changes to one resembling what Bratton called a "seamless web" (see Henry, 2000). In this seamless web—a structure that facilitates brainstorming, innovative problem solving, and the development of effective strategies and plans—every individual, every unit, and every function can communicate immediately and directly with every other individual, unit, or function. Once the

meeting concludes and decisions have been formulated, the structure reverts to one resembling a hierarchical bureaucratic organization—the kind of structure that is particularly well tailored to carrying plans through to fruition.

Intrinsic to grasping how and why the Compstat paradigm and process work so well is understanding that the paradigm and process work at every level of the agency and also beyond it. To remain prepared for the Compstat meetings at headquarters, most or all patrol borough and precinct commanders convene their own in-house Compstat-style meetings with key personnel. Accountability and responsibility for achieving results is not just placed on managers by executives but by managers upon supervisors and to some extent by supervisors upon rank-and-file members of the department. These formal or informal meetings within the agency have never been officially mandated, but the Compstat meetings set such a good example that other managers began to emulate them, and they have had a profound influence over the way other meetings are conducted. Because these formal and informal meetings emulate Compstat, a high degree of communication and a sharp focus on achieving results pervade the organization.

A HYBRID MANAGEMENT STYLE

The Compstat paradigm is a hybrid management style that combines the best and most effective elements of several organizational models as well as the philosophies that support them. Compstat retains the best practices of traditional policing, for example, but also incorporates insights and practices from community policing and problem-solving policing styles. It also utilizes the kind of strategic management approaches used by successful corporate entities that thrive in highly competitive industries. Because the Compstat paradigm is so flexible and because it emphasizes the rapid identification and creative solution of problems, it can be applied in virtually any goal-driven human organization.

Although Compstat police management draws on the strengths of the traditional professional model of policing as well as the community policing and problem-solving policing models, it also differs from each in important ways. The NYPD has based its approach to crime reduction and quality-of-life improvement on the "broken windows" theory articulated by Wilson and Kelling (1982)—an approach that many community policing theorists have also championed. In essence, this important theory takes the position that quality-of-life problems such as graffiti, public intoxication, loud radios, urban decay, and a host of other petty annoyances of modern urban life are in themselves criminogenic—when left untended, they subtly and overtly convey a message that disorder and incivility prevail, that social controls have broken down, and that no one really cares about the neighborhood in which they occur. This message often translates to the idea that such conditions are somehow acceptable and that because minor offenses are acceptable, more serious ones must be as well. Ultimately, if minor offenses are left unchecked they lead to more serious crime (Kelling, 1987, 1991, 1992, 1995; Kelling & Coles, 1996, 1997; Wilson, 1983; Wilson & Kelling, 1982, 1989).

The postulates of the broken windows theory are central to many community policing ideologies and practices, although many leading community policing theorists and practitioners place the burden for identifying and remedying a neighborhood's crime and quality-of-life problems on the beat officer. To empower the beat officer and support effective community policing, they say, the agency must be thoroughly decentralized so that power can be almost completely devolved to those at the bottom of the organizational pyramid. In the NYPD, though, the burden of identifying and solving problems has been placed squarely on the shoulders of middle managers—the agency's 76 precinct commanders. On the basis of its experience during the late 1980s and early 1990s in implementing a version of community policing that emphasized the primacy of the beat officer, the NYPD recognized that it is unfair and unreasonable to expect beat cops to disentangle and successfully address entrenched social problems whose solutions have confounded police executives, social scientists, and criminal theorists for years. Despite their best efforts and, in many cases, their skills and expertise,

beat-level police officers simply cannot muster the organizational resources needed to attack these problems.

REALIGNING ORGANIZATIONAL POWER, MOBILIZING EXPERTISE

Closely related to the NYPD's decentralization was the redistribution of power in the agency. The five bases of power operating within a police organization—coercive, legitimate, expert, reward, and referent power—need to be realigned if the agency and its members are to achieve their full potential. In traditionally managed agencies, the majority of power is concentrated among the executive cadre, and because others have almost no access to coercive, legitimate, or reward power, they cannot easily obtain or apply the agency's resources to address problems (see, generally, French & Raven, 1959).

The Compstat paradigm's effectiveness also derives from the emphasis it places on mobilizing expertise and good practice—in many cases, the expertise and good practices of experienced patrol officers—and making them the norm throughout the agency. This, too, is a tenet of problem-solving policing, but as an organizational reality it has often proven to be an illusory goal in American policing. The NYPD's executives gathered together experts from throughout the agency as well as from outside it and drew upon their knowledge and experience to develop a series of crime control and quality-of-life strategies. The strategies, specifically crafted to be flexible and adaptable to the local community's particular needs and conditions, addressed specific types of crime and disorder problems and were promulgated throughout the department. Every precinct commander was mandated to adapt and implement them, and Compstat meetings are one way to ensure that they have been implemented effectively.

As a practical matter, the crime control and quality-of-life strategies all proceeded from the broken windows theory's basic position that many serious crimes will be prevented and serious problems avoided if we attend to minor offenses as they occur or soon afterward. The strategies also primarily use enforcement tactics to suppress these lower-level offenses and quality-of-life problems. Some community policing adherents eschew enforcement as a means to reduce crime and disorder, or at best they express ambivalence about how effectively enforcement tactics work to reduce crime. Their position, in a nutshell, is that the police should become agents of change who empower and build communities to police themselves. Other community policing advocates rarely deal with the idea of *actually* reducing crime, preferring instead to emphasize that the perceived *fear* of crime in a community can be reduced through more positive police-community interactions. These community policing theorists have been criticized by more traditional thinkers for placing more emphasis on the *appearance* of public safety than upon substantive crime reduction. Highly focused enforcement activities were always a goal of the professional model that dominated American policing for most of this century, but they were nevertheless a rarely achieved goal. Through the Compstat paradigm and the Compstat meetings, the NYPD has had great success with the use of highly focused enforcement strategies in the broken windows context.

It is important to recognize that although the results achieved by the NYPD depended greatly on enforcement activities, the Compstat management paradigm can be used equally well to manage an agency in which enforcement has a lower priority. Another agency may or may not achieve the magnitude of crime reduction accomplished in New York City—indeed, some agencies may not face the compelling crime problems the NYPD did—but the agency will improve its performance on any criteria it deems important if it implements the paradigm cogently. If the agency's prevailing philosophy is that the number and quality of police-citizen encounters are of primary importance and that enforcement has little value, Compstat paradigm management can be adapted and used to tremendously improve the quality and number of those positive encounters. Regardless of the agency's specific goals and objectives, implementing the basic principles of the Compstat paradigm will dramatically increase the likelihood that they will be attained.

As practiced in New York City, the Compstat paradigm also articulated a bold new

philosophy—an unwavering belief in the capacity of police officers to make a difference and to reduce crime. Police officials in the NYPD and elsewhere have, of course, spoken to this philosophy for years, but in fact their words often amounted to mere platitudes and in many cases were betrayed by their actions. This perceived or real insincerity combines with other factors to foment cynicism, to drive a divisive wedge between street cops and management cops, and to undermine the legitimacy of managers as well as their capacity to effectively manage and direct their department's affairs.

At the heart of the Compstat paradigm is a realistic appreciation of the wealth of expertise and experience held by effective police officers. Expert officers of every rank worked together to create and implement the crime and quality-of-life strategies that helped reinvigorate the NYPD, and they worked together in the reengineering committees that restructured 12 important functional areas. In far too many agencies, including the NYPD, managers and executives have subtly or overtly devalued operational officers and their contribution to the agency. Good and effective cops exist in every agency, and the Compstat management paradigm insists that managers and executives capitalize on that expertise and use this essential resource effectively. Quite frankly, this often demands that executives and middle managers put aside their own egos to acknowledge that in many instances street cops may have greater expertise and greater knowledge than they.

In far too many agencies, including the pre-Compstat NYPD, some managers and executives have been too timid to take the kind of risks that might compromise their own careers, and they have concentrated a disproportionate share of their energies on restraining and controlling operational officers. This is not to say that executives, managers, and supervisors should exert no restraints on officers' behavior and conduct, but as often as not, the restraints have taken the form of broad policies and blanket practices that put unnecessary and burdensome restrictions on all cops without regard to their capabilities. Too often, managers and executives impose unnecessary impediments to good police work. Compstat management not only demands that these obstacles to performance be removed in order to let good cops flourish and influence others around them to do the same but helps identify and reward the officers who perform best.

Another central tenet of the new management philosophy is a belief in the idea that police *can* make a difference and that police *can* reduce crime. When it comes to crime, a great many criminologists, politicians, and police executives seem to equivocate about whether the police really make a difference. When a police officer performs a creditable act or when the agency performs well, executives laud it as an example of the kind of difference police can make, and certainly few police executives are reluctant to take credit when crime declines. When crime rises, though, many are unwilling to acknowledge their own managerial inadequacies or failures, and they begin looking about for other explanations. They may never directly articulate a *disbelief* in the capacity of police officers to make a difference by reducing crime and improving the quality of life enjoyed in their communities, but their failure to maintain a consistent approach often casts them as self-serving and undermines their legitimacy in the eyes of officers. In the world of policing, the disjunction between words and actions often breeds suspicion and distrust, and such subtleties rarely go unnoticed by cops. The Compstat paradigm rejects this pessimistic and cynical management view and optimistically asserts without question that conscientious police officers in a well-managed police organization can make a remarkable difference.

POSSIBILITIES AND OBSTACLES: THE FUTURE OF THE COMPSTAT PARADIGM

Paradoxically, the tremendous success of the Compstat paradigm raises a number of potential problems that may complicate its continued growth and future success as more and more agencies attempt to implement it. One of the primary difficulties might be called "cookie-cutter management." That is, there is a distinct tendency throughout policing to find some policy or practice that another agency has put to good use and to appropriate it. Agencies borrow these policies from another agency composed of

different people with a different organizational culture and structure and a different set of environmental and political forces working upon them, and then they press the borrowed policy down on their own agency as if it were a cookie cutter or template. In one fairly large police agency, for example, the newly appointed chief executive "borrowed" the NYPD's entire 800-plus page Patrol Guide of rules and regulations from several years before, had it retyped on a new letterhead with absolutely minimal changes, and adopted it lock, stock, and barrel. These unimaginative managers wind up trying to make the agency fit the policy or practice rather than the other way around. Experience and close observation of police agencies and systems in the United States and overseas show that in the vast majority of cases, a homegrown policy or practice will work much better than an imported policy or practice precisely because it was developed in conformance with the reality of the department. Such policies and practices also work better because they are developed by people who are intimately familiar with the agency, its history and culture, and the capabilities of its personnel. The same is true of the Compstat paradigm's adoption: Its general principles must be carefully tailored to the specific conditions, situations, and realities faced by other agencies in other contexts.

A related management practice that seems to affect American police management is what we might call "cargo cult management." The notion of "cargo cult management" derives from the millennial cults that developed in Melanesia and the South Pacific islands during and after the Second World War and continue to exist today. In essence, members of these primitive cultures had no exposure to the outside world, and as a function of their insularity the cultures were permeated with a deep strain of magical thinking and a propensity to attribute results to rituals rather than to their actual causes. These cultures had their first real exposure to outsiders during the war, when foreigners (Allied military personnel) arrived and began to carve out long flat strips of jungle. The foreigners engaged in such rituals as marching around in formation with unusual devices made of wood and metal over their shoulders. The foreigners built towers and spoke odd words into strange boxes, and shortly thereafter large birdlike flying machines came out of the sky laden with all sorts of good stuff the foreigners called "cargo." The foreigners eventually departed but left behind some of the cargo—Coca Cola, various ingenious machines, and building materials that were far superior to anything the tribes had known before. To this day, cargo cultists continue to carve out strips of jungle, march around with tree limbs over their shoulders, build bamboo towers or climb trees, and repeat "Roger, over and out" into coconuts as they await the return of the cargo-bearing magical flying machines.

The analogy of "cookie cutter management" and "cargo cult management" to the expanding use of Compstat-like programs is clear. Gootman (2000), for example, noted that 235 police agency representatives visited NYPD Compstat meetings in the first 10 months of 2000, whereas 221 visited in 1999 and 219 visited in 1998. There is no doubt that many of these representatives are highly experienced practitioners and fine managers, but on the basis of their too-brief exposure to the Compstat meetings, we can expect that many will return to their agencies with only a rudimentary and very superficial understanding of the Compstat process and even less knowledge of the Compstat paradigm as a whole. There is a distinct possibility that some proportion of these representatives will not fully grasp how and why the process works in terms of motivation, strategy development, the dissemination of knowledge and expertise, or organizational and cultural transformation and that they will not comprehend the important activities (such as reengineering and training) that were undertaken to support it. They may convene Compstat-type meetings where executives apply a heavy hand where a gentle touch is required, and in many cases they will not go beyond the statistical data to identify important qualitative issues that should be of concern to competent police executives.

Some police executives who see the wonderful things Compstat can bring to the organization (and to their careers) may engage in ritualistic repetition of the overt behaviors they've witnessed while the larger picture eludes them. This certainly may not be the case in all situations, but the overall pervasiveness of

"cookie cutter management" and "cargo cult management" in American police management certainly illuminates the potential harm that Compstat can do when unenlightened executives wield such a powerful management tool. Perhaps the greatest danger lies in the fact that the organizational and cultural damage they do remains submerged, creating a host of concealed difficulties with which future generations of police officers and executives will have to grapple. Moreover, once locked into the Compstat mind-set, future generations of managers in these agencies may lack the capacity or the experience to easily recognize or respond effectively to the subtleties and nuances of the problems they confront. Wittingly or unwittingly, far too many police chief executives seem willing to enter into the Faustian bargain of selling their souls to achieve immediate results without regard for the long-term organizational and social consequences or the management burdens that others will have to assume when they've moved on.

There are certainly useful templates or blueprints one might borrow to structure or design a policy, a practice, or an entire police organization, but the key to the success and applicability of the Compstat paradigm is the fact that its principles are eminently adaptable to *any* police agency. It is not necessary or desirable to create an NYPD Compstat-clone to achieve remarkable results—a far wiser and more effective approach is to adapt the Compstat paradigm management principles developed and refined by the NYPD, applying them to fit the agency's particular needs and objectives. As a management paradigm—a way of organizing our ideas about the nature of human organizations and their management—Compstat has proven its applicability throughout the public and private sectors. The paradigm has been successfully adapted, for example, to manage New York City's jail system (O'Connell & Straub, 1999), and the city of Baltimore has implemented a "Citistat" program based on the Compstat paradigm for managing the entire city and all its agencies (Clines, 2001; Swope, 2001).

Compstat continues to evolve and find broader application in other spheres of public sector administration, but its roots are firmly planted in policing. Compstat continues to evolve and to bring about remarkable changes in other spheres, but its most tangible, quantifiable, and dramatic impact continues to be in the area of crime reduction and quality-of-life improvement. American policing faces an array of unprecedented challenges in the coming years, and the Compstat paradigm represents an important opportunity for the kind of flexible and effective management style these challenges require.

NOTES

1. UCR data are based on slightly different offense definitions. These offense categories and definitions

are collected nationwide and provide a basis for comparing data across jurisdictions. The UCR definitions and the New York State Penal Law definitions vary slightly, but the two measures are roughly comparable.

2. At that time, the Transit Police was a separate agency under the jurisdiction of the Metropolitan Transportation Authority rather than the City of New York. Similarly, the Housing Police Department operated under the aegis of the New York City Housing Authority. The Transit and Housing Police Departments were merged with the NYPD in 1995.

3. For administrative purposes, the NYPD divides the five boroughs of the city into eight patrol boroughs. The 76 precincts, 12 transit division districts, and nine Housing Bureau Police Service Areas are about equally apportioned among the eight geographic patrol boroughs.

REFERENCES

Bratton, W. J. (1996, October 15). *Cutting crime and restoring order: What America can learn from New York's finest.* Heritage Foundation Lectures and Educational Programs, Lecture No. 573. Retrieved May 10, 2002, from Heritage Foundation Web site: www.heritage.com/ lectures.htm.

Bratton, W. J. (1998). *Turnaround: How America's top cop reversed the crime epidemic.* New York: Random House.

Buntin, J. (1999). *Assertive policing, plummeting crime: The NYPD takes on crime in New York City* (Case Study No. C16–99–1530.0). Cambridge, MA: Harvard University, John F. Kennedy School of Government.

Chetkovich, C. A. (2000a). *The NYPD takes on crime in New York City. (A)* (Case Study No. CR16–00–1557.3). Cambridge, MA: Harvard University, Kennedy School of Government.

Chetkovich, C. A. (2000b). *The NYPD takes on crime in New York City. (B) Compstat* (Case Study No. CR16–00–1558.3). Cambridge, MA: Harvard University, Kennedy School of Government.

Chetkovich, C. A. (2000c). *The NYPD takes on crime in New York City. (C) Short-term outcomes* (Case Study No. CR16–00–1559.3). Cambridge, MA: Harvard University, Kennedy School of Government.

City of New York, Mayor's Office of Operations. (1998, February). *The mayor's management report: Preliminary Fiscal 1998.* New York: Author.

City of New York, Mayor's Office of Operations. (2000, September). *The mayor's management report: Fiscal 2000.* New York: Author.

City of New York, Mayor's Office of Operations. (2001, February). *The mayor's management report: Preliminary Fiscal 2001.* New York: Author.

Clines, F. X. (2001, June 10). Baltimore uses a databank to wake up city workers. *New York Times,* p. A11.

Federal Bureau of Investigation. (1996). Uniform Crime Reports, Crime in the United States, 1996. Retrieved May 22, 2001, from Federal Bureau of Investigation Web site: www.fbi.gov/ucr/ 96cius.htm.

Federal Bureau of Investigation. (1999). Uniform Crime Reports, Crime in the United States, 1996. Retrieved May 22, 2001, from Federal Bureau of Investigation Web site: www.fbi.gov/ucr/99cius.htm.

French, J. P. R., & Raven, B. (1959). The bases of social power. In D. Cartwright (Ed.), *Studies in social power.* Ann Arbor: University of Michigan, Institute for Social Research.

Gladwell, M. (1995, June 3). The tipping point: What if crime really is an epidemic? *New Yorker.* Retrieved May 2002 from Malcolm Gladwell's Web site: www.gladwell.com/1996/ 1996_06_03_a_tipping.htm.

Gladwell, M. (2000). *The tipping point: How little things can make a big difference.* Boston: Little, Brown.

Gootman, E. (2000, October 24). Police department's allure is growing. *New York Times,* p. B1.

Henry, V. E. (2000). Interview with William J. Bratton. *Police Practice and Research: An International Journal, 1,* 397–434.

Henry, V. E. (2002). *The Compstat paradigm: Management accountability in policing, business and the public sector.* Flushing, NY: Looseleaf Law Publishers.

Innovations in American Government. (1996). Compstat: A crime reduction management tool: 1996 winner. Retrieved May 22, 2001, from Innovations in American Government Web site: http://ksgwww.harvard. edu/innovat/winners/cony96.htm.

Johnson, K., & Connelly, M. (1998, March 13). Americans have more positive image of New York City, poll finds. *New York Times,* p. B1.

Kelling, G. L. (1987). Acquiring a taste for order: The community and the police. *Crime and Delinquency, 33*(1), 90–102.

Kelling, G. L. (1991). Reclaiming the subway. *City Journal, 1*(2), 17–28.

Kelling, G. (1992). Measuring what matters: A new way of thinking about crime and public order. *City Journal, 2*(3), 21–33.

Kelling, G. L. (1995). How to run a police department. *City Journal, 5*(4), 34–45. Retrieved May 15, 2001, from City Journal Web site: www.city-journal.org/html/5_4_how-to_run.html.

Kelling, G. L., & Coles, C. M. (1996). *Fixing broken windows: Restoring order and reducing crime in our communities.* New York: Simon & Schuster.

Kelling, G., & Coles, C. (1997, January). The promise of public order (interview by Ryan Nally). *Atlantic Monthly,* pp. 22–30. Retrieved May 10, 2001, from City Journal Web site: www.theatlantic.com/unbound/bookauth/broken/broke.htm.

Krauss, C. (1994, January 26). Giuliani and Bratton start effort to shake up top police ranks: Reorganization will move more officers to street. *New York Times,* p. A1.

Kuhn, T. (1970). *The structure of scientific revolutions.* Chicago: University of Chicago Press.

Lester, T. (2000, March 29). Epidemic proportions. *Atlantic Monthly.* Retrieved May 10, 2001, from Atlantic Monthly Web site: www.theatlantic.com/unbound/interviews/ba2000-03-29.htm.

Maple, J., & Mitchell, C. (1999). *The crime fighter: Putting the bad guys out of business.* New York: Doubleday.

Marks, J., Egan, J., & Vogelstein, F. (1997, September 29). New York, New York: The Big Apple comes roaring back—and other cities wonder how it was done. *U.S. News and World Report, 123*(12), 44–48.

McKinley, J. C. (1994, January 26). Dispute flares over number of officers on patrol. *New York Times,* p. B3.

McQuillan, A. (1994, January 24). Bratton plan targets lazy cops: Says he'll "kick butt." *New York Daily News,* p. 18.

New York Police Department. (1990, October). *Staffing needs of the New York City Police Department.* New York: Author.

O'Connell, P. E., & Straub, F. (1999). Why the jails didn't explode. *City Journal, 9*(2), 28–37. Retrieved May 15, 2001, from City Journal Web site: www.city-journal.org/html/9_2_why_the_jails.htm.

Silverman, E. B. (1996, December 15). Mapping change: How the New York City Police department reengineered itself to drive down crime. *Law Enforcement News,* pp. 1–6. Retrieved May 10, 2001, from Lloyd Sealy Library Web site: www.lib.jjay.cuny.edu/len/96/15dec/12.html.

Silverman, E. B. (1998). Crime in New York: A success story. *Public Perspective, 8*(4), 3–5.

Silverman, E. B. (1999). *NYPD fights crime: Innovative strategies in policing.* Boston: Northeastern University Press.

Strong, O., & Queen, J. W. (1994, January 25). Where are all those extra cops? *Newsday,* p. 5.

Swope, C. (2001, April). Restless for results: Baltimore Mayor Martin O'Malley is tracking performance on a scale never seen before in local government. *Governing, 14,* 24–30.

Wilson, J. Q. (1983). Thinking about crime. *Atlantic Monthly, 252*(3), 72–88.

Wilson, J. Q., & Kelling, G. (1982, March). Broken windows. *Atlantic Monthly,* pp. 29–38.

Wilson, J. Q., & Kelling, G. (1989, February). Making neighborhoods safe. *Atlantic Monthly,* pp. 46–52.

Witkin, G. (1998, April 25). The crime bust. *U.S. News and World Report,* pp. 46–56.

9

COMMUNITY POLICING AND THE IMPACT OF THE COPS PROGRAM

A Potential Tool in the Local War on Terrorism

JOSEPH F. RYAN

In the first edition of this book, this chapter focused on describing the results of research sponsored by the National Institute of Justice to formulate a framework for defining community policing (Ryan, 1994). Although there are still skeptics that talk of the rhetoric of community policing (Mastrofski & Greene, 1991), to many practitioners the definition of community policing is quite clear. They would argue that they would be able to easily recognize community policing strategies in an agency trying to deal proactively with crime. The three components that they believe clearly define the essence of community policing are

1. *Partnerships* with various components of the community to engage their efforts in dealing with crime

2. *Problem-solving strategies* with identified community partners that attempt to address crime with a long-term focus on deterrence rather than a quick fix such as reliance only on arrest strategies

3. Use of these partnerships to develop joint *crime prevention strategies*

This chapter will share the results of the recent National Evaluation of the COPS (Community Oriented Policing Services) Program, of which I was the coprincipal investigator (Roth & Ryan, 2000), and will discuss how it furthered the development and growth of community policing.

THE COPS PROGRAM

The COPS Program was a key component of the 1994 Violent Crime Control and Law Enforcement Act, also known as the Crime Act. The act contained 33 different subtitles, dealing with issues such as prison construction, violence against women, and an assault weapons ban. This chapter deals only with Title 1, which was labeled "Public Safety and Policing." It shows how the COPS Program reshaped American policing: how it was started in various communities and how COPS grant funds proved to be seed money for efforts that ultimately enabled police to reverse the rising crime rates of the 1980s and 1990s. The chapter also shows how the implementation of the 911 emergency calling system impeded the growth of early experiments with community policing and how the events of September 11, 2001, have revealed that community policing is one of the key strategies needed by local law enforcement to deal with terrorism.

In his 1994 State of the Union address, President Clinton pledged to put 100,000 new police officers on America's streets. In response to this pledge, within 8 months, Congress presented the Crime Act to the president for his signature on September 13, 1994. The House of Representatives voted in favor of the legislation by 235 to 195, with the Senate voting 61 to 34.

Four specific purposes of the Crime Act were delineated: to increase the numbers of officers, provide training in problem-solving and community interaction skills, encourage innovation, and develop new technologies to assist officers in becoming proactive in dealing with crime (§ 10002). The task of administering the Crime Act was given to the Attorney General. In October 1994, the Attorney General, Janet Reno, formally established the Office of Community Oriented Policing Services (COPS Office) to assist in the police hiring grants program and to expand community policing (§ 1701).

Less than 2 months after the signing of the Crime Act, the COPS Office officially announced the first two grant programs. COPS AHEAD (Accelerated Hiring Education and Deployment) was designed for communities with populations greater than 50,000. COPS FAST (Funding Accelerated for Smaller Towns) was for those with populations less than 50,000. In the following month of December 1994, COPS MORE (Making Officer Redeployment Effective) was introduced. In 1995, two additional COPS programs were introduced: "Troops to COPS," a program that encouraged police departments to hire former members of the military by offering an additional $5,000 in reimbursement for training costs, and UNIHIRE (Universal Hiring), which superseded the FAST and AHEAD programs.

The key to the success of the COPS Program was its reengineering of the traditional cumbersome grant process. COPS also bypassed the block grant programs by sending a simple one-page letter directly to law enforcement agencies asking if they wanted additional officers. This was the first time that many law enforcement agencies had ever received a federal grant. Their communities were proud of finally being able to receive federal funds, which traditionally had been allocated to larger communities whose agencies had a grant-writing process in place.

The national evaluation conducted a census of all law enforcement agencies. A law enforcement agency was defined as one whose responsibilities clearly included "police"-type patrol duties; the definition therefore included sheriff, tribal, and public college agencies that were engaged in patrol activities. As of 2000, there were approximately 19,175 law enforcement agencies, whose personnel ranged from one sworn officer to, in the case of the New York City Police Department, almost 40,000.

The entire 1994 Crime Act allocated approximately $32 billion, of which almost $9 billion was earmarked for the hiring of 100,000 officers specifically mandated to perform community policing activities. The evaluation revealed that although Clinton's unofficial goal of 100,000 new hires was not fully realized, there were at the peak in 2000 approximately 84,000 more community police officers on the streets of America than in 1995. The evaluation concluded that the COPS Program was a positive example of how the federal government can address a national issue (i.e., increasing crime rates) with minimal interference in local affairs. The four purposes of the Crime Act were achieved in that numbers of officers were increased, training on problem-solving and community interaction skills was accomplished, innovation not only was encouraged but did occur, and new technologies to assist officers in becoming proactive in dealing with crime were developed (e.g., laptops in patrol cars to increase information sharing).

The 19,175 American law enforcement agencies vary a great deal in their standards and practices. However, as a result of the COPS Program, all American police departments can now be said to be at least aware of community policing. It would be hard to find a law enforcement agency that would not look to community policing strategies as a way to deal with a crime wave.

EVENTS THAT SHAPED THE COPS PROGRAM

It is difficult to isolate a single factor that led to the enactment of the COPS Program. But many police experts believe that the increasing crime rate in the 1980s, with homicides breaking all-time highs in some cities, was the sole factor. In 1991, homicides exceeded 2,200 for the first time in New York City's history. In response to the unprecedented levels of crime, the New York police commissioner, Lee Brown, often stated what he considered to be the underlying tenet of the philosophy of community policing: "It is time to stop blaming the police for the crime rate, it is everyone's responsibility."

Equally difficult to ascertain are those factors that led to the Crime Act's focus on increasing the number of police officers in the United States. Crime and fear of crime have historically been of concern to all citizens. Even in London in 1829, when Sir Robert Peel was developing the foundation for the American style of policing, police were attempting to address the concerns of their citizenry. Yet never before in the history of policing had there been an effort to hire as large a number of officers as that sought under the 1994 Crime Act.

Why was community policing selected as the strategy that the federal government would endorse in the Crime Act? One could speculate that persistent negative perceptions of the police in urban communities were a major factor. The civil unrest of the 1960s, highlighted by such events as the race riots in cities such as Newark and Watts and turmoil during the 1968 Democratic Convention in Chicago, left a lasting negative impression of American policing on the United States and on the world. Throughout the 1980s and 1990s, police continued to engage in conflicts with the communities they served, as evident in the MOVE bombing incident in Philadelphia and the Rodney King beating in Los Angeles (Ryan, 1994).

Obviously, a number of factors fostered the shift toward hiring community police officers. Three key factors that had a direct impact on the COPS Program were

1. The assumption, in law enforcement agencies and the federal government, that increasing the number of police officers would be widely viewed as an effective response to public concern over crime

2. Increasing attention by policing scholars and practitioners to new (or rediscovered) policing philosophies and strategies with such names as "community policing" and "problem-solving policing"

3. A growing interest in using federal grants as an instrument for both increasing the levels of policing and achieving widespread adoption of the new strategies

An article by George Kelling and Mark Moore, "The Evolving Strategy of Policing" (1988), provides a historical context for a discussion of these factors. Kelling and Moore divided American policing into three eras: the political era, the reform era, and the community problem-solving era. The political era, from the 1840s to the 1900s, was notorious for its local political patronage and corruption, which ultimately—or, some might say, logically—led to the reform movement, extending through the 1960s. The first two eras had been described by Robert Fogelson (1977); Fogelson's history ended with the 1970s with the assertion that reform was at a standstill. Kelling and Moore, however, saw the 1970s as the beginning of the third era of community problem solving. In this era, research began to challenge traditional policing strategies. The study that did so most dramatically was the Kansas City Preventive Patrol Experiment in 1972 (to be discussed later in this chapter).

As a former New York City police officer responsible for developing community policing strategies, I have attempted to provide a "practical police perspective" of those factors that pushed American policing into the community problem-solving era. That is, I have singled out events and studies that a police executive would have been especially likely to take into consideration when making decisions, such as those dealing with patrol allocation issues.

As indicated in a recent study by the Department of Justice (Sherman et al., 1997), the perspective of police practitioners and that of researchers can diverge widely, and there are significant misunderstandings as to "what works" in law enforcement. For example, studies dealing with one of the most popular national programs, D.A.R.E., have suggested that at best it has little to do with keeping children away from drugs and drug use. Yet communities are still vying for this program (National Institute of Justice [NIJ], 1994). This divergence suggests that despite the research findings, the program has been deemed valuable from a practical police perspective, probably because it offers police agencies an opportunity to present themselves to the community in a more positive light. The program can show youth that the officer is not only out there to arrest them but also there to help them understand that drug usage might lead to an arrest. It could be argued that this is a good way to improve police-community relations.

The divergence in perspectives seems to have caused friction between police practitioners and researchers. This was highlighted at the 1995 meeting of the American Society of Criminology, held in Boston. In a panel discussion during that meeting, a police practitioner and a researcher compared their perspectives on crime. The practitioner chosen was William Bratton, the police commissioner of New York City at that time and now a leading figure in helping police agencies worldwide to adopt strategies to deal with crime. Bratton forcefully argued, much to the dismay of academics, that the decrease in crime during his administration was due to the increase in the number of officers engaged in community policing strategies and not to some sociological or demographic factor such as the aging of the baby boomers.

The Impact of the Assumption That More Police Officers Means Less Crime

Earlier, it was proposed that one factor leading to the development of the COPS programs was the assumption that increasing the number of police officers in the United States would be widely viewed as an effective response to public concern over crime. Public opinion has always been a major force in pushing the police to do something. According to a 1993 Gallup Poll, more than 60% of Americans favor putting more police on the street as a way of dealing with crime (Bureau of Justice Statistics [BJS], 1994, p. 172).

Police have three goals: to protect life and property, to prevent crime, and to apprehend offenders. But they can often lose sight of these goals because they get caught up in the minutiae of all that they have to do on a daily basis (Ryan, 1994). Research has shown that "traditional" policing strategies (as contrasted to community policing) are only marginally effective in helping police achieve the goals set out for them.

Most people believe that the sheer presence of an officer deters crime. Thus, for both the police and the public, increasing the number of police means simply "seeing" more uniformed police officers on patrol. However, support for the deterrent effects of high police visibility, in and of itself, is scant. The most notable research that deals with the perception that police patrols are an effective response to crime is the Kansas City Preventive Patrol Experiment (Kelling, Pate, Dieckman, & Brown, 1974). It is also one that highlights the divergence in research and police perspectives. Police are reluctant to yield to the findings of this study, perhaps because they are convinced that police patrols "must be doing something."

The Kansas City study divided the city into three areas, each using a different patrol strategy. In the control group area, police patrols were maintained at normal levels and police engaged in their routine activities. In what was called the "experimental area," police patrols were removed, and the only police activity was in response to calls for service; officers would respond to the call and then immediately leave the area. Finally, in what was labeled the "proactive area," patrol levels were doubled.

The study found not only that the differences in patrol levels had no impact on crime but also that they had no impact on a number of other outcomes where police had traditionally believed that their efforts would produce results. When community residents were surveyed about a variety of perceptions such as fear of crime and observation of more or fewer police patrols, no differences were found across the three areas (Kelling et al., 1974).

Although the authors of the Kansas City study quite correctly advised against using their findings without further research, in 1975, with New York City facing a major budget deficit, the study was used as a justification for laying off 5,000 police officers. Unfortunately, the study probably could not be replicated in today's climate of civil liability and with today's federal guidelines on research involving human subjects. That is, no community would risk removing police patrols.

One limitation of the Kansas City study from a practitioner's perspective is that it dealt only with the amount of police patrolling and neglected to examine what the police officers were actually doing. The phrase *police patrolling* conveys the idea of police officers driving around and responding to calls for service. But police patrolling can also employ proactive measures to reduce crime (proactive police patrols will be discussed in the next section).

What do police officers do on patrol? In 1971, Al Reiss, examining the dispatch records of the Chicago Police Department, found that "non-criminal matters accounted for 83 percent of classified patrol calls and that service-related calls accounted for about 14 percent. . . . Reiss also suggested that about 85 percent of a patrol officer's time was spent in 'routine patrol'" (Greene & Klockars, 1991, p. 21). Stephen Mastrofski's (2000) recent update on Reiss's study similarly reported that most of officers' patrol time was taken up with non-crime-related activities and similarly described the undirected nature of police patrols.

These findings make it difficult to understand the clamor for more police patrols by police and the public. It appears that police departments have given in to the demands of citizens for more police patrols. Evidence for the public's willingness to pay an increase in their taxes can be found in New York City, where additional tax revenues and surcharges made it possible to develop the "Safe Streets, Safe City" initiative. Under this effort, the city was able to hire 6,479 additional officers during fiscal years 1991 and 1992 (Dinkins, 1991). In Milwaukee, two police budgets were passed, the first for the usual budget items and the second to support additional foot patrols (Kelling & Moore, 1988).

Some research has endeavored to answer the question of whether the police can do anything that will enable the public to "see" that they are getting something for their tax dollars. In Houston and Newark, police and researchers attempted to reduce fear of crime by increasing police visibility and improving communication with citizens through such efforts as police-community newsletters, community organizing response teams, citizen contact patrols, police community stations, and recontacting of victims. These measures did produce desirable results, but mostly in middle-income neighborhoods (Skogan, 1990).

From a researcher's perspective, the Kansas City study showed that there was no evidence that the police strategies used were effective in reducing crime or having an impact on a number of anticipated perceptions of police efforts. The Houston and Newark study suggested that some police strategies had a potential to increase citizens' awareness of police activities. These research findings did not imply that policing in a broad sense had no impact on crime or the citizen's fear of crime.

The findings of the Houston and Newark study did, however, provide the police with insight on alternative policing strategies that may have some impact. From a practitioner's point of view, the study provided insight on where police might have the potential to give citizens something they could "see" as a direct benefit from their tax dollars. It also suggested that the police needed to increase the quantity and quality of contacts between police and citizens.

Another strategy, besides increasing numbers of police officers on patrol, that the police and the public believed was helpful in deterring crime, as well as apprehending offenders, was rapid response to calls for help. In 1981, the Police Executive Research Forum (PERF) analyzed the Kansas City response time data. They found that even if the police were to travel to the scene faster than a "speeding bullet," they would increase their apprehension rate only from 3% to about 5%. Even in the best of circumstances, when the police are in front of the location just as the citizen is reporting the crime in progress, the police must still pick up the phone and talk to a dispatcher, and during those valuable moments all the offender has to do is walk away from the scene (Spelman & Brown, 1981).

Police and the public have wanted to believe that even if the offender cannot be captured at the scene, detectives can solve the crime and apprehend the offender. Unfortunately, there are few individuals in real life like Sherlock Holmes, the fictional detective who solves crime by the scientific method of deduction. As Chaiken, Greenwood, and Petersilia (1977) noted in the Rand Institute study of what detectives do, police have only three strategies for solving cases: obtaining the testimony of witnesses (including using informers), procuring physical evidence, or eliciting a confession (Klockars & Mastrofski, 1991).

Taken together, all of the above point to one fact: the traditional strategies that the police and the public had such strong beliefs in were faulty. But the Rand study does provide a valuable insight regarding potential police strategies. Two of the three methods detectives use to apprehend offenders rely to some extent on the community. Poor community relations will result in citizens' reluctance to cooperate with its police. Yet the community is the potential source of witnesses and of the information that could enable the police to begin their search for physical evidence. Detectives are in essence problem solvers and can solve problems best in cooperation with the community. In a discussion with a New York City detective about community policing, the detective commented, "I get it—community policing is stealing what we [i.e., detectives] have always been doing: getting to know the community so we can solve our cases. If we don't know who is who in the community, or where we can turn to for resources, we would never be able to solve a crime."

One of the more traditional aspects of American policing is the perception of police as involved in a "war" (Fogelson, 1977). During the 1960s and 1970s, as crime rates increased, the police, the media, the public, and Congress adopted the military language of the war in Vietnam and began to talk about the "war on crime" (and the "war on drugs"). With this frame of reference, the call for more troops became the call for more police.

The police believed that increasing the quality of contacts between police and citizens was not possible without first increasing the number of police officers. Concomitantly, law enforcement managers began to recognize that their police were now slaves to the 911 emergency telephone system (Sparrow, Moore, & Kennedy, 1990). They believed that there was no "free"

time for police officers to engage in "quality" contacts. To the police, quality contacts meant time-consuming activities. This simple reaction became a significant obstacle to police agencies that claimed they wanted to carry out community problem-solving strategies.

Research Pointing to the Effectiveness of New Policing Strategies

The second trend that led to the development of the COPS Program was the increasing attention by policing scholars and practitioners to new (or rediscovered) philosophies and strategies with such names as *team policing, problem-solving policing,* and *community policing.* Early among these experimental strategies was the concept of team policing. The idea for team policing was first put forward in 1967 by the President's Commission on Law Enforcement and the Administration of Justice. The commission recommended that "police departments should commence experimentation with a team-policing concept that envisions those officers with patrol and investigative duties combining under unified command with flexible assignments to deal with the crime problem in a defined sector" (Sherman, 1973, p. xiv).

One of the most important events shaping American policing was the advent of the emergency 911 telephone call system. As crime began to drastically increase in the late 1960s and 1970s, it was believed that the best strategy to meet the needs of the community was to make it easier for them to call the police. Unfortunately, the introduction of the 911 emergency call system brought with it an unforeseen increase in the volume of calls for service from the police. As the police quickly realized, citizens were treating every emergency situation as a situation for the police. Once I was called to a location with a gas leak. When I arrived there, the citizen asked me, "Do you smell gas?" I responded, "Have you called the emergency number for gas leaks?" Precious moments were passing, for the building had to be evacuated, and everyone was waiting for the gas utility to respond. While I was assigned to the Research and Planning Unit of the NYCPD, it was common to see the volume of calls to the police reach beyond 7 million, with the police being able to respond to only 3 million.

Unfortunately for American policing, the response to increased calls for service via 911 was to attempt to get officers to the scene faster. Officers were therefore taken out of the community to answer 911 calls, and their knowledge of their community began to diminish. As a former member of New York City's Neighborhood Police Teams (NPT) program, I can say from my own experience that 911 demands resulted in less time devoted to community concerns.

For most police practitioners, it was becoming obvious that this strategy was not working. Simultaneously, research was emerging that questioned what police were doing to reach their goals and that tried to find ways police could operate more efficiently and get back in touch with the community.

One of the earliest examinations of team policing was a set of case studies of team policing in seven cities of various sizes: Dayton, Detroit, Holyoke, Los Angeles, New York City, Richmond, and Syracuse (Sherman et al., 1973). The case studies summarized what occurred in each location and were not intended to be evaluations. They described team policing as a practice that varied from one city to the next but that generally redirected a group of officers from emergency call (i.e., 911) response to being "sensitive" to the community.

Although the seven case studies provided little assessment of the team policing projects, they show that the community liked this style of policing simply because the citizens believed that the police were listening to them and that the police viewed it as a good way to improve community relations. Unfortunately, the case studies did not provide insight as to why team policing did not catch on, other than to suggest that some chiefs viewed it as an expensive investment at a time when crime and calls for service were on the rise.

One factor that led to a move away from community policing was the report of the Knapp Commission investigating police corruption in New York City (Knapp, 1972). The commission found corruption among organized groups of officers who knew the community "too well." Given these findings, it is not difficult to understand why a police chief might see officers' intimacy with the community as more of a threat than a benefit. In some cases, close community relations can indeed affect officers' attitudes in ways that might interfere with their performance of their duty.

An incident from my own experience is suggestive. In 1987, I was the New York City police representative on a project that was assessing a multiagency attempt to deliver ameliorative services to families experiencing multiple problems. In one precinct where the project was being tested, community police officers played the crucial role of ensuring that representatives of the city's social service agencies would be able to deliver their services to the identified families. As part of the oversight, two community police officers hosted a visit to a large residential building with representatives from the mayor's office and commented on the nature of the problems there.

During this tour, the officers found an unsafe condition and sought the assistance of an individual who was walking down the hall of the building. The officers asked him how he was doing and appeared to know him well. The individual was extremely enthusiastic about helping the officers and left to get some tools. When he left, the officers informed us that he was "Fast Eddie," the number one cocaine dealer in the area, and then added, "But he actually is a really nice guy."

Hundreds of researchers have highlighted the potential merits of community policing, but those most readily recognized by police chiefs are Robert Trojanowicz, George Kelling, and Herman Goldstein. Trojanowicz was involved with evaluating the Flint, Michigan, "Foot Patrol Experiment" and was the first director of the National Neighborhood Foot Patrol Center, which in 1988 became the National Center for Community Policing (Trojanowicz & Bucqueroux, 1990). The center was best known for its series of publications dealing with community policing. One of its influential early publications, a booklet entitled *Job Satisfaction: A Comparison of Foot Patrol Versus Motor Patrol Officers* (Trojanowicz & Banas, 1985), suggested that community policing not only would be popular with the community but also would help increase officers' job satisfaction.

George Kelling's "broken windows" theory also stands out as one of the major influences to revive the idea of community policing. The theory suggested that visible evidence of a community's deterioration (e.g., broken windows, which suggest neglect by property owners) leads to increasing crime because criminals expect that social control in the community will be weak. Thus, the community is open to further decline because there is evidence that the community does not care. Ultimately, the community will be left to those who want to use the neighborhood for their own purposes, namely criminal activities. Kelling therefore suggested that the solution to neighborhood crime was simply to bring the officer back to the community to work proactively to prevent and fix "broken windows" before they became a sign that the community was in decline. This approach was in contrast to "traditional" policing, now synonymous with "reactive policing," or responding and dealing with crime only after it has occurred.

If traditional policing became synonymous with reactive policing, problem-oriented policing became synonymous with community policing. Although one chief said to me off the record that "the only difference between community policing and problem-solving policing is that with problem-solving policing you don't need the community," implying that the police can solve problems on their own, Herman Goldstein's landmark book *Problem-Oriented Policing* (1990) does not foster this perspective. Rather, it shows the community as an integral part of this process, whether in identifying the problem, participating in the solution, or providing insight toward an acceptable solution.

Goldstein's book combines the contributions of Trojanowicz and Kelling into one coherent strategy, providing a framework for the police to restructure their efforts to deal proactively with problems. It combines Trojanowicz's focus on community policing through foot patrols with Kelling's recognition that an officer's presence in the community on foot can lead to discussions on how to prevent problems.

Community policing has also gained attention as a tool that police administrators can use to address a growing issue—civil liability and accountability. Civil liability has been an issue for several decades and began to reach a new level of significance during the civil unrest of the 1960s. However, one of the most prominent examples occurred in 1984, in a case involving domestic violence where the victim sued the police in federal court for a violation of her civil rights. The police were found guilty and were assessed a fine of $2.3 million (*Thurman v. City of Torrington,* 1984). The police had been to the home almost 15 times before the near-fatal stabbing and had done nothing to help stop the violence in the home. In one instance, the victim went to the police station to find out what was happening with her case and was told that her file was lost; on another occasion, she was told that the assigned detective was on vacation. The lawsuit spurred police recognition of the need for improved interaction with the community, as well as the need for increased training of officers.

A major fact of policing has been, as Sherman et al. (1997) noted, the lack of evaluation of police activities. For years, police performed tasks at their own whim, inventing programs for soothing community demands or not doing anything. However, budget reductions have created an unprecedented demand on police to justify what they are doing for their taxpayers. Citizens, in response to never-ending increases in taxes, have begun to ask what they are getting for their money. It is now common to hear them ask such questions as "What are you doing to reduce crime?" The traditional responses of citing crime statistics or showing how summons activity has increased have not allayed community demands.

In one instance, a white middle-class citizen in New York City became concerned about crime in her community and demanded more police protection. When the local police commander responded that the community had already been assigned 150 officers on the basis of the crime rate and volume of 911 calls, she took the unprecedented step of verifying this number. For 1 month, she stood outside her police station at the beginning of each tour and counted the number of officers going on patrol. When she presented her finding that only 80 officers had in fact been assigned, the precinct commander had to acknowledge that she was right. Headquarters assigned 150 officers only on paper in order to use them for other needs around the city, such as mobile task forces.

Federal Grants to Address Crime

The last factor that contributed to the development of the COPS Program was the federal government's use of grants to promote and stimulate more efficient strategies in dealing with crime. Law Enforcement Assistance Administration (LEAA) programs represented the first attempt by the federal government to influence policing in the United States through block grants. Later these grants would be administered by the Office of Justice Programs through the Bureau of Justice Assistance (BJA) and NIJ. The federal government came to believe that it had "a unique responsibility to conduct research on criminal justice programs, and to develop creative programs based on research findings" (1981 Attorney General's Task Force Report on Violent Crime, quoted in Sherman et al., 1997) and that, with streamlining and simplification, federal grants could be an instrument for both increasing the level of policing and achieving widespread adoption of the new strategies.

LEAA was established in 1968 under the Omnibus Crime Control and Safe Street Acts. It immediately came under attack because of the inability of local governments to access funds. This was due in part to bureaucratic obstacles and underfunding of the sponsored programs. A number of studies were conducted to recommend improvements, but after 10 years Congress eliminated all funding for the implementation of the reorganized LEAA program (Lauer & Llewellyn, 1975).

LEAA was later reorganized into the Office of Justice Programs (OJP), with most of the different organizational units or functions being assumed by new agencies. The new agencies under the OJP were BJA, the Bureau of Justice Statistics, NIJ, the Office of Victims of Crime, and the Office of Juvenile Justice and Delinquency Prevention (which the Republicans have referred to as the Office of "Hug-a-Thug" for its failure to reduce juvenile violence). The functions of the OJP are administrative, such as providing financial review.

Two of OJP's units, NIJ and BJA, were instrumental in promoting both practitioner and researcher efforts to examine and change police strategies and can be viewed as especially influential in the development of community policing. Most of the following discussion will focus on the efforts of BJA through its direct funding for innovative police strategies. NIJ funded most, if not all, of the research cited earlier in this chapter. Some additional comments concerning NIJ's role will be discussed at the end of this section.

Funds from the BJA supported several programs that, according to BJA officials interviewed as part of the national evaluation, led to the development of community policing (interviews with BJA officials Ward and Shelko, 1997). They included the Systems Approach to Community Crime and Drug Prevention, Problem-Oriented Policing, the Drug-Impacted Small Jurisdictions Demonstration Program, Innovative Neighborhood-Oriented Policing, and Weed and Seed.

The Systems Approach to Community Crime and Drug Prevention was a BJA-funded demonstration program conducted in four cities beginning in 1986. It was law enforcement based and served as a bridge between traditional crime prevention and proactive policing. Its goal was to increase the capacity of law enforcement to address problems and to enable this program to become part of the daily work of other city agencies and community groups (BJA, 1993).

Problem-Oriented Policing was a project established in 1987 to test the application of analytical problem-solving strategies to community drug problems. Problem-oriented policing was described as a process for approaching the persistent problems in a community that create a need for a police response. It evolved from almost two decades of research showing that simply adding more officers and making more arrests would not solve community problems—especially durable and complex problems.

The goals of the problem-oriented approach were to increase the effectiveness of police in battling drug problems by addressing the underlying community problems; to increase reliance on the knowledge and creative approaches of line officers to analyze problems and develop solutions; to encourage police to tap diverse public and private resources in a cooperative effort to solve community problems; and to foster closer public involvement with the police to ensure that police would address the needs of citizens. This

The San Diego Police Department

VISION

We are committed to working together, within the Department, in a problem solving partnership with communities, government agencies, private groups and individuals to fight crime and improve the quality of life for the people of San Diego.

Values

The principles upon which we base our policing are:

- *Human Life*
  The protection of human life is our highest priority.

- *Crime Fighting*
  Our efforts to address neighborhood problems will be based on a partnership with the community.

- *Loyalty*
  We will be loyal to the community, to the Department and its members, and to the standards of our profession.

- *Fairness*
  Our decisions will be based on common sense, and will be balanced, moral, legal and without personal favoritism.

- *Ethics*
  We will demonstrate integrity and honor in all our actions.

- *Valuing People*
  We will treat each other with dignity and respect, protecting the rights and well-being of all individuals.

- *Open Communication*
  We will listen to one another's opinions and concerns.

- *Diversity*
  We appreciate one another's differences and recognize that our unique skills, knowledge, abilities and backgrounds bring strength and caring to our organization.

Mission

Our mission is to maintain peace and order through the provision of police services that are of the highest quality and responsive to the needs of the community. We will contribute to the safety and security of the community by apprehending those who commit criminal acts, by developing partnerships to prevent, reduce or eliminate neighborhood problems, and by providing police services that are fair, unbiased, judicious, and respectful of the dignity of all individuals.

Figure 9.1   Vision, Values, and Mission Statement of the San Diego Police Department

approach was, in essence, a restatement of Goldstein's "problem-oriented policing."

The Drug Impacted Small Jurisdictions Demonstration Program took the problem-solving strategy a step further toward community policing by giving the criminal justice system a primary role in mobilizing communities to develop comprehensive strategies for combating illegal drugs. Its focus was that police and communities must work together in a relationship of trust, cooperation, and partnership to promote safety.

The most insightful effort, not only from a BJA perspective but also from a practitioner's, was the Innovative Neighborhood-Oriented Policing Program (INOP), funded as part of the Edward Byrne Memorial State and Local Law Enforcement Assistance Program. INOP was initiated in 1990 to show how a neighborhood-oriented approach to policing could be applied to crime and drug problems in target communities to develop model strategies that would be effective in demand reduction. The cluster conferences involving the eight program sites provided much support for this effort. These meetings were extremely valuable because they provided the practitioners with the learning experiences of others who were encountering similar situations. Thus, the sites could adopt each other's promising ideas, be encouraged by each other's successes, and gain insight on how to overcome local obstacles (e.g., dealing with the bureaucratic processes of purchasing equipment).

An NIJ-funded evaluation of this effort by Sadd and Grinc (1995) described several problems and obstacles, such as the initial resistance of police officers to a community policing strategy and the difficulty of involving other public agencies and of organizing the community. But the largest and most persistent problem described was that although the police were weeding out identified problems, no one was following up with what everyone recognizes as the key to solving the situation—planting the seeds of positive community change that could prevent the problem from spreading. This recognition led to a new, broader-based program called Weed and Seed. Although the gardening analogy is perhaps simplistic, and although the program's title has been controversial, the BJA believes that Weed and Seed was a major contributor to the development of community policing.

Weed and Seed was designed to be a comprehensive, coordinated effort to control crime and to improve the quality of life in 19 targeted high-crime neighborhoods. In some of the target communities, a U.S. attorney for that state chaired the steering committees. This helped to bring the support of other federal agencies (Roehl et al., 1996). The first task, "weeding," involved using the resources of the criminal justice system, including intensive law enforcement suppression activities, to remove and incapacitate violent criminals and drug traffickers from targeted neighborhoods and housing developments. The second task, "seeding," involved attempting to revive the community by providing broad economic and social opportunities developed in cooperation with other federal, state, and local agencies, along with public and private organizations and community groups.

Community-oriented policing, with its emphasis on building police and community partnerships, was described as the bridge between the program's "weed" and "seed" components. Policing techniques such as officer foot patrols, citizen neighborhood watches, targeted mobile units, and community relations activities focused on increasing police visibility and developing cooperative relationships between the police and citizens in target areas. These strategies were aimed both at supporting suppression activities and at facilitating individuals' access to prevention, intervention, and treatment, as well as promoting the neighborhood reclamation and revitalization components.

The objective of Weed and Seed was to raise the level of citizen and community involvement in crime prevention activities and other partnership efforts, to help solve drug-related problems in neighborhoods, to enhance public safety, and to reduce fear in the community so that socioeconomic development and related services could be implemented successfully. However, as Roehl et al. (1996) noted, the "seeding" part of the program was the most difficult to implement because the targeted neighborhoods had deep-seated social problems. Most contained a large percentage of residents whose incomes were at or below the poverty line, complicating the effort. Weed and Seed did, however, address some of the shortcomings noted in the evaluation of INOP. According to Roehl et al. (1996), it was successful in that "people who ordinarily did not consult with one another—prosecutors, area residents, police officers, and social service personnel—were able to coordinate their efforts, share resources, and solve problems" (p. 44).

The role of NIJ in promoting community policing is difficult to summarize because NIJ has funded an enormous amount of research on policing strategies that could be applied in community policing programs.[1] Programs such as Hot-Spots, the Drug Market Analysis Project (DMAP), and Neighborhood Watch (Sherman et al., 1997) and their accompanying evaluations are crucial to understanding potential strategies that an agency engaging in community policing might use, but they did little to promote the philosophy of community policing. Thus, a discussion of NIJ's role will be limited to direct contributions to community policing.

From a practitioner's point of view, NIJ's most notable effort toward community policing was the funding provided to the John F. Kennedy School of Government to host the Executive Sessions on Community Policing. The focus of the sessions was to guide the development of community policing through the participation of top law enforcement executives such as Benjamin Ward of New York City; Willie Williams of Philadelphia; and Daryl Gates of Los Angeles. Seventeen books resulted from the sessions and were published in NIJ's Perspectives on Policing Series. The first of these, *Police and the Communities: The Quiet Revolution,* was by George Kelling (1988).

During my awarded Visiting Fellowship with NIJ, I was able to promote the idea for a national conference on community policing. A conference entitled "Community Policing for Safe Neighborhoods: Partnerships for the 21st Century" was held in August 1993. It was deemed a great success by all involved. Original projections were for 500 individuals, but over 700 attended. Nearly every major law enforcement executive and police researcher was present to talk about his or her interest and involvement with community policing. The entire 3-day conference was broadcast on C-SPAN. It included interviews by a reporter from a local television station (WJLA Television) with a panel of 18 police chiefs concerning their beliefs on the value of community policing.

CONCLUSION

The national evaluation of the COPS Program noted an overall success in achieving the purposes of the 1994 Crime Act. In the year 2002, there are significantly more community police officers on the streets of America than there were in 1995 before the program started. The purposes of the Crime Act in promoting community policing were realized. In visiting the sites as part of the national evaluation, I was left with a sense of excitement that American policing was changing. One bit of evidence that supports my excitement is the "Vision, Values and Mission" statement that the San Diego Police Department (SDPD) developed as part of its reengineering effort in implementing community policing (Figure 9.1). This statement is clearly a visionary effort toward changing how American law enforcement is policing communities. The statement clearly shows that the SDPD asked itself the simple question "Why do we exist?" and that the answer was "To serve the community." The creation of such a statement can help law enforcement managers inspire and lead their officers in policing strategies for the future. The statement, with the SDPD logo in the background, is conspicuously posted in each police facility, and a laminated 3 × 5 card is given to each officer.

I could see that community policing was changing at both the top and the bottom of the agencies we visited. Patrol officers would share how a problem that traditional officers had labeled as "unfixable" was now being analyzed and solved. Officers who were engaged in community policing strategies were clearly more excited about their daily work experiences than were traditional officers. However, there is more to be acknowledged about the legacy of the COPS Program.

As noted earlier in this chapter, William Bratton, the former police commissioner of New York City, asserted in his address to the 1995 annual meeting of the American Society of Criminology that New York City's hiring more police under the Safe Street Act was a major factor in beginning the decrease in the homicide rates. He argued that during all the years that crime was increasing, the police were faulted for not proactively fighting crime, so now, when crime was coming down, police should receive credit. At the time of the publication of the national evaluation, almost 5 years later in 2002, homicides were down to numbers not experienced since the early 1960s: They had dropped from approximately 2,200 to fewer than 600. Although from a researcher's point of view it would be hard to prove "cause and effect," it is difficult to ignore the correlation between the hiring of additional community police officers and the nationwide decrease in crime.

When the 911 emergency telephone system was first established, it pulled the police away from "knowing" their community. The COPS Program reversed this trend. Through its success in hiring more community police officers and thus giving officers more time to meaningfully communicate with their communities, it has begun to restore this wealth of community knowledge.

The terrorist attacks on this country on September 11, 2001, raise new issues in community policing. As Congress is trying to ascertain what truly happened, whether it could have been prevented, and what can we do in the future, they are guided by our Constitution's promise that our government will "provide for the common defense." One elementary question that we can begin to ask ourselves is "What role can our local law enforcement agencies play in providing for our common defense?"

A finding not fully examined in the national evaluation report was support found for my own experience of the value of community policing from walking a beat as a part of New York City's experiment with Neighborhood Police Teams in the early 1970s. Walking the beat and being responsible for the integrity of one's beat—that is, attempting to "know what is always going on" in your beat area[2]—facilitates all the stated purposes of the 1994 Crime Act: partnerships, problem solving, and crime

prevention. In each of the communities we visited as part of our national evaluation, it was clear that the officers who were successfully engaging in community policing were fulfilling the imperative of "beat integrity."

The challenge to those in American policing is to continue to examine the lessons learned as part of the legacy of the COPS Program. To proactively fight crime and keep crime rates down, we need to recognize two things. First, community policing is not soft on crime. In reality, it is a very difficult strategy that requires time and patience. It requires police to know their communities. And it requires the United States to keep the number of police on the streets at the 2002 level. Unfortunately, the current recession is threatening to push police levels back to the pre-1995 levels, and most experts fear the return of crime (Cooper, 2002). Second, police need to examine the reality of the value of community knowledge and hopefully put it into effect in our fight against terrorism. Community policing provides us with the tool to know both the good and bad side of "who's who in the community."

NOTES

1. For an excellent overview of federally sponsored research programs, see Sherman et al. (1997). See also the interviews that I conducted with Reed, Ross, and Lively of the NIJ staff as part of the national evaluation (Ryan, 1997).

2. Officers in New York City who were assigned to a beat in the early 1970s were required to know who each business owner was and to maintain what was called a "business house card file" on each location, to know and report all suspicious activity, and to interview each business owner and resident who was a victim of a crime on their beat. At times this was a stressful experience because these citizens had come to expect that the beat officer would be there for them; when the officer was not, they expected an answer as to why they had been victimized. In addition, if detectives assigned to headquarters found any suspicious activity that the beat officer had failed to report, the beat officer would be subject to disciplinary action, such as a loss of vacation time.

REFERENCES

Bureau of Justice Assistance. (1993, September). *The systems approach to crime and drug prevention: A path to community policing* (Bulletin of the Bureau of Justice Assistance, No. 1). Washington, DC: Author.

Bureau of Justice Statistics. (1994). *Sourcebook of criminal justice statistics—1994.* Washington, DC: U.S. Department of Justice, Office of Justice Programs.

Chaiken, J., Greenwood, P., & Petersilia, J. (1977). The Rand study of detectives. In C. B. Klockars & S. D. Mastrofski (Eds.), *Thinking about police* (2nd ed.). New York: McGraw-Hill.

Cooper, M. (2002, May 14). Kelly says police may be cut to lowest level in a decade. *New York Times,* p. B2.

Dinkins, D. (1991). *Safe streets, safe city: An omnibus criminal justice program for the city of New York (summary).* New York: City of New York, Office of the Mayor.

Fogelson, R. M. (1977). *Big-city police.* Cambridge, MA: Harvard University Press.

Goldstein, H. (1990). *Problem-oriented policing.* Philadelphia: Temple University Press.

Greene, J. R., & Klockars, C. B. (1991). What police do. In C. B. Klockars & S. D. Mastrofski (Eds.), *Thinking about police* (2nd ed.). New York: McGraw-Hill.

Kelling, George L. (1988). *Police and communities: The quiet revolution.* Washington, DC: National Institute of Justice.

Kelling, G. L., & Moore, M. H. (1988). *The evolving strategy of policing* (Perspectives on Policing, No. 4). Washington, DC: National Institute of Justice.

Kelling, G. L., Pate, T., Dieckman, D., & Brown, C. E. (1974). *The Kansas City Preventive Patrol Experiment.* Washington, DC: Police Foundation.

Knapp, W. (1972). *Commission to Investigate Allegations of Police Corruption and the City's Anti-Corruption Procedures.* New York: Bar Press.

Lauer, C.A., & Llewellyn, T. S. (1975). Reorganization of LEAA's grant programs. *Federal Grant Law,* White Paper, 40 pp.

Mastrofski, S. D., & Greene, J. R. (1991). *Community policing: Rhetoric or reality.* New York: Praeger.

National Institute of Justice. (1994, October). *The D.A.R.E. Program: A review of prevalence, user satisfactions, and effectiveness* (NIJ Update). Washington, DC: U.S. Department of Justice, Office of Justice Programs.

Roehl, J. A., Huitt, R., Wycoff, M. A., Pate, A., Rebovich, D., & Coyle, K. (1996). *National process evaluation of Operation Weed and Seed* (NIJ Research in Brief). Retrieved October 2002 from National Criminal Justice Reference Service Web site: http://ncjrs. org/.

Roth, J. A., & Ryan, J. F. (2000). *National evaluation of the COPS Program.* Washington, DC: National Institute of Justice.

Ryan, J. F. (1994). Community policing. In A. R. Roberts (Ed.), *Critical issues in criminal justice.* Thousand Oaks, CA: Sage.

Sadd, S., & Grinc, R.M. (1995). Innovative neighborhood-oriented policing: Descriptions of programs in eight cities. Retrieved October 2002 from

National Criminal Justice Reference Service Web site: http://ncjrs.org/.

Sherman, L. W., Gottfredson, D., MacKenzie, D., Eck, J., Reuter, P., & Bushway, S. (1997). Preventing crime: What works, what doesn't, what's promising. Retrieved October 2002 from National Criminal Justice Reference Service Web site: http://ncjrs.org/.

Sherman, L. W., et al. (1973). Team policing. Washington, DC: Police Foundation.

Skogan, W. G. (1990). *Disorder and decline: Crime and the spiral of decay in American neighborhoods.* New York: Free Press.

Sparrow, M. K., Moore, M. H., & Kennedy, D. M. (1990). *Beyond 911: A new era for policing.* New York: Basic Books.

Spelman, W. G., & Brown, D. K. (1981). Response time. In C. B. Klockars & S. D. Mastrofski (Eds.), *Thinking about police* (2nd ed.). New York: McGraw-Hill.

Thurman v. City of Torrington, 595 F. Supp. 1521 (1984).

Trojanowicz, R. C., & Banas, D. W. (1985). *Job satisfaction: A comparison of foot patrol versus motor patrol officers.* East Lansing: Michigan State University, National Neighborhood Foot Patrol Center.

Trojanowicz, R., & Bucqueroux, B. (1990). *Community policing: A contemporary perspective.* Cincinnati, OH: Anderson.

Violent Crime Control and Law Enforcement Act, Pub. L. No. 103-322, 108 Stat. 1796 (1994).

10

POLICE USE OF EXCESSIVE FORCE
Exploring Various Control Mechanisms

MARK BLUMBERG

BETSY WRIGHT KREISEL

The videotaped Rodney King beating in Los Angeles heightened Americans' sensitivity to the problem of police use of excessive force in the early 1990s. Civil unrest erupted when the police officers associated with the King beating were acquitted on state criminal charges. Since the King incident, the legitimacy of the police has been called into question in numerous instances. For example, in 2001 Cincinnati experienced the police shooting-death of a black unarmed man and subsequent community protests against police practices. In 1998, a New York City police officer was convicted of using an object to sexually assault Abner Louima, a Haitian immigrant. Another example of police misconduct is the Rampart scandal in 2000, when the Los Angeles Police Department (LAPD) went through one of the worst cases of police abuse of power. This misconduct stemmed from allegations of officers shooting innocent people and planting guns in order to effect an arrest for assault (LAPD, 2000). Some of these recent events led to community disorder. However, they are not the first instances of civil unrest precipitated by overzealous police action.

All police violence does not result in civil unrest. Further, it would be unfair to blame the police for the horrendous social and economic problems in many of our inner cities. In these communities, factors such as extensive unemployment, a high rate of poverty, poor schools, a declining industrial economy, limited opportunities for unskilled workers, racial discrimination, and hopelessness combine to create conditions that are conducive to the development of riots. Too often, though, an incident involving the use of excessive force by the police serves as the spark that triggers a violent reaction and is the catalyst for reforms of departmental policies and/or procedures.

The effects of police officer misconduct reach farther than the immediate subsequent civil disturbance. Misconduct has the potential to undermine police officers' effectiveness in performing their assigned duties. It can diminish the morale of police employees, and it harms the relations of citizens and police officers across the nation (Hunter, 1999). Because the effects of police use of excessive force can be extensive and because citizens across the nation are vigilantly watching police activities with a skeptical eye, policy makers must take the necessary steps to prevent this type of police activity.

This chapter explores the various institutional mechanisms that are available to accomplish the goal of police accountability and police misconduct avoidance. The discussion begins by defining what is meant by excessive police force and attempting to outline the scope of the problem. There follows an examination of the limitations inherent in a number of proposed external controls (i.e., those that originate outside the department) that seek to achieve this goal. Next, the potential for controlling excessive force through internal means is explored. Finally, the chapter discusses the most current attempts to control police officers' excessive use of force.

POLICE USE OF EXCESSIVE FORCE

Society has given police officers the unique authority to use force in situations where it is necessary for self-defense, to lawfully arrest a suspect, or to prevent criminal activity. According to Walker (1994), "Force includes the authority to take someone's life (deadly force), the use of physical force, and the power to deprive people of their liberty through arrest" (p. 10). This authority is one of the most important factors shaping the police role and is the defining feature of the police (Bittner, 1970; Walker, 1999). *Excessive force* is defined as that which exceeds the minimum amount needed to achieve the objective. Applying this definition to specific instances of use of force is difficult and often a matter of opinion. As Flanagan and Vaughn (1993) noted, this term "denotes a continuum of activities and interactions rather than a specific behavior on the part of the police" (p. 6). At one end of the spectrum, *excessive force* can refer to the behavior of an officer who intentionally handcuffs a suspect in a manner meant to cause discomfort. At the other end, it can be as severe as the beating administered to Rodney King or the torture suffered by Abner Louima at the hands of the police.

There are no national data on the incidence of the use of excessive force by police officers. There is no national reporting system on this subject, no state maintains a complete database,

and there is not even a standard for collecting use-of-force data at the agency level (McEwen, 1996). Although a number of researchers have relied on citizen interviews (Campbell & Schuman, 1969) or observational research (Reiss, 1971) to assess the extent of this problem in various communities, most of these studies are quite dated. In addition, the limited information on police use of force comes from a small number of agencies (Alpert & MacDonald, 2001).

Nonetheless, the available information indicates that police use of excessive force is considered a statistically infrequent occurrence. The best recent estimate of police misuse of force is a recent study of contacts between police and the public in a Bureau of Justice Statistics national survey (Langan, Greenfeld, Smith, Durose, & Levin, 2001). These researchers found that during 1999 the police used or threatened to use force against nearly 1% of the nearly 44 million people reporting face-to-face police contact with the police. This represented a total of 422,000 people during the year. About three quarters (76%) of those experiencing force in 1999 said the force used or threatened by the police was excessive (p. 26). With respect to racial and ethnic differences, about 59% of those experiencing force were white, 22.6% black, and 15.5% Hispanic. When respondents were asked to describe the type of force used against them, 72% involved in a force incident said the police pushed or grabbed them, and another 15.3% said the police pointed a gun at them. The remainder of those experiencing force from a police officer reported the use of chemical spray (9.8%) or the threat to fire a gun (5.4%) (Langan et al., 2001, pp. 26–27).

Despite the lack of a national reporting system, there is good reason to believe that this is a problem that varies greatly from department to department. Reiss (1971) indicated that use-of-force incidents accumulate over time and that a sizable minority of citizens will experience police misconduct at one time or another. Previous studies have found large variations in the rate of police shootings between various jurisdictions (Sherman & Cohn, 1986) that cannot be explained as a result of differences in the crime rate or the arrest rate (Milton, Halleck, Lardner, & Abrecht, 1977). There is no reason to believe that the pattern would be different for reports of excessive force.

The circumstances surrounding incidents of police use of excessive force are another issue that has received attention in the literature. In general, researchers have found that use of excessive force is a less common form of police misconduct than other forms such as verbal abuse or use of illegal methods of gathering evidence (Kerner Commission, 1968; Reiss, 1971). Nonetheless, allegations of excessive force account for the largest proportion of complaints filed against police officers in most cities (Walker, 1999). In addition, studies indicate that targets of abuse are almost always lower-class males and that the most common factor associated with this practice is citizen disrespect for the police (Reiss, 1971). Finally, it should be noted that the overwhelming majority of encounters between officers and offenders do not involve the use of any, let alone excessive, force (Friedrich, 1980; Langan et al., 2001).

EXTERNAL CONTROLS

Several institutional mechanisms are designed to assist in curbing the problem of excessive police force. Because these originate from outside the police department, they are referred to as "external" methods of controls. Included in this category are such potential remedies as (a) the U.S. Supreme Court, (b) criminal prosecution, (c) civil litigation, (d) appealing for redress to municipal officials, and (e) citizen oversight mechanisms. In this section, each of these methods is examined, with a particular focus on the shortcomings of the various approaches.

The Supreme Court

In the 1960s, the U.S. Supreme Court handed down a number of important decisions that were designed to curb various forms of police misconduct. The most far-reaching rulings were the cases of *Mapp v. Ohio* (1961) and *Miranda v. Arizona* (1966). In *Mapp,* the court ruled that evidence seized illegally by the police could no longer be used at trial in any state criminal procedure. In effect, the exclusionary rule was applied to state courts as well as the federal

judiciary. The *Miranda* decision dealt with the issue of police interrogations. The Supreme Court held that statements by suspects would no longer be admissible in a court of law unless certain procedural requirements had been followed. Part of the rationale for these decisions was to eliminate police abuses of important constitutional rights.

These rulings by the U.S. Supreme Court, although extremely controversial, have little effect on the use of excessive force by the police. The greatest impact of the *Mapp* decision was on police practices with respect to search and seizure. *Miranda* was intended to change the manner in which police conducted interrogations of suspects. Other decisions dealt with such law enforcement practices as wiretaps, lineups, and the right to consult with counsel when in police custody. If the police violated any of these rights, the evidence could not be used against the defendant; thus, the suppression of evidence would create a penalty for police misconduct. Presumably, this would serve to deter officers from acting in a manner inconsistent with the Constitution.

There are several explanations for the extremely limited impact of these decisions on police use of excessive force. For one thing, many of these rulings relate to legal and constitutional issues that are peripheral to the question of unjustified police force (e.g., search and seizure). Second, these decisions apply only in situations where the police have made an arrest. Many encounters between citizens and law enforcement officers do not result in an arrest, and citizens who have not been arrested file a substantial number of complaints. Third, the Court does not maintain daily supervision of police officers and cannot ensure that individual police officers are complying with its decisions (McGowan, 1972). Fourth, Walker (1999) contended that decisions imposed by the Court lead to evasion and lying. Finally, for these constitutional provisions to serve as a deterrent to misconduct, the police must be interested in convicting the suspect. In many cases where "street justice" is administered, this objective takes a back seat to teaching the complainant a lesson about "respect" for the law.

Obviously, these U.S. Supreme Court decisions have had an impact on many aspects of police behavior. It would be implausible to argue that police practices with respect to search and seizure have not improved significantly in the last four decades as a result of the *Mapp* decision. There is also merit in the argument that these rulings have stimulated efforts to upgrade the quality of policing in the United States, including improved police training, recruitment, and supervision (Walker, 1991). However, it is incorrect to suggest that the oversight exerted by the U.S. Supreme Court has had a major impact on police use of excessive force, except perhaps in the area of interrogation practices.

Criminal Prosecution

Police officers who use unnecessary force against citizens may face criminal prosecution either under state laws that prohibit assaultive behavior or in federal court under the U.S. civil rights statute. Although such prosecutions are possible, the reality is that criminal charges are rarely filed in these situations. For a variety of reasons, prosecutors are reluctant to initiate a criminal case against a police officer who has used excessive force against a citizen.

For one thing, the prosecutor must maintain a close working relationship with the police department. Without the cooperation of the police, the prosecutor is unable to perform his or her job. Prosecutors are dependent on the police to undertake an adequate investigation of criminal activity. If the police become reluctant to assist in this effort, it becomes much more difficult for the prosecutor to obtain convictions. As a consequence, the prosecutor must be careful not to antagonize the police by becoming overzealous in his or her pursuit of police misconduct.

Another problem that prosecutors must confront is the fact that obtaining a conviction against a police officer is very difficult. Two cases that demonstrate this are the Rodney King case, where officers were acquitted in the state criminal trial, and a recent Cincinnati case involving the shooting of an unarmed black man by a police officer. Juries are extremely reluctant to convict police officers who have been charged with assaulting citizens.

A variety of factors make these cases difficult to win. For one thing, persons who are targets of

police violence are not sympathetic figures to many jurors. As a general rule, police officers do not abuse corporate executives or doctors. Mostly these victims are poor and/or members of minority groups. In many cases, the individual has been involved in criminal activity. On the other hand, jurors do empathize with police officers. To the average citizen who sits on a jury, the police officer is a person who is risking his or her life in the defense of life and property. Because jurors see police as the "good guys," they are quite reluctant to convict even in situations where an officer may have stepped over the line of what is lawful conduct.

Adding to the prosecutor's burden is the fact that from an evidentiary standpoint, these are very difficult cases to win. Most allegations of police misconduct are ambiguous fact situations. Generally it is the officer's word versus that of the complainant. In most cases, there are no impartial observers who have witnessed the encounter. If there are any observers, these are likely to be other police officers or friends of the complainant. As a consequence, there is no way for the jury to determine who is telling the truth. Not surprisingly, they often take the word of the police officer over that of the complainant.

Ambiguous fact situations are not the only evidentiary problem in these cases. Achieving a successful prosecution is also made more difficult by the "blue curtain" of silence that is common among police officers. This term refers to the extreme reluctance of law enforcement personnel to testify against fellow officers. The independent commission established in the aftermath of the Rodney King incident noted that this is perhaps the greatest barrier to an effective investigation of police misconduct directed against citizens (Christopher, 1991, p. 14). Unfortunately, the unwritten code of silence is not limited to Los Angeles. It is pervasive throughout most, if not all, police departments.

Goldstein (1977) has advanced five reasons that account for the pervasive nature of this phenomenon among police officers: (a) The police see themselves as members of a group aligned against common enemies, (b) police officers are generally dependent upon one another for help in difficult situations, (c) the police see themselves as quite vulnerable to false accusations by citizens, (d) police officers understand that there is quite a disparity between formal departmental policies and actual police practices on the street, and (e) police officers have no occupational mobility (pp. 165–166).

There are other difficulties for prosecutors as well when they bring criminal charges against police officers. To gain a conviction, it is not enough to show that the officer in fact did the act in question. They must also prove beyond a reasonable doubt that the officer acted with criminal intent. If the case has been filed under the federal civil rights statute, the government must prove that the officer intended to deny the citizen his or her constitutional or federally protected rights. This is not generally an easy task. Juries must be persuaded that the alleged behavior is so serious that the officer should not only lose his or her job but be subjected to criminal penalties as well. Because of the difficulty of proving criminal intent by police officers, it is sometimes found that officers will be charged with lesser crimes or forms of negligence. For instance, a white police officer involved in a fatal shooting of an unarmed black man in Cincinnati in April 2001 was charged with negligent homicide and obstruction of official business as he allegedly gave misleading statements to investigators ("Cincinnati Officer," 2001). Even though the charges were lessened from what the public outcry demanded, this police officer was subsequently acquitted of all misdemeanor charges.

Finally, the prosecutor must be concerned about appearing too overzealous in the battle against police misconduct. There is a great deal of public anxiety regarding violent crime in our society. As a general rule, most citizens are far less worried about police brutality than about street crime. As a consequence, a prosecutor who spends what is perceived as an inordinate amount of time prosecuting errant police officers is likely to open him- or herself up to the charge that he or she is handcuffing the police. A local prosecutor who alienates the rank and file may also be inviting police retaliation at the ballot box. Although federal prosecutors do not have this concern because they are appointed, they still face all the other hurdles in cases of this nature.

Given these factors, it is not surprising that criminal prosecutions of police officers for excessive use of force are relatively infrequent. This is not to say that they do not occur. Prosecutors do sometimes file charges, especially in situations that involve serious misconduct and where there is some likelihood of obtaining a conviction. However, the difficulties inherent in pursuing this course of action lead many prosecutors to believe that administrative sanctions by police agencies are a more effective remedy against police violence than criminal prosecution. In addition, Walker (1999) contended that "criminal prosecution itself appears to have limited deterrent effect in departments where other effective controls do not exist" (p. 278). Take for example, the New York City Police Department, where police officers were prosecuted and convicted in scandals of the 1970s and 1980s, yet serious criminal law violations were found in the 1990s by the Mollen Commission (Walker, 1999).

Civil Litigation

The increasing number of lawsuits filed against the police and the litigation costs are an increasing concern of both individual officers and of the agencies for which they work (del Carmen & Smith, 2001; Walker, 2001). For example, New York City and its police union have been found liable in the police assault and sodomizing of Abner Louima and have been charged with a settlement of $8.75 million. This is the largest civil settlement the city has ever paid in a brutality claim ("Brutality's Hefty Price," 2001). To a large extent, the concern over increasing civil lawsuits has resulted from court decisions that have made it much easier for citizens to file lawsuits against both individual officers and police agencies. Del Carmen (1987) reported not only that the number of lawsuits had increased in recent years but also that there was some evidence that the average size of jury awards had risen (p. 398).

Citizens who have been the victims of police misconduct may either file a tort action in state court or initiate a lawsuit in federal court under Title 42, Section 1983 of the U.S. Civil Rights Code. A tort is defined "as a civil wrong in which the action of one person causes injury to the person or property of another, in violation of a legal duty imposed by law" (del Carmen,

1987, p. 400). Officers who employ excessive force may be liable under these statutes for assault and battery. In situations where the action results in a fatality, a wrongful death suit may be filed.

Despite the availability of state tort remedies, the majority of cases against public officials are filed in federal court under Section 1983 (del Carmen, 1991, p. 409). There are a number of reasons for this phenomenon, including more liberal discovery rules in the federal courts and the fact that since 1976, courts have been able to award attorney's fees to successful plaintiffs (Kappeler, 1993, p. 36). In addition, the U.S. Supreme Court ruled in 1978 that local municipalities are not protected by sovereign immunity and may under certain situations be held liable for the actions of their employees (*Monell v. Department of Social Services,* 1978). One analysis of published federal district court decisions indicated that almost half the cases involved an allegation of either excessive force or assault and battery (Kappeler & Kappeler, 1992, p. 71).

The evidentiary standard in a civil case is much lower (i.e., preponderance of the evidence) than in a criminal prosecution, where the state or federal government must prove guilt beyond a reasonable doubt. Nonetheless, citizens who sue the police face a number of very serious hurdles. For one thing, as already noted, juries tend to be quite sympathetic toward police officers and are not favorably disposed toward many of the individuals who file claims, especially members of minority groups or persons with a prior criminal record (New York City Council, 1979). If the case comes down to the word of a citizen versus that of the officer, the jury will generally be inclined to believe the latter. Second, there are a number of defenses available to police officers, such as the claim that they were acting in "good faith." These tend to make it more difficult for citizens to prevail. Third, many police officers are what lawyers would term "judgment proof." In other words, they do not earn a large salary and are unlikely to have much in the way of resources that can be attached if the lawsuit is successful. As a consequence, any sizable judgment that is awarded may ultimately be uncollectible.

Persons who file lawsuits under Section 1983 are not limited to seeking compensation from the individual. These cases may result in damage awards against the municipality as well. The rationale for holding the jurisdiction liable is that cities and counties are in a far better position to reimburse plaintiffs than are individual officers. After all, local government can always raise taxes if necessary to pay the judgment. In addition, it has been asserted that holding the municipality liable will create a financial incentive for the agency to correct the problem. The hope is that taxpayers and insurance carriers will not be anxious to pay repeated claims that result from misconduct on the part of their employees.

Despite the advantages of a Section 1983 lawsuit, there are a number of serious obstacles. For one thing, the plaintiff must prove that a serious violation of a constitutional or federally protected right occurred. Minor acts of police misconduct such as "mere words, threats, a push, a shove, a temporary inconvenience, or a single punch in the face" (del Carmen, 1991, p. 412) may not be grounds for recovery. In addition, to hold the municipality liable, the plaintiff must demonstrate that the officer's action was the product of an official policy or organizational custom that was either created or condoned by a high-level municipal policy maker (Kappeler, 1993, p. 41). This requires the plaintiff to establish not only a pattern of constitutional violations but an awareness of these violations by top administrators within the agency (Kappeler, 1993, p. 43). From an evidentiary standard, this may be a difficult burden to meet.

Despite the increased volume of litigation, the data indicate that most of the lawsuits filed against public officials do not succeed and that a great majority of cases do not even get to the trial stage (del Carmen, 1987, p. 398). Although Kappeler and Kappeler (1992) suggested that plaintiffs may be prevailing in a greater number of cases, it is unclear what impact these judgments are having on law enforcement agencies. One example of impact is the creation of the Los Angeles Sheriff's Department Special Counsel to law enforcement officers for the purpose of reducing civil litigation costs. In fact, during the initial 3 years of the Special Counsel's investigations and recommendations, costs of judgments and settlements dropped by one half

(Kolts, 1999). One the other hand, the Independent Commission on the Los Angeles Police Department (Christopher, 1991) noted that discipline against officers was frequently light or nonexistent even in cases that involved a settlement or judgment in excess of $15,000. This was so even though "a majority of cases involved clear and often egregious officer misconduct resulting in serious injury or death to the victim" (p. 5). Finally, the New York City Comptroller has asserted that the police department does not even regularly monitor lawsuits alleging police misconduct (New York City Council Committee on Public Safety, 1998).

Redress From Municipal Officials

Municipal executives (i.e., mayors and city managers) and members of the city council are responsible for exercising leadership with respect to issues that have an impact on their community. Unfortunately, they have few mechanisms at their disposal to deal with the problem of controlling police use of excessive force. Several factors account for their inability to discipline rogue officers.

For one thing, a tradition has developed in the United States that politicians should not become involved in police matters. During the 19th century and the early 20th century, there was a great deal of collusion between corrupt public officials and the police. In many cities, the police protected the interests of entrenched political machines at the expense of the citizenry. One of the reforms enacted to deal with this situation was the establishment of mechanisms that insulated the police from political influence. As a result, it became taboo for local municipal officials to become involved in the internal affairs of the police department. Any politician who violates this norm is likely to be the recipient of considerable negative publicity.

Second, there is generally little motivation for the municipal executive or city councilperson to take action in situations involving questionable police behavior. Although various Los Angeles municipal officials spoke out quite vigorously in the aftermath of the Rodney King incident, this response is the exception. Most allegations of police brutality are not videotaped, and the alleged conduct is rarely so offensive. As a general rule, community sympathy does not lie with the complainant. Because most citizens are much more concerned about crime than about the occasional use of excessive force by the police, public officials have little to gain and much to lose by becoming involved. Politicians must be concerned about winning the next election; charges that they are picking on the police can easily become the "kiss of death."

There is a final reason why municipal officials do not take action against officers who may have used excessive force or abused the rights of citizens. Basically, it is that they have no power to do so even if they are so inclined. As a result of civil service reforms enacted during the reform era in American policing, municipal executives lost the right to hire and fire police officers. Personnel decisions must now be made in accordance with civil service provisions, and municipal officials are not part of the discipline process. The goal is to ensure that departments are protected from political pressures and that positions are not filled through patronage appointments. However, it also has meant that municipal officials cannot discipline or terminate officers who engage in brutality or other forms of misconduct.

Although municipal officials cannot take action against individual police officers, they can have an indirect impact on police policy. As a general rule, the top municipal official (e.g., the mayor or city manager) or his or her appointees select the police chief. The chief not only sets the tone for the department with respect to such matters as policy, supervision, and training but is often a significant player in the disciplinary process as well. The appointment process therefore gives municipal officials some leverage over the police department over the long term.

Citizen Oversight of the Police

Police use of excessive force became an emotionally charged issue during the turbulent decade of the 1960s. This was a period marked by widespread political protests, numerous urban riots, a dramatic increase in violent crime, and general social and political upheaval. Because part of the task of the police was to maintain order, frequent confrontations developed between law enforcement officers and those who sought to challenge the status quo. Often there were complaints from protestors that the police were not impartial and that they used the threat of excessive force in dealing with demonstrators. In addition, there were allegations from civil libertarians and civil rights groups that the police frequently engaged in brutality, especially against members of minority groups. To deal with these concerns, critics demanded the establishment of citizen oversight mechanisms. These mechanisms are "procedures for providing input into the complaint process by individuals who are not sworn officers" (Walker, 2001, p. 5).

Allegations of police brutality and other complaints filed against police officers generally are investigated within the department. However, it is often asserted that the agency either cannot or will not adequately discipline officers who abuse the rights of citizens. For these reasons, there have been frequent suggestions dating back to the 1960s that some form of review by persons outside the department be established to hear complaints. Although the citizen oversight mechanism would hear evidence and make recommendations, it would not have the power to discipline officers. Nonetheless, the police were initially quite adamant in their opposition to the creation of any form of citizen review.

There were many reasons why the police were vigorously opposed to this concept. For one thing, they greatly distrusted its proponents, who they feared would gain control of the board. Second, spokespersons for the police organizations and unions asserted that only individuals with a law enforcement background could properly evaluate the propriety of police actions in various situations. Citizens could not perform this function because they lacked the necessary expertise and training. Third, the police felt that they were being singled out for criticism. After all, there is no citizen body to hear complaints directed against teachers, social workers, or other municipal employees. Why should things be different when it comes to the police? Finally, it was argued that the citizen review mechanism would handcuff the police in the battle against violent crime. It was suggested

that officers might not be as aggressive if they had to worry that their actions would be second-guessed by civilians (Walker & Kreisel, 2001).

The opposition of the police was so intense during the decade of the 1960s that not only did very few cities implement citizen review boards, but the boards were also disbanded in several jurisdictions that had initially established citizen oversight mechanisms. The most acrimonious battle took place in New York City, where the issue was put to a public referendum in November 1966 at the behest of the Patrolmen's Benevolent Association. Despite the support of the mayor and other prominent politicians, it was defeated by a margin of two to one. Furthermore, New York City was not alone. Police opposition also led to the abolition of the citizen complaint review board in Philadelphia, the other large city that had implemented this panel (Terrill, 1991; West, 1988).

Although most of the early boards met with failure, recent years have seen a renewal of interest in this concept and the emergence of a second generation of citizen oversight mechanisms (Alpert & Dunham, 1992; West, 1991). According to a recent study, over 100 citizen oversight agencies exist, covering law enforcement agencies that serve nearly one third of the American population (Walker, 1998). These researchers report that procedures vary considerably among communities: Some have subpoena power, whereas others do not. Some conduct public hearings, whereas others do not. Some have the power to make recommendations about general police policies, whereas most do not (Walker & Kreisel, 2001).

Second-generational citizen oversight mechanisms have not encountered the intense level of opposition that earlier proposals faced. To a large extent, this is the result of a less turbulent social and political climate than existed in the 1960s. In addition, many of the recent complaint mechanisms have been more carefully introduced than was the case with earlier efforts (West, 1991). Finally, police opposition in some communities has been diminished by oversight procedures that apply not just to law enforcement but to other municipal employees as well.

It is ironic that despite the acrimonious debate that had raged over this issue, there is evidence from a number of cities that citizen oversight mechanisms are actually less likely to find police officers guilty of misconduct than internal investigators from within the department. In addition, citizen reviewers are generally more lenient in their disciplinary recommendations when they do find that officers have misbehaved (Kerstetter, 1985). This has been attributed to several factors, including problems that reviewers have in obtaining access to information, the substantial procedural safeguards that are provided for officers accused of misconduct, and the citizen investigators' poor understanding of both the police subculture and various police practices (Kerstetter, 1985, pp. 162–163).

Despite public demands for citizen oversight of law enforcement agencies, it is clear that this approach is not a remedy to the problem of excessive police force. Not only are such mechanisms less aggressive in disciplining officers, but they are generally reactive instruments. In other words, the concern in most cases is to determine whether punishment is justified for past misconduct and not to implement the types of policies and training that can prevent future abuses from occurring in the first place.

INTERNAL CONTROL

None of the external controls reviewed offer a completely satisfactory remedy to the question of excessive police force. Clearly, this is a problem whose solution does not lie outside the department. Instead, what is required is a firm commitment at the highest level of the organization that misconduct will not be tolerated. To achieve this goal, the chief must be determined to curb brutality. Because the chief sets the tone for the entire agency, it is imperative that clear signals be sent to both supervisory personnel and rank-and-file members that excessive force will be punished.

In addition, "police administrators can do much of a positive nature to achieve conformity with desired standards of conduct" (Goldstein, 1977, p. 167). Departmental standards regarding selection, training, supervision, and policy are all relevant to this goal. Each of these should be reviewed with an eye toward preventing needless incidents from ever occurring in the first place.

Selection practices determine who can become a police officer. Many departments not only undertake an extensive background investigation to weed out unsuitable candidates but use some form of psychological screening as well. If properly performed, these tests can eliminate persons who have traits that are undesirable in a law enforcement officer. It is estimated that between 2% and 5% of applicants are rejected due to an emotional or mental dysfunction (Benner, 1989, p. 80).

Recruitment is only part of the process. Not only must good candidates be selected, they must also be properly trained. Specifically, officers must be taught how to deal with incidents in a manner that defuses the potential for violence instead of escalating potential conflicts to the point that force is required. In addition, departments must enhance the communication skills of officers. This training should focus on the difference between words and actions that are likely to lessen the likelihood for violence and those that are likely to anger citizens.

Supervision is another important means of preventing physical abuse. If the administration is really determined to eliminate the use of excessive force, supervisors must be held accountable for the actions of their subordinates. As Goldstein (1977) has noted, an overly aggressive officer who constantly offends citizens should be viewed as a serious administrative problem, not as a valuable employee who sometimes gets into trouble. Sergeants must be informed in no uncertain terms that they will be subject to discipline if they fail to take action against officers under their command who use unnecessary force against citizens. This is critical because if supervisors tolerate this type of conduct, patrol officers are likely to believe that the department condones their actions. Blalock (1992) supported this position by referring to examples of departments that due to mismanagement by incompetent and/or uncaring superiors actually promoted the incidence of police misconduct and abuse.

The chief police executive must also pay close attention to departmental policies. These guidelines should explicitly state the circumstances under which officers are authorized to use force. In addition, it must be stressed that anything beyond the minimum amount necessary is a serious violation of departmental rules that will result in swift punishment. Previous studies have demonstrated that a restrictive departmental policy can reduce the number of police shootings (Fyfe, 1979; Sherman, 1983). There is no reason to believe that policy cannot be an equally effective tool in curbing other types of police violence.

Finally, there is the issue of discipline. If officers are not penalized for abusing the rights of citizens, improvements in other areas will make little difference. Therefore, it is imperative that agencies not only welcome citizen complaints but investigate these cases in a fair and impartial manner. To achieve this goal, the overwhelming majority of large police departments have established an internal affairs or similar unit (West, 1988). There is much reason to believe that if the commitment is present, investigators can be far more effective in dealing with allegations of excessive force than other proposed mechanisms such as citizen oversight.

Internal control is more likely to be successful for a number of reasons (Kerstetter, 1985). For one thing, personnel from within the department do not have to establish their credibility with the police. Second, being trained law enforcement investigators, these individuals are in a much better position to gather the necessary information required to resolve complaints. Third, both supervisors and line personnel are much more likely to cooperate in a police investigation than in one commanded by outsiders. In the latter situation, an "us versus them" mentality is likely to develop. Finally, departmental personnel are likely to have greater procedural latitude in conducting investigations than a citizen oversight board would possess.

Despite the evidence that internal control is more effective, the public is often skeptical that departments can be trusted to police themselves. Internal affairs investigators basically do not have the credibility that outsiders possess. For this reason, police administrators must work diligently to convince the community that the agency does not cover up wrongdoing and that abusive officers receive swift punishment. Nonetheless, it is inevitable that some citizens will mistrust the findings of any investigation conducted within the department. To overcome this problem, Kerstetter (1985) has suggested that the police be given primary responsibility for investigating misconduct but that an outside review procedure also be established. This external review would focus "on the adequacy and integrity of the police department's response to complaints" (Kerstetter, 1985, p. 181). Citizens who were dissatisfied with the internal investigation would thus have an avenue of redress.

The Most Recent Efforts to Control Police Misconduct

With police integrity being questioned and a recent surge to improve police-community relations, different and somewhat new avenues of police accountability are developing. Three different means of addressing police abuse have evolved in recent literature: (a) police civil liability specialists or associations, (b) early warning systems, and (c) consent decrees between police departments and the Justice Department.

Police Civil Liability Specialists

In light of the increasing civil litigation against police departments, special associations and/or counsels are being created to improve risk management practices and reduce civil litigation costs. In California alone, there is the newly created position of the Los Angeles Sheriff's Department Special Counsel. This individual's purpose is to investigate claims of officer misconduct, investigate police officers accused of misconduct, and examine departmental policies and procedures in order to reduce civil litigation costs for the department and county (Kolts, 1999).

Similar to the Los Angeles Sheriff's Department Special Counsel is the Police Civil Liability Association, formed by law enforcement agencies in southern California to evaluate, among other things, complaints of police officers' excessive use of force. This organization is primarily a means of communicating ideas and information in hopes of improving their liability and risk management practices. It is also composed of three committees that oversee areas with potential problems. One of these is the Executive Force Review Committee, which reviews the tactics applied in excessive force incidents to determine if policy was followed, what civil liability risks exist, and what future improvements can be made (Ceniceros, 1998).

Although these committees and special counsels may be a concerted effort on the parts of the involved police departments, it is not certain what authority, if any, they carry. The best that can be hoped for from these efforts is some avenue of valuable positive information sharing.

Early Warning Systems

The literature indicates that a relatively small number of police officers ("rotten apples") generate a majority of citizen complaints (Christopher, 1991; Goldstein, 1977). Thus, early warning (EW) systems are a relatively new management strategy for identifying those "problem" officers (Walker, 2001). The EW system is a proactive approach to identifying police officers who have potential problems, and it is a tool for curbing police use of excessive force and achieving accountability. "EW systems are data-driven programs whose purpose is to identify police officers whose behavior is problematic and to subject those officers to some kind of intervention, often in the form of counseling or training" (Alpert & Walker, 2000, p. 59). Behavior-problematic officers are identified by an abnormally high number of citizen complaints or "other indicators of problematic performance" (Walker, 2001, p. 110).

The concept of EW systems was spawned by a recommendation by the U.S. Commission on Civil Rights (1981) to be more proactive at identifying problematic police officers. After that endorsement, by 1999, an estimated 39% of all municipal and county law enforcement agencies serving populations greater than 50,000 people either had an EW system in place or were planning to implement one (Walker, Alpert, & Kenney, 2001a, p. 1).

The majority of EW systems are still in their infancy, and their effectiveness is yet to be determined. However, there is early evidence that they appear to reduce problem behaviors significantly. In addition, EW systems encourage changes in the behavior of supervisors who

are responsible and accountable for monitoring problematic officers as well as serving as exemplary mentors (Walker, Alpert, & Kenney, 2001b).

Despite the positive findings from the various forms of EW systems, there are some limitations. The EW systems are expensive, high-maintenance operations that use a significant amount of administrative resources. Supervisors must routinely take the time and effort to evaluate each officer under their command as well as to follow up on problematic officers. Relevant to the limitation of expense is the cost of counseling or training for officers deemed "problematic." In addition, there are the probable costs of replacing that officer while he or she is in counseling or training. Walker (2001) also suggested that for the EW system to be effective, it should be viewed as one part of a system of accountability, not a panacea.

Consent Decrees

A relatively new reform tool in the form of a consent decree is being used across the country in a number of cities where patterns and practices of police misconduct have been found. City officials are signing consent decrees with the Justice Department in the face of federal civil rights litigation, and these are being touted as achieving notable success ("Bullied Cities," 2000). The consent decrees make demands of thorough investigations of police departments as well as recommendations for change. If the requirements of the consent decree are not met, the department and city face the threat of federal litigation. Los Angeles city officials recently signed a consent to investigate the Rampart scandal, which involved the city's elite antigang unit. In addition, the police department was to uncover and treat the root cause of the Los Angeles Police Department's history of corruption.

The consequence of not following through with the consent decree recommendations and orders is civil rights litigation. The problem with this ramification is that civil litigation has not always been successful for some departments. Police departments like New York City and Los Angeles have paid millions of dollars in liability cases, yet they continue to be plagued with officers accused of misconduct and abuse.

Conclusion

As long as police have the authority to use force, citizen complaints of excessive force will be inevitable. Therefore, it is crucial that these claims be investigated and evaluated in their proper context (Alpert & Smith, 1994). This chapter examined the issue of excessive police force. Various external methods of controlling this problem were examined. Each of these was found to be lacking in important ways that diminish its effectiveness. Internal control of unjustified police force was also explored. It was noted that departments have the capacity to ensure officer compliance with legal requirements if the administration is inclined to exercise this authority. Finally, the most recent innovations in controlling officer use of excessive force were explored, with some optimistic expectations of success from EW systems.

References

Alpert, G. P., & Dunham, R. C. (1992). *Policing urban America* (2nd ed.). Prospect Heights, IL: Waveland.

Alpert, G. P., & MacDonald, J. M. (2001). Police use of force: An analysis of organizational characteristics. *Justice Quarterly, 18,* 393–409.

Alpert, G. P., & Smith, W. C. (1994). How reasonable is the reasonable man? Police and excessive force. *Journal of Criminal Law and Criminology, 85*(2), 481.

Alpert, G. P., & Walker, S. (2000). Police accountability and early warning systems: Developing policies and programs. *Justice Research and Policy, 2*(2), 5–72.

Benner, A. W. (1989). Psychological screening of police applicants. In R. C. Dunham & G. P. Alpert (Eds.), *Criminal issues in policing: Contemporary readings* (pp. 72–86). Prospect Heights, IL: Waveland.

Bittner, E. (1970). *The functions of the police in modern society.* Washington, DC: National Institute of Mental Health.

Blalock, J. (1992). Mismanagement and corruption. *Police Studies, 15*(4), 184–187.

Brutality's hefty price. (2001). *U.S. News and World Report, 131*(3), 28.

Bullied cities finally gain tool to weed out bad cops. (2000, October 11). *USA Today,* p. 26A.

Campbell, A., & Schuman, H. (1969). *Racial attitudes in fifteen American cities* (Supplemental Study for the National Advisory Commission on Civil Disorders). Washington, DC: Government Printing Office.

Ceniceros, R. (1998). Police agencies look to peers for risk management advice. *Business Insurance, 32*(32), 16.

Christopher, W. (1991). *Report of the Independent Commission on the Los Angeles Police Department: Summary.* Unpublished paper.

Cincinnati officer faces minor charges in shooting; Ashcroft announces inquiry into police department. (2001, May 8). *St. Louis Post-Dispatch,* A1.

del Carmen, R. V. (1987). *Criminal procedure for law enforcement personnel.* Monterey, CA: Brooks/Cole.

del Carmen, R. V. (1991). Civil and criminal liabilities of police officers. In T. Barker & D. L. Carter (Eds.), *Police deviance* (2nd ed., pp. 405–426). Cincinnati: Anderson.

del Carmen, R. V., & Smith, M. R. (2001). Police, civil liability, and the law. In R. G. Dunham & G. P. Alpert (Eds.), *Critical issues in policing: Contemporary readings* (4th ed., pp. 181–198). Prospect Heights, IL: Waveland.

Flanagan, T. J., & Vaughn, M. S. (1993). Public opinion about police abuse of force. In W. A. Geller & H. Toch (Eds.), *Police use of excessive force and its controls: Key issues facing the nation.* Washington DC: Police Executive Research Forum/National Institute of Justice.

Friedrich, R. J. (1980). Police use of force: Individuals, situations and organizations. *Annals of the American Academy of Political and Social Science, 452,* 82–97.

Fyfe, J. J. (1979). Administrative interventions on police shooting discretion: An empirical examination. *Journal of Criminal Justice, 7,* 309–323.

Goldstein, H. (1977). *Policing a free society.* Cambridge, MA: Ballinger.

Hunter, R. D. (1999). Officer opinions on police misconduct. *Journal of Contemporary Criminal Justice 15*(2), 155–170.

Kappeler, S. F., & Kappeler, V. E. (1992). A research note on Section 1983 claims against the police: Cases before the federal district courts in 1990. *American Journal of Police, 11,* 65–73.

Kappeler, V. E. (1993). *Critical issues in police civil liability.* Prospect Heights, IL: Waveland.

Kerner Commission. (1968). *Report on the National Commission on Civil Disorders.* New York: Bantam.

Kerstetter, W. A. (1985). Who disciplines the police? Who should? In W. A. Geller (Ed.), *Police leadership in America* (pp. 149–182). New York: Praeger.

Kolts, J. G. (1999). *Special Counsel, 11th Semiannual Report.* Los Angeles: Los Angeles Sheriff's Department.

Langan, P. A., Greenfeld, L. A., Smith, S. K., Durose, M. R., & Levin, D. J. (2001). *Contacts between police and public: Findings from the 1999 National Survey.* Washington, DC: U.S. Department of Justice.

Los Angeles Police Department. (2000). *Board of Inquiry report into the Rampart area corruption incident.* Retrieved April 18, 2002, from LAPD Web site: www.lapdonline. org/whats_new/boi/boi_ report. htm.

Mapp v. Ohio, 367 U.S. 643 (1961).

McEwen, T. (1996). *National data collection of police use of force.* Alexandria, VA: Institute for Law and Justice.

McGowan, Carl. (1972, March). Rulemaking and the police. *Michigan Law Review, 70,* 659–694.

Milton, C. H., Halleck, J. W., Lardner, J., & Abrecht, G. L. (1977). *Police use of deadly force.* Washington, DC: Police Foundation.

Miranda v. Arizona, 384 U.S. 436 (1966).

Monell v. Department of Social Services, 436 U.S. 658 (1978).

New York City Council. (1979). Suing the police in federal court. *Yale Law Journal, 88,* 781–784.

New York City Council Committee on Public Safety. (1998). *Beyond community relations: Addressing police brutality directly.* New York: Author.

Reiss, A. J. (1971). *The police and the public.* New Haven, CT: Yale University Press.

Sherman, L. W. (1983). Reducing police gun use: Critical events, administrative policy and organizational change. In M. Punch (Ed.), *Control in the police organization* (pp. 88–125). Cambridge: MIT Press.

Sherman, L. W., & Cohn, E. G. (1986). *Citizens killed by big city police, 1970–84.* Washington, DC: Criminal Control Institute.

Terrill, R. J. (1991). Civilian oversight of the police complaints process in the United States: Concerns, developments and more concerns. In A. J. Goldsmith (Ed.), *Complaints against the police: The trend toward external review.* Oxford, UK: Clarendon Press.

U.S. Commission on Civil Rights. (1981). *Who is guarding the guardians?* Washington, DC: Author.

Walker, S. (1991). Historical roots of the legal control of police behavior. In D. Weisburd & C. Uchida (Eds.), *Police innovation and the rule of law* (pp. 32–55). New York: Springer.

Walker, S. (1994). *Taming the system: The control of discretion in criminal justice, 1950–1990.* New York: Oxford University Press.

Walker, S. (1998). *Citizen review of the police: 1998 update.* Omaha: University of Nebraska Press.

Walker, S. (1999). *The police in America: Introduction* (3rd ed.). Boston: McGraw-Hill College

Walker, S. (2001). *Police accountability: The role of citizen oversight.* Belmont, CA: Wadsworth.

Walker, S., Alpert, G. P., & Kenney, D. J. (2001a). *Early warning systems: Responding to the problem police officer.* Washington, DC: National Institute of Justice.

Walker, S., Alpert, G. P., & Kenney, D. J. (2001b). Early warning systems for police: Concept, history and issues. In R. G. Dunham & G. P. Alpert (Eds.), *Critical issues in policing: Contemporary readings* (4th ed., pp. 199–215). Prospect Heights, IL: Waveland.

Walker, S., & Kreisel, B. W. (2001). Varieties of citizen review: The relationship of mission, structure and procedures to police accountability. In R. G. Dunham & G. P. Alpert (Eds.), *Critical issues in policing: Contemporary readings* (4th ed., pp. 338–355). Prospect Heights, IL: Waveland.

West, P. (1988). *Police complaints procedures in the USA and in England and Wales: Historical and contemporary issues.* Unpublished MA thesis, Michigan State University, East Lansing.

West, P. (1991). Investigation and review of complaints against police officers: An overview of issues and philosophies. In T. Barker & D. L. Carter (Eds.), *Police deviance.* Cincinnati, OH: Anderson.

11

THE JUSTICE RESPONSE TO WOMAN BATTERING

The Evolution of Change

LISA A. FRISCH

Domestic violence and violence against women are universal problems of epidemic proportions. Peace begins at home, but millions of women do not have peace in their homes or in their lives. Domestic violence, rape, bride burning, genital mutilation, sex and slave trafficking, widow "suicides," the abortion of female fetuses, honor killings, the enslavement of women due to extremist religious beliefs, and rape and impregnation as an instrument of war. These are some of the horrors that women and girls are suffering around the world. While a husband in California assaults his pregnant wife with a gun, a husband in New Delhi burns his pregnant wife. While a father in Ohio rapes his daughter, incest is being committed to a 12-year-old girl in Peru. While a group of fraternity men in Vermont commits gang rape, a battalion of soldiers in Africa assaults a defenseless woman. We can no longer view this violence as a series of isolated incidents. . . . Our work needs to begin with the recognition that violence against women in all its forms is related, and driven by attitudes and cultural norms that devalue women's lives and deny women autonomy, dignity and equality. The same attitudes underlie all manifestations of violence against women. If we do not change these attitudes, we will be destined to live in a world in which the violence continues—a world in which no woman or girl is truly safe.

—Leni Marin, Managing Director, Family Violence Prevention Fund,
speaking at the Women's Funding Network 18th Annual Conference, April 26, 2002.

Author's Note: This chapter represents my own views and should not be construed as representing the opinions or positions of the New York State Office for the Prevention of Domestic Violence.

As we express our concern for oppressed women in Afghanistan and other parts of the world, we cannot forget the women who are beaten, murdered, and living in fear at the hands of their lovers and husbands in the United States. Today, after decades of struggle for equal protection for victims of intimate partner violence, we remain only midway through the process of the criminalization of domestic violence.[1] "Mandatory arrest" laws and policies are now the rule rather than the exception, and the federal Violence Against Women Act of 1994 (VAWA) changed the nature of funding and federal law in this area. Much has changed in the criminal justice and community responses to domestic violence, and much has stayed the same. Change is an elusive concept, particularly social change, and the slow and somewhat fragile changes in response to the battering of women are no exception. This chapter focuses on these change efforts and some of the critical issues in woman battering facing us today as we continue with our sometimes plodding efforts toward the institutionalization of the criminal justice response to this crime.

FITTING A SQUARE PEG INTO A ROUND HOLE: DOES EQUAL PROTECTION MEAN EQUAL TREATMENT?

The terms *domestic violence, woman battering, intimate partner violence, wife beating,* and *wife abuse* are often used interchangeably and are generally publicly recognized as violence between those engaged in an intimate relationship. The intimate nature of the relationship, however, is not the only distinction between domestic violence and stranger assault. Domestic violence, unlike stranger crime, almost always happens multiple times against the same victim. It is a crime that almost invites blaming the victim, at least in part, for the abuse by focusing on victim-offender dynamics and issues within the relationship. Conversely, justification for abuse is still all too common, inappropriately deflecting responsibility onto stress, loss of control, poor anger management, substance abuse, or unemployment to mitigate the abuse perpetrated by men. Domestic violence as a crime is perhaps most distinctive in that it goes well beyond the physical assault. This abuse embodies a continuum of coercive behaviors that include physical, sexual, psychological, emotional, and economic abuse that tend to escalate in frequency and severity over time (New York State Office for the Prevention of Domestic Violence [NYS OPDV], 1998). A range of tactics such as threats, physical beatings, isolation, and the manipulation of fear and other emotions are used to establish a pattern of behavior that maintains power and control over one or more persons (Pence & Lizdas, 1998). This imbalance of power is fixed and rigidly maintained, and the abuse often remains unidentified, owing to the private confines of the close, interpersonal relationships in which it takes place. Unlike the traditional understanding of stranger crime, *battering* therefore goes beyond physical beatings to include a pattern of coercive behaviors that effectively maintain the fixed imbalance of power over the victim. Because this pattern transcends the specific incident that may have brought police to the home, or the victim to court to request protection, the impacts and effects on victims are often underestimated and misunderstood (see Stark, 2002). If the pattern is not considered, the incident often seems of less consequence or seems "explained" by external factors.

A Critical Distinction: Battering Is Not a Gender-Neutral Crime

Unlike most other crimes except rape, domestic violence is not a gender-neutral crime. Although anyone can be a victim—heterosexual men and gay, lesbian, bisexual, and transgendered individuals—the most common victims by far are female victims of their male partners. This further complicates our responses to both victim and perpetrator. According to the National Institute of Justice (NIJ) and Centers for Disease Control (CDC) 1998 report *Prevalence, Incidence and Consequences of Violence Against Women: Findings From the National Violence Against Women Survey:*[2]

- Using a definition of physical assault that included a range of behaviors from slapping and hitting to using a gun, the survey found that 52% of women said they had been physically assaulted in their lifetimes. It is therefore estimated that 1.9 million women are physically assaulted annually in the United States.

- Women experience significantly more partner violence than men do: 25% of surveyed women, compared with 8% of surveyed men, said they had been raped or physically abused by a current or former spouse, live-in partner, or date in their lifetime.

- The more serious the assault, the wider the gap between women's and men's rates of physical assault. For example, women were 2 to 3 times more likely than men to report that an intimate partner had thrown something at them or pushed, grabbed, or shoved them. However, they were 7 to 14 times more likely to report that an intimate partner had beaten them up, choked or tried to drown them, threatened them with a gun, or actually used a gun on them. Women were also far more likely to be injured by their intimate partners.

- The vast majority of violence against women is partner violence: 76% of the women who had been raped or assaulted since age 18 had been harmed by a current or former intimate partner. Therefore, strategies to prevent violence against women should focus on ways of protecting women from risks by intimates.

- Intimate partner violence is primarily perpetrated by men. The survey found that 93% of women and 86% of the men who had been raped and/or physically assaulted since the age of 18 had been victimized by a male. Given these findings, the report recommended that adult violence prevention strategies should focus primarily on the risks posed by male perpetrators.

Victim Blaming Versus Victim Safety

Intimate violence against women is a social problem the awareness of which has evolved slowly and largely due to the efforts of women themselves. Lack of understanding of the dynamics of woman battering and a strong social tolerance for the behavior have been major obstacles to recognition of the problem, which is rooted in long-standing historical traditions that have permitted violence against women as a method of discipline and control. These traditions have been incredibly resistant to change over the centuries, but in recent years, social attitudes have slowly changed and have become less openly tolerant of abuse.

Early in my career, as a young, idealistic probation officer, I had little understanding of the context of abuse against women. I assumed that I could step in and somehow save the women I was dealing with, make decisions for them, and automatically stop the abuse. I often focused on the individual woman and her behavior in hopes of helping her somehow "avoid" being a victim. This was in the early 1980s, when the laws, as well as public attitude and awareness, were consistent with "walking the guy around the block" and using mediation and couples counseling to somehow solve "their" problem. Domestic violence was not considered a crime unless the harm was so great as to result in death or serious injury, and even then, justification and excuses for the behavior abounded.

Battered women continue to be "between a rock and a hard place." If they stay with their abuser, they are seen as weak or complicit in the abuse. If they attempt to end the relationship, especially if they have children, they are blamed for not keeping the family together or for provoking the attack and are forced to continue dealing with their tormentor through complex visitation and custody arrangements. There is no real *getting away* when children are involved.

Today, blaming women for men's violence and for their own victimization still happens, albeit less openly. There is particularly little sympathy for a woman who returns to her abuser. In a February 12, 2002, *News Flash* by the Family Violence Prevention Fund, there was a report of a judge in Kentucky who held two women in contempt and fined them for returning to their abusive partners after emergency stay-away protective orders had been issued.

"In my experience on the bench, I have found that there has been a number of petitioners who have chosen to come and get an order, and then ignore the order," said Judge Thornton at Hull's hearing "I think that both parties are obligated to follow through with the order. You can't have it both ways" ("Sometimes, Abuse Victim Goes to Jail," 2002, p. A1). At both women's hearings, Judge Thornton said she was offended by women who ask the court for protective orders and then invalidate the orders by contacting their abusers. "When these orders are entered, you don't just do whatever you damn well please and ignore them," said the judge during Harrison's hearing. "They are orders of the court. People are ordered to follow them, and I don't care which side you're on" (p. A1).

One thing that has changed is the public response to an openly articulated poor response. The judge in Kentucky was featured on national news stories and sparked debate between battered women's advocates who expressed grave concerns about the rulings and those, often within the justice system, who were frustrated by trying to deal with domestic violence cases as though they were the same as stranger crime situations. The fact is they are not the same. But rather than viewing this abuse as somehow less serious than other crimes, or as a "relationship" problem best solved by counseling rather than consequences, we should instead consider domestic violence as even *more* serious than most stranger crime. In abuse cases, we know who the next victim is and we can virtually predict that it will happen again. The abuser nearly always knows where his victim is and is keenly aware of her habits, her friends, her workplace, her family . . . and her fears.

INSTITUTIONALIZATION: THE COSTS AND BENEFITS

The way we have framed the successful "institutionalization" of the criminal justice response to domestic violence, a.k.a. *criminalization,* is by trying desperately to fit abuse against intimates into a traditional, adversarial model of law. Considering that our system of justice is not designed to protect individual victims but rather to protect the community, that premise alone fails to truly assist victims of intimate partner violence. The law was not originally constructed to protect women from their husbands and boyfriends. Instead, laws are framed to protect citizens from violations by strangers who are unlikely to ever see each other again. Even when laws are crafted to specifically respond to domestic violence, they are forced to fit within that traditional framework. To use a cliché, thinking "out of the box" was never encouraged in creating or enforcing the law.

Consider the experience of the so-called "typical" victim of abuse. The year is 2002. Susan is a 25-year-old woman who has been married for 4 years to a man she met in college. Since graduation, her husband Jim moved them far from Susan's family. He refused to allow her to work in her field of study and advance her career but insisted she stay home and start a family. Susan now has two children, ages 3 and 1. The physical abuse and fear began almost immediately upon their marriage, and in hindsight Susan can see how Jim subtly controlled her even during their courtship. It had been getting worse. Jim's threats were getting more vivid and more frequent. Only the month before, he had violently twisted her arm behind her back while screaming threats at her. The emergency room allowed the doting Jim to stay with Susan at the hospital, and domestic violence was never assessed. When Jim later strangled her until she passed out after accusing her of flirting with a neighbor, Susan called the police for the first time. Their 3-year-old was in the room with his terrified mother when the police arrived. Jim had left Susan passed out on the floor and had gone upstairs to bed. When the police went upstairs to speak with Jim, it was clear that he had been drinking, though he was not drunk. The police did end up arresting Jim, consistent with state law, but because the injury was not apparent, they filed a harassment charge that clearly did not reflect the seriousness of Susan's trauma or potential injury. They advised Susan of her right to go to a shelter and go to court for assistance, but because the police had no lockup at the station and could not get a judge to arraign him after hours, Jim was released on an appearance ticket, and he arrived back home before Susan even had time to consider her options. Jim alternately threatened Susan to drop the charges and pleaded with her to give him another chance, telling her that a criminal charge could affect his job status and thus harm the whole family. If he lost his job, the sole financial support for the family, Jim said, it would be "her own damned fault."

What exactly has the institutionalization of domestic violence done for Susan? She is still in grave danger, perhaps even more so since she called the police. Her fear, anger, confusion, and risk did not end with the arrest but are perhaps only just beginning once she enters the justice system. The case will be plea-bargained to an even lower charge or dismissed entirely. If the prosecutor does want to follow through with the case, Susan will often be compelled to testify against Jim, regardless of her terror and ambivalence. In some jurisdictions, if Susan refuses to testify, even if she fears it will put her at risk, *she* may be the one held in contempt by the court. If the court does take some action in the case, Jim is likely to be referred to an alcohol treatment program or a "batterers' intervention program," on the false assumption that either can and will stop his violence (Zubretsky, 2002; Zubretsky & Digirolamo, 1996). If he fails to follow through with either, little will happen to him as a result. Probation is rarely used as a way to provide ongoing monitoring or consequences for abusers, although it is the appropriate place in the criminal justice system to attempt to meet those goals (NYS OPDV & New York State Division of Criminal Justice Services [NYS DCJS], 2001). Child protective systems may be called in to intervene with the children, who are presumed to be at risk for having witnessed the violence. Should Susan fail to follow through on the criminal charges or "enforce" an order of protection, ironically, the child protection system may threaten *her* with charges of failure to protect and the loss of her children. Jim will have already filed for visitation and, probably, for custody of his children. Furthermore, Jim is almost certain to receive little or no sentence for his actions, with no more than a suspended sentence and perhaps an order of protection placed against him. As you consider Susan's experience, think about how this case would be treated if a strange man, instead of her husband, strangled her into unconsciousness.

Does this mean we should not provide equal protection to abused women and make arrests when crimes are committed? Some say yes. However, after nearly two decades of work on state policy and training on domestic violence, aimed at promoting a more effective response, it is clear that we must take action that changes the culture of tolerance for abuse against women. Laws are part of that change—clearly communicating that domestic violence is a crime that will be taken seriously. We do need to follow through with that promise, which in too many cases is an empty one. As Ellen Pence (1989) of the Duluth-based Domestic Violence Intervention Project has noted, we can't continue to attempt to make sweeping changes in our systemic responses while ignoring the safety of individual battered women. That is too high a price. To achieve both goals—systemic change and safety for victims—we also need to consider how we can adjust our traditional responses to meet the special needs of survivors of abuse so that our responses are no longer felt to be as punitive as the battering itself. We must recognize that the victim's safety is truly being compromised by her venture into the system or simply by trying to get away. Victim safety must now be our primary systems goal, although it is one that is virtually impossible to measure or guarantee. We must consider what must be done to truly promote safety for the victim, while ensuring that the abuser is given the message that this behavior will not be tolerated. We can no longer try to force victims' complex situations into our traditional paradigm of equal treatment. The cost of theoretical institutionalization is simply too high.

How Do We Really Effect Change?

Understanding the linkages and correlations between change efforts and their origins is crucial to promoting and maintaining this change. This awareness will also help us in moving forward in our efforts to achieve our collective goals of supporting and prioritizing safety for victims (i.e., let's at least make sure we don't make their situation worse) and promoting ongoing and consistent accountability for men who batter. It is clear that the significant changes in practice, policy, and law that have occurred are largely the result of the *interaction* of the various factors influencing change, not any one area. For example, the various external forces that worked together to inspire change in response to domestic violence must be examined. These include the early grassroots education and reform efforts of battered women's advocates (Pleck, 1987; Schechter, 1982);

successful lawsuits against law enforcement that claimed a violation of equal protection of the law and failure to protect victims of abuse; official responses, such as the formation of various commissions and task forces such as the U.S. Attorney General's Task Force on Family Violence (1984) and the offering of broad-based policy recommendations[3] for changes in practice; legislative reforms, such as mandatory arrest, designed to promote a consistent response by police who traditionally failed to arrest in domestic cases; and academic research conducted to determine the effectiveness of various interventions (Frisch & Caruso, 1996).

Police, unlike the other actors in the criminal justice system, are most vulnerable to external pressures for criminalization—changes in the law, research, and, most important, civil liability resulting in policies on arrest. Policies have the most opportunity to have impact when there is maximum participation in their formulation, implementation, and evaluation stages and when those who would be affected by the planned change were involved at some level in the change process. Swanson, Territo, and Taylor (1988, pp. 545–546) argued that successful organizational change generally includes a commitment by leadership to make the change, a thorough reexamination of past practices by all levels of the department, input from all levels into new practices and procedures, and consistent efforts to promote the change as a critical need for the organization until it is permanently absorbed into the organization's structure and culture. This can help alleviate the problem of overreliance on individual change agents within the systems, for although these people are critical to inspiring and supporting change, when they retire or are promoted, the change efforts often disappear with them.

It is important to recognize that it is not the actors within the systems that are necessarily the obstacle to change. Many want to do the best they can to help protect victims and hold offenders accountable. It is the *infrastructure* of the system—particularly the criminal justice system—that is unyielding. Consider the conundrum of the prosecutor who desperately wants to convict a violent batterer but, because of the nature of the laws and justice system, needs the victim to testify in order to do so. How does this prosecutor balance justice and safety? How do we know what will promote that safety? How do we weigh the victim's statement "If you prosecute, he'll kill me" against the prosecutor's belief, "If I *don't* prosecute, he'll kill her"? We are beginning to recognize that domestic violence cases do not fit into our overall adversarial justice system, but we are still struggling with what to do about it. The efforts toward "equal protection" have unfortunately led to us shoehorn abuse cases into this traditional response, often at the expense of victims. Our moves toward developing special units in police departments and prosecutors offices and establishing dedicated domestic violence courts are an important step, but we have been trying to make changes in response while maintaining that immovable infrastructure, without thinking creatively, from the bottom up, how to do things differently. Equal treatment doesn't need to mean *identical* treatment.

Critical Issues in Assessment and Evaluation

As more and more police agencies developed arrest policies, particularly in response to changes in law requiring them to do so, it became apparent that the definition of policy "success" was inconsistent and badly defined. Many of those police departments that had a written policy often automatically assumed "success," even when battered women's advocates confronted them on their individual officers' continued poor response to victims. Because the notion of success has been so unclear in this area, appropriately evaluating this change requires the development of assessment criteria to judge the effectiveness of policy outcomes as well as the overall community response.

Assessment strategies tend to be either quantitative or qualitative and generally try to measure changes in knowledge, attitude, or performance. Quantitative assessments of domestic violence policies may include the number of reports to police, the percentage of arrests made in domestic cases, the number of referrals made by police for victim services, and the increase in the number of training hours devoted to the problem. Evaluation plans that are purely quantitative are often narrow in focus and are conducted in isolation from the interorganizational field: for example, the effects on battered women. In contrast, qualitative assessments take a more holistic approach, assessing the policy change in relation to a variety of factors, such as attitude change and increased knowledge through training and supervision. In addition, this approach examines the impact of the policy on the other actors in the criminal justice system and other related agencies, as well as the victims' perception of increased safety and system access. The most effective evaluation strategy contains both quantitative and qualitative measures that focus on the assessment of the internal policy implementation program and on the external, community-based implementation process. The ongoing evaluation phase must include objective assessments by victim advocates and battered women in addition to others in the community, such as the prosecutor or judge. Assessment must be built into the program from its inception by including those affected by the change in defining "success," again creating active agents from passive targets. This participation helps reduce the alienation often felt by those who fear that evaluation can detect only the negative (Toch, Grant, & Galvin, 1975, p. 3). An effective evaluation design must look at the positive factors as well as the problems, and all members of the organization should be made aware of these varied goals. (See Fagan, 1996, for another perspective on research and evaluation in domestic violence.)

The trend is toward better evaluation of our activities, but there is a need to thoughtfully define our objectives so we know what to evaluate and how. Traditional cost-benefit analyses do not fit neatly when superimposed onto domestic violence responses. For example, lowering the rate of domestic violence may be an ideal goal, but in reality, increased arrests may show improved practice. On the other hand, significant increases in arrests of women should be viewed as a grave problem in instituting mandatory arrest laws and policies. Thus, our efforts to evaluate our responses to domestic violence need to be uniquely considered within the context of battering. To that end, the use of an innovative evaluation tool, the Safety and

Accountability Audit, has been replicated in a number of sites around the country, including New York State. It was developed by Ellen Pence as a doctoral dissertation. The Safety Audit, unlike its fear-inspiring name, which sounds like the IRS at the door, is a tool that operationalizes the philosophy that internal and external evaluation is necessary to implement and maintain positive change. It also brings in diverse partners to act as an interdisciplinary team to assess whether formal policies and forms serve to support the goal of promoting victim safety (Pence & Lizdas, 1998). These types of innovative evaluation strategies should be more fully explored as alternatives to more traditional, empirical studies, which often do not offer the necessary context and understanding of the dynamics of domestic violence. Many of the studies on the efficacy of arrest are good examples of this lack of context. Consider the problem of the changes that were initiated by Sherman's and Berk's initial research on arrest in domestic violence cases (Sherman & Berk, 1994). Although changes in police response and a trend, slow that it was, toward criminalization were in process before Sherman and Berk's research blanketed the country, the timing of research that seemed to support the push of advocates toward arrest gave it even more power to influence change. Changes in law enforcement were further prompted by research on the effectiveness of various techniques for handling domestic violence calls.

Research can both prematurely promote and dismantle change efforts. For example, one of the five NIJ-funded replication studies of the original Minneapolis Police Experiment, based in Milwaukee (Sherman et al., 1992), determined that unemployed, high-risk batterers were more likely to reoffend after an arrest than those with more to lose. The researchers therefore reasoned that arrest strategies must somehow be tailored to fit the offender, suggesting that some batterers would again be violent as a "result" of the arrest and that perhaps "punishment should be made less severe in order to reduce an escalation effect" (Sherman et al., 1991, pp. 158–159). When these batterers assaulted their victims after having been arrested, the researchers presumed that the assault was the direct result of the arrest, rather than the choice and responsibility of the offender. This analysis ignores the reality that batterers tend to increase their control and abuse over time, with violence becoming more severe and more frequent over time. Further, attempting to leave or otherwise hold the batterer accountable is also correlated with escalated abuse and homicide. It is no surprise, then, that arrest, particularly with little other consequence, fails to deter and may even exacerbate the escalation of these obviously high-risk offenders. Thus, battering will naturally escalate, and these studies seem to show simply that arrest did not stop that progression in these particularly challenging offenders, who had little to lose by arrest. The lack of deterrence here mirrors our overall lack of deterrence resulting from arrest for other crimes, such as burglary and drug offenses. Does a lack of deterrence mean we should abandon the provision of equal protection under law? Would we arrest one burglar but not another? Additionally, because deterrence alone was an unreasonably high standard for "success" and little focus was placed on how the rest of the criminal justice system did or did not support ongoing offender accountability, these studies were conducted in a systemic vacuum. Even the researchers themselves have cautioned against the use of the various studies on arrest to unduly influence policy (Berk, 1994; Sherman, 1992). However, the research on the presumed "escalation" effects of arrest, for example, has prompted some to rethink the use of arrest in domestic violence cases, even though other research studies provide more positive results of "success" of arrest. Ironically, this is as much of a knee-jerk reaction as the response to the original Minneapolis Experiment.

The problem remains—research alone can't define our policies. Policies must be driven by both proactive and reactive needs. It cannot simply be a reaction to a crisis; instead, it must be a considered way to guide practice so that problems can better be avoided. Good practice can be encouraged and supported by research, but it must not be driven by research alone.

Critical Legal Changes

Laws, particularly in the area of criminal justice response, are the most concrete and resistant to change of all intimate partner violence planned-change outcomes. Unlike policy, when laws do change, the new laws often become their own structural institution. Though they are created with the intent to significantly reform institutions and practices, in reality laws can exist only through political compromise. They generally end up looking quite different from their original incarnations and are not always a product of critical thinking as to the various, unintended consequences of their passage. Mandatory arrest laws are a good example of this challenge: They were passed with good intent, but alone they do not provide the necessary guidance or assessment tools to support ongoing institutionalization of a change in response.

As a result of the myriad of criminalization efforts, states passed mandatory arrest legislation all through the 1990s, paralleling VAWA in 1994. Mandatory arrest law was based on the reality that battered women have historically been given short shrift by the justice system and that consequently clear guidelines within law were now required if women were to achieve equal protection. In New York, many, if not most, police agencies had voluntarily developed "proarrest" policies during the mid- to late 1980s (Frisch & Caruso, 1996). These policies were instigated by a concern for civil liability but evolved into a desire for good practice in many departments. Because most police in New York had some sort of proarrest policy on domestic violence before the legislation, many saw the subsequent legal change as reinforcing the trend toward criminalization. Therefore, unlike some other states, most police in New York did not actively resist the move toward mandatory arrest legislation. What resulted is the most comprehensive piece of legislation to date affecting domestic violence in New York State. The Family Protection and Domestic Violence Intervention Act, which was signed into law in June 1994, is far-reaching, amending numerous sections of the Family Court Act, the Domestic Relations Law, and the Criminal Procedure Law. Most relevant to the criminal justice response is language that attempts to clarify the law enforcement role in domestic violence cases. The law requires police to make arrests in cases in which there is reasonable cause to believe that a felony, misdemeanor, or violation of an order of protection has been committed by one "family or household member" against another.[4] The sole exception to the arrest requirement is in misdemeanor cases in which the victim, without prompting by the officer, requests that an arrest not be made. In such cases, the officer still has the authority to make the arrest but is not required by the law to do so. In addition, the law requires police to complete a written report of the incident, called a Domestic Incident Report (DIR), whether or not an arrest is made or a crime committed. These laws provide a context for change in police response but do not, in and of themselves, provide the means toward institutionalization of a change in practice. NYS OPDV and NYS DCJS, in coordination with the School of Criminal Justice at the State University of New York at Albany, conducted an evaluation of the mandatory arrest provisions of the law. The framework of the evaluation took into account that deterrence alone could not be the standard for "success" in the implementation of the law, as deterrence is difficult to measure in these cases and the overall criminal justice response has at least as much to do with implementing mandatory arrest as the police do. For example, what good would an arrest be if the district attorney failed to prosecute the case or plea-bargained it into oblivion, or if the judge ordered mutual orders of protection or dismissed the case the moment he or she sensed ambivalence in the victim? This evaluation was conducted with these broader questions in mind and stepped "out of the box" of traditional deterrence-based evaluation studies as a result (NYS OPDV & NYS DCJS, 2001). After considering all of the factors involved, in light of victim safety and offender accountability objectives, this evaluation concluded that mandatory arrest should continue but that concerns regarding implementation and differential responses must be addressed.

Unlike the New York study, most mandatory arrest laws (if they are being assessed at all) are still being evaluated in light of a myopic and unrealistic standard of deterrence. Again, this blurs the intended outcomes of the law. We need to begin to develop more comprehensive assessments with more informed and attainable definitions of "success" that fit the framework of woman battering rather than stranger violence. We must also clearly recognize that holding offenders accountable can inadvertently increase victims' risks, and we can no longer describe our "dual goals" of offender accountability and victim safety in the same breath without acknowledging those risks and concerns. We must insist that the other actors within the system besides police do their jobs, and we must not presume that arrest, with no other sanctions, can inspire change in a man who batters.

Although the police have been the major focus nearly everywhere, the overall goal should be to promote a consistent, coordinated community response to victims of woman battering. Duffee (1980) suggested that when the community's response is the context for planned change involving criminal justice agencies, then measures of those community functions affected by the change, such as social control (crime prevention) and mutual support (services provided to victims), must become part of the assessment plan. This means examining the procedures of prosecutors and probation, court-mandated programs for batterers, judicial sentencing trends, and family court intake proceedings. Furthermore, the community's response to domestic violence goes beyond the criminal justice system to include health care, social services, substance abuse, mental health, clergy, and other areas of the human service system—which battered women call on whether or not they seek help from the police (Hart, 1995). Most important, the impact of the police policy and the community's response must be evaluated in light of the ultimate goal—safety for victims of battering (Jones, 1994).

Although we can't presume that having a proarrest policy or mandatory arrest law will in any way ensure victims' safety, it is hoped that such changes will at least not *increase* their danger. Concern for battered women's safety in light of these new policies is real. Since 1994, when such policies and laws began proliferating, an increased number of victims began to feel that they now had more options to help them. More women began to make official reports of these crimes to the police—59%

reported to police across the country in 1998 compared to 48% in 1993. It is troubling that publicizing these legal and policy "changes" will increase battered women's expectations of assistance when the actual level of change in the community is far more modest. For example, victims may assume that the police will surely arrest their batterer when they call for help, but the arrest rate for batterers is startlingly low, especially when suspects have left the scene (NYS OPDV & NYS DCJS, 2001; Roberts, 2002). Given the messages about domestic violence now being treated as a crime, victims may assume that they can readily obtain an order of protection from the court, even though the process is still arduous in many jurisdictions and victims can be understandably discouraged from following through, particularly in criminal court. And they may assume that their batterer will be held accountable for his violence, even though only a tiny percentage are actually convicted and an even smaller number are punished in any way, let alone incarcerated. These higher expectations for assistance have resulted, at least in part, in increased safety—*for men.* According to the Bureau of Justice Statistics (BJS, 2001), the number of men murdered by intimates has dropped by a whopping 69% since 1976. Unfortunately, the report found no such drop in the deaths of women at the hands of their male partners, indicating that women will try to exercise these identified options before getting to the terrible place of feeling compelled to kill or be killed. Women, however, are still dying at the hands of their partners as a direct result of our inability to make all of our promises for change much more real.

Clearly, without long-term monitoring and supervision, there can be no true accountability for offenders and no ability to promote safety for victims. There is a slow but growing national effort to better use probation supervision in the response to batterers, in an attempt to provide some ongoing monitoring and accountability as well as to focus on the safety and well-being of the victims. New York has a unique multiyear project, funded by the Office of Justice Programs, Grants to Encourage Arrest Policies, that develops the skills and domestic violence awareness of probation departments statewide and is intensively training representatives from the local departments to be Probation Domestic Violence Liaisons (PDVLs) or "point people" for probation on this issue in the community. The development of probation and domestic violence advocate linkages is a large focus of the project, which is administered collaboratively by the New York State Coalition Against Domestic Violence, the Division of Probation and Correctional Alternatives, and NYS OPDV. Probation departments in the state are being provided new tools for supervising these difficult cases, such as a specialized guide for preparing presentence reports, and an increasing number of departments are redeploying resources to domestic violence cases. This project, in addition to new laws, policies, training, and public education about domestic violence, has helped send the message that domestic violence is a crime and will be treated seriously. The reality in communities, however, even after more than 5 years since mandatory arrest became law and after nearly two decades of the "criminalization" process, is that only a very small number of offenders receive any sanction beyond arrest and an order of protection. Many offenders are being offered opportunities to attend batterers' intervention programs as a way to avoid other consequences, though efficacy of these programs is inconsistent at best (NYS OPDV, 2000). The majority of offenses are low level in official charge but are often experienced by the victim as increasingly controlling and threatening. She continues to see her offender's violence diminished and minimized by the structure and system of laws as much as by official attitudes (NYS OPDV & NYS DCJS, 2001).

What Have We Really Achieved?

Many questions remain. Does the development of police policy or the passage of mandatory arrest laws reflect the achievement of criminalization? Have we achieved a semblance of coordination of responses? Do we listen as a rule to the voices of survivors and advocates as we continue with our planned change efforts? Have we found ways to provide better economic support for victim programs so they will have the luxury of providing input and developing community partnerships while still providing emergency services? Clearly, the answer to these questions is no. It seems clear from the experiences in New York and other states that the focus for further study must go well beyond the police response and must assess changes in the community—looking at the entire criminal justice and legal systems as well as the larger community. Proactive police policies and clearer laws that foster equal protection are integral parts of this change, but moving toward positive changes in response to domestic violence requires the coordination of the community, both within the criminal justice system and beyond, with agreement on all levels that the changes must take place. Fagan (1996) stated that "social control is most effective when legal controls interact reciprocally with extralegal social controls" (p. 28). Success as it must be reframed cannot be achieved until there is agreement within the broader community that such violence will not be tolerated, that violent persons will be held accountable, and that the goal of safety for victims is paramount. Understanding how we came to this current place of mandatory arrest, evidence-centered prosecution, and proliferation in specialized domestic violence courts requires that we remember where we have been. If we care to look, the past also helps point us to where we need to go. If we finally begin to frame the future with this experience, and plan for change not in isolation but in partnership, we will be able to better define "success" and meet the needs of victims as well as the larger community, rather than our individual institutions—a challenge, but one well worth overcoming.

Systems Accountability: The Missing Link

Possibly the most important and overriding issue in the criminal justice response to domestic violence is that of systems accountability (see NYS OPDV, 1998). This is the missing link in inspiring true institutional change in the response to woman battering. As a society, we should have no tolerance for lack of policies, unenforced laws, and victim-blaming attitudes or strategies that endanger or disempower victims. Perhaps most of all, we should not allow institutional goals to take precedence over victim safety. We must listen to the voices of survivors and their advocates with at least the same respect and consideration that we give to academic researchers. In fact, research goals and methodology should be as informed as our laws and policies should be by their experiences. Equally important, analyses of research should not be done in a vacuum but should also have the benefit of input and review by those affected most directly by the results of the policy (Yllo, 1988). This input would not diminish the relevance or academic perspective of the research; rather, it would provide a context for analyses that would consider the impact on the victim and prioritize her safety in both the long and the short term. NIJ has recognized the importance of the need for such partnerships, but finding ways to foster ongoing, egalitarian relationships between these two diverse groups is challenging at best (Travis, 1998). Perhaps most critical to supporting these partnerships is finding ways to provide better economic support to programs that serve victims. These programs are seriously understaffed, undertrained, and underpaid, though immensely dedicated. As even the federal government recommends that policy changes and program plans be made with the input of survivors and their advocates, we need to remain mindful that this is one more thing to add to their overflowing plate. With increased recognition, awareness, and demand, domestic violence advocacy has by necessity shifted away from social change and toward individual services just when we need informed support for this change the most.

CRITICAL ISSUES IN WOMAN ABUSE: WHERE DO WE GO FROM HERE?

The criminalization process has triggered a number of other critical issues in domestic violence that should be considered for further study and consideration. A nonexhaustive list includes

- The aforementioned need for a more relevant role for probation and parole supervision to inspire longer term accountability

- The limitations of programs for men who batter and the negative consequences of using these programs as diversion from the system
- The increase in arrests of women for domestic violence offenses without appropriate assessment of their own victimization
- The need to develop domestic violence screening tools and protocols for drug courts to avoid inappropriate referrals and mandates
- The pros and cons of evidence-based police and prosecutorial responses; assessments of mandatory prosecution (no-drop) policies
- The proliferation of specialized domestic violence courts and newly developed "integrated" or "consolidated" courts, and the need for evaluation of these initiatives
- Troubling trends toward filing "failure to protect" charges against the mother when children witness domestic violence
- The dearth in policy development and evaluation of the criminal justice process *post arrest*
- Critical evaluation of community justice strategies used in domestic violence and sexual assault cases that mirror tribal justice initiatives that favor conferencing and agreements rather than formal justice responses
- The spread of the use of electronic monitoring, cell phones, and other security systems and the need for evaluation of the effects of their use and impacts on victim safety
- The use of community coordination activities to reduce fragmentation and inconsistency in response
- Response to the needs of communities of color, gay/lesbian/bisexual and transgendered communities, and disabled victims, and prioritizing of cultural competency in our practices and research
- Recognition of abuse in teen relationships and the need for additional services for teen victims and accountability strategies for young offenders
- Efforts to ensure that our legal definition of who is a victim of domestic violence is consistent and reaches all victims, such as same-gender partners and those who have had a dating relationship
- Improvement of service delivery to victims, who are increasingly in need of assistance as they become more involved in the legal system
- Development of better quality control strategies to help ensure that increased federal and state dollars spent on domestic violence programs are dollars well spent
- Assessment of the impacts of newly proposed federal "fatherhood initiatives" that promote marriage as a way to lower public assistance costs, perhaps at the expense of women and children's safety

These and other issues must be critically analyzed within the context of intimate partner violence and in collaboration between policy makers, victim advocates, survivors, researchers, and practitioners. Sometimes due to lack of resources, lack of information, or simply too much to do with too little time in a crisis-oriented environment, we continue to do our criminal justice planning in domestic violence in compartmentalized segments, with little sense of the whole. Institutionalization of the justice response to domestic violence, though something we have all worked tirelessly for, can unwittingly become a compromise for the benefit of the system at the expense of victim safety. It can be the proverbial case of "be careful what you wish for."

Developing holistic assessment strategies, with informed research strategies and respectful partnerships with advocates and practitioners, can assist us to fill these information gaps, and hopefully—finally—help us save lives and improve the quality of life of women and children. In this post-September 11 world, we all know what it feels like to live with a level of fear in our hearts every day. Battered women have known this forever. We owe it to survivors of this terror in the home to help them live free, just as we all wish to do in our lives.

Notes

1. The term *criminalization* describes the process(es) by which substantive and procedural criminal laws are written, amended, and applied to specific behaviors that were previously socially tolerated, for the purposes of general and specific deterrence, punishment, and social control (Beirne & Messerschmidt, 1991, pp. 24–25). Thus, to criminalize a set of behaviors is to assign them a criminal status in written law based on perceived social harms and to enforce such laws when violations become known. Written laws describe behaviors that are considered harmful and authorize an official response. The enforcement or lack of enforcement of these laws plays a critical role in

shaping public attitudes toward the seriousness of particular actions. Predictable law enforcement conveys the message of wrongfulness and social rejection and educates victims, offenders, children, neighbors, and the public at large that such behavior will not be tolerated. Criminalization focuses the legal authority and coercive power of the criminal justice system on a set of behaviors now perceived as socially dangerous and currently not manageable using informal means of social control. Criminalization is thus more than the manipulation of law. The process includes the planned change of individual perceptions, organizational policies and procedures, and ultimately, the status quo of everyday life (Frisch & Caruso, 1996).

2. It should be noted that domestic violence also seriously affects same-gender partners and in rare cases can involve male victims of female partners, but the vast majority of victims are women by male intimate partners.

3. Calls for treating domestic violence as a crime have come from organizations that one would hardly call feminist. As far back as 1976, the International Association of Chiefs of Police (1976a, 1976b) recommended to their police constituency in Training Keys and policy statements that there be equal treatment of victims of what was called "wife beating"—a stand that was quite progressive in its time. Since then, that national organization has been remarkably conscientious in addressing domestic violence and policing by issuing and widely disseminating model domestic violence policies on arrest, as well as policy language relating to the sensitive issue of officer-involved abuse (see www.theiacp.com for more information on these policies).

4. In New York State, "family or household" members are those who are related by blood, marriage, or having a born child together. Only these individuals may access Family Court for civilly based legal relief; all others must go to the criminal justice system for assistance. The majority of police departments, however, apply a much broader definition of who is a victim of abuse. For example, police recognize that at least as many cases they respond to involve boyfriends/girlfriends and that they see same-gender intimate violence cases as well. Throwing a wide blanket of protection over domestic violence cases is generally deemed the best for both liability protection *and* victim protection, at least theoretically. Thus police will now at least consider making an arrest in a domestic violence offense, as opposed to the earlier trend toward arrest avoidance in virtually all cases.

REFERENCES

Beirne, P., & Messerschmidt, J. (1991). *Criminology*. San Diego: Harcourt Brace Jovanovich.

Berk, R. (1994, Summer). What the scientific evidence shows: On average, we can do no better than arrest. *Domestic Violence Project Research Update*, pp. 1–4.

Berk, R., & Newton, P. (1985). Does arrest really deter wife battery? An effort to replicate the findings of the Minneapolis Spouse Abuse Experiment. *American Sociological View, 50*, 253–262.

Bureau of Justice Statistics. (2001). *Homicide trends in the U.S.: Intimate partner homicide*. Washington, DC: U.S. Department of Justice.

Duffee, D. (1980). *Explaining criminal justice*. Cambridge, MA: Oelgeschlager, Gunn & Hain.

Fagan, J. (1996, January). *The criminalization of domestic violence: Promises and limits*. Washington, DC: National Institute of Justice.

Family Protection and Domestic Violence Intervention Act, L. 1994, cc. 222 & 224.

Frisch, L., & Caruso, J. (1996). The criminalization of woman battering: The New York State experience. In A. Roberts (Ed.), *Helping battered women: New perspectives and remedies*. New York: Oxford University Press.

Hart, B. J. (1995, March 31). *Coordinated community responses to domestic violence*. Paper presented at the Strategic Planning Workshop on Violence Against Women, National Institute of Justice, Washington, DC.

International Association of Chiefs of Police. (1976a). *Investigation of wife-beating* (Training Key 246). Gaithersburg, MD: Author.

International Association of Chiefs of Police. (1976b). *Wife-beating* (Training Key 245). Gaithersburg, MD: Author.

Jones, A. (1994). *Next time she'll be dead: Battering and how to stop it*. Boston: Beacon.

National Institute of Justice & Centers for Disease Control. (1998). *Prevalence, incidence and consequences of violence against women: Findings from the National Violence Against Women Survey*. Washington, DC: National Institute of Justice.

New York State Office for the Prevention of Domestic Violence. (1998). *Governor's model domestic violence policy for counties.* Albany: Author.

New York State Office for the Prevention of Domestic Violence. (2000, Fall). Programs for men who batter: What have we learned? (OPDV Bulletin). Retrieved September 2002 from www.opdv.state.ny.us.

New York State Office for the Prevention of Domestic Violence & New York State Division of Criminal Justice Services. (2001, January). *Evaluation of mandatory arrest law in New York State: Final report to the governor and legislature.* Albany: Author.

Pence, E. (1989). *The justice system's response to domestic violence.* Duluth: Minnesota Program Development.

Pence, E., & Lizdas, K. (1998). *The Duluth Safety and Accountability Audit: A guide for assessing institutional responses to domestic violence.* Duluth: Minnesota Program Development.

Pleck, E. (1987). *Domestic tyranny: The making of American social policy against family violence from colonial times to the present.* New York: Oxford University Press.

Roberts, A. R. (Ed.). (2002). *Handbook of domestic violence intervention strategies: Policies, programs, and legal remedies.* New York: Oxford University Press.

Schechter, S. (1982). *Women and male violence: The visions and struggles of the battered women's movement.* Boston: South End.

Sherman, L. (1992). *Policing domestic violence: Experiments and dilemmas.* New York: Free Press.

Sherman, L., Schmidt, J., Rogan, D., Garten, P., Cohn, E., Collins, D., & Bacich, A. (1991). From initial deterrence to long-term escalation: Short-term custody arrest for poverty ghetto domestic violence. *Criminology, 29,* 821–841.

Sherman, L., Schmidt, J., Rogan, D., Smith, D., Gartin, P., Cohn, E., Collins, D., & Bacich, A. (1992). The variable effects of arrest on criminal careers: The Milwaukee Domestic Violence Experiment. *Journal of Criminal Law and Criminology, 83,* 137–169.

Sometimes, abuse victim goes to jail. (2002, April 20). *Lexington Herald-Leader,* p. A1.

Stark, E. (2002). Preparing for expert testimony in domestic violence cases. In A. R. Roberts (Ed.), *Handbook of domestic violence intervention strategies* (pp. 216–252). New York: Oxford University Press.

Swanson, C. R., Territo, L., & Taylor, R. W. (1988). *Police administration structures, processes and behavior* (2nd ed.). New York: Macmillan.

Toch, H., Grant, J., & Galvin, R. (1975). *Agents of change: A study in police reform.* Cambridge, MA: Schenkman.

Travis, Jeremy. (1998, November 30). Can we talk? Yes, let's!" *Law Enforcement News, 24*(1), 3.

U.S. Attorney General's Task Force on Family Violence. (1984). *Final report.* Washington, DC: U.S. Department of Justice.

Violence Against Women Act, 18 U.S.C. § 40101; 42 U.S.C. § 40211; 42 U.S.C. § 40401 (1994).

Yllo, K. (1988). Political and methodological debates in wife abuse research. In K. Yllo & M. Bograd (Eds.), *Feminist perspectives on wife abuse.* Newbury Park, CA: Sage.

Zubretsky, T. (2002). Promising directions for helping chemically-involved battered women get safe and sober. In A. R. Roberts (Ed.), *Handbook of domestic violence intervention strategies* (pp. 321–340). New York: Oxford University Press.

Zubretsky, T. M., & Digirolamo, K. M. (1996). The false connection between adult domestic violence and alcohol. In A. R. Roberts (Ed.), *Helping battered women.* New York: Oxford University Press.

ADDITIONAL SUGGESTED READINGS

Alexander, F. (1985, July 17). Municipal negligence: Failure to provide police protection. *New York Law Journal,* p. 17.

Bachman, R. (1994). *Violence against women: A national crime victimization report.* Washington, DC: Bureau of Justice Statistics.

Beck, L. (1987). Protecting battered women: A proposal for comprehensive domestic violence legislation in New York. *Fordham Urban Law Journal, 15,* 999–1048.

Browne, A. (1987). *When battered women kill.* New York: Free Press.

Bureau of Justice Statistics. (1986). *National Crime Survey.* Washington, DC: U.S. Department of Justice.

Bureau of Justice Statistics. (1991). *Violent crime in the United States.* Washington, DC: U.S. Department of Justice.

Bureau of Justice Statistics. (1998). *National Crime Survey*. Washington, DC: U.S. Department of Justice.

Buzawa, E. S., & Buzawa, C. (1990). *Domestic violence: The criminal justice response*. Newbury Park, CA: Sage.

Carter, J., Heisler, C., & Lemon, N. (1991). *Domestic violence: The crucial role of the judge in criminal court cases. A national model for judicial education*. San Francisco: Family Violence Prevention Fund.

Cohn, E., & Sherman, L. (1987). Police policy on domestic violence, 1986: A national survey. In Crime Control Institute (Ed.), *Crime control reports*. Washington, DC: Crime Control Research Corporation.

Community Council of Greater New York. (1988, August). Arrest and domestic violence: Trends in state laws and findings from research. *Research Utilization Update*, pp. 1–9.

Dobash, R. E., & Dobash, R. P. (1979). *Violence against wives: A case against the patriarchy*. New York: Free Press.

Erez, E., & Belknap, J. (1998). In their own words: Battered women's assessment of the criminal processing system's responses. *Violence and Victims, 13*(3), 3–20.

Evans, M. (1987). Domestic violence: A proactive approach. In Victim Services Agency (Ed.), *The law enforcement response to family violence* (pp. 1–4). New York: Victims Services Agency.

Feeley, R., & Sarat, L. (1980). *The policy dilemma*. Minneapolis: University of Minnesota Press.

Felson, R. B., & Ackerman, J. (2001, August). Arrest for domestic and other assaults. *Criminology, 39*, 655–675.

Fields, M. D. (1987, September). *Municipal liability for police failure to arrest in domestic violence cases*. Albany: New York State Governor's Commission on Domestic Violence.

Fields, M. D. (1991, October). *Rape and domestic violence legislation: Following or leading public opinion?* Unpublished paper.

Ford, D. A., & Regoli, M. J. (1993). The criminal prosecution of wife batterers: Process, problems and effects. In Z. Hilton (Ed.), *Legal responses to wife assault*. Newbury Park, CA: Sage.

Frisch, L. (1992, Spring). Research that succeeds, policies that fail. *Journal of Criminal Law and Criminology, 83*, 209–216.

Garner, J., & Clemmer, E. (1986). *Danger to police in domestic disturbances: A new look*. Washington, DC: National Institute of Justice.

Gordon, L. (1988). *Heroes of their own lives*. New York: Penguin.

Hart, B., Stuehling, J., Reese, M., & Stubbing, E. (1990). *Confronting domestic violence: Effective police response*. Reading: Pennsylvania Coalition Against Domestic Violence.

Haviland, M., Frye, V., Rajah, V., Thukral, J., & Trinity, M. (2001). *The Family Protection and Domestic Violence Intervention Act of 1995: Examining the effects of mandatory arrest in New York City*. New York: Urban Justice Center.

Klaus, P., & Rand, M. (1984). *Family violence*. Washington, DC: Bureau of Justice Statistics.

Langan, P., & Innes, C. (1986). *Preventing domestic violence against women*. Washington, DC: Bureau of Justice Statistics.

Loving, N. (1980). *Responding to spouse abuse and wife beating: A guide for police*. Washington, DC: Police Executive Research Forum.

Nakamura, R., & Smallwood, F. (1980). *The politics of policy implementation*. New York: St. Martin's.

New York State Governor's Commission on Domestic Violence. (1988). *Domestic violence: A curriculum for probation and parole*. Albany, NY: Author.

Newman, D. J., & Anderson, P. R. (1989). *Introduction to criminal justice*. New York: Random House.

Pennsylvania Attorney General's Task Force on Family Violence. (1989). *Domestic violence: A model protocol for police response*. Harrisburg, PA: Office of the Attorney General.

President's Task Force on Victims of Crime. (1984). *Final report*. Washington, DC: U.S. Department of Justice.

Stark, E., & Flitcraft, A. (1988). Violence against intimates: An epidemiological review. In V. D. Van Hasselt, A. Bellack, & M. Hersen (Eds.), *Handbook of family violence* (pp. 159–199). New York: Plenum.

Stark, E., & Flitcraft, A. (1997). *Women at risk: Domestic violence and women's health*. Thousand Oaks, CA: Sage.

Steinman, M. (Ed.). (1991). *Women battering: Police responses*. Cincinnati, OH: Anderson.

Toch, H., & Grant, J. (1982). *Reforming human services*. Beverly Hills, CA: Sage.

U.S. Commission on Civil Rights. (1982). *Under the rule of thumb: Battered women and the administration*. Washington, DC: Author.

Victim Services Agency. (1989). *The law enforcement response to family violence: The training challenge*. New York: Victim Services Agency.

Walker, L. E. (1989). *Terrifying love: Why battered women kill and how society responds*. New York: Harper & Row.

Zorza, J., & Woods, L. (1994). *Mandatory arrest: Problems and possibilities*. New York: National Center on Women and Family Law.

12

VICTIM PERCEPTIONS OF THE UTILITY OF DOMESTIC VIOLENCE ARRESTS AND TEMPORARY RESTRAINING ORDERS

A Qualitative Study

GINA PISANO-ROBERTIELLO

We have begun to witness a commitment to ending family violence through comprehensive intervention strategies which ensure the safety of the victim and the complete account-ability of the offender. However, the elimination of family violence throughout society will only become a reality after responsive communities work toward an integrated response from the police, the courts, health care providers, and social service agencies. This will require a commitment to continued research, community-wide collaboration and well-coordinated community responses, policy development and improved access to services, and an investment in technological advancements such as cellular phones, panic alarms, emergency response pendants, and the Juris monitors.

Albert R. Roberts, *Handbook of Domestic Violence Intervention Strategies* (2002)

Suzette is a 31-year-old white woman from a New Jersey suburb. The first time she was victimized was in 1995, when her boyfriend hit her because she found him using heroin. Her abuse escalated thereafter and contin-ued for 5 years. She called the police 20 to 25 times during this relationship. It finally ended on December 6, 2000, when "he came home high and proceeded to beat me in the head after I ques-tioned him about charges on my credit cards."

Suzette's interaction with the police was quite positive. They responded every time she called them, and her abuser was arrested on three occasions, including an arrest for the above-mentioned beating, which brought Suzette to the shelter.

Jessica obtained some assistance from the police. She is a 21-year-old Hispanic woman, also from urban New Jersey, who was first beaten two and a half years into her relationship. As with many battered women, her boyfriend began physically abusing her once he found out she was pregnant. According to Jessica, "When he beat me while I was pregnant, I left him. I came back after I had the baby and left for good when he hit me in front of our child."

Jessica called the police three to four times, but the police only responded once. Her boyfriend was arrested on this occasion, but this did not deter him from abusing her upon his release.

Keiko was somewhat dissatisfied with the police response to her plight. She is a 22-year-old woman from suburban New Jersey. Her abuse started right after the birth of their child and lasted 2 years. The abusive incident that brought Keiko to the shelter was when "he came after me with a bat, tried to choke me, hit me with his fists on the back of my head, and threw me into our door."

The police were called three times during their relationship. Her husband was arrested twice, and she was arrested once when she grabbed him while being beaten and left nail marks on him. At that time, she did not display any visible signs of abuse on her.

These are just a few examples of different women with different stories. Their age, race, ethnicity, marital status, and socioeconomic status may vary. The severity and type of victimization may vary, and the number and type of contact with the police may also vary. But these women are all survivors of domestic violence.

In 2000, I began an exploratory study to examine the impact of police arrest and temporary restraining orders (TROs) on domestic violence victims. I examined two critical issues:

1. Do victims perceive mandatory arrest as helpful or harmful to them?

2. Do victims feel that a TRO is helpful (i.e., that it will deter or has deterred the abuser)?

I developed a set of interview questions that were administered to battered women currently living in two different shelters in New Jersey: one located in a city (Newark, in Essex County) and the other located in a suburban neighborhood (Morristown/Morris Plains, in Morris County). I expected large differences between the two groups on account of the lower socioeconomic status of Essex County residents. The knowledge gained from these interviews was helpful in assessing victims' perceptions of the law and should be useful in assisting police with combating domestic violence, as well as addressing the important critical issues facing police and victims today.

SCOPE OF THE PROBLEM

Estimates from the National Crime Victimization Survey indicate that in 1998 about 1 million violent crimes were committed against persons by current or former spouses, boyfriends, or girlfriends (Bureau of Justice Statistics [BJS], 1999). Although rates have declined in the last 5 years and are far lower than in the last 25 years, the percentage of female murder victims killed by intimate partners has remained at about 30% of all homicides since 1976 (BJS, 1999).

According to the BJS, the number of men murdered by intimates dropped 69% since 1976, and the number of women murdered by intimates declined after 1993. In the year 2000, 11% of murder victims were killed by intimates. In addition, the crime index rate fell for the ninth straight year in 2000 (down 3.3% from 1999, 18.9% from 1996, and 30.1% from 1991 (Federal Bureau of Investigation [FBI], 2001).

New Jersey is one of the 20 states that requires an arrest if probable cause exists. According to the section of the *New Jersey Police Manual* on domestic violence entitled "Arrest of Alleged Perpetrator":

> When a person claims to be a victim of domestic violence, and where a law enforcement officer responding to the incident finds probable cause to believe that domestic violence has occurred, the police officer shall arrest the person who is alleged to be the person who subjected the victim

to domestic violence and shall sign a criminal complaint if:

1. The victim exhibits signs of injury caused by an act of domestic violence;

2. A warrant is in effect;

3. There is probable cause to believe that the person has violated N.J.S. 2C:29–9, and there is probable cause to believe that the person has been served with the order alleged to have been violated. If the victim does not have a copy of a purported order, the officer may verify the existence of an order with the appropriate law enforcement agency; or

4. There is probable cause to believe that a weapon as defined in N.J.S. 2C:39–1 has been involved in the commission of an act of domestic violence. (New Jersey State Police, 1998, 2C: 25-21)

In summary, New Jersey requires that when a complaint is made to police, the officer shall arrest the person if it is felt that probable cause exists that a crime of domestic violence occurred.

In New Jersey, "warrantless" arrests are allowed for both felony and misdemeanor domestic violence as long as the woman has been beaten, even if no visible signs of injury exist (Buzawa & Buzawa, 1996). However, allowing "warrantless" arrests has not guaranteed that police officers will make that arrest. In fact, many arrest policies only encourage arrest (Hoctor, 1997). Officers can still decide to do nothing regardless of the law or policy in a particular state. However, according to Roberts (2002), it is a myth that police never arrest the batterer because they see domestic violence calls as a private matter. For the years 1992 to 1996, the National Crime Victimization Survey demonstrated that only 20% of these cases resulted in an arrest. But more recently, proarrest policies have been changing those figures. For example, Jones and Belknap (1999) reported a 57% arrest rate for domestic violence offenses in one progressive jurisdiction. Nationwide, arrest rates for domestic crimes increased from 1995 to 2000 even while arrest rates for violent crimes in general decreased

(FBI, 2002). As of 2001, all 50 states had employed warrantless arrest policies, and mandated domestic violence police training, specialized police domestic violence units, crisis response teams, and collaboration among victim advocates, police, and prosecutors are all becoming more common. Thus, we have seen an increase in offender accountability in the last few years (Roberts, 2002).

According to Roberts (2002), it is a myth that TROs are rarely effective. With improved technology, such as advanced photographic techniques, DNA profiling, computer-aided investigation and dispatch, and automated case study tracking of battering offenders and victims' reported victimizations, TROs can indeed be effective. The evidence suggests a significant decline in the probability of abuse following the issuance of a protective order (Carlson, Harris, & Holden, 1999). This topic will be examined in the interviews with battered women who have obtained TROs and have assessed their success.

The mandatory arrest requirement for domestic violence is unique in that no other class of offenses requires arrest. Although mandatory arrest may appear to simplify the work of police officers confronted with gray areas and to offer protection from civil liability for officers, many officers historically did not see things this way. Zorza (1994), for example, in a study conducted relatively early in the trend toward mandatory arrest, found that police did not like mandatory arrest laws because they felt that these laws would interfere with their exercise of discretion. Hence, police could find ways to keep from making an arrest regardless of the law. However, some information indicates that police now prefer having the guidance of a formalized arrest policy or law. In New York State, one study surveyed police about whether the mandatory arrest law should be renewed or allowed to expire. A startling 95% of the respondents said they wanted mandatory arrest to continue (New York State Office for the Prevention of Domestic Violence [NYS OPDV], 2001).

Studies are mixed as to whether mandatory arrest of the batterer is effective. Maxwell, Garner, and Fagan (2001), in a study of the deterrent effects of arrest, found that arrest of batterers was consistently related to reduced subsequent aggression against female intimate partners, although results were not statistically significant and although the overall size of the relationship between arrest and repeat offending was modest when compared to the size of the relationship between recidivism and prior criminal record. No association between arresting the offender and increased risk of subsequent aggression against women was found.

METHODOLOGY

In this study, I interviewed 30 women who had recently been victims of domestic violence: 19 of them were residing in a domestic violence shelter in the urban community of Newark in Essex County, New Jersey, and 11 were residing in a shelter in suburban Morris County. The number was higher for the urban facility because turnover rates were higher at this shelter. At the Essex County shelter, but not at the Morris County shelter, I was allowed to sit in on support group meetings and could build rapport with subjects through group and individual interaction before interviewing. At both shelters, I set up an interview schedule for each week and interviewed, on average, two victims per day. Interviews lasted approximately 90 minutes. Occasionally, I scheduled follow-up appointments to clarify information after review of the answers given by the subjects. The suburban shelter, located in a residential area, had relatively new furniture. The urban shelter, on the other hand, was much more cramped and furnished with older, worn materials. However, its administration was also much more open and cooperative than that of the suburban location. Nevertheless, the volunteer residents at both locations were extremely helpful.

The character and intensity of the interview allowed me to gain a good grasp of perceptions. The questions concerned both the victimization itself and recollections of police response. Interviews were structured in a manner that encouraged in-depth responses and individual comments. Each woman had something unique and interesting to contribute. The results of this study should be useful in helping society and police understand the perceptions of victims and can assist in evaluating and revamping current policies and practices.

RESULTS

Mandatory Arrest

Most (14) of the 16 urban shelter victims who said the police had been called (by them or someone else) regarding an incident or incidents reported that the police had showed up every time they were called. However, 2 respondents claimed that the police had not showed up every time they were called. All of the suburban shelter victims claimed that the police had showed up every time they were called (11 out of 11).

All of the suburban victims had called the police at least once during an abusive incident. One victim said she had called the police 20 to 25 times. Eight suburban victims thought the police had responded quickly to their call (within 5 minutes). Two victims said officers had taken 5 to10 minutes, and one victim said officers had taken over 10 minutes to respond. In contrast, 12 of the 19 urban victims stated that they had called the police, 4 stated that others had called the police for them, and 3 stated that neither they nor others had called the police. Of the 12 urban victims who stated that they had called the police themselves, 6 called one time, 2 called three times, 1 called two times, 1 called four times, and 2 called five times. Of the 16 urban victims who stated that the police had been called, by either themselves or others, 9 thought police had responded quickly and 7 thought they had taken too long.

Three of the urban battered women reported a failure by police to consistently implement mandatory arrest policy. Only one of the women in the suburban shelter reported this.

One victim claimed that her boyfriend sometimes knew the responding officer and that the officer would warn him of an impending arrest if the circumstances occurred again or just tell him to hurry up and leave the premises before he would be arrested. One urban victim claimed that police just told her boyfriend to "take a walk." He would then go get some alcohol and come back in 30 minutes. She also said,

He was not scared by the threat of an arrest, and when my boyfriend was not arrested [and was instead just told to leave], he did not listen and returned immediately upon the police leaving the scene. The abuse actually increased after the police left the scene. One time in particular, the officer said there was "nothing he could do."

The victim knew where the perpetrator had fled to, but the police did not want to hear it. A few women heard "There is nothing we can do" when police could have acted and did not, due to either ignorance or inconvenience.

This is a perfect example of how the law says one thing on paper but is enforced differently. Most of the victims interviewed, however, thought the police handled their situation professionally. They arrived in a timely manner (within minutes) and followed procedure by seeing signs of abuse and making an arrest.

The 14 urban victims who had called police reported 29 occasions on which the police had responded. On 6 of those 29 occasions, the parties were separated; on 5 occasions, the parties were told to talk it out; on 3 occasions, both parties were arrested; and on 14 occasions only the abuser was arrested.

The 11 suburban victims who had called police reported 22 occasions on which the police had responded. On 9 occasions, the parties were separated; on 10 occasions, the abuser was arrested; on 1 occasion, both parties were arrested; and on 1 occasion, the abuser was given a warning.

Thus, arrest of the abusers happened more often than any other outcome for both sets of victims. However, in incidents that involved the suburban women, the abusers were almost as likely to be separated from the suburban victims as arrested (in 10 incidents they were arrested and in 9 they were separated). On one occasion, nothing was done, and on two occasions, the female victim was arrested along with her battering counterpart.

Dual arrest is more common then most authorities realize, especially when police arrive at a scene and both parties show signs of injury. Whether dual arrest rates reflect true mutual violence or a failure of police to assess for the "primary" or "predominant" aggressor, increased arrest of women becomes a deterrent

to victims seeking police assistance. Martin (1997a) found that arrest of battered women may reinforce their isolation and belief that no resources are available for them.

When urban victims were asked what they had expected from the encounter, 13 responded that they expected an arrest. Three were unsure, and the other three had never called, so the question was not applicable to them. When asked why they did not contact the police, two reasoned that the abuse was minor, and one thought there was nothing police (or others) could do to help them. Eight urban victims felt there was no discrepancy between what they expected and what the police did, and three victims could not respond to this question because the police were never contacted. One victim did not know why there was a discrepancy between what she expected and what the outcome was. However, eight victims had numerous explanations for the discrepancy, including

- "The police did not take the situation seriously enough and they saw him as a victim."
- "The police were friends with him."
- "I hit him back."
- "They did not want to fill out the paperwork."

When suburban victims were asked what they had expected from the encounter, five said they had expected an arrest. Three expected a warning or separation, one just expected to be taken out of the abusive situation, and two did not know what they expected from the encounter. Eight suburban victims felt there was no discrepancy between what they expected and what the police did, and one did not know why there was a discrepancy. Those who thought there was a discrepancy and had an opinion why presented the following explanations:

- "There were no visible signs of battering on either of us."
- "I did not show them any signs of battering."
- "The lease was in his name."

When the urban victims were asked whether they thought that in general police followed current law in handling domestic violence cases, nine said yes and six said no. Four victims said they did not know if police were following the current law. Ten suburban victims thought that in general police followed current domestic violence law and one did not.

Although training aimed at breaking down common myths and stereotypes about domestic violence is given to all police recruits, it is obvious that neither policy nor training translates directly into action. Police officers' response is shaped not only by their training and by procedural law but also by officers' styles of policing, values, and beliefs, as well as cultural norms.

I found that the police did not make an arrest every time they were called or every time they witnessed signs of abuse. Thirteen of the 16 urban victims who called the police and 10 out of the 11 suburban victims who called the police claimed that police did what they should have done and that there was no occasion where they witnessed signs of abuse and took no action. For example, one informant, Sharon, stated that police "were looking for every sign to try to make an arrest" of her abuser.

The interview results demonstrate that very few women blamed the police for their situation. Only three urban victims and one suburban victim claimed that police had not made an arrest when there were visible signs of abuse.

Both urban and suburban groups were divided on the question of how helpful mandatory arrest was. The urban victims were more negative on this question: 6 said mandatory arrest keeps the batterer away, 12 said it doesn't, and 1 was unsure. Of the suburban victims, five said mandatory arrest works and six said it does not.

Temporary Restraining Orders

This pilot study also evaluated victims' perceptions of the utility of TROs. Sixteen urban victims had obtained TROs and three had not. When the urban shelter victims were asked whether they felt that in general TROs were effective, five said yes and six said no. Five victims had come directly to the shelter after obtaining the TRO, so they considered themselves unable to determine whether TROs worked. According to Monique, obtaining the restraining order was not effective: "He still came near me, and I never went

to court to finalize the restraining order." Tara said, "The TRO was dismissed because the judge said I could not substantiate the domestic violence against me." For Jessica, the TRO was "effective for a while, but then I dropped it."

All 11 suburban victims obtained TROs. Three felt they were effective and seven did not. One did not know if the TRO was effective because she came directly to the shelter after obtaining it. Even after obtaining a TRO, some victims continued to be harassed. Upon release from jail, many victims claimed that their abuser came directly back to them. Keiko said: "He violated the TRO right away—outside the courtroom." Suzy said: "He still came in the house, but I did not call the police because he cut off the phone."

Of the 16 urban victims who obtained the TRO, 6 claimed that the abuser did not subsequently violate it. Ten claimed that their abuser continued to contact them but that they did not tell the police. Margie said he did not contact her but did contact her friends and family. Other victims stated:

- "The TRO is just a piece of paper—he returned, I changed the locks, called the police, but he always left before the police arrived."
- "He thought he was invincible. He would come over, get arrested, but [get] out the next day and harass me again."

Twelve urban victims claimed that the TRO was easy to obtain, and three said it was difficult to obtain. Jessica got the runaround because "some of the abuse occurred in another town, and Newark did not want to issue the TRO due to this fact."

Of the 11 suburban victims who obtained TROs, 2 said the abuser never violated the TRO, and 1 said that he did violate the TRO but that she did not tell the police. Seven victims said the TRO was effective (i.e., it kept the abuser away), and one said it was effective some of the time. Nine said it was easy to obtain, and two said it was not.

In this study, almost all of the victims obtained TROs (16 of 19 urban victims and 11 out of 11 suburban victims). However, over one third of the New Jersey victims did not report

TRO violations if they occurred (11 of 30 victims). Thus, the frequency of violations is probably much higher, and their effectiveness must be questioned.

Of the 27 victims in the New Jersey study who obtained TROs, only five urban and three suburban victims said the TRO stopped the abuser from coming around. These victims claimed,

- "The restraining order was effective because he knew all I had to do was call the police."
- "He has had no contact with me after I got the TRO."

Yet Forell (1990) found that protective orders give women a false sense of hope. The orders have often been ignored by batterers and by the police. Hoctor (1997) found that law enforcement officials often fail to arrest and prosecute batterers even when the victim has a TRO. Erez and Belknap (1998) found TROs to be largely unenforced and ineffective. On the basis of these results, they appear to be less effective than anticipated.

Balos and Trotzky (1988) examined the effectiveness of orders of protection in Minnesota. They found that 22% of those who obtained an order of protection were named in subsequent police reports of domestic violence. In every misdemeanor case where a not-guilty plea was entered, the case was dismissed. Yet the authors concluded that for most individuals, an order of protection was an effective means of reducing domestic violence. Holmes (1993) found that police officers were more apt to arrest if a TRO was violated, witnesses were present, the assault took place in public, and/or the offender was black.

Chaudhuri and Daly (1992) found that TROs increased police responsiveness. However, the chance of a victim's being battered again depended upon numerous other characteristics such as the batterer's prior history, employment status, and substance abuse problems. They also found that the victims most likely to obtain TROs were younger, had achieved a higher level of education and a higher income status, and were in the relationship for a shorter time than the victims who did not obtain the TROs.

Thus, it appears that most victims in this exploratory study did not believe TROs would deter their abuser.

CONCLUSION

What does this pilot study show about the effectiveness of mandatory arrest and TROs as based on victims' experience? Most abusers were not deterred by a TRO or an arrest. The urban victims were slightly less likely to call the police, obtain TROs, or tell the police when their abuser violated the TRO. They were slightly less likely to say police responded every time they were called and slightly less satisfied with the police response than suburban victims. They were also more likely to say that they had never left their abuser before and more likely to believe that mandatory arrest did not work (while a slight majority of suburban victims also did not believe that mandatory arrest worked). Again, the definition of success needs more exploration when considering what, in fact, "works" to address domestic violence.

The urban victims in this study were slightly different from the suburban victims. All suburban victims said they easily obtained TROs (not all of the urban victims did), most of them had less severe injuries, and the abuse that they experienced lasted a shorter time for most and began later into the relationship. All suburban victims had called the police at least once. Some urban victims had never called the police at all.

This was a pilot study, and the results cannot be generalized. However, they do fit in with several findings in the literature such as negative perceptions of police by poor inner-city residents and positive views of the police by suburban middle-class residents.

The key findings of this pilot qualitative study were (a) that many victims of domestic violence believed that the police who responded to them had followed the law with respect to domestic violence; and (b) that most believed mandatory arrest and TROs were not panaceas.

Further research is needed on a representative sample of the 2,000 shelters nationwide (at least five shelters from each of the 12 federal regions of the United States). The perceptions of the shelter directors, residents, and police should all be included.

REFERENCES

Balos, B., & Trotzky, K. (1988). Enforcement of the Domestic Abuse Act in Minnesota: A preliminary study. *Law and Inequality, 6,* 83–125.

Bureau of Justice Statistics. (1999). *Domestic violence: Crime in New Jersey, homicide trends in the U.S.* Washington, DC: U.S. Department of Justice.

Buzawa, E. S., & Buzawa, C. G. (1993). Introduction: The impact of arrest on domestic violence. *American Behavioral Scientist, 36,* 558–574.

Buzawa, E. S., & Buzawa, C. G. (1996). (Eds.). *Do arrest and restraining orders work?* Thousand Oaks, CA: Sage.

Chaudhuri, M., & Daly, D. (1992). Do restraining orders help? In E. S. Buzawa & C. G. Buzawa (Eds.), *Domestic violence: The changing criminal justice response.* Westport, CT: Greenwood.

Erez, E., & Belknap, J. (1998). In their own words: Battered women's assessment of the criminal processing system's response. *Violence and Victims, 13,* 251–268.

Federal Bureau of Investigation. (2001). *Crime in the United States, 2000* (Uniform Crime Reports). Washington, DC: Government Printing Office.

Federal Bureau of Investigation. (2002). *Crime in the United States, 2001* (Uniform Crime Reports). Washington, DC: Government Printing Office.

Forell, C. (1990). Stopping the violence: Mandatory arrest and police tort liability for failure to assist battered women. *Berkeley Women's Law Journal, 6,* 215–262.

Hoctor, M. (1997). Domestic violence as a crime against the state: The need for mandatory arrest in California. *California Law Review, 85,* 643–700.

Holmes, W. M. (1993). Police arrests for domestic violence. *American Journal of Police, 12*(4), 101–125.

Jones, D. A., & Belknap, J. (1999). Police response to battering in a progressive pro-arrest jurisdiction. *Justice Quarterly, 16,* 249–273.

Martin, M. E. (1997a). Double your trouble: Dual arrest in family violence. *Journal of Family Violence, 12,* 139–157.

Maxwell, C. D., Garner, J. H., & Fagan, J. A. (2001). *The effects of arrest on intimate partner violence: New evidence from the Spouse Assault Replication Program.* Washington, DC: National Institute of Justice.

New Jersey State Police. (1998). *New Jersey Police Manual.* Newark, NJ: Gann Law Books.

New York State, Office for the Prevention of Domestic Violence and Division of Criminal Justice Services. (2001). *Evaluation of mandatory arrest provisions: Final report to the governor and legislature.*

Roberts, A. R. (Ed.). (2002). *Handbook of domestic violence intervention strategies.* New York: Oxford University Press.

Zorza, J. (1994). Women battering: High costs and the state of the law. *Clearinghouse Review, 28,* 382–395.

13

Homicide Investigations, Trends, and Forensic Evidence in the 21st Century

Norman B. Cetuk

The homicide investigator, the first police officer who enters the scene, the detective who responds to investigates the circumstances surrounding the death, or the medical examiner who is responsible for determining the cause of death enters a world that the ordinary person seldom experiences or even thinks about. The exposure that individuals ordinarily have to violent death investigations is through television, books, newspaper accounts, or limited personal experiences. When investigators enter a violent death scene or homicide scene, they must be prepared to confront not only a gruesome sight of human suffering but also their private thoughts, feelings, and fears about their own death and the death of their loved ones. Investigators need to put these fears and questions aside and objectively examine the evidence that is and is not at the scene. It is essential that they have a clear understanding of their role in the investigation. They need to be trained in the latest forensic sciences and either have these services available or know where to obtain them. Thorough investigators need to be able to draw on past experiences. They must have an understanding of human behavior. They must have knowledge and understanding of causes of death. Finally, they need to understand

homicide's nature and current trends. Not only does this aid them in reconstructing the scene and help them formulate a motive, but they are required by the courts to possess a current knowledge and understanding of homicide and to be able to articulate this knowledge at the time of the trial.

Most of the violent deaths in the United States are caused by accident or suicide. But of these violent deaths in 1999, 5.7 per 100,000 inhabitants in the United States were due to murder (Bureau of Justice Statistics [BJS], 2000, p. 279). When murder investigations are examined, certain patterns develop and probabilities begin to take form. The patterns and the probability of the victim-suspect relationship, types of weapons used, cause and manner of death, time of day, season, past police contact, victim-suspect sex, race, and ethnicity are but a few of the homicide trends that an investigator needs to consider during an investigation. Homicides frequently fit into predictable patterns that, if recognized and understood by the investigator, may give understanding and insight into the suspect's mentality and motive. Perhaps this information will protect the police officer and investigator during the investigation and shorten the process of apprehending and bringing the suspect before the criminal justice system.

This chapter will present, discuss, and analyze the current trends of homicides in the United States. Investigators will be able to use it to gain a general understanding of the patterns and trends of this crime. Two case studies and an analysis of the trends of homicide will show how this information can be practical and useful in helping the criminal investigator, in any given instance, to understand how and why a crime was committed and to identify possible suspects.

CASE STUDY 1: STATE V. REED

At approximately 8:00 a.m. on a Monday, the local police received a telephone call from a woman named Fran. She told the police that her live-in boyfriend, John, age 26, had found the dead body of one of John's coworkers, Susan, age 24. The victim's body was reportedly in her apartment, where she lived by herself. The police met the caller, Fran, and her boyfriend, John, at Susan's apartment. After speaking with Fran and John, the police entered Susan's apartment, where her body was found on the living room floor. She had been stabbed, and her pants and underwear had been pushed down around her knees. At the trial, the medical examiner testified that Susan had been stabbed 53 times. Twenty-three of the wounds were potentially fatal, penetrating her abdomen, liver, lungs, and heart. In addition, a strong blow to her head had fractured her skull.

At Susan's apartment, John told the police that Susan had called him the previous Friday, terrified that a "black man" was pounding on the window. John told the police that when he arrived at Susan's apartment, no one was in sight. He spoke with Susan for a short time and left. The following day, Saturday, John was to meet Susan at her apartment for a dinner date at 5:00 p.m. However, no one answered the door. John attempted to reach Susan by telephone the remainder of the weekend but could not contact her. John decided to stop at Susan's apartment on Monday before going to work. John found the front door of the apartment unlocked, so he entered the apartment and found Susan's body on the living room floor. He called Fran and told her what had happened, and Fran called the police.

John suffered from a speech impediment due to a harelip and a cleft palate, and he stuttered severely when nervous. Detectives attempted to continue the interview with John at his apartment, but John could not be understood. The detectives then asked John if he would accompany them back to their office. John voluntarily agreed to accompany the detectives back to the police department, where he would give a statement and provide elimination fingerprints. Because John was very upset, Fran informed the police that she would drive John to the police department.

When they arrived at the police department, the detectives separated John and Fran. The detectives took John into the interview room. Fran was asked to remain in the waiting room. At this time, Fran called a family member and received the name of an attorney. Fran then asked the police not to question John until she was able to contact an attorney. A while later, the attorney arrived to represent Fran and John. At this time, the police moved John to another location for further questioning and never informed John that Fran had retained an attorney for him.

While at police headquarters, John was advised of the Miranda warnings. John acknowledged that he understood his rights and signed a waiver form indicating so. John never requested to speak with an attorney, and the detectives never informed John that Fran had retained an attorney for him. At this time, John gave a second account of his actions on Monday that was somewhat different than his first account. John now claimed that on Monday morning he had entered Susan's apartment, discovered her body, covered it with a jacket and pillow, and then gone to work. At the conclusion of the second statement, John was asked to take a polygraph examination, to which he agreed.

During this time, the attorney that Fran had retained contacted the county prosecutor. The prosecutor informed the attorney that John was a witness and not a suspect and that "he [the attorney] had no right to walk into an investigation." The attorney gave the prosecutor a business card and asked that John call him if and when he wanted an attorney.

At about noon, John was scheduled for a polygraph examination. John was again given the Miranda warning, and he stated that he understood his rights. In preparation for the examination, John was again questioned. His story was now markedly different from his previous statements. In this third version, John was looking through Susan's front window when he observed a "black man" repeatedly stabbing her. After hearing John's new statement, police left him alone to think about what had happened.

After a short time, John was again advised of the Miranda warning, and he indicated he understood his rights. During questioning, John admitted that he had killed Susan. According to John, Fran had gone away for the weekend, and he felt "depressed and weak." After Fran left, John drank three cans of beer and went over to Susan's apartment. John stated that they had talked for a while; then Susan had become seductive and started to disrobe. John told Susan he did not want to have sex, and as a result Susan started calling him names. As they were arguing, Susan brandished a knife. John asked her to put it down and to "back off." Susan asked John to leave but suddenly grabbed a board and came toward him with it. John stated that he then "freaked out and stabbed her." John could not remember where he had gotten the knife from or how many times he had stabbed her.

John was convicted in the Superior Court of murder and aggravated criminal sexual contact. However, the verdict was appealed by the defendant. The Supreme Court concluded that the trial court erred in admitting John's confession into evidence because the police failed to inform John that an attorney was present and had asked to speak with him. This failure by the police violated John's state privilege against self-incrimination. As a result, the Supreme Court of New Jersey reversed the defendant's conviction and remanded the case for retrial. Because John had served 7 years in state prison between the time of his first conviction and the Supreme Court's ruling, and because John was now eligible for parole, the state elected not to retry this case. John has been released from prison.

Case Study 2:
State v. Demarco

On a clear, cold, Saturday in December, a local hunter found the dead, nude body of a young woman lying in an open field. The body lay at the top of a hill adjacent to a dirt road. The victim had been stabbed numerous times in the chest, and there were many deep cuts around the neck, almost severing her head from her body. Numerous defensive wounds were visible on the victim's hands and arms. At autopsy, the medical examiner documented the following injuries:

Eight anterior, lateral, and posterior incisions on the neck

Six stab wounds to the left breast and left lateral chest

One stab wound to the abdomen

Incision, right anterior chest

Stab wound to right lower abdomen

Two incisions to the right forearm

Stab wound to the right back

Stab wound to the central back

Stab wound to the lower back

12 superficial and deep stab wounds to the left lateral chest and back

Five superficial incisions to the left upper arm

Four incisions to the right thumb and the second and third fingers, consistent with defense wounding

Four superficial incisions to the left forearm and left thumb, consistent with defense wounding

The only items on her body were two rings, a necklace, bracelet, and one knee-high stocking. A short distance away from the body was a pile of bloody clothing. This pile contained dark blue pants, a dark blue jacket, a gold shirt, one knee-high stocking, one black high-heeled shoe, a bra, and underwear. There was no personal identification of any kind either on the body or in the clothing.

During a crime scene search, a yellow crumpled-up piece of paper was discovered approximately 50 feet from the body. Close examination of this paper revealed that it was a blood-stained receipt from a gas station dated 2 days earlier. Besides the clothing and the receipt, nothing else was found at the scene.

A photograph of the victim was taken to the gas station in an attempt to identify who the victim was. The gas station attendant identified the victim as a 26-year-old woman named Karen who lived in the neighboring town. Karen's parents were contacted and positively identified her. At this time, it was learned from her parents that on the preceding night Karen had called her 29-year-old boyfriend, John, and had a very upsetting argument with him. At two o'clock in the morning, contrary to her father's advice, Karen decided to travel the 25 miles to John's apartment to retrieve her personal possessions from his apartment. She was never seen alive again.

As the crime scene investigation was continuing and attempts were being made to identify Karen, John was up and traveling around the town. During the morning, John called the local police and reported that when he had left his apartment early that morning he had observed Karen's car parked on the street outside his apartment. He asked the police to be on the "lookout" for her because he did not know where she was, and he was concerned for her and thought something might have happened to her. John decided to leave a note on Karen's car windshield for her to come up to his apartment when she returned. After placing the note on Karen's windshield, John got into his car and traveled to a flea market 30 miles away to look around for the afternoon.

Learning that Karen had an argument with John, the police went to John's apartment to interview him. When they got there, they discovered a black high-heeled shoe at the base of the staircase leading to John's apartment. This shoe later turned out to be the matching shoe to the one found in the pile of clothing at the crime scene. John returned home late that afternoon and was questioned by the police. John stated that he and Karen had talked on the phone the preceding night but had not had an argument. He had been unaware that Karen was coming over to his apartment after the phone conversation and was surprised to find her car parked outside his apartment that morning. John stated that he had no knowledge of Karen's death. When asked what he thought might have happened, John said he feared that when Karen had gotten out of the car in the early morning hours she had been approached by an unknown individual who forced her into his car, abducted her, and killed her.

John was asked to take a polygraph examination concerning the death of Karen, but he refused and indicated that he wanted to speak with an attorney prior to any further questioning. John was given the opportunity to call his attorney and was accompanied by a detective to a private office to use the telephone. The detective was not working on this investigation and had no knowledge of the facts of the investigation.

John was unable to make contact with anyone on the telephone. As he sat at the desk, John put his head in his hands and said to the detective who had accompanied him to the office, "I bet I know what happened." The detective did not respond. John continued, "Karen must have driven over to see me, and when she was getting out of her car, she was approached by a guy who asked her if she wanted to have a good time. She must have told him no, and the guy grabbed her, forced her into his car, drove her to the field, where he tried to rape her. Karen probably fought back, jumped out of the car, ran up the hill, where he stabbed her and killed her." The next day, the detective reported John's description (statement) of what had occurred. As the detective recounted John's story, the detective assigned to the case recognized that John's account matched the facts of the investigation that John should not have known. This was John's confession.

A few days later, a weapon was found lying on the side of the road some distance from where Karen's body had been found. Forensic examination identified fibers on the knife as coming from John's car seat, Karen's pants, Karen's jacket, Karen's gold shirt, and a pair of green leather gloves recovered from John's car. Other fiber evidence that was identified included blue acrylic fibers that were found on John's blue jeans, Karen's pants, and John's car seat, but the origin of this fiber evidence could not be determined. In addition, the knife's blade width, shape, and length matched a bloodstained pattern of a knife that was on Karen's gold blouse.

Transferred fiber evidence between two subjects who know each other is not very significant. But in this case, Karen had bought the blue pants, blue jacket, gold blouse, and shoes the day before her murder. John told the police that he had not been with Karen and had not seen Karen for the 3 days preceding her death. If John had no contact with Karen in her new clothing, then how did the fiber evidence transfer between the knife, the car, John's clothing, and Karen's clothing?

At trial, the jury heard testimony as to the cause and manner of Karen's death and the events concerning John's alibi and claim of innocence, and they examined the evidence. After deliberation, John was found guilty of the murder of Karen. He is currently serving a life sentence in prison.

ENTERING THE DEATH SCENE

When new investigators walk into a violent death scene the first couple of times, they are totally focused on the sights, sounds, and smells that are before them. The violent scene is nothing like what they have ever experienced before. Often inexperienced investigators want to get to the scene as quickly as possible and then jump right in and "busy" themselves with securing the scene, preserving evidence, documenting the scene in diagram, photography, and video, recording the details of the scene in the notebook, logging and properly collecting the evidence, working with the medical examiner, helping the funeral home remove the body, and at times even trying to clean the area to lessen the impact of the scene on those present. While they try to detach themselves from the emotions that the scene stimulates, it is easy for them to become absorbed in detailing the position of the body, cataloging evidence, interviewing witnesses, and responding to the many questions of the other investigators and supervisors. But at times they miss the key element of the investigation—observation of not only what is present but also, just as important if not more important at times, what is not present.

By contrast, experienced investigators at the time of dispatch begin to pull themselves together. They organize their thoughts either mentally or on a notepad that is on the car seat next to them. As they drive to the scene, they take mental and written notes of the area they are driving into, the neighborhood, the people standing on the street corners, the vehicles leaving the area, the weather conditions. As they pull up to the scene, they take special note of the lighting in the area and the sounds and

odors that are present. After ensuring that the victim exhibits no signs of life, they secure the area around the victim.

The veteran investigator then steps back from the area to note the obvious and question the details. If the death scene is outside, the investigator must turn his or her back to the victim and survey the surrounding area. How did the victim get there? Where did the suspect come from? Where did the suspect go? And as Edmond Locard would ask, what did the suspect bring to the scene and what did the suspect take with him? Locard's "exchange principle" is: "It is not possible to come into contact with an environment without changing it in some small way, whether by adding to it or by taking something away."

When entering a dwelling or a business, the investigator needs to note any signs of forced entry and identify the exit paths that could have been taken by the perpetrator. If no forcible entry is detected, was the perpetrator known to the victim? Was the perpetrator invited in? If so, why was he or she invited in? Did the victim know the perpetrator?

The experienced investigator will not use the phone or the bathroom or turn light switches on or off. Evidence may be destroyed if these items are touched before they are examined for latent prints or DNA. The investigator will walk in and out of the scene on a path that has been checked for evidence and that will not further alter or destroy anything of evidential value. After the area is photographed and a video is taken, the investigator then examines the area or every room in a slow, methodical, deliberate way, constantly repeating a methodical search as he or she moves from area to area or room to room.

First, the investigator needs to note how the area is organized. Was this an area that few people visited? Was this a room that was well kept, clean, and orderly? Only when the investigator understands how the area or room was can he or she begin to understand what is out of place, what was touched or moved, and, just as important, what was not touched or not moved.

If the house is an orderly home and the living room desk has been torn apart and papers scattered on the floor while the rest of the room looks clean and well kept, the investigator must ask, What was the purpose? What was being sought? Could it be bank records, money, PIN numbers, or credit cards? If the pillowcases have been removed from the pillows in the bedroom, were they used to carry away jewelry or proceeds from a burglary? Were the bedroom closets entered and searched? Were guns or hidden monies taken? Who knew these items were there? Has the victim's pocketbook been emptied? Everyone carries bank cards and credit cards; are they missing? As soon as it can be determined that credit cards or bank cards are missing, the investigator must gather all the information that is available—card account number, PIN number (if available), name on the card, and bank. It is essential that the investigator immediately contact the financial institution, not closing the account but allowing the account to remain open so that account activity can be documented. The investigator must assure the financial institution that the investigating agency is prepared to accept any loss. If this assurance is not given, the financial institution will cancel the account in an attempt to minimize their loss. Working closely with the financial institution, the investigator can track the movement of whoever is using the stolen cards.

The investigator must also understand that people are creatures of habit. Persons using a stolen credit card taken from a dead person are certainly under stress. Under stress, people tend to repeat behaviors that have been successful in the past. If an individual using a stolen credit card has been successful at a particular ATM machine or at a particular type of store or location, then it is advisable to assign a detail to monitor that location. Individuals who have stolen jewelry or property revisit the same location time and time again to sell these items. Repeatedly, this type of investigation, the identification of ATM machines or locations that will buy stolen property, has proven successful.

Another consideration for the investigator is the wounds on the body. What violence occurred between the victim and the perpetrator? Was there violence, and what occurred before death? How was the person killed? What was done to the victim after death? Are there signs that the victim fought back? Are there defensive wounds? Was there "overkill"? Was the victim stabbed numerous times or beaten with an object repeatedly? Repeated violence or "overkill" could

indicate that the perpetrator was lashing out in emotional frustration. It could indicate that an argument took place between two people who knew each other and this death was not a random act of violence or the result of another crime, such as a robbery or theft. Was the weapon brought to the scene of the killing and used in the commission of another crime, or was the weapon already at the scene and used by the perpetrator during an uncontrollable emotional outburst? Examination of the wounds will help the investigator narrow down the list of suspects and focus on particular suspects.

An investigator needs to stand back and examine the entire scene. The investigator must also understand that what is not at the scene is just as important as what is at the scene. He or she must not jump to any conclusions but must examine all of the evidence before drawing a conclusion. At times what appears obvious at the beginning of an investigation later turns out to be the opposite. Such investigations are referred to as equivocal death investigations.

For example, in one case, when a man returned home from work, he found his wife, a woman in her late 40s, dead in her bed. The woman was lying on her bed, on her back, and the bedsheets were pulled up to her neck. A large kitchen knife was impaled in her neck. Only the knife's handle could be seen. Next to the body was a second knife. Rigor mortis indicated that the women had died about 6 to 8 hours earlier. The husband had left for work 10 hours earlier and had remained at work until he returned home that evening. A check with his employer substantiated his alibi. An examination and search of the house showed that there was no forcible entry and that all windows and doors were locked and secured.

Examination of the victim revealed that both her wrists and inner elbows had been cut and had bled. There were 11 stab wounds to her lower abdomen, but none penetrated the abdominal muscle. Examination of her neck revealed numerous superficial horizontal cuts, cuts that penetrated only the first layer of skin. Blood spatter analysis showed that as the victim lay in bed as she bled, her blood pooled on the sheets and mattress. Also observed on her legs were low-velocity blood drops that had dropped from the vertical position. Examination of the bedsheet indicated that the victim had been sitting on the side of the bed when she was bleeding from her wrists and inner elbows. When one first entered the scene, it was easy to make the assumption that the woman lying in bed had been the victim of a brutal attack.

The bedroom as well as the entire house showed no signs of violence. No one of the wounds to the victim was fatal. Even the knife that was plunged into her throat did not strike any vital organs. She had bled to death as a result of the neck injury and the cuts to her wrists and arms. The knives were taken from the kitchen counter. But when you looked at the scene, something was still missing. Examination of the victim revealed that she was not wearing any jewelry. She was not wearing her wedding ring or any other rings, bracelets, or necklaces. The victim's jewelry box was closed and on her dresser. When the jewelry box was opened, under her wedding ring was a handwritten suicide note.

What was important in this investigation was not the violence in the bedroom but where the violence was. The only violence was to the victim's body. There were no signs of a struggle. There were no defensive wounds to the hands or arms. The entire house was neat and orderly. There was no theft, sexual assault, or burglary; there was no other crime. The only violence was to her body, self-inflicted torture and pain. Later, a check with her doctor revealed that the victim was anorexic (a form of self-torture) and depressed. She had attempted suicide on a number of previous occasions. There was no need to pursue any further investigation.

Determining the Cause of Death

Determining the manner and cause of death is the responsibility of the medical examiner. However, the death investigation is a joint investigation between the medical examiner, the crime scene investigator, and the criminalist. All parts of the team must work together to successfully piece together the circumstances and events of a particular incident. The medical examiner is responsible for determining the cause of death and the identity of the victim. The investigator is responsible for the identification, preservation,

documentation, and collection of evidence associated with the cause and manner of death. The criminalist is responsible for the examination, analysis, and comparison of evidence as it relates to the scene and investigation.

The cause of death of an individual loosely falls into one of two broad categories, termed natural or unnatural. Natural causes of death cover everything from stillbirth to the degenerative processes of advanced age and can be linked to a fatal disease or illness. Unnatural or premature deaths are deaths that occur under unusual or suspicious circumstances. Unnatural deaths can be accidental or intentional, but in either case they require an official investigation. Often it is not easy to distinguish an accidental from an intentional death. The medical examiner and the criminal investigator must work together and understand each other to properly investigate the cause and manner of death. For example, in one case a police officer on the job for a number of years attempted to take his own life by ingesting a large quantity of prescription medication. However, family members discovered him alive, unconscious, and unresponsive. He was transported to a local hospital, where he was admitted and remained for 5 days in a coma and on life support. On the sixth day he died. The death certificate listed the cause of death as renal failure, a natural cause of death. But the death certificate did not report that the renal failure was the result of the officer's intentional ingestion of a dose of medication, which caused his kidneys to fail. Nowhere was there the mention of suicide.

Homicide Trends Overview

It has been my experience that approximately 20% of all people that die in America die under circumstances that require some type of official investigation into the cause of death. This investigation can range from a routine autopsy that determines the death was one from natural causes, such as a heart attack or Alzheimer's, to a multijurisdictional investigation involving hundreds or even thousands of investigators, as in the cases of the 1993 World Trade Center bombing or the bombing of the Federal Building in Oklahoma City. Approximately 10% of all deaths are caused by or the result of violence. These violent deaths are usually a result of motor vehicle accidents, industrial accidents, home accidents, or sporting accidents. Another 10% of all deaths occur under unusual or suspicious circumstances that require an official law enforcement investigation—for example, suicide, self-defense, or cases where no cause of death can be immediately determined. Included in this 10% is less than one death in 20,000 that is the result of murder (BJS, 2000).

It must be understood that this murder rate, reported by the U.S. Department of Justice's Bureau of Justice Statistics (BJS), is a national number and does not apply to every city or town. The murder rate in parts of the rural United States varies greatly from that of the inner cities, where a concentrated population, gang violence, and drug activity are more common. Also, in my experience, sometimes no murders may occur in an area for a long time and then, over a short time, this same area may experience an unusual high number of violent deaths or murders. For example, a small residential community located in the central part of New Jersey, populated by approximately 2,000 families, had not experienced a murder for over a quarter of a century. But during a 3-year period in the 1990s, this community experienced six murders: three that were the result of domestic or family disputes, one that may have been the result of a serial murder, and a double murder that was the result of a home intrusion and robbery.

During the past 25 years, the homicide rate in the United States has declined. Homicide rates have decreased to levels last seen in the 1960s. In 1991, 24,700 murders occurred, setting the high mark for violence in the United States. Every year following 1991 has shown a steady decrease in the number of murders. The latest report by BJS reported that 15,530 murders occurred in the United States in 1999. There were 5.7 murders per 100,000 citizens in 1999 compared to 9.8 murders per 100,000 citizens in 1991 (BJS, 2000, pp. 309–316).

The significance of studying the murder rate in the United States is that it is a fairly reliable barometer of all violent crime at a national level. The higher murder rates of the late 1970s, the 1980s, and the first two thirds of the 1990s parallel higher rates in the other crimes of the Violent Crime Index (forcible rape, robbery, aggravated assault, burglary, larceny-theft, and motor vehicle theft).

When one examines the ages of those committing murder in the 1980s, one can see an increase in gun violence by juveniles and young adults (aged 13–24) but a decrease in gun violence by adults aged 25 and older. Disturbingly, even though there has been a decrease in the U.S. murder rate, the level of gun homicides by juveniles and young adults today is well above that reported in the early and mid-1980s (Bureau of Justice Statistics, 2000, p. 309). Is this increase in youth violence attributable to more guns available to U.S. youth? Are U.S. youth today more prone to use a gun than they were in the 1980s? Do youth who commit murder view it as an acceptable answer to a dispute or an acceptable way to get what one wants? Have the motives for murder changed between the 1980s and the present? Is murder viewed as an acceptable resolution of a dispute? Is it now acceptable to murder an individual who has shown you or your gang "disrespect"? Has there been a change in interventions targeting troubled youth? No final answers can be given to explain murder. But if we examine and try to understand who the offenders and the victims are, then maybe as concerned citizens and as a nation we can attack this epidemic.

The statistics on youth violence are disturbing. The following information summarizes the BJS (2000) and the Federal Bureau of Investigation (2000) concerning offenders committing murder and non-negligent manslaughter who are known to police. White males aged 14 to 24 years account for 18% of the murders in the United States. Black males aged 14 to 24 years account for 27% of the murders. Together they account for 45% of all murders in the United States.

The statistics on the victims of murder and non-negligent manslaughter known to police are just as troubling. If you examine the total number of murders between 1964 and 1990, the sex of the victims has remained relatively constant. The proportion of male victims of murder for this time period overall was 74%. In 1971, this proportion reached a high of 79%; in 1998, it was 75%; and in 1999, it was 76%. This figure has remained within this range, even though the total number of homicides varied (a high of over 23,271 in 1993, a low of 7,990 in 1964, and 12,658 in 1999). White males between the ages of 14 to 24 account for 10.3% of the victims. Black males in the same age range account for 14.9%. This age range accounts for one quarter, or over 3,900, of the victims of murder.

Black female victims between the ages of 18 to 24 years are 4 times more likely to be the victim of murder than white females. Black males are about 10 times more likely to be the victim of murder than white males within the same age range.

According to the National Commission on the Causes and Prevention of Violence in the United States (BJS, 2000, p. 316), during recent years 45.7% of the murders in which the murderers were known and convicted involved black persons killing black persons, 44% involved white persons killing other white persons, 7.2% involved black persons killing white persons, and 2.6% involved white persons killing black persons. Thus, persons of the same ethnic group kill others in the same ethnic group more often than they kill persons of different ethnic groups.

Murders tend to take place at certain hours of the day, on certain days of the week, and in certain months of the year. Generally, most murders occur during nonworking hours, between 8:00 p.m. and 2:00 a.m. There are more murders on Saturdays and Sundays. There are fewer murders in February and March and more murders in July and December. Could a possible explanation be that social events and circumstances that bring people together, such as holidays, vacations, and family events, combined with increased alcohol consumption and drug abuse, increase the intensity of interpersonal interaction and the potential for confrontation and crisis? There are no data to support this supposition, but during my past 30 years in law enforcement, I have seen that alcohol consumption or drug use was almost always present and a contributing factor in domestic violence and social confrontations.

Most murder victims were familiar with or at least knew their assailant. Statistics on the relationship of the victims and the offenders are disturbing. Of the 12,658 murders in the United States in 1999, 6,037 murders (48%) were by persons who knew or were in a relationship with the victim. In 1,743 murders (14%), the offender was a family member; in 3,213 (25%) of the murders, the victim and the offender were not family members but knew each other. For every age group, female murder victims are more likely than male victims to have been killed by someone who was known to them. Of the 3,085 female murder victims in 1999, 1,323 or 43% were murdered by someone identified as either a family member or an acquaintance of the victim. In 1,506 (12%) of the murders where there was a female victim, the murder was committed by a stranger. (Some of these numbers might vary a little from year to year.). In 5,115 (40%) of the 12,658 murders of male and female victims that occurred in 1999, the offenders could not be identified.

White females account for the most victims killed by an intimate (someone in a friendship or romantic relationship with the victim) during each of the past 15 years. In 1999, white females accounted for a little over 800 victims, black females accounted for a little below 400 victims, and white males and black males accounted for about 200 victims each. Women between the ages of 25 and 49 were the most likely to be killed by someone they had been friends with or intimately involved with—a spouse, ex-spouse, or boyfriend. However, during the past 15 years, murders of women by their spouse have steadily declined to a low of 800 in 1999. The numbers of women who were killed by their ex-spouse or boyfriend remained relatively constant during this same period, with under 100 murders being committed by ex-spouses and approximately 750 murders being committed by a boyfriend or girlfriend.

During 1999, in 855 (7%) of the murders, the offender used his or her physical strength—that is, his or her fist, feet, and/or hands—to kill the victim. In 8,259 (65%) of the murders, a firearm was used. In 1,667 (13%) of the murders, a knife or other cutting instrument was used, and in 736 (9%) of the murders, a blunt instrument was used. The age group of 20 to 24 years more frequently used a firearm than other age groups. The age group of 40 to 44 years used their hands, their feet, or a blunt instrument more frequently than did other age groups.

It is apparent that the deliberately planned killing is not too common. In the majority of murders, the perpetrator has not planned out the consequences of his or her actions, and as a result the perpetrator makes mistakes and takes action that is incriminating if the investigator can identify it or recognize it.

What can be done to stop family members from killing loved ones? What can be done to stop friends and acquaintances from killing each other? Are harsher criminal penalties, more gun control legislation, or a police officer at every corner going lessen to the rage that people feel when they kill someone they love or have a relationship with? Do we need to teach and reinforce in our children that rage, anger, and frustration need to be talked about, not acted out?

Conclusion

Homicide investigation is, generally speaking, an investigation of a crime, such as assault, sexual assault, robbery, theft, burglary, or a domestic disturbance, that results in the intentional ending of a life. Patterns of homicide can be observed and studied, such as victim-suspect relationship, types of weapons used, causes and manner of death, time of day, season of the year, past police contact, victim-suspect sex, race, and ethnicity. Homicide frequently falls into patterns that, if recognized by the investigator, may give the investigator insight into who the suspect is and the suspect's mentality. Knowing that the murderer frequently calls the police to report the murder and remains at the scene not only can aid the police in the investigation but can also protect the police officer or investigator who is at the scene.

Understanding the trends and predictable patterns of homicide helps the investigator bring the perpetrator to a swift arrest and facilitate the prosecution of murder. In addition, the courts require the investigator to have more than a basic understanding and knowledge of homicide trends and investigative procedures.

Finally, the study of homicide trends is a reliable barometer of all violent crime across America. No other crime is measured more accurately and precisely. This aids social agencies and legislatures in funding initiatives, developing programs, and designing interventions in crime prevention and enforcement. Understanding the trends of homicide has also aided in aggressive enforcement and prosecution of offenders during the past 10 years. As the statistics indicate, this enforcement and prosecution has resulted in a reduction of crime and homicides across the United States.

References

Bureau of Justice Statistics. (2000). *Source book of criminal justice statistics—2000.* Washington, DC: U.S. Department of Justice.

Federal Bureau of Investigation. (2000). *Crime in the United States.* Washington, DC: Government Printing Office.

Additional Suggested Readings

Adams, T. F., & Krutsinger, J. L. (2000). *Crime scene investigation.* Upper Saddle River, NJ: Prentice Hall.

DiMaio, V. J. M. (1985). *Gunshot wounds, practical aspects of firearms, ballistics, and forensic techniques.* New York: Elsevier Science.

Geberth, V. J. (1996). *Practical homicide investigation* (3rd ed.). Boca Raton, FL: CRC.

Goode, E. (2001). *Deviant behavior* (6th ed.). Upper Saddle River, NJ: Prentice Hall.

Kaci, J. H. (2000). *Criminal evidence* (4th ed.). Incline Village, NV: Copperhouse.

Lee, H., Palmbach, T., & Miller, M. T. (2001). *Henry Lee's crime scene handbook.* San Diego, CA: Academic Press.

Saferstein, R. (2001). *Criminalistics: An introduction to forensic science* (7th ed.). Upper Saddle River, NJ: Prentice Hall.

Saferstein, R. (2002). *Forensic science handbook* (2nd ed.,

Vol. 1). Upper Saddle River, NJ: Prentice Hall.

State of New Jersey v. John Robert Reed, 133 N.J. 237 627 A.2d 630, 62 USLW 2130, Supreme Court of New Jersey.

Wetli, C. V., Mittlemean, R. E., & Rao, V. J. (1988). *Practical

forensic pathology.* New York: Igaku-Shoin Medical Publishers.

Wilson, C. (1991). *Clues! A history of forensic detection.* London: Warner Communications.

Wilson, D. (2000). *The mammoth book of the history of murder.* New York: Carroll & Graf.

PART IV

CRITICAL ISSUES IN JUVENILE JUSTICE

14

AN OVERVIEW OF JUVENILE JUSTICE AND JUVENILE DELINQUENCY

Cases, Trends, Policy Shifts, and Intervention Strategies*

ALBERT R. ROBERTS

This overview chapter provides an orientation to the critical issues, trends, policies, programs, and intervention strategies of the juvenile justice system. It lays the groundwork for this part of the text on juvenile delinquency and the juvenile justice process by describing the types, functions, and legal responsibilities of the various juvenile justice agencies and institutions, discussing the legal definitions of juvenile status offenses and juvenile delinquency, outlining the scope of the problem in terms of official and unofficial delinquency statistics, and describing punishment-oriented and rehabilitation-oriented policies and programs.

Generally speaking, youths can be charged with two types of wrongdoing: delinquency offenses, which are criminal acts (e.g., auto theft, breaking and entering) for which they would be held accountable if they were adults, and status offenses (e.g., truancy, incorrigibility, and running away from home), which are illegal only for juveniles.

The following are some case illustrations of a variety of juvenile criminal offenses and responses of the juvenile justice system:

- A 15-year-old male runaway was before the juvenile court judge for possession of a concealed, unloaded .38-caliber handgun. The juvenile probation officer who conducted the predispositional investigation recommended detention for the youth on the handgun charge. The public defender stated that the 3 days the youth had already spent in detention had had a profound effect on him and requested

Author's Note: Adapted and updated from A. R. Roberts (1998). An Introduction and Overview of Juvenile Justice. In A. R. Roberts (ed.). *Juvenile Justice: Policies, Programs and Services* (Published by Nelson-Hall, Chicago), pp. 3–20. The third edition is In Press with Oxford University Press. Copyright held by A. R. Roberts. Reprinted by permission.

leniency. The judge ruled that the youth and his parents had to attend 12 sessions of family counseling provided through the probation department and that the stepfather, who worked as a security guard, had to keep his revolver in a lockbox so that the boy would not be able to take it again.

- A 16-year-old girl named Suzie fought with another girl when it was learned that Suzie had become pregnant by the other girl's boyfriend. During the argument, Suzie cut the other girl with a broken bottle. The juvenile court judge suspended commitment to the girl's training school and placed Suzie in a group home for adolescent girls for 1 year.

- A 15-year-old boy broke into the home of an elderly woman to search for cash and valuables. The woman returned home to find the youth ransacking her house; in a fit of rage, he brutally assaulted her. The judge ruled that the boy, although only 15 years old, was to be tried as an adult. Found guilty by a jury, he was sentenced to 2 years of incarceration.

- Candace, a 14-year-old girl, stole a car and had repeatedly run away from the group home where she was placed because her mother could not handle her at home. Candace's mother had two jobs; the girl had never met her biological father. Candace had been suspended from school three times in the past year for possession of marijuana. Her case record reflected that she had been sexually abused by her mother's boyfriend. The judge adjudicated Candace to the girl's training school until she would reach age 16.

Violent juvenile crimes receive the most media attention and serve to intensify the fear and outrage of concerned citizens. This fear and outrage, in turn, frequently influence prosecutors, juvenile court judges, and correctional administrators to subject more juvenile offenders to harsher penalties. But juveniles who commit status offenses or nonviolent property crimes are far more prevalent than those who commit violent crimes. The following are examples of the latter type of offense:

- A 12-year-old boy was brought to the county juvenile detention center after his mother complained that he was on drugs and was uncontrollable. The social worker's investigation for the court revealed that the mother had been released from the state hospital 3 months earlier. She had a history of psychotic episodes. The judge ruled that the boy should live with the aunt and uncle with whom he had lived while his mother was confined in the state hospital.

- A 16-year-old boy whose mother was an alcoholic had been reared in a home in which he was neglected and received no discipline or limit setting. He had a history of 12 arrests for petty theft and shoplifting starting at the age of 11. Following the most recent arrest, he was sent to a rehabilitation-oriented juvenile training school that provided group therapy 3 days a week, a behavior modification program, and vocational training. The youth was learning to be an auto mechanic and proudly demonstrated his knowledge of automobile repair.

Students and practitioners working in juvenile justice agencies on either a volunteer or a paid basis encounter the discretionary, deficient, flawed, and often overwhelming system of juvenile justice. Although the goal of justice-oriented agencies is to care for and rehabilitate our deviant children and youth, in actuality the juvenile justice system all too often labels, stigmatizes, and reinforces delinquent patterns of behavior. The controlling and punitive orientation of many juvenile justice officials has led to a revolving-door system in which we find an overrepresentation of children from neglectful and/or abusive homes.

DEFINING JUVENILE JUSTICE

The juvenile justice system consists of the agencies and institutions whose primary responsibility is caring for, punishing, and rehabilitating juvenile offenders. These agencies and their programs concern themselves with delinquent youths as well as those children and youths labeled incorrigible, truant, and/or runaway. Juvenile justice focuses on the needs of over 2 million youths who are taken into custody, diverted into special programs, processed through the juvenile court and adjudicated, placed on probation, referred to a community-based day treatment program, or placed in a group home or a secure facility.

The history of juvenile justice concerns the development of policies, programs, and agencies for dealing with youths involved in legal violations. The juvenile justice system today comprises the interrelated, yet different, functions of several agencies and programs: the police, pretrial diversion projects, the juvenile court, children's shelters and detention facilities, juvenile correctional facilities, group homes, wilderness programs, family counseling programs, restitution programs, and aftercare programs.

What has been done during the past 100 years to provide juvenile offenders with equal opportunities for justice? Which policies and program alternatives are currently prevalent within the juvenile justice subsystems (police, courts, and juvenile corrections)? What are the latest trends in processing and treating juvenile offenders? What does the future hold? This chapter will explore the answers to these questions.

DEFINING JUVENILE DELINQUENCY

Juvenile delinquency is a broad, generic term that includes many diverse forms of antisocial behavior by a minor. In general, most state criminal codes define juvenile delinquency as behavior that is in violation of the criminal code and is committed by a youth who has not reached adult age. The specific acts on the part of the juvenile that constitute delinquent behavior vary from state to state. A broad definition that is commonly used was developed by the U.S. Children's Bureau (1967):

> Juvenile delinquency cases are those referred to courts for acts defined in the statutes of the State as the violation of a state law or municipal ordinance by children or youth of juvenile court age, or for conduct so seriously antisocial as to interfere with the rights of others or to menace the welfare of the delinquent himself or of the community. (p. 4)

Other agencies define as delinquent those juveniles who have been arrested or contacted by the police, even though many of these

individuals are merely reprimanded or sent home when their parents pick them up. Slightly less than half of the juveniles handled by the police are referred to the juvenile court. These are the children and youth whom the Children's Bureau would classify as delinquents.

The primary concern of juvenile corrections officials is delinquency offenses: acts that are considered illegal whether committed by an adult or a juvenile. Such illegal acts include aggravated assault, arson, homicide, rape, burglary, larceny, auto theft, and drug-related crimes. According to the national and local statutes on juvenile criminality, also known as Juvenile Code Statutes, burglary and larceny are the most frequently committed offenses. The brutal crimes of homicide and forcible rape constitute only a small percentage of the total number of crimes committed by juveniles.

Generally, the definition of what constitutes juvenile delinquency according to the legal statutes holds that any person, usually under 18 years of age, who commits an illegal act is considered a delinquent when he or she is officially processed through the juvenile or family court. A juvenile does not become a delinquent until he or she is officially labeled as such by the specialized judicial agency (i.e., the juvenile court). For example, Ohio defines a "delinquent child" as any child:

a. Who violates any law of this State, the United States or any ordinance or regulation of a political subdivision of the state, which would be a crime if committed by an adult . . .
b. Who violates any lawful order of the court (Page's Ohio Revised Code Annotated, 1994, Ch. 2151.02)

A Montana statute specifies that a delinquent youth is one who has either committed a crime or violated the terms of his or her probation, and the Mississippi statute defines a delinquent child broadly as a youth (10 years of age or older) "who is habitually disobedient, whose associations are injurious to his welfare or to the welfare of other children (Mississippi Criminal Code Annotated, 1997). Therefore, a youth who could be defined as a "delinquent" under the Mississippi statute in many situations would not be so considered under the Ohio or Montana codes.

Police officers and prosecutors handle juveniles differently depending upon the nature, severity, and frequency of the juvenile's acts. The frequency and repeated nature of an illegal act provides evidence of whether an antisocial pattern of breaking the law exists or whether the act is isolated and impulsive. Prosecutorial and judicial discretion with juvenile offenders and beliefs about punishment versus rehabilitation explain why two different youths will be handled differently even if they commit the same offense. Four major factors usually contribute to the disparate treatment, diversion, or formal processing, and adjudication of juveniles as delinquents: (a) the juvenile court judge's belief in traditional incarceration versus community treatment; (b) the *frequency* and *repeated* nature of the acts; (c) the *seriousness* or *severity* of the act; and (d) the attitude (e.g., overt hostility, explosive anger) expressed by the juvenile toward parents and other authority figures.

Definitions of juvenile delinquency may or may not include status offenses such as truancy, incorrigibility, curfew violations, and running away from home: forms of misbehavior that would not be considered crimes if engaged in by an adult. Approximately half of the states include status offenses in their definition of juvenile delinquency offenses. Other states have passed separate legislation that distinguishes juveniles who have committed criminal acts from those who have committed status offenses. In those states, status offenders are viewed as individuals "in need of supervision" and are designated as CHINS, CINS, MINS, PINS, or JINS. The first letter of the acronym varies, depending on whether the initial word is *children, minors, persons,* or *juveniles,* but the rest of the phrase is always the same: "in need of supervision."

AGE LIMITS FOR JUVENILE COURT JURISDICTIONS

All 50 states and the District of Columbia have legal statutes that define an upper age limit for juvenile court jurisdictions. But states differ on the age at which a juvenile's wrongdoing is handled as a criminal (adult) offense rather than a juvenile offense. The highest age at which an offender is still treated as a juvenile ranges from 15 to 18. In the overwhelming majority of states and the District of Columbia, an individual is under the jurisdiction of the juvenile court up through age 17 (see Figure 14.1). However, youths in the states of Connecticut, New York, North Carolina, and Vermont who violate their state's criminal code when they are age 16 (or older) are within the jurisdiction of the criminal court (Hamparian et al., 1982, pp. 18–21).

Although all states specify maximum ages for juvenile court jurisdiction, only a few designate a specific minimum age in their juvenile code. Most state juvenile and criminal codes implicitly follow the English common-law position that a child under the age of 7 is incapable of criminal intent. Therefore, a child below the age of 7 who commits a crime is not held morally or criminally responsible for that act.

An increasing number of state legislatures have determined that juvenile offenders accused of brutal crimes should be processed by the criminal court rather than the juvenile court. Many states have authorized a waiver of jurisdiction that automatically grants the adult criminal courts jurisdiction over certain violent juveniles. For example, in 1978, New York State passed legislation creating a classification of juvenile offenses called "designated felonies." Under this state law, 14- and 15-year-olds charged with committing the crimes of murder, kidnapping, arson, manslaughter, rape, and/or assault were tried in Designated Felony Courts. If the accused offenders were found guilty of the charges, they could be imprisoned for up to 5 years (Prescott, 1982, p. 1). This is an increase over the typical sentence for a juvenile offender, which averages 12 months. (For further information on the criminal courts' handling of and intervention with repeat violent juvenile offenders, see Roberts, 1998, chaps. 10 and 12.)

TRENDS AND SCOPE OF THE DELINQUENCY PROBLEM

Data on juvenile delinquency trends and estimates of the scope of the delinquency problem come from both official and unofficial studies. The three major sources of data on delinquency and victimization are

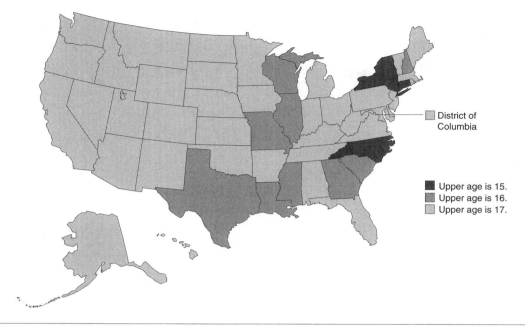

Figure 14.1   Upper Age of Original Juvenile Court Jurisdiction, 1997
SOURCE: Adapted from Torbet and Szymanski (1997).

1. Official police and juvenile court statistics, such as the FBI Uniform Crime Reports; these data are based on crimes reported to the police

2. Self-report studies, such as the National Youth Survey, which involve asking youths whether they have engaged in one or more delinquent behaviors in the past year (e.g., damaging or destroying school property, making obscene phone calls, running away from home)

3. Victimization studies, such as the National Crime Survey (NCS), which involve asking a sample of individuals and household heads whether they have been victims of criminal acts

Official Statistics

One set of official statistics on juvenile delinquency is the number of cases handled by juvenile courts (see Figure 14.2). Juvenile court cases rapidly increased from 405,000 in 1961 to 1.25 million in 1977, with a slight decline to 1.1 million in 1984 and 1985 and then a gradual year-by-year increase to 1.83 million cases by 1996. In 1960, there were an estimated 405,000 juvenile court cases in the United States. By 1975, the number had more than doubled at 1,051,000. The all-time peak is 1,827,621, in 1996. Since the peak in 1996, there was a decline in juvenile court cases for the next 5 years (National Center for Juvenile Justice, 2002).

Since 1972, the U.S. Children's Bureau has issued periodic estimates on the number of juvenile delinquents in the United States. Their figures, based on reports from a sampling of juvenile courts across the country, show a significant increase in the number of crimes committed by juveniles. In 1930, they estimated that there were 200,000 delinquents; in 1950, the figure climbed to 435,000 (Robinson, 1960). In 1966, over 1 million individuals under the age of 18 were arrested. Alfred Blumstein (1967) estimated that 27% of all male juveniles would probably be arrested before reaching their 18th birthday. By 1984, the total number of arrests for youths under 18 years of age had exceeded 2 million (2,062,448). Juveniles over 15 years of age were almost 3 times as likely to be arrested (regardless of the type of offense) as juveniles aged 15 or younger (1,537,688 to 524,760).

The FBI's Uniform Crime Reports give data on juvenile arrests broken down by offense

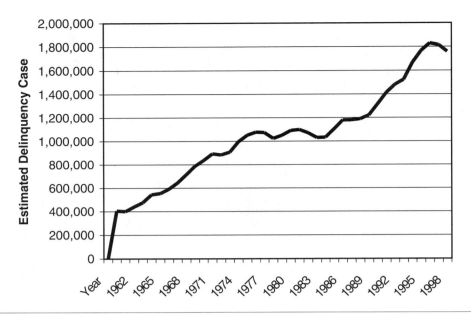

Figure 14.2 Juvenile Court Cases, 1960–1998
SOURCE: Adapted from Puzzanchera et al. (2001).

Table 14.1 Most Frequent Offenses by Juveniles Aged 15 to 18, 1984

Offense Charged	No. of Juveniles Arrested
Larceny/theft	338,785
Burglary	127,708
Running away	114,275
Liquor law violations	101,904
Vandalism	87,135
Disorderly conduct	73,552
Curfew and loitering law violations	67,243
Drug abuse violations	67,211
Motor vehicle theft	33,838

SOURCE: Federal Bureau of Investigation (1985).

category. Table 14.1 lists the most frequent offenses for older juveniles between the ages of 15 and 18. As shown, the top categories are larceny/theft, burglary, running away, and liquor law violations.

We now turn our attention to the eight major index crimes. As shown in Table 14.2, a vast majority of juvenile arrests were for property offenses rather than violent crimes. Property crimes include burglary, larceny/theft, motor vehicle theft, and arson. Violent crimes include murder, forcible rape, robbery, and aggravated assault.

Table 14.2 shows that juvenile crime increases dramatically with age. Juveniles under age 15 were over twice as likely as those aged 15 to 17 to be arrested for property crimes and over three times as likely to be arrested for violent crimes.

The table also shows that of the more than 1.6 million crimes committed by juveniles in 1984, 88% were property crimes, and 12% were violent crimes. Although most of the crimes were property related, there was a noticeable trend toward violent crime as the juveniles become older. In the under-15 population, the proportion of property crimes was 92%, whereas the proportion of violent crimes was 8%. Violent crime increased by 3 percentage points for the 15- to 17-year-olds, and another 3 points for the 18- to 20-year-olds, whose proportion of violent crimes was 14%. These data on juvenile arrests represent only the tip of the iceberg. They do not include status offenses, nor do they reflect the number of juvenile lawbreakers who were not apprehended or who

Table 14.2   Total Arrests of Juveniles for Property and Violent Crime, 1984

Offense Charged	Number of Under Age 15	Juveniles Aged 15–17	Arrested Aged 17–18	Any Age
Property crimes	218,894	506,575	743,801	1,469,270
Violent crimes	18,791	64,344	122,543	205,678
Crime Index total	237,685	570,919	866,344	1,674,948

SOURCE: U.S. Department of Justice (1984).

Table 14.3   Person Offenses Cases Handled by U.S. Juvenile Courts, 1989–1998

Cases Disposed	1989	1994	1996	Percentage 1989–98	Change 1994–98
Total person offenses*	214,300	360,900	403,600	88	12
Violent Crime Index	77,300	131,700	102,600	33	-22
Criminal homicide	1,900	3,100	2,000	6	-36
Forcible rape	4,800	6,600	6,000	26	-9
Robbery	22,900	38,300	29,600	29	-23
Aggravated assault	47,800	83,700	65,100	36	-22
Simple assault	115,000	197,800	262,400	128	33
Case rate**	8.5	13.0	13.9	64%	7%
Total person offenses*	3.1	4.7	3.5	15	-26
Criminal homicide	0.1	0.1	0.1	-8	-39
Forcible rape	0.2	0.2	0.2	10	-13
Robbery	0.9	1.4	1.0	12	-26
Aggravated assault	1.9	3.0	2.2	18	-26
Simple assault	4.5	7.1	9.0	98	27

NOTE: Percent changes are calculated using rounded numbers.
*Total includes others person offense categories not listed.
**Per 1,000 youth aged 10 through the upper age of juvenile court jurisdiction.
SOURCE: Office of Justice Programs (2001a).

committed offenses that were never reported to the police.

Table 14.3 sheds a useful perspective on person offenses cases handled by the juvenile court. Between 1989 and 1998, we see an increase of 88% in the number of cases handled. However, the biggest increase took place between 1989 and 1994. The period between 1994 and 1998 saw an increase of only 12%, minimal compared to the increase during the overall span. Although the total number of juvenile court cases seemed to continually increase over this 10-year period, over the last 5 years of the chart (1994–1998), we see a decrease in the number of juvenile court cases for violent crimes. This 22% drop shows a promising trend with regard to juvenile violent crimes handled

by the juvenile court. It suggests that in recent years there has been a decrease among juveniles committing the most violent crimes, whereas the rates of the lesser violent offenses such as simple assault (e.g., slapping or punching, with no severe injuries) and property crimes have been increasing.

Table 14.4 clearly indicates that violent and property crime among juveniles under 18 years of age peaked in 1994 and has been declining each year since then. During any given period of time, the crime rate may well shift, decreasing, increasing, or fluctuating up and down. Over the 20-year period between 1980 and 1999, we see fluctuations in the juvenile crime rate.

In the first half of the 1980s, mainly from 1980 to 1984, we saw a gradual decline in the

Table 14.4   Juvenile (Ages 10–17) Arrest Rates for 1980–1999 (per 100,000)

Year	Violent Crime	Property Crime	All Crimes
1980	334.09	2562.16	7414.28
1981	322.64	2442.85	7384.78
1982	314.48	2373.32	7344.97
1983	295.98	2244.41	6750.79
1984	297.46	2220.73	6765.81
1985	302.97	2370.74	7245.20
1986	316.70	2427.07	7505.04
1987	310.55	2451.39	7527.47
1988	326.48	2418.73	7599.86
1989	381.56	2433.80	7730.89
1990	428.48	2563.65	8032.85
1991	461.06	2611.51	8381.43
1992	482.14	2522.62	8239.23
1993	504.48	2431.01	8437.49
1994	526.64	2545.56	9274.63
1995	517.74	2445.62	9312.64
1996	460.27	2381.78	9477.16
1997	442.46	2265.12	9441.76
1998	369.64	1959.62	8567.91
1999	339.14	1750.75	7928.68

NOTE: Rates are arrests of persons ages 10 to 17 per 100,000 persons in the resident population. The Violent Crime Index includes the offenses of murder and non-negligent manslaughter, forcible rape, robbery, and aggravated assault. The Property Crime Index includes burglary, larceny/theft, motor vehicle theft, and arson.
SOURCE: Adapted from Snyder (2000).

Table 14.5   Drug Cases in Juvenile Court: 1989, 1994, and 1998

	1989	1994	1998
Male	86%	86%	84%
Female	14	14	16
Age (Years) at Time of Referral			
14 and younger	18%	20%	19%
15	22	23	21
16	30	30	31
17 and older	30	28	29
Race/Ethnicity			
White	58%	61%	68%
Black	40	37	29
Other	2	2	3

SOURCE: Office of Justice Programs (2001b).

and age, showed practically no change at all over the 10-year period from 1989 to 1998. The only significant change occurred on the demographic variable of race. Between 1989 and 1998, we saw a slow but steady 10% increase in the number of juvenile cases involving whites and a concomitant 13% decrease, from 40% to 27%, in the number of juvenile cases involving blacks.

Although the official statistics may be appropriate in examining the extent of the labeling process, law enforcement and juvenile court statistics are of limited usefulness in measuring the full extent and volume of delinquent behavior. Because not all delinquent behavior is detected (and thus cannot be officially recorded), the acts that are officially recorded do not represent a random sample of all delinquent acts. In other words, official statistics provide only a limited index of the total volume of delinquency.

Unofficial Measures of Delinquency

Researchers have been persistent in their attempts to identify juvenile delinquents and to measure juvenile delinquency. As mentioned previously, the primary sources of data have been self-reports, victimization surveys, and observational studies in gang hangouts and schools. The major limitation of these methods relates to the representativeness of the individuals reported as delinquent.

overall crime rate for juveniles. This decline was due to the decrease in both property and violent crimes. However, from the mid-1980s through the mid-1990s, we began to see an increase in both violent and property crimes. The peak for juvenile property crimes was reached in 1991, and the peak for juvenile crimes of violence was 1995. Gradually by the end of 1999, the overall juvenile crime rate had shifted back to close to the level it was in 1980. In large part, this was due to a major decrease in the number of both violent and property crimes, especially in the period between 1997 and 1999. This trend bodes well for the future because not only did the violent crime rate fall back to its 1980 annual rate, but we saw a considerable decrease in the total number of property crimes.

According to Table 14.5, drug cases in the juvenile court have remained virtually steady. Two of the three demographic variables, gender

Self-Report Delinquency Studies

Self-report questionnaires were first introduced as a method of measuring delinquency by Short and Nye (1957). Several other prominent sociologists and criminologists used the self-report approach: Empey and Erickson (1966), Gold (1966, 1970), Hindelang, Hirschi, and Weis (1981), and Hirschi (1969). Some self-report studies focus on measuring the proportion of youths who have engaged in delinquent acts, asking such questions as "Have you ever stolen something?" or "Have you stolen something since you were 9 years old?" These types of studies gather data on the extent of participation in specific delinquent behaviors. Other self-report delinquency studies make inquiries about the frequency of individual involvement in delinquent behaviors over a specific period, such as within the past year or two. Studies having time bounds of 4 or more years have only limited usefulness because of memory decay and filtering of recall.

Most self-report surveys indicate that the number of youths who break the law is much greater than official statistics report. These surveys reveal that the most common juvenile offenses are truancy, alcohol abuse, shoplifting, using a false ID, fighting, using marijuana, and damaging the property of others (Farrington, 1973; Hindelang, 1973).

Some of the more striking findings of self-report studies relate to the extent of delinquency and status offenses. According to estimates of the President's Commission on Law Enforcement and Administration of Justice (1967), self-report studies indicate that the overwhelming majority of all juveniles commit delinquent and criminal acts for which they could be adjudicated.

Erickson and Empey (1963) studied self-reported delinquency behavior among high school boys aged 15 to 17 in Provo, Utah. Using a list of 22 criminal and delinquent offenses to question the youths about their illegal activities, the researchers found that in 90% of the cases, minor offenses (such as traffic violations, minor thefts, liquor law violations, and destruction of property) went undetected. They compared the self-reported delinquency of boys who were officially nondelinquent with a group of official

delinquents—those who had been processed by the court only once, those on probation, and those who were incarcerated. The results were surprising: the nondelinquents admitted to an average of almost 158 delinquent acts per youth. However, official delinquents' self-reports greatly exceeded the reports of nondelinquents: Youths with a one-time court appearance had an average of 185 delinquent acts; repeat offenders who had been on probation had an average of 855 delinquent acts; and incarcerated delinquents had an average of 1,272 delinquent acts. Williams and Gold (1972) found that "eighty-eight percent of the teenagers in the sample confessed to committing at least one chargeable offense in the three years prior to their interview" (p. 213).

The "dark figures of crime"—the unreported delinquent acts—are difficult to determine. Self-report studies indicate that the dark or unknown figure of delinquency may be more than 9 times greater than the official estimate, since about 9 of every 10 juvenile law violations are either undetected or not officially acted upon. The overwhelming majority of juveniles have broken the law, even though their offenses are usually minor. Yet a small group are chronic, violent offenders; these are the youths who habitually violate the law and, as a result, are more likely to be apprehended and formally adjudicated.

Empey (1982) offered an analogy comparing juvenile offenders to fish being caught in a net in a large ocean:

> The chances are small that most fish will be caught. And even when some are caught, they manage to escape or are released because they are too small. But because a few fish are much more active than others, and because they are bigger, they are caught more than once. Each time this occurs, moreover, the chances that they will escape or be thrown back decrease. At the very end, therefore, they form a very select group whose behavior clearly separates them from most of the fish still in the ocean. (p. 113)

The most important and enlightening developments in research on hidden delinquency came from the national youth surveys. Beginning in 1967, the National Institute of Mental Health

(NIMH) implemented the first national survey of a representative sample of adolescents, focusing on their attitudes and behaviors, including delinquent behavior. In 1972, the national youth survey (NYS; originally NAYS) was repeated; in 1976, the National Institute of Juvenile Justice and Delinquency Prevention became the co-sponsor of what developed into an annual survey of self-reported delinquent behavior, based on a national probability panel of teenage youths from 11 to 17 years of age (Weis, 1983). These longitudinal national surveys were based on a carefully drawn sample of 1,725 adolescents from throughout the United States who were asked to report on their delinquent behavior for 5 consecutive years.

The major survey findings indicated that the overwhelming majority of American youth (11 to 17 years old) in the sample admitted that they had committed one or more juvenile offenses. In agreement with other self-report studies, the NAYS has found that the majority of youths had committed minor offenses, especially as they grew older. Less than 6% of the youths in the survey (Elliot & Huizinga, 1983; Huizinga & Elliot, 1984) admitted to having committed one of the more serious index offenses that are listed in the FBI Uniform Crime Reports. The trends in self-reported delinquency revealed that youths in the 1980s seemed to be no more delinquent than youths in the early 1970s.

The survey found that youths living with both natural parents reported lower delinquency rates than those who came from single-parent or reconstituted families. Youths who stayed in school reported less involvement in delinquency than those who dropped out. In addition, school dropouts admitted that they had participated in crimes such as felony assault and theft, hard drug use, disorderly conduct, and general delinquency to a much greater extent than in-school youths. In contrast, higher percentages of in-school youth indicated they had participated in minor assaults, vandalism, and school delinquency (Thornberry, Moore, & Christenson, 1985). Finally, the 1978–1980 national youth surveys asked about attendance at religious services. Those youths who reported regular attendance at religious services also reported less involvement in virtually all types of delinquent

behaviors (U.S. Department of Justice, 2001) (see Table 14.6).

Victimization Surveys

Recently, victimization studies conducted as joint efforts by the Bureau of Justice Statistics and the Bureau of the Census have been completed in a large number of cities throughout the United States. The most well-known victimization survey is the National Crime Survey (NCS), a massive, annual, house-to-house survey of a random sample of 60,000 households and 136,000 individuals (Flanagan & McCloud, 1983). The NCS provides annual estimates of the total number of crimes committed by both adult and juvenile offenders. The six types of crime measured are rape, robbery, assault, household burglary, personal and household larceny, and motor vehicle theft. On the basis of the survey data, it has been estimated that 40 million serious crimes occur each year in the United States. The NCS estimate is 4 times greater than that reported in the FBI Uniform Crime Reports. These data indicate that underreporting is far more pervasive than was generally recognized before the completion of the national victimization surveys.

This survey also provided estimates of the juvenile offense rates for a limited number of crimes: rape, robbery, assault, and personal larceny. The analysis of Laub (1983) indicated that the rates of delinquency for the years 1973–1981 remained relatively stable. In addition, there has been very little change in the types of persons who are victimized by juveniles. For the most part, juvenile perpetrators victimized other youths, rather than adults; males were the victims twice as frequently as females.

Figure 14.3 shows the percentage of victimization committed by juveniles from 1973 to 1997. Over the 25-year-period, robbery was the most common victimization committed by juveniles. From 1973 to 1987, we saw a slow decrease in the percentage of robbery. However, this trend was short lived, for the percentage slowly and inconsistently increased once again. Although we have seen the robbery percentage increase, the rate in 1997 was still lower than the initial 1973 rate.

Table 14.6   Percentages of Students Reporting Problem Behaviors by Grade Level of Respondent, 1999–2000*

	Never	*Seldom*	*Sometimes*	*Often*	*A Lot*
Have you been in trouble with the police?	77.6	13.2	5.4	1.9	2.0
Grades 6 to 8	81.0	10.7	4.7	1.7	1.8
Grades 9 to 12	73.9	15.8	6.1	2.0	2.2
12th grade	72.5	17.5	6.3	1.7	2.1
Do you take part in gang activities?	90.8	4.1	2.3	1.0	1.7
Grades 6 to 8	90.8	4.2	2.4	1.0	1.5
Grades 9 to 12	90.7	4.0	2.3	1.0	1.9
12th grade	92.0	3.2	2.0	0.8	2.0
Have you thought about committing suicide?	72.6	13.6	7.9	2.8	3.1
Grades 6 to 8	77.6	11.0	6.1	2.3	2.9
Grades 9 to 12	67.2	16.4	9.8	3.3	3.3
12th grade	67.3	17.6	9.7	2.9	2.6
Do you drink alcohol at home?	74.5	13.4	8.0	2.2	1.8
Grades 6 to 8	83.9	9.6	4.2	1.2	1.1
Grades 9 to 12	64.5	17.4	12.1	3.4	2.6
12th grade	59.8	18.2	14.8	4.0	3.2
Do you use drugs at home?	89.9	3.9	2.9	1.5	1.8
Grades 6 to 8	94.3	2.4	1.6	0.7	1.0
Grades 9 to 12	85.1	5.6	4.4	2.3	2.7
12th grade	82.9	6.4	4.8	2.6	3.3
Have you threatened to harm a teacher?	93.4	3.8	1.5	0.5	0.9
Grades 6 to 8	94.3	3.3	1.3	0.4	0.7
Grades 9 to 12	92.4	4.3	1.6	0.6	1.1
12th grade	92.7	4.0	1.5	0.5	1.3
Have you threatened to harm one or both of your parents, guardian, etc.?	90.7	5.5	2.2	0.7	1.0
Grades 6 to 8	92.0	4.6	1.9	0.6	0.9
Grades 9 to 12	89.3	6.4	2.4	0.8	1.0
12th grade	90.2	5.8	2.2	0.7	1.1

NOTE: These data are from a survey of 6th- through 12th-grade students conducted between September 1999 and June 2000 by Parent Resources Institute for Drug Education (PRIDE) Surveys. Participating schools are sent the PRIDE questionnaire with explicit instructions for administering the anonymous, self-report survey. Schools that administer the PRIDE questionnaire do so voluntarily or in compliance with a school district or state request. For the 1999–2000 academic year, survey results are based on students from 24 states. The following states participated in the 1999–2000 PRIDE survey: Alabama, Arizona, Arkansas, California, Florida, Georgia, Idaho, Illinois, Kentucky, Massachusetts, Mississippi, Montana, New Jersey, New Mexico, New York, Ohio, Oklahoma, Pennsylvania, Tennessee, Texas, Virginia, Washington, West Virginia, and Wisconsin. To prevent any one state from having a disproportionate influence on the summary results, random samples of students were drawn from those states where disproportionately large numbers of students were surveyed. Therefore, no state makes up more than 10% of the sample. The results presented are based on a sample consisting of 114,318 students drawn from juniors and seniors in public high schools throughout the United States.
*Percentages may not add to 100 because of rounding.
SOURCE: U.S. Department of Justice (2001, Tables 6.15, 6.24, 6.25, & 6.26).

Violent crime and aggravated assault saw similar trends to the robbery percentage. Both of these variables underwent slow and choppy declines in their rates only to once again increase back to their original starting level. This graph shows how victimizations committed by juveniles, although fluctuating frequently, overall stay fairly consistent, with

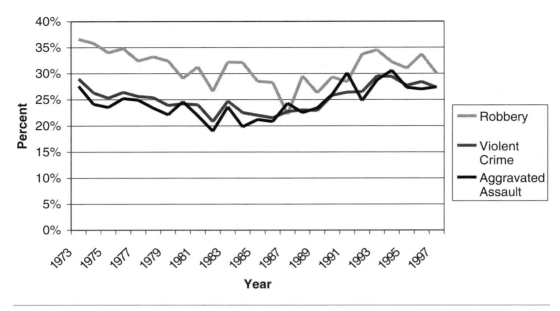

Figure 14.3   Percentage of Victimizations Committed by Juveniles

robbery being the most frequent and aggravated assault and violent crime staying fairly equal in their percentage rates.

Three important findings relate to the extent of violence among juvenile offenders. First, there was very little change in the use of guns, knives, and other weapons during the 9-year period. Second, only 27% of the juvenile crimes involved weapons use. This finding contradicts the view that most juvenile offenders use a deadly weapon. Finally, the number of victims who required hospitalization in the aftermath of a delinquent act seemed to remain stable over the 9-year period.

Although juvenile delinquency rates stabilized during the 1980s, millions of adolescents continue to become involved in juvenile delinquency each year, and it continues to be a complex and difficult social problem with no easy solution.

DISPOSITION OF JUVENILE OFFENDERS

According to the U.S. Department of Justice (2001), probation is the most common form of sanction given by the juvenile court. Since 1989, of the almost 1 million juvenile court cases per year, a probation sanction accounted for almost half of all of the total sanctions given. Figure 14.4 shows an increasing trend in the numbers of overall sanctions given out by juvenile courts, particularly probation. Noteworthy is the percentage increase in juvenile probation sanctions, an increase of over 50% over the 10-year period from 1989 to 1998. This is a vast increase when compared to the other possible juvenile court sanctions, which saw increases of about 5% to 20% since 1989. The other two forms of sanctions show very slight increases, if any at all. The number of cases waived to the adult criminal courts seemed to stay at virtually the same level throughout the entire period.

With the increase in the number of overall juvenile court cases from 1989 to 1998, we should have expected to see rises in the number of sanctions given out by the court. But the trend we are beginning to witness is that juvenile court judges seem to be adjudicating to some form of probation (e.g., intensive probation supervision, restitution and community service, or traditional probation) as a preferred form of sanction. This is evidenced by the percentage increase compared to the overall rise in all forms of sanctions.

Probation was ordered in 58% of the more than 1 million cases that received a juvenile court sanction

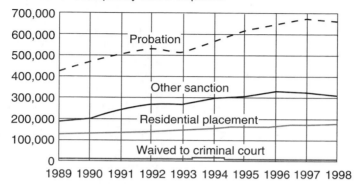

Number of delinquency cases disposed

Probation cases accounted for 59% of the increase in delinqency cases that received a juvenile court sanction (those that were not dismissed or otherwise released).

Figure 14.4 Juvenile Court Case Dispositions
SOURCE: Office of Justice Programs (2001a).

There has been considerable debate over the appropriate way to handle status offenders. The major issue is whether the juvenile court should retain authority over them. Those in favor of the court's continuing authority believe that youth's habitual misbehavior will eventually lead to more serious delinquent acts; therefore, it is wise for the court to retain its jurisdiction over status offenders. An opposing view (often advanced by deviance theorists with a societal-reaction or labeling-theory perspective) holds that when one defines status offenders as delinquents, the youths may actually become delinquents as the result of a self-fulfilling prophecy, leading to secondary deviance (Becker, 1963; Lemert, 1971; Schur, 1973).

Another belief held by a number of social workers is that the needs of status offenders can be better met within the community social service and child welfare service systems (Boisvert & Wells, 1980; Roberts, 1987; Springer, 2001). For example, my own research documented the need for the full range of social services, including 24-hour telephone hotlines, short-term runaway shelters, family treatment programs, education and treatment services for abusive parents, and vocational training and placement services.

At issue is the type of treatment status offenders should receive. Should they be sentenced to a secure juvenile facility or referred to a community social service agency for counseling? For many years, it was common for juvenile status offenders to be sentenced to juvenile training schools, where they were confined in the same institution with youths convicted of serious crimes. The practice of sending status offenders to juvenile correctional institutions has become much less common in recent years due to the deinstitutionalization of status offenders. However, in certain circumstances a minor who has committed no crime can still be sent to a juvenile training school. For example, a youth with a history of chronic runaway behavior is placed in a group home by the court. If the juvenile then runs away from that facility, the court views that act as a delinquency offense, and the youth is usually sent to a secure juvenile institution.

Probation officers often believe that although most status offenders do not pose a danger to others, they frequently exhibit destructive behavior patterns such as drug abuse, alcohol abuse, or suicidal ideation. They often come from dysfunctional, conflict-ridden families where physical, sexual, and/or emotional abuse is prevalent. Thus, the social work perspective urges that a continuum of services be provided for status offenders and their families through a social service agency, a family service agency, or a juvenile court-based program. Available services should include family counseling, individual and group counseling, addiction treatment, alternative education programs, and vocational evaluation, education, and training. (For an in-depth analysis of the nature and types of social service and counseling programs to which status offenders are referred, see Roberts, 1998, chaps. 5 and 9.)

CONCLUSION

Every youth who violates the law is not labeled a juvenile delinquent. Many either escape detection entirely or, when apprehended by the local police, receive a strict lecture or warning and then are taken home. Most authorities do not consider law-breaking youths to be juvenile delinquents unless they are officially processed through the juvenile court and adjudicated delinquent.

Youths can be referred to the juvenile court for two types of offenses: delinquency offenses—illegal acts that are considered crimes whether committed by an adult or a juvenile (e.g., aggravated assault, arson, burglary, drug-related offenses, theft, and rape); and status offenses—deviant acts or misbehavior that, if engaged in by an adult, would not be considered crimes (e.g., truancy, incorrigibility, and running away from home).

For status offenders, many states have separate legislation that views these juveniles as individuals "in need of supervision." An array of crisis intervention services, runaway shelters, youth service bureaus, and family counseling programs have been developed to serve these youths.

This chapter laid the groundwork for an examination of the field of juvenile justice. This was done by presenting legalistic, sociological, and criminological information focusing on three areas: the nature of the juvenile justice system and its subsystems, the definition of the terms *juvenile delinquency* and *status offenses,* and official, as well as unofficial, trends and statistics on the extent of juvenile delinquency in the United States.

REFERENCES

Becker, H. S. (1963). *Outsiders: Studies in the sociology of deviance.* New York: Free Press.

Blumstein, A. (1967). Systems analysis and the criminal justice system. *Annals of the American Academy of Political and Social Science, 474,* 3–24.

Boisvert, M. J., & Wells, R. (1980). Toward a rational policy on status offenders. *Social Work, 25,* 230–234.

Elliot, D. S., Ageton, S. S., Huizinga, D., Knowles, B. A., & Canter, R. J. (1983). *The prevalence and incidence of delinquent behavior: 1976–1980.* Boulder, CO: Behavioral Research Institute.

Elliot, D. S., & Huizinga, D. (1983, May). Social class and delinquent behavior in a national youth panel: 1976–1980. *Criminology, 21,* 149–177.

Empey, L. T. (1982). *American delinquency: Its meaning and construction.* Chicago: Dorsey.

Empey, L. T., & Erickson, M. L. (1966). Hidden delinquency and social status. *Social Forces, 44,* 546–554.

Erickson, M. L., & Empey, L. T. (1963). Court records, undetected delinquency, and decision making. *Journal of Criminal Law, Criminology, and Police Science, 54,* 456–469.

Farrington, D. P. (1973, March). Self-reports of deviant behavior. Predictive and stable?

Journal of Criminal Law and Criminology, 44, 99–111.

Federal Bureau of Investigation. (1985). *Crime in the United States, 1984* (Uniform Crime Reports). Washington, DC: Government Printing Office.

Flanagan, T. J., & McCloud, M. (1983). *Sourcebook of criminal justice statistics, 1982.* Washington, DC: U.S. Department of Justice.

Gold, M. (1966, January). Undetected delinquent behavior. *Journal of Research in Crime and Delinquency, 3,* 27–46.

Gold, M. (1970). *Delinquent behavior in an American city.* Belmont, CA: Brooks/Cole.

Hamparian, D., et al. (1982). *Youth in adult courts: Between two worlds*. Washington, DC: U.S. Department of Justice.

Hindlelang, M. J. (1973). Causes of delinquency: A partial replication and extension. *Social Problems, 20,* 471–487.

Hindelang, M. J., Hirschi, T., & Weis, J. (1981). *Measuring delinquency*. Beverly Hills, CA: Sage.

Hirschi, T. (1969). *Causes of delinquency*. Berkeley: University of California Press.

Huizinga, D., & Elliot, D. S. (1984). *Self-reported measures of delinquency and crime: Methodological issues and comparative findings*. Boulder, CO: Behavioral Research Institute.

Laub, J. H. (1983). *Juvenile criminal behavior in the United States: An analysis of an offender and victim characteristics*. Albany: State University of New York at Albany, Michael J. Hindelang Criminal Justice Research Center.

Lemert, E. M. (1971). *Instead of court: Division in juvenile justice*. Chevy Chase, MD: National Institute of Mental Health, Center for the Studies of Crime and Delinquency.

National Center for Juvenile Justice. (2002). *National Juvenile Court Data Archive: Juvenile court case records*. Washington, DC: Office of Juvenile Justice and Delinquency Prevention.

Office of Justice Programs. (2001a, August). *Delinquency cases in juvenile courts* (Fact Sheet). Washington, DC: U.S. Department of Justice.

Office of Justice Programs. (2001b, September). *Drug offences juvenile court, 1989–1998* (Fact Sheet). Washington, DC: U.S. Department of Justice.

Prescott, P. S. (1982). *The child savers*. New York: Alfred A. Knopf.

President's Commission on Law Enforcement and Administration of Justice. (1967). *Task force report: Juvenile delinquency and youth crime*. Washington, DC: Government Printing Office.

Puzzanchera, C., Stahl, A., Finnegan, T., Snyder, H., Poole, R., & Tierney, N. (2001). *Juvenile court statistics, 1998*. Washington, DC: Office of Juvenile Justice and Delinquency Prevention.

Roberts, A. R. (1987). *Runaways and nonrunaways*. Chicago: Dorsey.

Roberts, A. R. (1998). Juvenile justice (2nd ed.). Chicago: Nelson-Hall.

Robinson, S. M. (1960). *Juvenile delinquency: Its nature and control*. New York: Holt, Rinehart & Winston.

Schur, E. (1973). *Radical nonintervention: Rethinking the delinquency problem*. Englewood Cliffs, NJ: Prentice Hall.

Short, J. F., Jr., & Nye, F. I. (1957). Extent of unrecorded delinquency: Tentative conclusions. *Journal of Criminal Law, 49,* 296–302.

Snyder, H. (2000). *Juvenile arrests, 1999*. Washington, DC: Office of Juvenile Justice and Delinquency Prevention.

Springer, D. W. (2001). Runaway adolescents: Today's Huckleberry Finn crisis. *Brief Treatment and Crisis Intervention, 1*(2), 131–152.

Thornberry, T. P., Moore, M., & Christenson, R. L. (1985, February). The effect of dropping out of high school on subsequent criminal behavior. *Criminology, 23,* 3–18.

Torbet, P., & Szymanski, L. (1997). *State legislative responses to violent juvenile crime: 1996–97 update* (Juvenile Justice Bulletin). Washington, DC: Office of Juvenile Justice and Delinquency Prevention.

U.S. Children's Bureau. (1967). *Juvenile court statistics, 1966* (Statistical Series 90). Washington, DC: Government Printing Office.

U.S. Department of Justice. (2001). *Sourcebook on criminal justice statistics 2000*. Washington, DC: Government Printing Office.

Weis, J. G. (1983). Crime statistics: Reporting systems and methods. In *Encyclopedia of crime and justice* (Vol. 1). New York: Free Press.

Williams, J. R., & Gold, M. (1972). From delinquent behavior to official delinquency. *Social Problems, 20,* 209–229.

Youth Development and Delinquency Prevention Administration. (1973). *The challenge of youth service bureaus*. Washington, DC: Government Printing Office.

15

YOUTH HOMICIDE

A Critical Review of the
Literature and a Blueprint for Action*

KATHLEEN M. HEIDE

The issue of juvenile homicide has been headlined repeatedly in the news all over the United States and abroad since the early 1990s. Though it is difficult to assess the exact number of murders committed by juveniles because the age of the killer is not specified by the arresting authority in as many as a third of the cases (Snyder, 2001), there is no question from available data that murders by young people have risen over the last two decades. Dramatic increases in youths being arrested for homicide beginning in the early 1980s are apparent whether the frame of reference being examined is youths under 18 or those in their middle to late teenage years (Heide, 1999).

Youth is a broad term that encompasses both juveniles and adolescents. Although these words are often used interchangeably in the media and in the professional literature, they can be distinguished from each other. Juvenile or minority status is determined on the basis of age and is a legislative decision (Butts & Snyder, 1997). The federal government and the majority of the states, for example, designate youths under 18 as juveniles (Bortner, 1988; Sickmund, 1994). The Federal Bureau of Investigation (2001) classifies arrests of "children 17 and under" as juvenile arrests.

Adolescence, in contrast to juvenile status, is based on human development and varies across individuals. It is a stormy period characterized by hormonal changes, growth spurts, psychological changes, and enhancement of intellectual abilities and motor skills. According to child development experts, adolescence begins with puberty, which typically commences by age 12 or 13 but may start earlier (Solomon, Schmidt, & Ardragna, 1990; see also Lee, Lee, & Chen, 1995). Adolescence extends through the teen years to ages 19 or 20 (Solomon et al., 1990). The term *children* is commonly used to refer to prepubescent youth.

Analysis of crime patterns clearly indicates that youth involvement in homicide remains a serious problem in the United States in the 21st century. Homicide arrests of juveniles rose every year from 1984 through 1993. The dramatic escalation in murders by juveniles during this time frame put the United States in the grip of fear. In 1993, the number of juveniles arrested for murder—3,284—was three times higher than the number arrested in 1984 and had reached an all-time high (Heide, 1999). The rate at which

---

*Author's Note: Some of this material has been previously published in *Young Killers* (1999).

juveniles were arrested for murder also increased substantially during this time frame. The juvenile murder rate peaked in 1993 at 14 per 100,000 and was more than twice the level of the early 1980s (6 per 100,000) (Snyder, 1997). The significant rise in murders committed by those under 18 during this time frame cannot be attributed to an increase in the juvenile population. In fact, the percentage of young Americans during this time frame had generally been declining (Blumstein, 1995; Ewing, 1990; Fox, 1996).

Although the numbers of minors arrested for murder decreased over the period 1994 through 2000, it would be wrong to conclude that the crisis in lethal violence by youth is over. The percentage of all homicide arrests involving juveniles in 2000, after 7 years of decline, is still higher than it was in 1984, when juvenile homicide was just beginning to increase. In 1984, 7.3% of all homicide arrestees were juveniles; in 2000, 9.3% of those arrested for murder were under 18. The mean percentage of homicide arrestees who were juveniles during the 10-year period of escalation (1984–1993) was 11.6%; the comparable mean percentage during the 7-year period of decline (1994–2000) was 13.0%.

The seriousness of the youth homicide problem is underscored when the unit of analysis is 15- to 19-year-olds. Smith and Feiler (1995) computed the *absolute* and *relative* levels of youth involvement in arrests for murder during the period 1958–1993. Examination of trends in arrest rates revealed a very striking escalation in homicides by youths in their late teens beginning in the mid-1980s. The arrest rates for 15- to 19-year-olds recorded in 1992 (42.4 per 100,000) and 1993 (42.2 per 100,000) were the highest recorded for *any* age group during the 36 years covered by the data. Similarly, the researchers concluded that "the ratio for 15- to 19-year-olds in 1993 marks the greatest relative involvement in murder arrests for any age group during the period under study" (p. 330). Recent follow-up data revealed that youth involvement remained highest even when the analyses were extended to encompass data through 1999 (Feiler & Smith, 2000).

Today adults ask why youths kill just as they have for centuries. The question, however, has become more urgent because available data suggest that young people in the United States are killing more since the mid-1980s than in previous

generations. This chapter first synthesizes the literature on clinical and empirical findings related to youth homicide. Thereafter, it reviews the literature with respect to the treatment of juvenile homicide offenders. The chapter concludes with recommendations to guide future research efforts, with the aim of increasing understanding of etiological factors associated with juvenile homicide and designing effective intervention strategies.

FACTORS ASSOCIATED WITH YOUTH HOMICIDE

Many clinicians and researchers have examined cases of youths killing over the last 50 years in an effort to determine the causes of juvenile homicide. Two excellent critiques of the literature by University of Virginia Professor Dewey Cornell (1989) and SUNY Professor Charles Patrick Ewing (1990) have been published. Both scholars cited a number of methodological problems with most of the studies conducted on juvenile homicide through the late 1980s and suggested that reported findings be viewed with caution.

Much of the difficulty with the available literature stems from the fact that most published accounts of young killers consist of case studies. The cases reported were often drawn from psychiatric populations referred to the authors for evaluation and/or treatment after the youth committed a homicide. The conclusions drawn from these cases, though interesting and suggestive, cannot provide us with precise explanations regarding why youths kill because it is unknown to what extent the youths examined are typical of the population of juvenile murderers. In addition, in the absence of control groups of any kind, it is unknown in what ways these young killers differ from nonviolent juvenile offenders, violent juvenile offenders who do not kill, and juveniles with no prior records.

Research on juvenile murderers has been primarily descriptive. Not surprisingly, psychogenic explanations (e.g., mental illness, defective intelligence, childhood trauma) have largely predominated in the literature, given the professional background of many of the authors. Biopsychological explanations (e.g., neurological impairments, brain injury) have been investigated by some scientists. Data on important sociological variables (e.g., family constellation, gang involvement, drug and alcohol use, participation in other antisocial behavior, peer associations) have been reported by some researchers. Sociological theories of criminal behavior (e.g., strain/anomie, subcultural, social control, labeling, conflict, and radical theories; see Bynum & Thompson, 1999, for discussion), however, have not been systematically investigated in the literature on youth homicide.

Statements about juvenile murderers in the professional literature are typically about male adolescents who kill. Although some studies of adolescent homicide have included both females and males (e.g., Dolan & Smith, 2001; Labelle, Bradford, Bourget, Jones, & Carmichael, 1991; Malmquist, 1990), most research has focused on male adolescents because they constitute the overwhelming majority of juvenile homicide offenders.

Girls Who Kill

A few publications report case studies of girls who have murdered (e.g., Benedek & Cornell, 1989; Ewing, 1990; Gardiner, 1985; Heide, 1992; McCarthy, 1978; Medlicott, 1955; Russell, 1986). These studies reveal that girls are more likely than boys to kill family members and to use accomplices to effect these murders. Girls are also more likely to perform secondary roles when the killings are gang related or occur during the commission of a felony, such as robbery. Their accomplices are generally male. Pregnant unmarried girls who kill their offspring at birth or shortly thereafter, in contrast, often appear to act alone. Girls' motives for murder are varied. Instrumental reasons include ending abuse meted out by an abusive parent, eliminating witnesses to a crime, and concealing a pregnancy. Expressive reasons include acting out psychological conflict or mental illness, supporting a boyfriend's activities, and demonstrating allegiance to gang members (Ewing, 1990).

Homicides Involving Young Children as Offenders

Research specifically investigating "little kids" who kill has also been sparse, partly because of its low incidence and the difficulty of obtaining access to these youths (e.g., Bender, 1959; Carek & Watson, 1964; Ewing, 1990; Goetting, 1989, 1995; Petti & Davidman, 1981; Pfeffer, 1980; Shumaker & Prinz, 2000; Tooley, 1975). The importance of distinguishing between preadolescents and adolescents in understanding what motivates youths to kill and in designing effective treatment plans was recognized by clinicians as early as 1940 (Bender & Curran, 1940). Subsequent investigators, however, have not consistently used age as a criterion in selecting samples of youths who kill (Easson & Steinhilber, 1961; Goetting, 1989; Myers, Scott, Burgess, & Burgess, 1995; Sargent, 1962). In one frequently cited report, for example, the young killers ranged in age from 3.5 to 16 years (Sargent, 1962).

Though recent research suggests that preteen and adolescent murderers may share some characteristics (Shumaker & Prinz, 2000), the differences between the two groups cannot be ignored. Physically healthy children under 9 who kill, in contrast to older youths, typically do not fully understand the concept of death (Bender & Curran, 1940; Cornell, 1989; Heide, 1992; O'Halloran & Altmaier, 1996). They have great difficulty comprehending that their actions are irreversible (Bender & Curran, 1940). Prepubescent children who kill often act impulsively and without clear goals in mind (e.g., Adelson, 1972; Carek & Watson, 1964; Goetting, 1989). Preadolescent murderers are also more likely to kill than older youths in response to the unstated wishes of their parents (e.g., Tooley, 1975; Tucker & Cornwall, 1977). In addition, the incidence of severe conflict (e.g., Bernstein, 1978; Paluszny & McNabb, 1975) or severe mental illness (Bender, 1959; Heide, 1992; "Incompetency Standards," 1984; Pfeffer, 1980; Tucker & Cornwall, 1977; Zenoff & Zients, 1979) tends to be higher among younger children who kill than among their adolescent counterparts. Adolescent killers are more likely to kill because of the lifestyles that they have embraced or in response to situational or environmental constraints that they believe to be placed upon them (Heide, 1984, 1992, 1999; Sorrells, 1977; Zenoff & Zients, 1979).

CASE STUDY RESEARCH PERTAINING TO ADOLESCENT MURDERERS

The literature below highlights findings from studies of youths who typically ranged in age from 12 to 17 years of age, although some research included adolescents in their late teen years. In recognition of reporting practices in the professional literature, the terms *juvenile* and *adolescent* are treated as equivalent in the discussion that follows. Similarly, the terms *murder* and *homicide* are used synonymously, though the intended legal meaning is that of *murder.* Accordingly, terms such as *juvenile homicide offender, adolescent murderer,* and *young killer* are used interchangeably throughout the remainder of this chapter. The following sections address various areas covered in case studies of adolescent murderers. It is important to keep in mind the caveats discussed above regarding the shortcomings of this body of literature.

Psychological Disorder and Youth Homicide

Several scholars have synthesized existing scientific publications relating to various types of juvenile homicide offenders (Adams, 1974; Busch, Zagar, Hughes, Arbit, & Bussell, 1990; Cornell, Benedek, & Benedek, 1987b, 1989; Ewing, 1990; Haizlip, Corder, & Ball, 1984; Lewis et al., 1985; Lewis, Lovely, et al., 1988; Myers, 1992; with respect to preadolescent murderers, Shumaker & Prinz, 2000), including youths who commit sexual murders (Myers, 1994; Myers et al., 1995; Myers, Burgess, & Nelson, 1998). Much of the literature, particularly during the 1940s, 1950s, 1960s, and 1970s, suggested that psychodynamic factors propelled youths to kill (Lewis et al., 1985). These factors included impaired ego development, unresolved Oedipal and dependency needs, displaced anger, the ability to dehumanize the victim, and narcissistic deficits (Cornell, 1989; e.g., Bender & Curran, 1940; Mack, Scherl, & Macht, 1973; Malmquist, 1971, 1990; McCarthy, 1978; Miller & Looney, 1974; Scherl & Mack, 1966; Smith, 1965; Washbrook, 1979).

Many studies have investigated the extent of severe psychopathology, such as psychosis,

organic brain disease, and neurological impairments (see Cornell, 1989; Ewing, 1990). The findings, particularly with respect to the presence of psychosis among juvenile homicide offenders, are mixed and may be the result of how the samples were generated. Individuals who are diagnosed as psychotic have lost touch with reality, often experience hallucinations (seeing or hearing things that are not occurring) and delusions (bizarre beliefs), and behave inappropriately. Most studies report that juvenile homicide offenders are rarely psychotic (e.g., Bailey, 1994; Corder, Ball, Haizlip, Rollins, & Beaumont, 1976; Cornell, 1989; Cornell et al., 1987b, 1989; Dolan & Smith, 2001; Ewing, 1990; Hellsten & Katila, 1965; Kashani, Darby, Allan, Hantke, & Reid, 1997; King, 1975; Labelle et al., 1991; Malmquist, 1971; Myers & Kemph, 1988, 1990; Myers et al., 1995; Patterson, 1943; Petti & Davidman, 1981; Russell, 1965, 1979; Shumaker & McGee, 2001; Sorrells, 1977; Stearns, 1957; Walshe-Brennan, 1974, 1977; Yates, Beutler, & Crago, 1983; with respect to adolescent mass murderers, see Meloy, Hempel, Mohandie, Shiva, & Gray, 2001). Some studies, however, do posit a high incidence of psychosis (e.g., Bender, 1959; Lewis, Pincus, et al., 1988; Rosner, Weiderlight, Rosner, & Wieczorek, 1978; Sendi & Blomgren, 1975), or episodic psychotic symptomatology (Lewis et al., 1985; Lewis, Lovely, et al., 1988; Myers et al., 1995; Myers & Scott, 1998), and other serious mental illness, such as mood disorders (Lewis, Pincus, et al., 1988; Malmquist, 1971, 1990).

Several case reports have suggested that young killers suffered from a brief psychotic episode that remitted spontaneously after the homicides (e.g., Cornell, 1989; McCarthy, 1978; Miller & Looney, 1974; Mohr & McKnight, 1971; Sadoff, 1971; Smith, 1965). This phenomenon, initially introduced by the renowned psychiatrist Karl Menninger and one of his colleagues 40 years ago, is known as *episodic dyscontrol syndrome* and is characterized by incidents of severe loss of impulse control in individuals with impaired ego development (Menninger & Mayman, 1956). Diagnosing psychosis in homicide offenders who kill impulsively, brutally, and apparently senselessly, in the absence of clear psychotic symptoms, has been strongly challenged by some of the leading experts on juvenile homicide (e.g., Cornell, 1989; Ewing, 1990).

Examination of the literature indicates that there is considerable variation in diagnoses given to adolescent murderers within studies (e.g., Labelle et al., 1991; Malmquist, 1971; Myers & Scott, 1998; Rosner et al., 1978; Russell, 1979) as well as across studies. Personality disorders and conduct disorders rank among the more common diagnoses (Bailey, 1994; Dolan & Smith, 2001; Ewing 1990; Labelle et al., 1991; Malmquist, 1971; Myers & Kemph, 1988, 1990; Myers et al., 1995, 1998; Rosner et al., 1978; Russell, 1979; Santtila & Haapasalo, 1997; Schmideberg, 1973; Sendi & Blomgren, 1975; Sorrells, 1977; Yates et al., 1983; with respect to preteen murderers, Shumaker & Prinz, 2000). Attention deficit hyperactive disorder (ADHD) has also been noted with some frequency (Myers et al., 1995; Myers & Scott, 1998; Santtila & Haapasalo, 1997; with respect to preteen murderers, Shumaker & Prinz, 2000).

Neurological Impairment and Youth Homicide

Significant disagreement also exists with respect to the prevalence of neurological problems in juvenile killers (Cornell, 1989; Ewing, 1990), which may be partly due to differences in assessment and reporting practices used by various clinicians (Restifo & Lewis, 1985; e.g., Podolsky, 1965; Thom, 1949). Neurological impairment may be indicated by brain or severe head injuries, past and present seizure disorders, abnormal head circumferences or electroencephalogram (EEG) findings, soft neurological signs, and deficits on neurological testing (Myers, 1992). Several researchers have found significant neurological impairment or abnormalities among young killers (e.g., Bailey, 1996a; Bender, 1959; Busch et al., 1990; Lewis et al., 1985; Lewis, Lovely, et al., 1988; Michaels, 1961; Myers, 1994; Myers et al., 1995; Woods, 1961; Zagar, Arbit, Sylvies, Busch, & Hughes, 1990), particularly those on Death Row (Lewis, Pincus, et al., 1988). Others maintain that neurological difficulties are absent or rare among the juvenile murderers assessed

in their studies (e.g., Dolan & Smith, 2001; Hellsten & Katila, 1965; Labelle et al., 1991; Petti & Davidman, 1981; Russell, 1986; Scherl & Mack, 1966; Walshe-Brennan, 1974, 1977).

The Intelligence of Young Homicide Offenders

The findings with respect to intelligence are also mixed. Several studies reported that there were mentally retarded youths (IQ score below 70) among their samples of adolescent homicide offenders (Bender, 1959; Busch et al., 1990; Darby, Allan, Kashani, Hartke, & Reid, 1998; Labelle et al., 1991; Lewis, Pincus et al., 1988; Patterson, 1943; Solway, Richardson, Hays, & Elion, 1981; Zagar et al., 1990). There is a consensus across many studies, however, that few young killers are mentally retarded.

In contrast, there is disagreement regarding the intelligence of the majority of young killers (e.g., Ewing, 1990). Some researchers reported that the average IQ scores of the juvenile homicide offenders in their samples were in the below-average range (70–99) (Busch et al., 1990; Darby et al., 1998; Hays, Solway, & Schreiner, 1978; Labelle et al., 1991; Lewis, Pincus, et al., 1988; Petti & Davidman, 1981; Solway et al., 1981; Zagar et al., 1990). Others, however, have found that the IQ scores were typically in the average to above-average ranges (100–129) (Bender, 1959; Kashani et al., 1997; King, 1975; Patterson, 1943).

The literature on juvenile homicide offenders indicates that, regardless of intelligence potential, many struggle in educational settings. As a group, they tend to perform poorly academically (Bernstein, 1978; Hellsten & Katila, 1965; Myers et al., 1995; Myers & Scott, 1998; Scherl & Mack, 1966; Sendi & Blomgren, 1975; Shumaker & McGee, 2001; Stearns, 1957), have cognitive and language deficits (King, 1975; Myers & Mutch, 1992), experience severe educational difficulties (Bailey, 1994; Busch et al., 1990; Zagar et al., 1990), suffer from learning disabilities (Bender, 1959; Darby et al., 1998; Dolan & Smith, 2001; Hardwick & Rowton-Lee, 1996; King, 1975; Lewis, Pincus, et al., 1988; Myers et al., 1995; Myers & Scott, 1998; Patterson, 1943; Sendi & Blomgren, 1975) and engage in disruptive behavior in the classroom (Bailey, 1996a; Myers et al., 1995).

Home Environments of Youths Who Murder

In-depth analyses of the families of young killers have been lacking (Crespi & Rigazio-DiGilio, 1996). Case studies of adolescents who killed biological parents and stepparents have appeared far more often in the professional literature than case studies of other types of juvenile homicide offenders (Ewing, 1990; Zenoff & Zients, 1979; e.g., Anthony & Rizzo, 1973; Cormier, Angliker, Gagne, & Markus, 1978; Duncan & Duncan, 1971; Heide, 1992; Kalogerakis, 1971; Kashani et al., 1997; McCully, 1978; Mones, 1985, 1991; Mouridsen & Tolstrup, 1988; Post, 1982; Russell, 1984; Sadoff, 1971; Sargent, 1962; Scherl & Mack, 1966; Tanay, 1973, 1976; Wertham, 1941). These studies have indicated that youths who killed parents or stepparents, particularly fathers or stepfathers, were typically raised in homes where child abuse, spouse abuse, and parental chemical dependency were common (Heide, 1992). Research on the "adopted child syndrome" suggests that adopted youths who kill their fathers may be driven by other psychodynamic factors, including unresolved loss, extreme dissociation of rage, hypersensitivity to rejection, and confusion about their identity (Kirschner, 1992).

With few exceptions (e.g., Fiddes, 1981; King, 1975), published research and case studies report that the majority of adolescent homicide offenders are raised in broken homes (Ewing, 1990; e.g., Darby et al., 1998; Easson & Steinhilber, 1961; Labelle et al., 1991; McCarthy, 1978; Patterson, 1943; Petti & Davidman, 1981; Rosner et al., 1978; Russell, 1986; Scherl & Mack, 1966; Smith, 1965; Sorrells, 1977; Woods, 1961). Recent studies suggest that the majority are likely to come from criminally violent families (Busch et al., 1990; Zagar et al., 1990). Parental alcoholism, mental illness, and other indicators of parental psychopathology are commonly found in the histories of juvenile murderers (Ewing, 1990; e.g., Bailey, 1994, 1996a; Corder et al., 1976; Dolan & Smith, 2001; Heide,

1992; Hellsten & Katila, 1965; Labelle et al., 1991; Lewis et al., 1985; Lewis, Lovely, et al., 1988; Lewis, Pincus, et al., 1988; Myers et al., 1995, 1998; Petti & Davidman, 1981; Santtila & Haapasalo, 1997; Sorrells, 1977). Child maltreatment and spouse abuse are also repeatedly encountered in the homes of adolescent homicide offenders (Dolan & Smith, 2001; Ewing, 1990; Myers et al., 1995, 1998). Young killers as a group (e.g., King, 1975; Myers & Scott, 1998; Woods, 1961) and youths who kill parents in particular (Corder et al., 1976; Duncan & Duncan, 1971; Heide, 1992, 1994; Malmquist, 1971; Patterson, 1943; Post, 1982; Russell, 1984; Sargent, 1962; Tanay, 1976) have frequently witnessed one parent, typically the mother, being abused by the other parental figure. Juvenile murderers (e.g., Bailey, 1994, 1996a; King, 1975; Lewis et al., 1985; Lewis, Pincus, et al., 1988; Myers et al., 1995; Myers & Scott, 1998; Santtila & Haapasalo, 1997; Sendi & Blomgren, 1975), especially adolescent parricide offenders (e.g., Corder et al., 1976; Duncan & Duncan, 1971; Heide, 1992, 1994; Malmquist, 1971; Scherl & Mack, 1966; Tanay, 1976), have often been physically abused. Sexual abuse has also been documented in the lives of juvenile murderers (e.g., Bailey 1994, 1996a; Corder et al., 1976; Dolan & Smith, 2001; Lewis, Pincus, et al., 1988; Sendi & Blomgren, 1975), including those who kill parents (Heide, 1992, 1994).

Involvement in Other Antisocial Behavior

Juvenile homicide offenders engaged in several types of deviant behavior prior to committing homicide (Ewing, 1990). Several studies have reported that the majority of adolescent murderers have had a prior arrest or offense history (e.g., Bailey, 1996a; Cornell, Benedek, & Benedek, 1987a; Darby et al., 1998; Dolan & Smith, 2001; Ewing, 1990; Fiddes, 1981; Labelle et al., 1991; Myers et al., 1995, 1998; Rosner et al., 1978; Sorrells, 1977). Findings regarding whether young killers have had a lengthy history of fighting and other violent or antisocial behavior have been mixed. Some researchers have reported extensive antisocial behavior (e.g., Darby et al., 1998; Lewis et al., 1985; Lewis, Lovely, et al., 1988; McCarthy, 1978; Myers et al., 1995, 1998); others have uncovered little or none (Malmquist, 1971; Patterson, 1943; Walshe-Brennan, 1974); and still others found that previous delinquency varied significantly by the type of juvenile homicide offender (e.g., Zenoff & Zients, 1979) or the nature of the relationship between the offender and the victim (Corder et al., 1976). Gang participation has also been found among juvenile homicide offenders (Busch et al., 1990; Darby et al., 1998; Zagar et al., 1990).

Substance Abuse

The literature on substance abuse among juvenile homicide offenders has been sparse (Ewing, 1990). Examination of available studies reveals that the percentages of juvenile homicide offenders who reported abusing substances or were substance dependent have increased over the last 20 to 30 years. Earlier studies indicated that between 20% to 25% of young killers abused alcohol or drugs (Corder et al., 1976; Malmquist, 1971). Cornell et al. (1987a) reported that more than 70% of the 72 juvenile murderers in their Michigan sample reportedly drank alcohol or used drugs. Zagar et al. (1990) compared alcohol abuse among 101 juvenile murderers and 101 matched nonviolent delinquents in Cook County, Illinois. They reported in 1990 that juvenile murderers were significantly more likely to abuse alcohol than the control group (45% vs. 28%). Myers and Kemph (1990) reported that half of the 14 homicide youths in their study were diagnosed as substance dependent. In a later study of 18 juvenile murderers, Myers and Scott (1998) found that 50% were substance dependent. Psychiatrist Susan Bailey (1996a) reported in 1996 that, of the 20 juvenile murderers in the United Kingdom whom she treated, 75% abused alcohol and 35% abused drugs. Dolan and Smith (2001) found that 50% of the 46 juvenile homicide offenders referred to an adolescent forensic unit in Britain during 1986 to 1996 had a history of alcohol abuse, manifested in binge drinking, and 39.1% had a history of illicit drug use. Researchers in Finland reported in 1997 that 10

of the 13 young homicide offenders in their study were dependent on alcohol (Santtila & Haapasalo, 1997).

In addition to increases in substance abuse and dependence, there is evidence that the percentage of those who indicated that they were "high" at the time of the murder has also risen since the 1970s. Sorrells's (1977) study of juvenile murderers in California indicated that approximately 25% (8 of 31) of juvenile homicide offenders were under the influence of drugs and alcohol at the time of the homicidal event. Cornell et al. (1987a) noted 10 years later that more than 50% (38 of 72) of their sample of juvenile killers had killed while they were intoxicated. The U.S. Department of Justice (1987) indicated that 42.5% of juvenile murderers were under the influence of either alcohol, drugs, or both at the time of the incident. Fendrich, Mackesy-Amiti, Goldstein, Spunt, and Brownstein (1995) compared substance involvement among 16- and 17-year-old juvenile murderers with four different age groups of adult murderers incarcerated in New York state prisons. The groups were compared in terms of regular lifetime use, use during the week preceding the homicide, and use at the time of the crime. In general, the juvenile murderers had relatively "lighter" use and lower levels of drug involvement than adults in the sample. Of the 16 juvenile homicide offenders, 8 indicated that they were "substance affected" (intoxicated, crashing, or sick or in need of a substance) at the time of the murder. Of these 8, 5 acknowledged using alcohol and 3 marijuana. Only 3 young killers reported using cocaine, heroin, or psychedelics. The research team cautioned against concluding that substance use does not present a special risk for violent, homicidal behavior among juveniles.

Analysis of respondent substance use attribution patterns suggests that when 16–17 year old perpetrators use substances, the substances they use tend to have considerably more lethal effects than they do on perpetrators in older age groups. Thus, our study suggests that a focus on ingestion and involvement rates may underestimate the risk posed by substances for homicidal behavior among juveniles. (Fendrich et al., 1995, p. 1363)

Other Social Difficulties

Studies have indicated that a significant proportion of juvenile murderers do not attend school regularly (Ewing, 1990) due to truancy (e.g., Bailey, 1994, 1996a; Dolan & Smith, 2001; Myers et al., 1998; Shumaker & McGee, 2001; Smith, 1965), dropping out, or suspension/expulsion (e.g., Cornell et al., 1987a; Myers et al., 1998; Shumaker & McGee, 2001). Running away is a common response of adolescent parricide offenders (e.g., Heide, 1992; Sadoff, 1971; Scherl & Mack, 1966; Tanay, 1976). Enuresis (bedwetting) (e.g., Dolan & Smith, 2001; Easson & Steinhilber, 1961; Michaels, 1961; Myers et al., 1995; Russell, 1986; Sendi & Blomgren, 1975) and difficulties relating to peers (e.g., Corder et al., 1976; Marten, 1965; Zenoff & Zients, 1979) have been found in the histories of youths who kill.

Summary: A Case Study Portrait of Adolescents Who Kill

Given the methodological problems in the literature cited earlier, generalizations from many of these studies to the population of juvenile murderers must be made with caution. Some consensus among the studies reported, however, suggests that a portrait of the typical adolescent murderer can be drawn. Before sketching his profile, it is important to note that many youths who possess these characteristics do not commit murder. With this caveat in mind, available data suggest that today's young killer tends to be a male who is unlikely to be psychotic or mentally retarded, to do well in school, or to come from a home where his biological parents live together in a healthy and peaceful relationship. Rather, he is likely to have experienced or to have been exposed to violence in his home and to have a prior arrest record. He is increasingly more likely to use/abuse drugs and alcohol than juvenile homicide offenders in the past.

EMPIRICAL STUDIES OF JUVENILE HOMICIDE OFFENDERS

Well-designed empirical studies of juvenile murderers that attempt to compensate for the weaknesses of case studies do exist. However, they are relatively few and typically also suffer from methodological limitations related particularly to sample selection and size. Hence, generalizations from these studies also must be made with caution.

In one study, for example, 71 adolescent homicide offenders were matched with 71 nonviolent delinquents with respect to age, race, gender, and socioeconomic class. Both groups were selected from a group of 1,956 juveniles referred for evaluation by the juvenile court. The sample of juvenile murderers represented all youths convicted of homicide in the referral sample. The control group was a subset of the larger sample. Accordingly, the samples were retrospective and nonrandom and reflected the selection bias in the referral and adjudication process.

Both groups were assessed on numerous educational, psychiatric and psychological, social, and physical dimensions. Four significant differences were found between the two groups. Compared to nonviolent delinquents, juvenile homicide offenders were more likely to come from criminally violent families, to participate in a gang, to have severe educational deficits, and to abuse alcohol (Busch et al., 1990). The same results were obtained when the study was repeated using different groups of juveniles who were obtained and matched in the same way (Zagar et al., 1990).

Dorothy Otnow Lewis, a psychiatrist at New York University School of Medicine, and her colleagues have also conducted several investigations of juvenile homicide offenders (Lewis, Shanok, Grant, & Ritvo, 1983; Lewis et al., 1985; Lewis, Lovely, et al., 1988; Lewis, Pincus, et al., 1988). Their research, although groundbreaking, often consisted of small samples of cases referred to the senior author and her team for evaluation. In one study, Lewis, Lovely, et al. (1988) compared 13 juvenile murderers evaluated after the homicide with 14 violent delinquents and 18 nonviolent youths. All three groups were incarcerated at the time of the evaluation and were compared with respect to a set of neurological, psychiatric, psychological, and social variables. Analyses revealed that the adolescent homicide offenders did not differ from violent delinquents. However, the juvenile murderers were significantly more likely than the nonviolent delinquents to be neuropsychiatrically impaired, to have been raised in violent homes, and to have been physically abused.

A Finnish study (Santtila & Haapasalo, 1997) was similarly designed to assess whether selected risk factors would differentiate young murderers from other violent offenders and nonviolent youths. Sample subjects were recruited from among all Finnish male prisoners born after a certain year and ranged in age from 18 to 22 at the time of the study. The pool of volunteers was asked to indicate their prior criminal involvements on a questionnaire. The subjects were then divided into groups from which study participants were randomly chosen. The groups were small: 13 had committed murder or attempted to do so, 13 had committed less serious assaults, and 11 reported no violent crimes in their offending history. Several reliable differences were found. Members of the homicide group were significantly more likely to admit having been cruel to animals than members of the two other groups. The age at which members in the murderer group began abusing alcohol was significantly younger than those in the nonviolent group. Relative to the nonviolent offenders, the homicide offenders were significantly more likely to have been physically abused and dependent on harder drugs, such as cocaine, speed, stimulants, and tranquilizers. Although other differences were observable among the groups, the findings did not reach significance, a result that could have been due to the small sample sizes.

Two studies published in 2001 explored differences between a group of juvenile murderers and control groups of violent offenders (Shumaker & McGee, 2001) and fire setters (Dolan & Smith, 2001), using retrospective case analyses. In the first study, 30 juvenile males charged with murder were compared to 62 boys charged with other violent offenses. These individuals were all referred for pretrial psychiatric evaluation between 1987 and 1997. This study was methodologically superior to earlier studies in several ways. It had a relatively large number of murderers, explored many clinical and offense-related variables, and used a control group. In addition, the authors assessed potential differences as well as differences between youths charged with murder and those charged with battery with intent to kill. Comparisons of the two groups made from existing case files on 33 demographic, historical, clinical, offense, and forensic variables yielded only three differences. Juvenile murderers were significantly less likely to have an Axis I diagnosis of mental disorder than other violent juveniles (63.3% vs. 81.4%). Group differences in the types of diagnoses were discernible. Half of the homicide group were diagnosed as having an adjustment disorder or a substance abuse disorder, whereas 69% of the violent group were diagnosed with a chronic or organically based disorder such as conduct disorder, attention deficit disorder, psychosis, or mood disorder. The homicide offenders were also significantly more likely than the other violent offenders to have acted alone (46.7% vs. 8.1%) and to have committed their crimes in a domestic setting (40% vs. 6.5%) (Shumaker & McGee, 2001).

In the second study, 46 juvenile murderers were compared with 106 fire setters who were referred to an adolescent forensic center for evaluation in England between 1986 and 1996. The matching of the fire setters to the young killers on age, ethnicity, socioeconomic data, and criminal history data was an obvious strength of this study. The two samples were also matched in terms of referral to the same unit for assessment on the basis of court adjudications during the same period to ensure that the two groups were evaluated by the same team. Extensive data on demographics, personal/family history, and medical/psychiatric history were extracted from case files; offense-related characteristics were culled from legal depositions and newspaper articles. Juvenile killers were significantly more likely than the arsonists to be male and to have histories of frequent changes of school, alcohol abuse, and alcohol intoxication at the time of the murder. The homicide group was significantly less likely than the fire-setting group to be diagnosed as psychotic, to have had delayed developmental milestones, to have a history of being in care, and to have had previous contact with social/psychological and psychiatric services. A discriminant function analysis found that psychotic illness, care history, prior psychology contact, and alcohol abuse at the time of the crime successfully differentiated the two groups. Of these four, only alcohol abuse at the time of the offense was more prevalent in the homicide group (Dolan & Smith, 2001).

Other empirical studies have looked for distinguishing characteristics among youths who commit murder (e.g., Corder et al., 1976; Cornell et al., 1987b; Cornell, Miller, & Benedek, 1988). Corder et al. (1976) compared 10 youths charged with killing parents to 10 youths charged with killing other relatives or close acquaintances and 10 youths charged with killing strangers. The groups were not only small but also not randomly generated. Individuals in the three groups had been sent to the hospital for evaluation and were matched with one another by age, gender, intelligence, socioeconomic status, and date of hospital admission. The three groups differed significantly from one another on several variables. Those who killed parents, for example, were significantly more likely than those who killed others to have been physically abused, to have come from homes where their mothers were beaten by their fathers, and to have amnesia for the murder.

TYPOLOGIES OF JUVENILE MURDERERS AND CRIME CLASSIFICATION

Several other researchers, as highlighted earlier, have proposed typologies of youths who kill. Attempts to validate typologies of juvenile homicide offenders, however, have often failed because they consisted of small samples or lacked control groups. In contrast, the typology proposed by Cornell et al. (1987b) has shown remarkable promise. This scheme classifies juvenile homicide offenders into three categories based on circumstances of the offense: psychotic (youths who had symptoms of severe mental illness such as hallucinations or delusions), conflict (youths who were engaged in an argument or dispute with the victim when the killing occurred), and crime (youths who killed during the commission of another felony, such as rape or robbery).

The Cornell et al., typology was tested using 72 juveniles charged with murder and a control group of 35 adolescents charged with larceny.

Both groups were referred for pretrial evaluation and were assessed with respect to eight composite categories: "family dysfunction, school adjustment, childhood problems, violence history, delinquent behavior, substance abuse, psychiatric problems, and stressful life events prior to the offense" (Cornell et al., 1987b, p. 386). On the basis of information pertaining to the offense, 7% of the juvenile homicide offenders were assigned to the psychotic subgroup, 42% to the conflict subgroup, and 51% to the crime subgroup.

Analyses revealed significant differences on all eight composite categories between the homicide group and the larceny group. In addition, a number of significant differences emerged among the three subgroups of juvenile homicide offenders. Psychotic homicide offenders were significantly more likely to score higher on the psychiatric history composite and lower on the index of criminal activity than the nonpsychotic groups. In relation to the conflict group, the crime group scored significantly higher on school adjustment problems, substance abuse, and criminal activity and lower on stressful life events. This study provided preliminary support that juvenile homicide offenders could be distinguished from other groups of offenders and from one another. The authors correctly advised that further studies are needed to determine whether the differences among the homicide subgroups will hold up when group assignment is not determined by offense circumstances (Cornell et al., 1987b, 1989).

Subsequent research has found significant differences between the crime and conflict groups. The crime group youths had higher levels of psychopathology on the Minnesota Multiphasic Personality Inventory (an objective measure of personality) (Cornell et al., 1988) than the conflict group youths. The crime group killers also had more serious histories of substance abuse and prior delinquent behavior than the conflict group murderers (Cornell, 1990). The crime group adolescents were more likely to act with others and to be intoxicated on drugs at the time of the murder than the conflict group youths. The crime group homicide offenders also showed poorer object differentiation and more of a victim orientation in responses to the Rorschach (a projective measure of personality) than their conflict group counterparts. The Rorschach responses suggest that crime group youth are more likely to dehumanize other people, to respond violently when frustrated, and to have more severe developmental deficits than conflict group youths (Greco & Cornell, 1992).

Distinctions also emerged within the conflict group between youths who murdered parents and those who killed other victims, none of whom were family members. Juvenile parricide offenders scored lower on school adjustment problems and prior delinquent history than those who killed others but higher on a family dysfunction measure. Cornell's findings with respect to youths who kill parents are similar to conclusions reached in clinical case studies and provide further empirical support that these youths may represent a distinct type of homicide offender (Cornell, 1990; Heide, 1992).

Myers et al. (1995) classified 25 juvenile homicide offenders using the FBI Crime Classification Manual (CCM). The murderers involved in their study included children and adolescents. The sample size was too small to test differences among the four categories of motives in the CCM. Accordingly, the cases were classified into only two categories: "criminal enterprise" and "personal crime."

Important findings from this study included characteristics common to the young killers, as well as those that differentiated the two groups from each other. Ten profile characteristics applied to more than 70% of the total sample. These consisted of family dysfunction, previous violent acts toward others, disruptive behavior disorder, failing at least one grade, emotional abuse by family member, family violence, prior arrests, learning disabilities, weapon of choice, and psychotic symptoms (Myers et al., 1995, Table 2, p. 1488).

Statistical analyses compared the two crime classification groups with respect to psychiatric diagnoses, psychotic symptoms, biopsychosocial variables, and crime characteristics. Statistically significant differences were found between the criminal enterprise and personal cause groups with respect to victim age, victim relationship, and physical abuse. Youths in the criminal group were more likely than those in the personal group to have been abused and to have killed an adult or elderly victim whom they

did not know. Personal group murderers tended to select a child or adolescent victim whom they knew (Myers et al., 1995).

THE TREATMENT OF YOUNG KILLERS

The literature on treating adolescent murderers is sparse and suffers from the same problems as the general literature on juvenile homicide (Benedek, Cornell, & Staresina, 1989; Myers, 1992) and violent juvenile delinquents (Tate, Reppucci, & Mulvey, 1995). Most of the treatment results are based on clinical case reports of a few cases referred to the author for evaluation and/or treatment (e.g., Agee, 1979; Myers & Kemph, 1988; Petti & Wells, 1980; Washbrook, 1979). The extent to which these cases of juvenile murderers are representative of the population of young killers is unknown (Cornell, 1989; Ewing, 1990). In addition, the interventions used are often not based on established therapeutic principles or empirically documented successes (Benedek et al., 1989; Tate et al., 1995). Programs are also frequently not tailored to the type of juvenile murderer.

Although most young killers will be released back into society, few receive any type of mental health treatment following the homicide. In fact, the likelihood of juvenile murderers receiving intensive psychiatric intervention appears to diminish as they enter adolescence (Myers, 1992; e.g., Rosner et al., 1978). University of Florida psychiatrist Wade Myers (1992) summarized the literature on the treatment of homicidal youths by focusing on four main areas: psychotherapy, psychiatric hospitalization, institutional placement, and the use of psychopharmacologic agents.

Psychotherapy with aggressive youths has generally been viewed with pessimism (Myers, 1992; Tate et al., 1995). The overriding assumption among many clinicians has been that juvenile homicide offenders are antisocial and hence are not good candidates for psychotherapeutic interventions. It is important to remember, however, that not all young killers have extensive delinquent or violent histories or antisocial character structures. Available evidence does indicate that psychotherapy can be effective with some adolescents who have engaged in violence (Keith, 1984), even murder (Bailey, 1996a; McCarthy, 1978; Myers & Kemph, 1988; Scherl & Mack, 1966; Smith, 1965). Preliminary data suggest that psychotherapy may be an effective treatment for conduct-disordered youths who meet the diagnostic criteria for the undifferentiated type, have prior emotional relationships with their victims, and are suicidal (Myers & Kemph, 1988).

Among those who have worked with juvenile homicide offenders, psychotherapy, including art therapy (Bailey, 1996a, 1996b; e.g., McGann, 1999), is generally considered to be an important component of treatment with this population. Offenders likely to benefit from interventions of this nature are higher-maturity youths, particularly those who are capable of self-examination and introspection and of forming emotional relationships with others. Youths unlikely to do well with this approach include those with low intelligence, limited insight, and aggressive behavioral response patterns. Therapeutic gains typically do not come quickly even among those who are amenable to psychotherapy because these youths generally have been raised in chaotic and abusive environments and are slow to trust the therapist (Bailey, 1996b; Myers, 1992).

Psychiatric hospitalization, although commonly used for "little kids" who kill, is rarely used for adolescent murderers. Unlike the homicidal child who is typically viewed as psychologically disturbed (Carek & Watson, 1964; Mouridsen & Tolstrup, 1988; Pfeffer, 1980), the adolescent killer is generally regarded as antisocial and is likely to be institutionalized in a facility for juvenile delinquents or adult criminals. Adolescents are more likely to be hospitalized if they appear psychotic, remain homicidal, or need intensive psychopharmacological management. Inpatient treatment can be particularly helpful in stabilizing the youth, redirecting his homicidal impulses, and reducing his internal conflict (e.g., Haizlip et al., 1984). In addition, it can provide an optimal setting for evaluating the youth, assessing his potential for continued violent behavior, and understanding the family system of which he is a member (Myers, 1992).

Institutional placement in juvenile offender programs is a more typical disposition if the

adolescent murderer is retained in the juvenile justice system rather than transferred to the adult criminal justice system to stand trial. Mental health care in juvenile facilities, as well as adult prisons, is typically minimal due to financial constraints and limited awareness of the psychological needs of this population. Despite the lack of treatment, institutional placement appears to have been effective in many cases, as measured by the lack of commission of serious crimes after release (Myers, 1992; e.g., Russell, 1965; Gardiner, 1985).

Myers (1992) discussed four reasons for the apparent success of these "preventive detention" programs. Two of these reasons involve normal maturational processes. First, while the youth is institutionalized, "further neurodevelopmental, cognitive, and emotional growth" may occur, thus enabling the adolescent to acquire "better control of his emotions and aggressive impulses" (p. 53). Second, youths who are contained in a safe and prosocial environment may simply "outgrow" their antisocial behavior over time. Third, for some youths, the homicidal violence was an atypical and isolated event, largely the result of extreme circumstances and/or psychological difficulties, and "would never be repeated" regardless of the court disposition or treatment provided. Fourth, for other youths, the program components, treatment agents, and therapeutic setting had a positive effect on their character structure and behavioral responses. For example, Muriel Gardiner (1985), a psychiatrist who worked extensively with homicidal youths, found that those who made successful adjustments to society when released had learned a vocation, had strong social support systems and had developed meaningful relationships, and did not return to the unhealthy environments where they lived prior to the killing.

Several researchers have expressed concern that institutional programs emphasize behavioral control and conformity to the institutional regime as a measure of progress and success rather than individualized and specialized treatment of the youthful offenders (Fiddes, 1981; Myers, 1992; Sorrells, 1981). Myers has argued persuasively for the development of a "corrective emotional experience" for a subgroup of juvenile murderers who have killed as a result of interpersonal conflict (Cornell et al.'s conflict group) as opposed to furtherance of another crime (Cornell et al.'s crime group) (Cornell et al., 1989; Myers, 1992). This subgroup consists "primarily of youths with some degree of psychological problems (e.g., adjustment disorders, depression), disturbed family functioning, and concomitant stressful life events" (Myers, 1992, p. 55). Myers recommended placement of these youths in a "therapeutically designed institution" staffed by sincerely interested, empathic, and supportive adults who would function as "prosocial role models" and set appropriate behavioral limits. He argued that the program should be tailored to ensure that each youth receives quality mental health care and educational and vocational programs that are consistent with his ability.

Psychopharmacological management of some juvenile murderers holds promise, although empirical studies are lacking (Myers, 1992; Tate et al., 1995). Researchers have hypothesized that several different neurological processes and biological conditions are linked to violent behavior. These include genetic influences, neurophysiological abnormalities, and malfunctioning of neurotransmitter systems and steroid hormones (Reiss & Roth, 1993; Roth, 1994). Psychotropic medication as a component of treatment, rather than the sole type of treatment, is an appropriate intervention for youths who have an associated mental disorder that is likely to respond to medication, such as ADHD (Scott, 1999; Yeager & Lewis, 2000). Myers (1992), noting that many juvenile homicide offenders are conduct disordered, reviewed studies that evaluated the effectiveness of various drugs in reducing aggressive symptoms among this population. There is some evidence that haloperidol (an antipsychotic drug), methylphenidate ("Ritalin," a psychostimulant), imipramine (a tricyclic antidepressant), and propanolol (an antihypertensive drug), as well as lithium and carbamazepine (both mood stabilizers), may be useful in treating certain conduct-disordered and aggressive children and adolescents (Myers, 1992; see Campbell et al., 1984; Kafantaris et al., 1992; Kaplan, Busner, Kupietz, Wassermann, & Segal, 1990; Post, Rubinow, & Uhde, 1984; Puig-Antich, 1982; Scott, 1999; for description of the types of

drugs, see *Physicians' Desk Reference,* 2001). The newer class of antidepressants known as selective serotonin reuptake inhibitors (SSRIs; e.g., Prozac, Zoloft, Paxil) also have been used with good results with violence-prone individuals (Coccaro, 1995), although some caution is advised (Myers & Vondruska, 1998).

Benedek et al. (1989) discussed four classes of drugs that might be considered in the treatment of the homicidal adolescent, depending upon the previous psychiatric history of the youth and his current clinical functioning. These consist of antidepressant, antianxiety, antipsychotic, and antimanic (mood-stabilizing) medications. The authors advised that these medications must be carefully monitored for occasional paradoxical effects and for possible side effects. Dorothy Otnow Lewis, a psychiatrist who has studied violent juveniles for more than two decades, and her colleagues reported that their "greatest successes" occurred when they "targeted underlying specific psychopathology rather than aggression per se. . . . The more specifically directed the medication, the better the outcome will be" (Yeager & Lewis, 2000, p. 807).

Benedek et al. (1989) indicated that other medications, including "beta- and alpha-adrenergic blockers, anticonvulsants, calcium-channel blockers, and antiandrogen hormones," have not been proven to be effective in the treatment of violent adolescents or adults (p. 234; e.g., Tupin, 1987). These authors advised that long-term use of psychotropic medication is most appropriate for youths who are severely mentally ill. Short-term use of antianxiety drugs may be correctly prescribed for youths who have killed due to interpersonal conflict. Use of medication for youths who have killed during the commission of a felony should be carefully considered in the context of a possible history of drug abuse and addiction.

Myers (1992) maintained that each of these four different interventions can play an important role in the treatment of young killers. He advised that effective treatment planning for this population should include all of the possible factors that lead to murder. The family system, as well as the adolescent homicide offender, needs to be thoroughly evaluated. Intervention needs to target chemical abuse/dependency and neuropsychiatric vulnerabilities (e.g., language disorders, learning disabilities, psychomotor seizures) where indicated.

LONG-TERM OUTCOMES OF JUVENILE HOMICIDE OFFENDERS

Benedek et al. (1989) summarized the pre-1990 literature with respect to long-term outcomes of juvenile murderers (Corder et al., 1976; Cormier & Markus, 1980; Duncan & Duncan, 1971; Foster, 1964; Gardiner, 1985; Hellsten & Katila, 1965; Medlicott, 1955; Tanay, 1976). They concluded that "with few exceptions the limited follow-up information is surprisingly positive" (p. 239). Young killers tend to make a satisfactory adjustment in prison and in the community after release from custody and to relate well to their families.

The authors advised caution in extrapolating from these studies to the entire population of juvenile homicide offenders. These case studies frequently report on criminally unsophisticated youths who were involved in what appear to be isolated acts of violence, often involving intense interpersonal conflict with the victim (Benedek et al., 1989). Many specifically involved family members as victims. Perusal of follow-up reports on adolescent parricide offenders indicates that they typically make a successful re-entry into society (Heide, 1992; Hillbrand, Alexandre, Young, & Spitz, 1999). Accordingly, Benedek and her colleagues hypothesized that the outcomes for chronic delinquents who killed in the course of committing other crimes would be much less favorable in terms of recidivism and readjustment to society than the outcomes for youths who killed in response to interpersonal conflict. Extensive follow-up research on Canadian adolescent homicide offenders by Toupin (1993) confirmed that the 18 subjects in the "crime group" committed significantly more offenses, violent offenses, and serious offenses than the 23 youths in the "conflict group" per year of stay since return to the community.

Hagan (1997) conducted a follow-up study on 20 juveniles who had been convicted of homicide or attempted homicide and were sentenced to state custody. He compared the homicide offenders (experimental group) to a control group of 20 juveniles who had been convicted of other nonhomicide offenses. Recidivism was very similar for the two groups. Over half (60%) of the homicide offenders and 65% of the control group were reconvicted of a crime after their release, although none were involved in another homicide. The results indicated that a conviction for attempted homicide or homicide as a juvenile did not show that a person was likely to be involved in another homicide upon return to the community. Of the 60% of the experimental group who recidivated, 25% were convicted of property crimes, and 35% were convicted of crimes against persons. Hagan did not explore the variables that were correlated with success or failure for either the homicide group or the nonhomicide group.

The Texas Youth Commission (2002) evaluated the postrelease outcomes of youths who were placed in the Capital Offender Treatment Program (COTP) located at the Giddings State Home and School in Texas. This program was specifically designed for juveniles who had been convicted of murder. Recidivism was measured by comparing rearrest and reincarceration rates of COTP participants with those of a control group of untreated juvenile homicide offenders (who were not treated due to space limitations) at 1- and 3-year intervals. Known initial differences in recidivism propensities were statistically removed to ensure that differences between the groups were due to treatment effects. The Texas Youth Commission reported in its first annual report, released in December 1996, that specialized treatment for juvenile homicide offenders reduced the likelihood of capital offenders being arrested for a violent crime within a year after release by 52.9%.

Subsequent data indicated that juvenile homicide offenders who received treatment were less likely to be rearrested and reincarcerated than their nontreated counterparts at 1- and 3-year follow-up periods. Compared to controls, treated capital offenders were 16% less likely to be rearrested at both 1-year and 3-year intervals. Juvenile homicide offenders in the treated group were 70% and 43% less likely to be reincarcerated within 1-year and 3-year periods, respectively, than untreated capital offenders (Heide, 1999; Texas Youth Commission, 2002).

The Texas Youth Commission has evaluated personality changes of the youths while in the program, as well as postrelease outcomes. Youths became significantly less hostile and aggressive, assumed more responsibility, and had more empathy for their victims during program involvement.

A 2001 study provided follow-up data on 59 juveniles who were committed to adult prison during a 2-year period in the early 1980s for one or more counts of murder, attempted murder, or, in a few cases, manslaughter. These youths, unlike those in the Hagan study and COTP, were incarcerated in adult prisons, where they received little or no treatment. Although many of these adolescents received lengthy prison sentences, 73% had been released from prison at the time of follow-up conducted 15 to 17 years later. Results indicated that 58% of the 43 subjects released from prison were returned to prison, and most of those who failed did so within the first 3 years of release (Heide, Spencer, Thompson, & Solomon, 2001).

CONCLUSIONS AND FUTURE DIRECTIONS

This chapter synthesizes studies on juvenile homicide offenders that span more than 50 years. Though a large body of literature exists, many questions regarding etiology, associated risk factors, intervention strategies, and long-term outcomes remain unanswered. The established body of literature has been successful in elucidating the phenomenon of youth homicide and in shedding light on the demographic, medical/psychiatric, educational, social, familial, and behavioral characteristics often associated with young killers. Greater advances in knowledge will follow with the implementation of enhanced methodological designs that examine juvenile homicide across four distinct time frames: the years preceding the homicide, the time period immediately following the homicide, the incarcerative or treatment period, and the postrelease period. Each of these areas are discussed below.

The Years Preceding the Homicide

The available literature on juvenile homicide offenders is retrospective in nature. To say something definitive about etiological factors associated with youth murder requires longitudinal studies of children. Three ongoing longitudinal studies, known collectively as the Program of Research on the Causes and Correlates of Delinquency, are studying the developmental pathways to delinquency. This research effort, supported by the Office of Juvenile Justice and Delinquency Prevention (OJJDP), represents the largest and most comprehensive effort undertaken to date to identify the development of violence, other delinquency, and substance use. These studies, which have collected many of the same variables and are well coordinated with each other, began with a total of 4,500 inner-city youths across three sites located in Rochester, New York; Pittsburgh; and Denver. The youth ranged from 7 to 13 years of age at the beginning of data collection in 1988. Youth who were considered at high risk for delinquency were intentionally overrepresented in the three samples. Sample retention has been high across all sites. Interviews have been conducted with the sample subjects and their primary caretakers at either 6-month or 12-month intervals, depending on the site location. In addition to interviews, data have been collected from schools, police, courts, and social service agencies (Thornberry, Huizinga, & Loeber, 1995).

Youths who murdered have been identified across the three groups (Loeber, 2002). Although the numbers appear to be small, the prospective nature of this study may enable the researchers to say something about the developmental pathways to murder by young people living in inner cities, where the largest percentage of juvenile homicide offenders occur. The strength of this methodological design is that researchers can determine which factors preceded the homicide for youths in these three cities.

Recent accounts of youth murderers, particularly those involved in school shootings, have underscored a fact that researchers, clinicians, and law enforcement personnel have known for years: Not all young killers are poor kids from inner-city areas. During the 10-year period from 1991 through 2000, for example, 10.6% of the homicide arrestees in suburban areas and 7.8% of those in rural areas were under 18. Analysis of the young killers in the Program of Research on the Causes and Correlates of Delinquency, though uncovering the developmental pathways to murder for inner-city youth, may have little relevance to more affluent youth living in sub-urban and rural areas. A longitudinal study employing a design such as the one used by the National Longitudinal Study of Adolescent Health (Carolina Population Center, 1998) would seem to be appropriate in this regard. This study consisted of a nationally representative sample of youth who were in Grades 7 through 12 as of 1994. Data were collected from the adolescents themselves, from their parents, siblings, friends, and romantic partners, and from school administrators. One of the areas of health investigated was the youth's participation in violent behavior. Several waves of data collection at different periods involving subsamples have been done. Juvenile murderers probably exist within this sample. Identifying their developmental pathways would be a first step to determining whether the findings from the OJJDP sites can be replicated and, if the numbers are sufficient, whether different pathways can be identified for different types of murderers (see Carolina Population Center, 1998; Resnick et al., 1997).

The Time Period Immediately Following the Homicide

The second period that warrants systematic investigation is the homicidal incident. To the extent possible, researchers should thoroughly investigate the circumstances surrounding the murder by speaking to the youth about what happened as soon as possible. Questions should be designed to uncover the youth's mental state, conscious decisions he was making, conditions of which he was aware, and circumstances that evolved or changes in the moments preceding the homicide. It may be that researchers should think in terms of charting probabilities of lethal outcomes rather than looking for precise causes of murder (Cheatwood, 2002). Many murders committed by juveniles, for example, occur in groups and are associated with other criminal behavior. When a particular youth went out with his peers or gang members, did he intentionally take his gun? For what purpose? Did he know that he was going to participate in a robbery?

Had he participated in a robbery before? With what result to the victim or victims? When did he pull out his gun in the instant robbery? What was behind his pulling out his gun? Was it in response to something his friends did? The victim did? When he pulled out his gun, what was he thinking or intending? Had the youth been drinking or using drugs prior to the shooting? Was the juvenile intoxicated or high? Had he ever pointed a gun at someone before? With what result? Did he decide to pull the trigger? What was he thinking or intending? What happened in the incident? Did the victim die? Subsequent investigation needs to determine whether the victim died on the scene, police response time, timeliness of emergency response teams, and so forth.

Future research involving youth may wish to include those who attempted murder among its sample of murderers. Screening of cases using police reports and follow-up interviews with youth charged with attempted murder would seem useful from the standpoint of better understanding the phenomenon of murderous behavior and increasing sample size (see Heide, 1999; Heide et al., 2001). Frequently what distinguishes a completed murder from an attempted murder is the marksmanship of the offender and the timeliness and quality of the medical response (Block, 1977). If an offender fired five shots at a robbery victim and hit him in the shoulder, one could argue that the victim survived because the youth was a lousy shooter. Similarly, if the youth fired multiple shots into the victim and the emergency response team went to work on the victim until he arrived at a trauma center, one could argue that the victim recovered because he received immediate, life-saving surgery. In both instances, the juvenile's actions, consistent with his intention, appear to kill. From the standpoint of research on the dynamics and pathways to youth murder, it may make little sense to exclude youths convicted of attempted murder from the pool of murderers.

The Incarcerative or Treatment Period

Youths need to be followed up at periodic intervals after the homicide, whether they are incarcerated in prison or sent to a treatment facility. Follow-up at intervals of 6 or

12 months may shed light on which factors are related to successful adjustment in prison and to postrelease success. Information may be gathered, for example, on contacts with family and significant others over time, program participation, mental heath and drug-related services received, physical and mental illnesses suffered during the period of confinement, disciplinary incidents, victimizations, changes in perception, and personality and behavioral changes. Control groups are needed to determine which sanctions or treatments, and which life experiences during this period of incarceration or treatment, are correlated with postrelease success and failure. Random assignment of juvenile homicide offenders to different types of treatment programs would help determine which treatment programs are successful with this population. Random assignment of young killers to prison versus intensive treatment would allow researchers to determine which sanctions hold more promise among youths involved in murderous activities.

Very few treatment programs target juvenile homicide offenders. Recidivism data discussed earlier indicated that COTP has shown remarkable promise in treating juveniles committed for violent crimes, many of whom have been involved in homicide. Further programs modeled on this one may be in order, particularly if they build on components known to be effective with serious, violent, and chronic offenders. A recently published meta-analysis of 200 treatment programs is worth noting in this regard. This study specifically examined whether intervention programs could reduce recidivism among serious delinquents and explored which programs were most successful with this population (Lipsey & Wilson, 1999; Lipsey, Wilson, & Cothern, 2002). The study found an overall 12% decrease in recidivism among treated delinquents when compared to their control groups. The treatment programs that were most effective in reducing recidivism varied to some extent between noninstitutionalized and institutionalized serious delinquents. The most effective treatment programs for noninstitutionalized serious juvenile offenders included individual counseling, interpersonal behavioral programs, and multiple services; for their institutionalized counterparts, they included interpersonal skills,

teaching family homes, behavioral programs, community residential programs, and multiple services. Reduction in recidivism was dramatic for both types of serious juvenile offenders. The researchers estimated that the most effective treatment programs would reduce recidivism for the noninstitutionalized offenders by 40%; for the institutionalized group, by 30% to 35%. Lipsey and his colleagues concluded that the results of their study provided compelling data that serious juvenile offenders could be helped and that intervention can reduce recidivism. They maintained that new and better programs could be developed by skills, studying the characteristics of effective programs, implementing them, and evaluating them.

In recent years, governmental leaders and policy makers have displayed a heightened awareness of the need for effective treatment for violent youths, even as they have argued for sanctions for wrongdoing and protection of the public. OJJDP has maintained that effective programs for rehabilitating violent juvenile offenders must be developed. Its "Comprehensive Strategy for Serious, Violent, and Chronic Offenders" incorporated both prevention and intervention components in an effort to reduce juvenile delinquency and to manage juvenile crime more effectively (Howell, 1995). The experts who devised this initiative conceptualized an effective model for treating juvenile offenders as combining accountability and sanctions with increasingly intensive treatment and rehabilitation.

Postrelease Adjustments of Juvenile Homicide Offenders

Youths involved in homicidal incidents must be followed beyond the incarcerative or treatment period. The high recidivism found by Heide et al. (2001) indicates that more must be known about what occurs after individuals are released from prison that leads to their reoffending. Follow-up interviews at 6-month intervals, particularly during the first 3 years, seem critical. Charting the life course of individuals who make good adjustments to society and understanding the pathways that take offenders back to prison put society in a position where its agents and members can be proactive in reducing recidivism and future violence.

Some might argue that, in light of the recent decreases in juvenile murder, such an extensive investigation of young homicide offenders hardly seems warranted. But two recently released government reports concluded that, despite the decreasing trends, youth homicide and youth violence have not significantly abated. The U.S. Surgeon General in a January 2001 Report on Youth Violence made the following observations:

> Since 1993, when the epidemic peaked, youth violence has declined significantly nationwide, as signaled by downward trends in arrest records, victimization data, and hospital emergency room records. But the problem has not been resolved. Another key indicator of youth violence–youths' confidential reports about their violence behavior–reveals no change since 1993 in the proportion of young people who have committed physically injurious and potentially lethal acts. (American Psychological Association, 2001, pp. 1–2)

The Bureau of Justice Statistics in a March 2000 report noted that "despite the encouraging improvement since 1993, the levels of gun homicide by juveniles and young adults are well-above those of the mid-1980's" and that "the levels of youth homicide remain well-above those of the early and mid-1980's" (Fox & Zawitz, 2000, pp. 1, 2). The report of the U.S. Surgeon General ended with the following warning:

> This is no time for complacency. The epidemic of lethal violence that swept the United States from 1983 to 1993 was funneled in large part by easy access to weapons, notably firearms. If the sizable numbers of youths still involved in violence today begin carrying and using weapons as they did a decade ago, this country may see a resurgence of the lethal violence that characterized the violence epidemic. (American Psychological Association, 2001, p. 2)

REFERENCES

Adams, K. A. (1974). The child who murders: A review of theory and research. *Criminal Justice and Behavior, 1,* 51–61.

Adelson, L. (1972). The battering child. *Journal of the American Medical Association, 222,* 159–161.

Agee, V. L. (1979). *Treatment of the violent incorrigible adolescent.* Lexington, MA: Lexington.

American Psychological Association. (2001, June). Youth violence: Report from the Surgeon General (executive summary). *Child, Youth, and Family Services, 34*(2), 1–7.

Anthony, E. J., & Rizzo, A. (1973). Adolescent girls who kill or try to kill their fathers. In E. J. Anthony & C. Koupernik (Eds.), *The impact of disease and death* (pp. 330–350). New York: Wiley Interscience.

Bailey, S. (1994). Critical pathways of child and adolescent murderers. *Chronicle of the International Association of Juvenile and Family Court Magistrates, 1*(3), 5–12.

Bailey, S. (1996a). Adolescents who murder. *Journal of Adolescence, 19,* 19–39.

Bailey, S. (1996b). Current perspectives on young offenders: Aliens or alienated? *Journal of Clinical Forensic Medicine, 3,* 1–7.

Bender, L. (1959, December). Children and adolescents who have killed. *American Journal of Psychiatry, 116,* 510–513.

Bender, L., & Curran, F. J. (1940). Children and adolescents who kill. *Criminal Psychopathology, 1,* 297–321.

Benedek, E. P., & Cornell, D. G. (1989). Clinical presentations of homicidal adolescents. In E. P. Benedek & D. G.

Cornell (Eds.), *Juvenile homicide* (pp. 37–57). Washington, DC: American Psychiatric Press.

Benedek, E. P., Cornell, D. G., & Staresina, L. (1989). Treatment of the homicidal adolescent. In E. P. Benedek & D. G. Cornell (Eds.), *Juvenile homicide* (pp. 221–247). Washington, DC: American Psychiatric Press.

Bernstein, J. I. (1978). Premeditated murder by an eight year old boy. *International Journal of Offender Therapy and Comparative Criminology, 22,* 47–56.

Block, R. L. (1977). *Violent crime: Environment, interaction, and death.* Lexington, MA: Lexington.

Blumstein, A. (1995, August). Violence by young people: Why the deadly nexus? *National Institute of Justice Journal, 229,* 2–9.

Bortner, M. A. (1988). *Delinquency and justice.* New York: McGraw-Hill.

Busch, K. G., Zagar, R., Hughes, J. R., Arbit, J., & Bussell, R. E. (1990). Adolescents who kill. *Journal of Clinical Psychology, 46,* 472–485.

Butts, J. A., & Snyder, H. N. (1997, September) *The youngest delinquents: Offenders under age 15* (Juvenile Justice Bulletin). Washington, DC: U.S. Department of Justice, Office of Juvenile Justice and Delinquency Prevention.

Bynum, J. E., & Thompson, W. E. (1999). *Juvenile delinquency: A sociological approach* (4th ed.). Boston: Allyn & Bacon.

Campbell, M., Small, A. M., Green, W. H., Jennings, S. J., Perry, R., Bennett, W. G., & Anderson, L. (1984). Behavioral efficacy of haloperidol and lithium carbonate: A comparison in hospitalized aggressive children with conduct disorder children. *Archives of General Psychiatry, 41,* 650–656.

Carek, D. J., & Watson, A. S. (1964). Treatment of a family involved in fratricide. *Archives of General Psychiatry, 11,* 533–542.

Carolina Population Center, University of North Carolina, Chapel Hill. (1998). The National Longitudinal Study of Adolescent Health. Retrieved April 14, 2002, from www.cpc.unc.edu/projects/lifecourse/adhealth.html.

Cheatwood, D. (2002, June). *A proposed model to better integrate theory and policy on homicide.* Paper presented at the annual meeting of the Homicide Research Working Group, St. Louis, MO.

Coccaro, E. F. (1995, January/February). The biology of aggression. *Scientific American, 273,* 38–47.

Corder, B. F., Ball, B. C., Haizlip, T. M., Rollins, R., & Beaumont, R. (1976). Adolescent parricide: A comparison with other adolescent murder. *American Journal of Psychiatry, 133,* 957–961.

Cormier, B. M., Angliker, C. C. J., Gagne, P. W., & Markus, B. (1978). Adolescents who kill a member of the family. In J. M. Eekelaar & S. N. Katz (Eds.), *Family violence: An international and interdisciplinary study* (pp. 466–478). Toronto: Butterworth.

Cormier, B. M., & Markus, B. (1980). A longitudinal study of adolescent murderers. *Bulletin of the American Academy of Psychiatry and the Law, 8,* 240–260.

Cornell, D. G. (1989). Causes of juvenile homicide: A review of the literature. In E. P. Benedek & D. G. Cornell (Eds.), *Juvenile homicide* (pp. 3–36). Washington, DC: American Psychiatric Press.

Cornell, D. G. (1990). Prior adjustment of violent juvenile offenders. *Law and Human Behavior, 14,* 569–577.

Cornell, D. G., Benedek, E. P., & Benedek, D. M. (1987a). Characteristics of adolescents charged with homicide. *Behavioral Sciences and the Law, 5,* 11–23.

Cornell, D. G., Benedek, E. P., & Benedek, D. M. (1987b). Juvenile homicide: Prior adjustment and a proposed typology. *American Journal of Orthopsychiatry, 57,* 383–393.

Cornell, D. G., Benedek, E. P., & Benedek, D. M. (1989). A typology of juvenile homicide offenders. In E. P. Benedek & D. G. Cornell (Eds.), *Juvenile homicide* (pp. 59–84). Washington, DC: American Psychiatric Press.

Cornell, D. G., Miller, C., & Benedek, E. P. (1988). MMPI profiles of adolescents charged with homicide. *Behavioral Sciences and the Law, 6,* 401–407.

Crespi, T. D., & Rigazio-DiGilio, S. A. (1996). Adolescent homicide and family pathology: Implications for research and treatment with adolescents. *Adolescence, 31*(122), 353–367.

Darby, P. J., Allan, W. D., Kashani, J. H., Hartke, K. L., & Reid, J. C. (1998). Analysis of 112 juveniles who committed homicide: Characteristics and a closer look at family abuse. *Journal of Family Violence, 13,* 365–375.

Dolan, M., & Smith, C. (2001). Juvenile homicide offenders: 10 years' experience of a adolescent forensic psychiatry service. *Journal of Forensic Psychiatry, 12,* 313–329.

Duncan, J. W., & Duncan, G. M. (1971). Murder in the family. *American Journal of Psychiatry, 127,* 74–78.

Easson, W. M., & Steinhilber, R. M. (1961, January). Murderous aggression by children and adolescents. *Archives of General Psychiatry, 4,* 27–35.

Ewing, C. P. (1990). *When children kill.* Lexington, MA: Lexington.

Federal Bureau of Investigation. (2001). *Crime in the United States 2000* (Uniform Crime Reports). Washington, DC: Government Printing Office.

Feiler, S. M., & Smith, M. D. (2000, November). *Absolute and relative involvement in homicide offending: An update.* Paper presented at the annual meeting of the American Society of Criminology, San Francisco.

Fendrich, M., Mackesy-Amiti, M. E., Goldstein, P., Spunt, B., & Brownstein, H. (1995). Substance involvement among juvenile murderers: Comparisons with older offenders based on interviews with prison inmates. *International Journal of the Addictions, 30,* 1363–1382.

Fiddes, D. O. (1981). Scotland in the seventies: Adolescents in care and custody: A survey of adolescent murder in Scotland. *Journal of Adolescence, 4,* 47–58.

Foster, H. H. (1964). Closed files on juvenile homicides: A case report. *Journal of Offender Therapy, 8,* 56–60.

Fox, J. A. (1996). *Trends in juvenile violence.* Washington,

DC: U.S. Department of Justice, Bureau of Justice Statistics.

Fox, J. A., & Zawitz, M. W. (1996). *Homicide trends in the United States: 1998 update.* Washington, DC: U.S. Department of Justice, Bureau of Justice Statistics.

Gardiner, M. (1985). *The deadly innocents: Portraits of children who kill.* New Haven, CT: Yale University Press.

Goetting, A. (1989). Patterns of homicide among children. *Criminal Justice and Behavior, 16,* 63–80.

Goetting, A. (1995). *Homicide in families and other special populations.* New York: Springer.

Greco, C. M., & Cornell, D. G. (1992). Rorschach object relations of adolescents who committed homicide. *Journal of Personality Assessment, 59,* 574–583.

Hagan, M. P. (1997). An analysis of adolescent perpetrators of homicide and attempted homicide upon return to the community, *International Journal of Offender Therapy and Comparative Criminology, 41,* 250–259.

Haizlip, T., Corder, B. F., & Ball, B. C. (1984). Adolescent murderer. In C. R. Keith (Ed.), *The aggressive adolescent* (pp. 126–148). New York: Free Press.

Hardwick, P. J., & Rowton-Lee, M. A. (1996). Adolescent homicide: Toward assessment of risk. *Journal of Adolescence, 19,* 263–276.

Hays, J. R., Solway, K. S., & Schreiner, D. (1978). Intellectual characteristics of juvenile murderers versus status offenders. *Psychological Reports, 43,* 80–82.

Heide, K. M. (1984, November). *A preliminary identification of types of adolescent murderers.* Paper presented at the annual meeting of the American Academy of Criminology, Cincinnati, OH.

Heide, K. M. (1992). *Why kids kill parents: Child abuse and adolescent homicide.*

Columbus: Ohio State University Press.

Heide, K. M. (1994). Evidence of child maltreatment among adolescent parricide offenders. *International Journal of Offender Therapy and Comparative Criminology, 38,* 151–162.

Heide, K. M. (1999). *Young killers.* Thousand Oaks, CA: Sage.

Heide, K. M., Spencer, E., Thompson, A., & Solomon, E. P. (2001). Who's in, who's out, and who's back: Follow-up data on 59 juveniles incarcerated for murder or attempted murder in the early 1980s. *Behavioral Sciences and the Law, 19,* 97–108.

Hellsten, P., & Katila, O. (1965). Murder and other homicide, by children under 15 in Finland. *Psychiatric Quarterly, 39*(Suppl. 1)*,* 54–74.

Hillbrand, M., Alexandre, J. W., Young, J. L., & Spitz, R. T. (1999). Parricide: Characteristics of offenders and victims, legal factors, and treatment issues. *Aggression and Violent Behavior, 4*(2), 179–190.

Howell, J. C. (Ed.). (1995). *Guide for implementing the comprehensive strategy for serious, violent, and chronic juvenile offenders.* Washington, DC: U.S. Department of Justice, Office of Juvenile Justice and Delinquency Prevention.

Incompetency standards in death penalty and juvenile cases. (1984). *Mental and Physical Disability Law Reporter, 8*(2), 92–93.

Kafantaris, V., Campbell, M., Padron-Gayol, M. V., Small, A., Locascio, J., & Rosenberg, C. R. (1992). Carbamazepine in hospitalized aggressive conduct-disordered children: An open pilot study. *Psychopharmacology Bulletin, 28*(2), 193–199.

Kalogerakis, M. G. (1971). Homicide in adolescents: Fantasy and deed. In J. Fawcett (Ed.), *Dynamics of violence* (pp. 93–103). Chicago: American Medical Association.

Kaplan, S. L., Busner, J., Kupietz, S., Wassermann, E., & Segal, B. (1990). Effects of methylphenidate on adolescents with aggressive conduct disorder and ADDH: A preliminary report. *Journal of the American Academy of Child and Adolescent Psychiatry, 29,* 719–723.

Kashani, J. H., Darby, P. J., Allan, W. D., Hantke, K. I., & Reid, J. C. (1997). Intrafamilial homicide committed by juveniles: Examination of a sample with recommendations for prevention. *Journal of Forensic Sciences, 42,* 873–878.

Keith, C. R. (1984). Individual psychotherapy and psychoanalysis with the aggressive adolescent: A historical review. In C. R. Keith (Ed.), *The aggressive adolescent* (pp. 191–208). New York: Free Press.

King, C. H. (1975). The ego and the integration of violence in homicidal youth. *American Journal of Orthopsychiatry, 45,* 134–145.

Kirschner, D. (1992). Understanding adoptees who kill: Dissociation, patricide, and the psychodynamics of adoption. *International Journal of Offender Therapy and Comparative Criminology, 36,* 323–334.

Labelle, A., Bradford, J. M., Bourget, D., Jones, B., & Carmichael, M. (1991). Adolescent murderers. *Canadian Journal of Psychiatry, 36,* 583–587.

Lee, A. S., Lee, E. S., & Chen, J. (1995). Young killers. In M. Reidel & J. Boulahanis (Eds.), *Proceedings of the 1995 meeting of the Homicide Research Working Group* (pp. 15–20). Washington, DC: National Institute of Justice Report.

Lewis, D. O., Lovely, R., Yeager, C., Ferguson, G., Friedman, M., Sloane, G., Friedman, H., & Pincus, J. H. (1988). Intrinsic and environmental characteristics of juvenile murderers. *Journal of the*

American Academy of Child and Adolescent Psychiatry, 27, 582–587.

Lewis, D. O., Moy, E., Jackson, L. D., Aaronson, R., Restifo, N., Serra, S., & Simos, A. (1985). Biopsychosocial characteristics of children who later murder: A prospective study. *American Journal of Psychiatry, 142,* 1161–1167.

Lewis, D. O., Pincus, J. H., Bard, B., Richardson, E., Feldman, M., Prichep, L. S., & Yeager, C. (1988). Neuropsychiatric, psychoeducational, and family characteristics of 14 juveniles condemned to death in the United States. *American Journal of Psychiatry, 145,* 584–589.

Lewis, D. O., Shanok, S. S., Grant, M., & Ritvo, E. (1983). Homicidally aggressive young children: Neuropsychiatric and experimental correlates. *American Journal of Psychiatry, 140,* 148–153.

Lipsey, M. W., & Wilson, D. B. (1999). Effective intervention for serious juvenile offenders. In R. Loeber & D. P. Farrington (Eds.), *Serious and violent juvenile offenders* (pp. 313–345). Thousand Oaks, CA: Sage.

Lipsey, M. W., Wilson, D. B., & Cothern, L. (2002). *Effective intervention for serious juvenile offenders.* Washington, DC: U.S. Department of Justice, Office of Juvenile and Delinquency Prevention.

Loeber, R. (2002, June). *The prospective prediction of homicide in two community samples.* Paper presented at the annual meeting of the Homicide Research Working Group, St. Louis, MO.

Mack, J., Scherl, D., & Macht, L. (1973). Children who kill their mothers. In A. J. Anthony & C. Koupernik (Eds.), *The child in his family: The impact of disease and death* (pp. 319–332). New York: Wiley Interscience.

Malmquist, C. P. (1971). Premonitory signs of homicidal aggression in juveniles. *American Journal of Psychiatry, 128,* 461–465.

Malmquist, C. P. (1990). Depression in homicidal adolescents. *Bulletin of the American Academy of Psychiatry and the Law, 18,* 23–36.

Marten, G. W. (1965). Adolescent murderers. *Southern Medical Journal, 58,* 1217–1218.

McCarthy, J. B. (1978). Narcissism and the self in homicidal adolescents. *American Journal of Psychoanalysis, 38,* 19–29.

McCully, R. S. (1978). The laugh of Satan: A study of a familial murderer. *Journal of Personality Assessment, 42*(1), 81–91.

McGann, E. P. (1999). Art therapy assessment and intervention in adolescent homicide. *American Journal of Art Therapy, 38*(2), 51–62.

Medlicott, R. W. (1955). Paranoia of the exalted type in a setting of folie à deux: A study of two adolescent homicides. *British Journal of Medical Psychology, 28,* 205–223.

Meloy, J. R., Hempel, A. G., Mohandie, K., Shiva, A., & Gray, B. T. (2001). Offender and offense characteristics of a nonrandom sample of adolescent mass murderers. *Journal of the American Academy of Child and Adolescent Psychiatry, 40,* 719–728.

Menninger, K., & Mayman, M. (1956). Episodic dyscontrol: A third order of stress adaptation. *Bulletin of the Menninger Clinic, 20,* 153–165.

Michaels, J. J. (1961). Enuresis in murderous aggressive children and adolescents. *Archives of General Psychiatry, 5,* 94–97.

Miller, D., & Looney, J. (1974). The prediction of adolescent homicide: Episodic dyscontrol and dehumanization. *American Journal of Psychoanalysis, 34,* 187–198.

Mohr, J. W., & McKnight, C. K. (1971). Violence as a function of age and relationship with special reference to matricide. *Canadian Psychiatric Association Journal, 16,* 29–32.

Mones, P. (1985). The relationship between child abuse and parricide. In E. H. Newberg & R. Bourne (Eds.), *Unhappy families: Clinical and research perspectives on family violence* (pp. 31–38). Littleton, MA: PSG.

Mones, P. (1991). *When a child kills: Abused children who kill their parents.* New York: Pocket.

Mouridsen, S. E., & Tolstrup, K. (1988). Children who kill: A case study of matricide. *Journal of Child Psychology and Psychiatry, 29,* 511–515.

Myers, W. C. (1992). What treatments do we have for children and adolescents who have killed? *Bulletin of the American Academy of Psychiatry and the Law, 20,* 47–58.

Myers, W. C. (1994). Sexual homicide by adolescents. *Journal of American Academy of Adolescent Psychiatry, 33,* 962–969.

Myers, W. C., Burgess, A. W., & Nelson, J. A. (1998). Criminal and behavioral aspects of juvenile sexual homicide. *Journal of Forensic Science, 43,* 340–347.

Myers, W. C., & Kemph, J. P. (1988). Characteristics and treatment of four homicidal adolescents. *Journal of the American Academy of Child and Adolescent Psychiatry, 27,* 595–599.

Myers, W. C., & Kemph, J. P. (1990). DSM-IIIR classification of homicidal youth: Help or hindrance. *Journal of Clinical Psychiatry, 51,* 239–242.

Myers, W. C., & Mutch, P. A. (1992). Language disorders in disruptive behavior disordered homicidal youth. *Journal of Forensic Sciences, 37,* 919–922.

Myers, W. C. & Scott, K. (1998). Psychotic and conduct disorder symptoms in juvenile murderers. *Journal of Homicide Studies, 2*(2), 160–175.

Myers, W. C., Scott, K., Burgess, A. W., & Burgess, A. G. (1995). Psychopathology, biopsychosocial factors, crime characteristics, and classification of 25 homicidal youths. *Journal of the American Academy of Child and Adolescent Psychiatry, 34,* 1483–1489.

Myers, W., & Vondruska, M. A. (1998). Murder, minors, selective serotonin reuptake inhibitors, and the involuntary intoxication defense. *Journal of the American Academy of Psychiatry and Law, 26,* 487–496.

O'Halloran, C. M., & Altmaier, E. M. (1996). Awareness of death among children: Does a life-threatening illness alter the process of discovery? *Journal of Counseling and Development, 74,* 259–262.

Paluszny, M., & McNabb, M. (1975). Therapy of a six-year-old who committed fratricide. *Journal of the American Academy of Child Psychiatry, 14,* 319–336.

Patterson, R. M. (1943). Psychiatric study of juveniles involved in homicide. *American Journal of Orthopsychiatry, 13,* 125–130.

Petti, T. A., & Davidman, L. (1981). Homicidal school-age children: Cognitive style and demographic features. *Child Psychiatry and Human Development, 12,* 82–89.

Petti, T., & Wells, K. (1980). Crisis treatment of an adolescent who accidentally killed his twin. *American Journal of Psychiatry, 3,* 434–443.

Pfeffer, C. R. (1980). Psychiatric hospital treatment of assaultive homicidal children. *American Journal of Psychotherapy, 34*(2), 197–207.

Physicians' desk reference (52nd ed.). (1998). Montvale, NJ: Medical Economics Data Production.

Podolsky, E. (1965). Children who kill. *General Practitioner, 31,* 98.

Post, R. M., Rubinow, D. R., & Uhde, T. W. (1984). Biochemical mechanisms of action of carbamazepine in affective illness and epilepsy. *Psychopharmacology Bulletin, 20,* 585–590.

Post, S. (1982). Adolescent parricide in abusive families. *Child Welfare, 61,* 455–455.

Puig-Antich, J. (1982). Major depression and conduct disorder in prepuberty. *Journal of the American Academy of Child Psychiatry, 21,* 118–128.

Reiss, A. J., & Roth, J. A. (Eds.). (1993). *Understanding and preventing violence.* Washington, DC: National Academy Press.

Resnick, M. D., Bearman, P. S., Blum, R. W., Bauman, K. F., Harris, K. M., Jones, J., Tabor, J., Beuhring, T., Sieving, R. E., Shew, M., Ireland, M., Bearinger, L., & Udry, J. R. (1997). Protecting adolescents from harm: Findings from the National Longitudinal Study on Adolescent Health. *Journal of the American Medical Association, 278,* 823–832.

Restifo, N., & Lewis, D. O. (1985). Three case reports of a single homicidal adolescent. *American Journal of Psychiatry, 142,* 388.

Rosner, R., Weiderlight, M., Rosner, M. B. H., & Wieczorek, R. R. (1978). Adolescents accused of murder and manslaughter: A five year descriptive study. *Bulletin of the American Academy of Psychiatry and the Law, 7,* 342–351.

Roth, J. A. (1994). *Understanding and preventing violence.* National Institute of Justice Research in Brief. Washington, DC: U.S. Department of Justice, Office of Justice Programs.

Russell, D. H. (1965). A study of juvenile murderers. *Journal of Offender Therapy, 9*(3), 55–86.

Russell, D. H. (1979). Ingredients of juvenile murder. *International Journal of Offender Therapy and Comparative Criminology, 23,* 65–72.

Russell, D. H. (1984). A study of juvenile murderers of family members. *International Journal of Offender Therapy and Comparative Criminology, 28,* 177–192.

Russell, D. H. (1986). Girls who kill. *International Journal of Offender Therapy and Comparative Criminology, 30,* 171–176.

Sadoff, R. L. (1971). Clinical observations on parricide. *Psychiatric Quarterly, 45*(1), 65–69.

Santtila, P., & Haapasalo, J. (1997). Neurological and psychological risk factors among young homicide, violent, and nonviolent offenders in Finland. *Homicide Studies, 1,* 234–253.

Sargent, D. (1962, January). Children who kill: A family conspiracy? *Social Work, 7,* 35–42.

Scherl, D. J., & Mack, J. E. (1966). A study of adolescent matricide. *Journal of the American Academy of Child Psychiatry, 5,* 569–593.

Schmideberg, M. (1973). Juvenile murderers. *International Journal of Offender Therapy and Comparative Criminology, 17,* 240–245.

Scott, C. L. (1999). Juvenile violence. *Child and Adolescent Psychiatric Clinics of North America, 22,* 71–83.

Sendi, I. B., & Blomgren, P. G. (1975). A comparative study of predictive criteria in the predisposition of homicidal adolescents. *American Journal of Psychiatry, 132,* 423–427.

Shumaker, D. M., & McGee, G. R. (2001). Characteristics of homicidal and violent juveniles. *Violence and Victims, 16,* 401–409.

Shumaker, D. M., & Prinz, R. J. (2000). Children who murder: A review. *Clinical Child and Family Psychology Review, 3*(2), 97–115.

Sickmund, M. (1994, October). *How juveniles get to criminal court* (OJJDP Update and Statistics). Washington, DC: U.S. Department of Justice, Office of Justice Programs, Office of Juvenile Justice and Delinquency Prevention.

Smith, M. D., & Feiler, S. M. (1995). Absolute and relative involvement in homicide offending: Contemporary youth and the baby boom cohorts. *Violence and Victims, 10,* 327–333.

Smith, S. (1965). The adolescent murderer: A psychodynamic interpretation. *Archives of General Psychiatry, 13,* 310–319.

Snyder, H. N. (1997, November). *Juvenile arrests 1996* (Juvenile Justice Bulletin). Washington, DC: U.S. Department of Justice, Office of Juvenile Justice and Delinquency Prevention.

Snyder, H. N. (2001). *Law enforcement and juvenile crime.* Washington, DC: U.S. Department of Justice, Office of Justice and Delinquency Prevention.

Solomon, E., Schmidt, R., & Ardragna, P. (1990). *Human anatomy and physiology.* Philadelphia: W. B. Saunders.

Solway, I. S., Richardson, L., Hays, J. R., & Elion, V. H. (1981). Adolescent murderers: Literature review and preliminary research findings. In J. R. Hays, T. K. Roberts, & K. Solway (Eds.), *Violence and the violent individual* (pp. 193–210). Jamaica, NY: Spectrum.

Sorrells, J. M. (1977). Kids who kill. *Crime and Delinquency, 23,* 313–320.

Sorrells, J. M., Jr. (1981). What can be done about juvenile homicide? *Crime and Delinquency, 16,* 152-161.

Stearns, A. (1957). Murder by adolescents with obscure motivation. *American Journal of Psychiatry, 114,* 303–305.

Tanay, E. (1973). Adolescents who kill parents: Reactive parricide. *Australian and New Zealand Journal of Psychiatry, 7,* 263–277.

Tanay, E. (1976). Reactive parricide. *Journal of Forensic Sciences, 21*(1), 76–82.

Tate, D. C., Reppucci, N. D., & Mulvey, E. P. (1995). Violent juvenile delinquents: Treatment effectiveness and implications for future action. *American Psychologist, 50,* 777–781.

Texas Youth Commission. (2002, April). Capital Offender Treatment Program, reports December 1996 and February 27, 1997, Texas Youth Handbook. Retrieved April 14, 2002, from http://austin.tyc.us/cfinternet/handbook/youth_e9.html.

Thom, D. (1949). Juvenile delinquency and criminal homicide. *Journal of the Maine Medical Association, 40,* 176.

Thornberry, T. P., Huizinga, D., & Loeber, R. (1995). The prevention of serious delinquency and violence: Implications from the program of research on the causes and correlates of delinquency. In J. C. Howell et al. (Eds.), *A sourcebook: Serious, violent, and chronic juvenile offenders* (pp. 213–237). Thousand Oaks, CA: Sage.

Tooley, K. (1975). The small assassins. *Journal of the American Academy of Child Psychiatry, 14,* 306–318.

Toupin, J. (1993). Adolescent murderers: Validation of a typology and study of their recidivism. In A. V. Wilson (Ed.), *Homicide: The victim/offender connection* (pp. 135–156). Cincinnati, OH: Anderson.

Tucker, L. S., & Cornwall, T. P. (1977). Mother-son "folie à deux": A case of attempted patricide. *American Journal of Psychiatry, 134,* 1146–1147.

Tupin, J. (1987). Psychopharmacology and aggression. In L. Roth (Ed.), *Clinical treatment of the violent person* (pp. 79–94). New York: Guilford.

U.S. Department of Justice. (1987). *Bureau of Justice Statistics special report: Survey of youth in custody.* Washington, DC: Government Printing Office.

Walshe-Brennan, K. S. (1974). Psychopathology of homicidal children. *Royal Society of Health, 94,* 274–276.

Walshe-Brennan, K. S. (1977). A socio-psychological investigation of young murderers. *British Journal of Criminology, 17*(1), 53–63.

Washbrook, R. A. H. (1979). Bereavement leading to murder. *International Journal of Offender Therapy and Comparative Criminology, 23,* 57–64.

Wertham, F. (1941). *Dark legend: A study in murder.* New York: Duell, Sloan, & Pearce.

Woods, S. M. (1961). Adolescent violence and homicide: Ego disruption and the 6 and 14 dysrhythmia. *Archives of General Psychiatry, 5,* 528–534.

Yates, A., Beutler, L. E., & Crago, M. (1983). Characteristics of young, violent offenders. *Journal of Psychiatry and Law, 11*(2), 137–149.

Yeager, C. A., & Lewis, D. O. (2000). Mental illness, neuropsychologic deficits, child abuse, and violence. *Child and Adolescent Psychiatric Clinics of North America, 9,* 793–813.

Zagar, R., Arbit, J., Sylvies, R., Busch, K., & Hughes, J. R. (1990). Homicidal adolescents: A replication. *Psychological Reports, 67,* 1235–1242.

Zenoff, E. H., & Zients, A. B. (1979). Juvenile murderers: Should the punishment fit the crime? *International Journal of Law and Psychiatry, 2*(4), 533–553.

16

SENTENCING GUIDELINES AND THE TRANSFORMATION OF JUVENILE JUSTICE IN THE 21ST CENTURY

DANIEL P. MEARS

As we enter the 21st century, many states have introduced fundamental changes to their juvenile justice systems. The changes focus on jurisdictional authority, especially transfer to adult court; sentencing guidelines and options; correctional programming; interagency information sharing; offender confidentiality; and victim involvement. At the same time, attention has turned increasingly to prevention, early intervention, rehabilitation, and the use of specialized courts. Because of their special significance in the historical context of the juvenile court, this article focuses on the emergence of sentencing guidelines to identify underlying trends and issues in the transformation of juvenile justice. In so doing, the article argues that the considerable attention given by policy makers and researchers to transfer rather than other changes provides a distorted picture of current juvenile justice practice.

The past decade witnessed dramatic changes to juvenile justice in America, changes that have altered the focus and administration of juvenile justice as it enters the 21st century (Butts & Mitchell, 2000; Feld, 1991; Harris, Welsh, & Butler, 2000). In contrast to the philosophical foundation and practice of the first juvenile courts, punishment and due process today constitute central features of processing. These emphases, which run counter to the rehabilitative *parens patriae* ("state as parent") foundation of the first juvenile courts, emerged in the 1960s with a series of U.S. Supreme Court decisions. In cases such as *In re Gault*, the Supreme Court recognized that juvenile courts served not only a rehabilitative function but also a punishment function, and that consequently due process rights and procedures should figure more prominently in juvenile proceedings (Feld, 1999). In recent years, the transition has become more pronounced, with states enacting

Mears, D. P. (2002, February). Sentencing Guidelines and the Transformation of Juvenile Justice in the 21st Century. *Journal of Contemporary Criminal Justice, 18*(1), 6-19. Reprinted with permission.

sweeping legislative changes affecting all aspects of the juvenile justice system (National Criminal Justice Association, 1997; Torbet et al., 1996; Torbet & Szymanski, 1998).

It is important to recognize, however, that the changes have not been entirely or even primarily focused on punishment. One would not know this from a review of research, the bulk of which has examined patterns, correlates, and effects of transfer (for a review, see Butts & Mitchell, 2000). The focus is understandable—transfer provides an easily identifiable symbol for debates about the merits of maintaining two separate juvenile and adult systems (Feld, 1999; Hirschi & Gottfredson, 1993). Indeed, why have a juvenile justice system if youth are being sent into adult courts? But the fact is that only about 1% of all formally processed delinquency cases ultimately are transferred (Snyder & Sickmund, 1999, p. 171).

Focusing solely on transfer ignores the fact that other equally, if not more significant, transformations have occurred in juvenile justice. These include enactment of sentencing guidelines; creation of blended sentencing options for linking the juvenile and criminal justice systems; enhanced correctional programming, with an increasing emphasis on treatment; greater interagency and cross-jurisdiction cooperation and information sharing; reduced confidentiality of court records and proceedings; and increased participation of victims in juvenile justice processing (Fagan & Zimring, 2000; Guarino-Ghezzi & Loughran, 1996; National Criminal Justice Association, 1997; Torbet et al., 1996). In addition, states increasingly are turning their attention to prevention, early intervention, rehabilitation, and the use of specialized courts to address juvenile crime (Butts & Harrell, 1998; Butts & Mears, 2001; Cocozza & Skowyra, 2000; Coordinating Council on Juvenile Justice and Delinquency Prevention, 1996; Cullen & Gendreau, 2000; Howell, 1995; Lipsey, Wilson, & Cothern, 2000; Rivers & Anwyl, 2000).

It is apparent that juvenile justice has been evolving along many dimensions. With all of these changes, the question arises: What, if any, are the common trends and issues underlying these different changes? To answer this question, I examine sentencing guidelines, showing that they reflect many of the major trends and issues in juvenile justice. I focus on guidelines because typically they apply to all juvenile offenders and embody a range of goals, thus reflecting many of the conflicts and tensions inherent in attempts to modify the focus and administration of juvenile justice. By contrast, transfer laws, which have received much more attention in the research literature, focus only on select age groups and offenders and have the delimited purpose of punishing and deterring offenders.

The primary goal of this article, in short, is to use an analysis of sentencing guidelines to highlight a range of critical underlying trends and issues in juvenile justice. A secondary goal is to show that research on transfer laws provides little insight into juvenile justice as it is practiced today and, in the absence of research on or attention to other reforms, can provide a distorted picture of current practice. To achieve these goals, I begin by briefly describing the history of the juvenile court and the emergence of juvenile sentencing guidelines. I then use this discussion to identify key trends and issues in juvenile justice.

FOUNDATION OF THE JUVENILE JUSTICE SYSTEM

Juveniles have not always been viewed the same way throughout U.S. history. For example, in the 18th century, juvenile offenders were treated as adults and received the same types of punishments. During the 19th century, a movement began that focused on the unique, less-than-adult capacities and needs of youth. This movement highlighted the need for a specialized sanctioning process, one that emphasized rehabilitation and deemphasized punishment.

The result of this movement was the development of the first U.S. juvenile court in Cook County, Illinois, in 1899. By 1925, juvenile courts were established in all but two states, with most courts defining juveniles as individuals who were aged 17 years or younger. (For histories of the juvenile court, see Platt [1977], Bernard [1992], Feld [1999], and Butts and Mitchell [2000].)

These new youth-centered courts were grounded in the doctrine of parens patriae. The guiding rationale was that states had an obligation to intervene in the lives of children whose parents provided inadequate care or supervision. Juvenile court interventions were to be benevolent and in the "best interests" of the child.

For this reason, court processing entailed fundamentally different notions of procedural and substantive justice. Unlike adult court proceedings, juvenile court proceedings were to be informal and conducted on a case-by-case basis, with the aim of improving the lives of children through individualized treatment and varying dispositional options, ranging from warnings to probation to confinement.

The basis for intervening in the lives of juvenile offenders derived not from criminal law but civil law, further highlighting the focus on helping youth rather than sanctioning them for their crimes. Similarly, the philosophy of parens patriae clearly suggested that the courts had an obligation to help youth who committed crimes or who clearly needed help. As a result, juvenile courts could use coercive means to help youth, even when relatively minor crimes had been committed or when there was insufficient basis for determining that a crime in fact was committed.

The potential for abuse of this discretionary authority is evident in critiques of the juvenile court (see Feld, 1999). Indeed, as many scholars have shown, the transition to establishing a juvenile justice system was not motivated entirely by benevolent concerns. Under the guise of providing social services and crime control, juvenile courts could, for example, be used instead to provide a form of social control over "undesirable classes," including minorities, immigrants, and indigents (Butts & Mitchell, 2000).

By the 1960s, deep-rooted concerns arose about the procedural and substantive unfairness of juvenile court proceedings, leading the U.S. Supreme Court, through a series of decisions, to emphasize greater procedural parity with criminal court proceedings. The result was an increasingly criminal-like juvenile court. This trend, coupled with tougher transfer provisions in the 1990s, led to considerable debate about the merits of having two separate court systems, one for juveniles and one for adults (Feld, 1999).

JUVENILE SENTENCING GUIDELINES: AN OVERVIEW

The early juvenile court emphasized individualized, offender-based treatment and sanctioning. Indeed, almost every justification of the juvenile court rests on the notion that the most appropriate and effective intervention for youth is one that takes into account their particular needs and resources. Ironically, despite the establishment of this view more than 100 years ago, recent research provides considerable empirical support for it—the most effective interventions are those premised on addressing the specific risk, needs, and capacities of youth (Cullen & Gendreau, 2000; Lipsey, 1999).

Under the Office of Juvenile Justice and Delinquency Prevention's (OJJDP's) Comprehensive Strategy for Serious, Violent, and Chronic Juvenile Offenders (Howell, 1995; Wilson & Howell, 1993), states have been encouraged to adopt individualized sanctioning and to emphasize risk and needs assessment. Many have responded by enacting guideline systems that are modeled to a considerable extent on the Comprehensive Strategy.

In some states, these guideline systems are voluntary, in others there are incentives to use them, and in still others they are required. In each instance, the guidelines typically are offense-based and outline a sequence of increasingly tougher sanctions, while at the same time emphasizing rehabilitative interventions when appropriate.

In 1995, for example, Texas enacted what it termed the Progressive Sanctions Guidelines. The Guidelines outline seven tiers of sanctioning, with each linked to the instant offense and the offender's prior record. Once the appropriate level of sanctioning is established, courts are encouraged to include additional, nonpunitive interventions. Although the Guidelines are voluntary, Texas documents the extent to which county-level sanctioning deviates from the recommendations of the Guidelines (Texas Criminal Justice Policy Council, 2001). Similar approaches have been implemented in other states, including Illinois, Kansas, Nebraska, New York, Utah, Virginia, and Washington (Corriero, 1999; Demleitner, 1999; Fagan &

Zimring, 2000; Lieb & Brown, 1999; National Criminal Justice Association, 1997; Torbet et al., 1996).

State guideline systems often identify their goals explicitly. In Texas, for example, the Progressive Sanction Guidelines are used to "guide" dispositional decision making in providing "appropriate" sanctions and to promote "uniformity" and "consistency" in sentencing (Dawson, 1996). At the same time, the Guidelines are seen as furthering the newly established and explicitly stated goal of the Texas Juvenile Justice Code—namely, punishment of juveniles. But they also promote rehabilitative sanctioning by encouraging appropriate treatment and interventions for each recommended sanction level. In addition, the Guidelines implicitly promote certain goals, including public safety through incapacitation of the most serious or chronic offenders and reduced crime through get-tough, deterrence-oriented sanctioning.

Other states have followed similar paths. For example, Washington established sentencing guidelines aimed directly at reducing the perceived failings of a system founded on practitioner discretion (Lieb & Brown, 1999). The guidelines focus not only on offense-based considerations but also on the juvenile's age, with younger offenders receiving fewer "points" and thus more lenient sanctions. Similarly, Utah has enacted sentencing guidelines focusing on proportionate sentencing, early intervention, and progressively intensive supervision and sanctioning for more serious and chronic offenders (Utah Sentencing Commission, 1997).

Because many states increasingly are adopting sentencing guidelines and because the guidelines focus on all youth rather than simply those who may be transferred, an examination of them can help to identify underlying trends and issues emergent in juvenile justice. By contrast, a focus on transfer, typical of most research on recent reforms, provides relatively little leverage to do so. Transfer laws typically focus on "easy cases," those in which the seriousness of the offense largely vitiates, rightly or wrongly, concerns many would have about individualized or rehabilitative sanctioning. Any resulting debate therefore centers on extremes: Should we retain or eliminate the juvenile court?

But a broader issue in juvenile justice is how to balance individualized, offender-based sanctioning with proportional and consistent punishment. These issues, among several others, are a consideration in almost every case coming before the juvenile court. It is appropriate, therefore, to focus on a recent reform, such as sentencing guidelines, that typically target, in one manner or another, all youth and that reflects attempts to shape the entire juvenile justice system. For this reason, the remainder of this article uses a focus on sentencing guidelines to identify key trends and issues in the transformation of juvenile justice.

Juvenile Sentencing Guidelines: Trends and Issues in the Transformation of Juvenile Justice in the New Millennium

Balancing Multiple and Conflicting Goals

The motivation for transforming juvenile justice has come from many sources. Scholars cite a range of factors, including the desire to address violent crime, inconsistency and racial/ethnic disproportionality in sentencing, financial burdens faced by counties versus states, and public support for get-tough and rehabilitative measures (Bazemore & Umbreit, 1995; Bishop, Lanza-Kaduce, & Frazier, 1998; Butts & Mitchell, 2000).

As suggested by the different motivations for reform, a key trend in juvenile justice is the move toward balancing multiple and frequently competing goals, only one of which includes the punitive focus associated with transfer (Bazemore & Umbreit, 1995; Guarino-Ghezzi & Loughran, 1996; Mears, 2000). Today, many juvenile justice codes and policies focus on retributive/punitive sanctioning (through get-tough sanctions generally), incapacitation, deterrence, rehabilitation, individualized as well as consistent and proportional sentencing, and restorative sanctioning.

Reduced crime is a broad goal underlying many but not all of these more specific goals. For example, get-tough sanctions are viewed as a primary mechanism to instill fear and achieve specific or general deterrence (i.e., reduced offending among sanctioned or would-be offenders) or to reduce crime through temporary incapacitation of offenders. In many instances, retribution serves as the primary focus of sanctioning, irrespective of any potential crime control impact.

Some goals, like rehabilitation, serve as steps toward enhancing the lives of juveniles, not simply reducing their offending. Others, such as restorative sanctioning, focus on reintegrating offenders into their communities while at the same time providing victims with a voice in the sanctioning and justice process. Still others, including proportional and consistent sentencing, focus primarily on fairness rather than crime control. That is, the motivation is to provide sanctions that are proportional to the crime and that are consistent within and across jurisdictions so that juveniles sanctioned by Judge X or in County X receive sanctions similar to those administered by Judge Y or in County Y.

Historically these different goals, including what might be termed intermediate goals leading to reduced crime, have overlapped considerably with those of the criminal justice system (Snyder & Sickmund, 1999, pp. 94-96). In general, though, criminal justice systems have given greater weight to punishment than rehabilitation, whereas juvenile justice systems generally have favored rehabilitation more than punishment.

In reality, the goals in each system are diverse, as are the weightings given to each goal. Indeed, the diversity of goals and their weightings can make it difficult to determine how exactly the two systems differ, especially if we focus only on new transfer laws (see, however, Bishop & Frazier, 2000). But one major difference between the two is that juvenile justice systems—as is evident in their sentencing guideline systems—are actively struggling to balance as wide a range of goals as possible. By contrast, most criminal justice systems have veered strongly toward retribution and incapacitation (Clark, Austin, & Henry, 1997).

Giving Priority to Punishment Through Offense-Based Guidelines and Changes in Discretion

Most state guideline systems use offense-based criteria for determining which types of sanctions to apply (Coolbaugh & Hansel, 2000). Once the punishment level has been established, the court is supposed to consider the needs of the offender and how these may best be addressed. However, these needs frequently are only vaguely specified and rarely assessed. One result is that priority implicitly and in practice may be given to punishment.

This priority can be reinforced through various mechanisms that place greater discretion in the hands of prosecutors rather than judges. For example, laws that stipulate automatic sanctions for certain offenses do not eliminate discretion; instead, they shift it to prosecutors, who can determine whether and how to charge an offense (Feld, 1999; Mears, 2000; Sanborn, 1994; Singer, 1996). Consequently, in practice, many guideline systems make punishment a priority not just for youth who may be transferred but for all youth referred to juvenile court.

Sentencing guidelines have not gone unopposed. For example, research on the Texas Progressive Sanction Guidelines indicates that many judges resisted enactment of the Guidelines and then, once they became law, resisted using them (Mears, 2000). One reason is their belief that offense-based criteria provide too limited a basis for structuring decision making. Thus, even though compliance with the Guidelines is voluntary, some judges feel that the Guidelines symbolize too narrow a focus, one that draws attention from factors they believe are more important, such as the age and maturity of the youth and their family and community contexts. Such concerns have been expressed about adult sentencing guidelines (e.g., Forer, 1994). One difference with juvenile sentencing guidelines is that, despite the views of opponents, they generally state explicitly that there are multiple goals associated with sanctioning and that practitioners should consider a range of mitigating factors (Howell, 1995).

Balancing Discretion Versus Disparity and Consistency, and Procedural Versus Substantive Justice

In stark contrast to the early foundation of the juvenile court, many states today are intent on eliminating disparity and inconsistency in sentencing (Feld, 1999; Torbet et al., 1996). The widespread belief, evident in many sentencing guidelines, is that (a) judicial discretion causes disparity and inconsistency and (b) that offense-based systems can eliminate or reduce these problems. Both beliefs prevail despite the fact that little empirical evidence exists to support them (Mears & Field, 2000; Sanborn, 1994; Yellen, 1999).

But the fact that such strategies may not work does not belie the underlying trend toward discovering ways to promote fairness and consistency in sentencing. Nor does it belie the fact that, as with adult sanctioning, there likely will continue to be an ongoing tension between the use of discretion and the need to have sanctions that are relatively similar for different populations and within and across jurisdictions.

This tension is captured in part by the distinction in the sociology of law between procedural and substantive justice. From the perspective of procedural justice, fairness emerges from decisions that are guided by established rules and procedures for sanctioning cases that exhibit specific characteristics. By contrast, from the perspective of substantive justice, fairness emerges from decisions that are guided by consideration of the unique situational context and characteristics of the defendant (Gould, 1993; Ulmer & Kramer, 1996).

In recent years, and as exemplified by the creation of offense-based sentencing guidelines, juvenile justice systems increasingly are focusing on procedural justice. In the case of transfer particularly, the Supreme Court and state legislatures have attempted to ensure that there is procedural parity with adult proceedings. Yet despite the increased proceduralization, for most cases facing the juvenile courts, substantive justice also remains a priority, especially when sanctioning first-time and less serious offenders. In these instances, states have devised strategies, outlined in their guidelines, that promote diversion, rehabilitation, and treatment.

Maintaining the View That Most Youth are "Youth," Not Adults

Public opinion polls show that whereas most people consistently support rehabilitative sanctioning of youth, they also support punitive, get-tough measures for serious and violent offenders (Roberts & Stalans, 1998). Moreover, even when the public supports transferring youth to the adult system, they generally prefer youth to be housed in separate facilities and to receive individualized, rehabilitative treatment (Schwartz, Guo, & Kerbs, 1993).

The apparent contradiction likely constitutes the primary reason that wholesale elimination of the juvenile justice system has not prevailed. In the debate about abolishing the juvenile court, this fact frequently is omitted, perhaps because so much attention has centered on changes in transfer laws. Indeed, were one to focus solely on recent trends in transfer, one might conclude that an eventual merging of juvenile and adult systems is inevitable (Feld, 1999).

Yet the focus and structure of juvenile sentencing guidelines, which explicitly call for rehabilitation and early intervention, suggest otherwise. In contrast to get-tough developments in the criminal justice system (Clark et al., 1997), most states—even those without guideline systems—have struggled to maintain a focus not only on the most violent offenders but also on efficient and effective intervention with less serious offenders.

This trend is reflected in the proliferation of alternative, or specialized, courts, including community, teen, drug, and mental health courts (Butts & Harrell, 1998; Office of Justice Programs, 1998; Santa Clara County Superior Court, 2001). These courts focus on timely and rehabilitative sanctioning that draws on the strengths of families and communities and the cooperation and assistance of local and state agencies.

Some authors suggest that these courts threaten the foundation of the juvenile court (Butts & Harrell, 1998). But specialized courts can be viewed as symbolic of the reemergence of the juvenile justice system as historically conceived—namely, as a system designed to intervene on an individualized, case-by-case basis, addressing the particular risks and needs of offenders (Butts & Mears, 2001). Indeed, to this end, many guidelines promote diversion of first- and second-time, less serious offenders from formal processing to informal alternatives available through specialized courts.

Limited Conceptualization and Assessment of the Implementation and Effects of Changes in the Juvenile Justice System

One last and prominent trend in juvenile justice bears emphasizing—the lack of systematic attention to conceptualizing and assessing the implementation and effects of recently enacted laws. A focus on sentencing guidelines illustrates the point: Few states have systematically articulated precisely what the goals of the guidelines are, how specifically the guidelines are expected to achieve these goals, or what in fact the effects of the guidelines have been (Coolbaugh & Hansel, 2000; Fagan & Zimring, 2000; Mears, 2000).

One example common to many guidelines is the focus on consistency. Several questions illustrate the point. What exactly does *consistency* mean? Is it identical sentencing of like offenders within jurisdictions? Across jurisdictions? Does it involve similar weighting of the same factors by all judges or judges within each jurisdiction in a state? Across states? Apart from definitional issues, does consistency lead to reduced crime or increased perceptions of fairness? If so, how? What precisely are the mechanisms by which increased consistency would lead to changes in crime or perceptions of fairness? The failure to address these questions means that it is impossible to assess whether there has been more or less consistency resulting from guideline systems.

Similar questions about many other aspects of recent juvenile justice reforms remain largely unaddressed, with two unfortunate consequences. First, as noted above, it is impossible to assess the effects of the reforms without greater clarity concerning their goals and the means by which these goals are to be reached. As a result, it is difficult if not impossible to

make informed policy decisions, including those focusing on maintaining or eliminating the juvenile justice system (Schneider, 1984; Singer, 1996). Second, without conceptualization and assessment of the effects of recent reforms, there is an increased likelihood that research on delimited aspects of juvenile justice systems will be generalized into statements about entire systems, even though there may be little to no correspondence between the two.

Conclusion

Recent changes to juvenile justice systems throughout the United States indicate a trend toward developing more efficient and effective strategies for balancing different and frequently competing goals. This trend is evident in recent juvenile sentencing guidelines. As the above discussion demonstrates, guidelines focus on more than transferring the most serious offenders to the criminal justice system. They also focus on balancing competing goals, reducing discretion and promoting fair and consistent sanctioning, and tempering procedural with substantive justice. More generally, guidelines aim to preserve the notion that youth are not adults.

One result of such trends is increasing interest in alternative administrative mechanisms for processing youthful offenders. Specialized "community," "teen," "drug," "mental health," and other such courts have been developed to do what the original juvenile court was supposed to do—provide individualized and rehabilitative sanctioning. But the "modern" approach involves doing so in a more timely and sophisticated fashion, and in a way that draws on the cooperation and assistance of local and state agencies as well as families and communities.

In the new millennium, juvenile justice thus involves more than an emphasis on due process and punishment. It also involves substantive concerns, including a range of competing goals, a belief in the special status of childhood, and the desire to develop more effective strategies for preventing and reducing juvenile crime.

By focusing on sentencing guidelines, these types of issues become more apparent, highlighting the need for researchers to look beyond transfer laws in assessing recent juvenile justice reforms. Indeed, there is a need for research on many new and different laws, policies, and programs in juvenile justice, most of which remain unassessed. As we enter the new millennium, it will be critical to redress this situation, especially if we are to move juvenile justice beyond "juvenile" versus "adult" debates and to develop more efficient and effective interventions.

REFERENCES

Bazemore, G., & Umbreit, M. (1995). Rethinking the sanctioning function in juvenile court: Retributive or restorative responses to youth crime. *Crime & Delinquency, 41,* 296-316.

Bernard, T. J. (1992). *The cycle of juvenile justice.* New York: Oxford University Press.

Bishop, D. M., & Frazier, C. E. (2000). Consequences of transfer. In J. Fagan & F. E. Zimring (Eds.), *The changing boundaries of juvenile justice: Transfer of adolescents to the criminal court* (pp. 227-276). Chicago: University of Chicago Press.

Bishop, D. M., Lanza-Kaduce, L., & Frazier, C. E. (1998). Juvenile justice under attack: An analysis of the causes and impact of recent reforms. *Journal of Law and Public Policy, 10,* 129-155.

Butts, J. A., & Harrell, A. V. (1998). *Delinquents or criminals? Policy options for juvenile offenders.* Washington, DC: The Urban Institute.

Butts, J. A., & Mears, D. P. (2001). Reviving juvenile justice in a get-tough era. *Youth & Society, 33,* 169-198.

Butts, J. A., & Mitchell, O. (2000). Brick by brick: Dismantling the border between juvenile and adult justice. In C. M. Friel (Ed.), *Criminal justice 2000: Boundary changes in criminal justice organizations* (Vol. 2, pp. 167-213). Washington, DC: National Institute of Justice.

Clark, J., Austin, J., & Henry, D. A. (1997). *"Three strikes and you're out": A review of state legislation.* Washington, DC: National Institute of Justice.

Cocozza, J. J., & Skowyra, K. (2000). Youth with mental health disorders: Issues and emerging responses. *Juvenile Justice, 7,* 3-13.

Coolbaugh, K., & Hansel, C. J. (2000). *The comprehensive strategy: Lessons learned from the pilot sites.* Washington, DC: Office of Juvenile Justice and Delinquency Prevention.

Coordinating Council on Juvenile Justice and Delinquency Prevention. (1996). *Combating violence and delinquency: The national juvenile justice action plan.* Washington, DC: Office of Juvenile Justice and Delinquency Prevention.

Corriero, M. A. (1999). Juvenile sentencing: The New York youth part as a model. *Federal Sentencing Reporter, 11,* 278-281.

Cullen, F. T., & Gendreau, P. (2000). Assessing correctional rehabilitation: Policy, practice, and prospects. In J. Horney (Ed.), *Criminal justice 2000: Policies, processes, and decisions of the criminal justice system* (Vol. 3, pp. 109-175). Washington, DC: National Institute of Justice.

Dawson, R. O. (1996). *Texas juvenile law* (4th ed.). Austin: Texas Juvenile Probation Commission.

Demleitner, N. V. (1999). Reforming juvenile sentencing. *Federal Sentencing Reporter, 11,* 243-247.

Fagan, J., & Zimring, F. E. (Eds.). (2000). *The changing borders of juvenile justice.* Chicago: University of Chicago Press.

Feld, B. C. (1991). The transformation of the juvenile court. *Minnesota Law Review, 75,* 691-725.

Feld, B. C. (1999). *Bad kids: Race and the transformation of the juvenile court.* New York: Oxford University Press.

Forer, L. (1994). *A rage to punish: The unintended consequences of mandatory sentencing.* New York: Norton.

Gould, M. (1993). Legitimation and justification: The logic of moral and contractual solidarity in Weber and Durkheim. *Social Theory, 13,* 205-225.

Guarino-Ghezzi, S., & Loughran, E. J. (1996). *Balancing juvenile justice.* New Brunswick, NJ: Transaction.

Harris, P. W., Welsh, W. N., & Butler, F. (2000). A century of juvenile justice. In G. LaFree (Ed.), *Criminal justice 2000: The nature of crime: Continuity and change* (Vol. 1, pp. 359-425). Washington, DC: National Institute of Justice.

Hirschi, T., & Gottfredson, M. R. (1993). Rethinking the juvenile justice system. *Crime & Delinquency, 39,* 262-271.

Howell, J. C. (1995). *Guide for implementing the comprehensive strategy for serious, violent, and chronic juvenile offenders.* Washington, DC: Office of Juvenile Justice and Delinquency Prevention.

Lieb, R., & Brown, M. E. (1999). Washington state's solo path: Juvenile sentencing guidelines. *Federal Sentencing Reporter, 11,* 273-277.

Lipsey, M. W. (1999). Can rehabilitative programs reduce

the recidivism of juvenile offenders? An inquiry into the effectiveness of practical programs. *Virginia Journal of Social Policy and Law, 6,* 611-641.

Lipsey, M. W., Wilson, D. B., & Cothern, L. (2000). *Effective intervention for serious juvenile offenders.* Washington, DC: Office of Juvenile Justice and Delinquency Prevention.

Mears, D. P. (2000). Assessing the effectiveness of juvenile justice reforms: A closer look at the criteria and impacts on diverse stakeholders. *Law and Policy, 22,* 175-202.

Mears, D. P., & Field, S. H. (2000). Theorizing sanctioning in a criminalized juvenile court. *Criminology, 38,* 101-137.

National Criminal Justice Association. (1997). *Juvenile justice reform initiatives in the states: 1994-1996.* Washington, DC: Office of Juvenile Justice and Delinquency Prevention.

Office of Justice Programs. (1998). *Juvenile and family drug courts: An overview.* Washington, DC: Author.

Platt, A. M. (1977). *The child savers: The invention of delinquency.* Chicago: University of Chicago Press.

Rivers, J. E., & Anwyl, R. S. (2000). Juvenile assessment centers: Strengths, weaknesses, and potential. *The Prison Journal, 80,* 96-113.

Roberts, J. V., & Stalans, L. J. (1998). Crime, criminal justice, and public opinion. In M. Tonry (Ed.), *The handbook of crime and punishment* (pp. 31-57). New York: Oxford University Press.

Sanborn, J. A. (1994). Certification to criminal court: The important policy questions of how, when, and why. *Crime & Delinquency, 40,* 262-281.

Santa Clara County Superior Court. (2001). *Santa Clara County Superior Court commences juvenile mental health court.* San Jose, CA: Author.

Schneider, A. L. (1984). Sentencing guidelines and recidivism rates of juvenile offenders. *Justice Quarterly, 1,* 107-124.

Schwartz, I. M., Guo, S., & Kerbs, J. J. (1993). The impact of demographic variables on public opinion regarding juvenile justice: Implications for public policy. *Crime & Delinquency, 39,* 5-28.

Singer, S. I. (1996). Merging and emerging systems of juvenile and criminal justice. *Law and Policy, 18,* 1-15.

Snyder, H. N., & Sickmund, M. (1999). *Juvenile offenders and victims: 1999 national report.* Washington, DC: Office of Juvenile Justice and Delinquency Prevention.

Texas Criminal Justice Policy Council. (2001). *The impact of progressive sanction guidelines: Trends since 1995.* Austin, TX: Author.

Torbet, P., Gable, R., Hurst, H. IV, Montgomery, I., Szymanski, L., & Thomas, D. (1996). *State responses to serious and violent juvenile crime.* Washington, DC: Office of Juvenile Justice and Delinquency Prevention.

Torbet, P., & Szymanski, L. (1998). *State legislative responses to violent juvenile crime: 1996-97 update.* Washington, DC: Office of Juvenile Justice and Delinquency Prevention.

Ulmer, J. T., & Kramer, J. H. (1996). Court communities under sentencing guidelines: Dilemmas of formal rationality and sentencing disparity. *Criminology, 34,* 383-407.

Utah Sentencing Commission. (1997). *Juvenile sentencing guidelines manual.* Salt Lake City, UT: Author.

Wilson, J. J., & Howell, J. C. (1993). *Comprehensive strategy for serious, violent, and chronic juvenile offenders: Program summary.* Washington, DC: Office of Juvenile Justice and Delinquency Prevention.

Yellen, D. (1999). Sentence discounts and sentencing guidelines. *Federal Sentencing Reporter, 11,* 285-288.

17

"WHAT DO WE DO WITH GIRLS?"

The Dimensions of and Responses to Female Juvenile Delinquency

SCOTT K. OKAMOTO

MEDA CHESNEY-LIND

Tosha is a 15-year-old African American girl who has been arrested for assault, terroristic threatening, and drug possession. During her sentencing, she received a "slap on her wrist" by the court, was put on probation for 2 years, and received no services. Consequently, she began cutting classes and smoking marijuana after school. Although she stated to a school counselor that she "didn't want to get high anymore," her family, particularly her mother and mother's boyfriend, continued to support her drug use by using and offering her drugs. Eventually, she got picked up by the police for truancy and was sentenced to a local residential treatment center for her substance abuse. However, because of a lack of "bed space" in the girls' unit, she ended up sitting in detention for 6 weeks.

Britney is a 17-year-old Caucasian female arrested for drug possession with the intent to sell and assault. During her sentencing, she was given probation for 1 year and was referred to a day treatment center for substance abuse. Her mother also agreed to pay for individual psychotherapy in order to address Britney's anger management issues. Although the court was somewhat concerned with the mother's ability to supervise Britney, they ultimately felt that case management from the probation officer would allow Britney to reside with her mother until a space opened up in the day treatment program. Britney returned home, began psychotherapy, and was admitted into the day treatment program 2 weeks later.

Historically, female juvenile delinquency has been "ignored, trivialized or denied" (Chesney-Lind & Okamoto, 2001, p. 3). This response has gradually changed, as statistics consistently illustrate the increasing involvement of female youth in the juvenile justice system (Budnick & Shields-Fletcher, 1998). Throughout the past decade, an

244

increasing amount of literature has focused on the etiology, prevalence, and treatment of female juvenile delinquency (see Belknap, Dunn, & Holsinger, 1997; Chesney-Lind & Okamoto, 2001; Chesney-Lind & Shelden, 1998) and has highlighted the unique patterns of female juvenile offending (Poe-Yamagata & Butts, 1996). Despite this attention in the literature, relatively little progress has been made to address the unique needs of this population from a political and practice perspective. This chapter examines the dimensions of female juvenile delinquency and the legal, political, and practice responses to this phenomenon. The nature of girls' aggression, violence, and status offenses to date is explored, and practice principles evolving from the treatment literature are proposed and examined.

DIMENSIONS OF FEMALE JUVENILE DELINQUENCY

Throughout the past decade, female juvenile delinquency has undergone substantial changes compared to boys' delinquency. For example, between 1991 and 2000 in the United States, girls' arrests increased 25.3% while arrests of boys actually decreased by 3.2% (Federal Bureau of Investigation, 2001). Within this same time period, arrests of girls for serious violent offenses and "other assaults" increased by 27.9% and 77.9%, respectively (Federal Bureau of Investigation, 2001). What do statistics like these suggest about the prevalence of, and our subsequent responses to, female juvenile delinquency? This section will examine the changes in this phenomenon, focusing specifically on the current research examining trends in girls' aggression and violence, trends in enforcement practices and arrests, the increasing use of detention for girls, and the different ways that girls of color are processed through the juvenile justice system.

Trends in Girls' Aggression and Violence

Contrary to arrest data, *self-report* data of youthful involvement in violent offenses suggest that girls' violence and aggression are decreasing

at a substantially greater rate than those of boys. For example, data collected by the Centers for Disease Control (CDC) over the last decade support the decrease in girls' aggression. The CDC has been monitoring youthful behavior in a national sample of school-aged youth in a number of domains (including violence) at regular intervals since 1991 in a biennial survey entitled the Youth Risk Behavior Survey. Data collected over the 1990s reveal that, whereas 34.4% of girls surveyed in 1991 said that they had been in a physical fight in the last year, by 1999 that figure had dropped to 27.3% (a 21% decrease in girls' fighting). Boys' violence also decreased during the same period but only slightly—from 44.0% to 42.5% (a 3.4% drop; Brener, Simon, Krug, & Lowry, 1999; CDC, 2000). Further, the number of girls who reported carrying weapons and guns declined substantially. Summarizing these trends, Brener et al. (1999) concluded that, although there was a significant linear decrease in physical fighting for both male and female students, females seemed to have had a steeper decline. These findings are consistent with earlier research that revealed significant *decreases* in girls' involvement in felony assaults, minor assaults, and hard drugs and no change in a wide range of other delinquent behaviors—including felony theft, minor theft, and index delinquency (Huizinga, 1997).

The trends in girl's lethal violence illustrate even more dramatic decreases over time. Whereas girls' arrests for all forms of assault skyrocketed in the 1990s, girls' arrests for robbery fell by 45.3% and murder by 1.4% between 1991 and 2000 (Federal Bureau of Investigation, 2001). Further, recent research on girls' violence in San Francisco that used vital statistics maintained by health officials (rather than arrest data) found that there had been a 63% drop in teen-girl fatalities between the 1960s and the 1990s and that girls' hospital injuries were dramatically underrepresented among those reporting injuries, including assaults (Males & Shorter, 2001).

Enforcement Practices and Arrest Trends

Why is it that arrest data suggest an increase in female juvenile delinquency, whereas self-report data suggest a decrease in it? Research suggests that it is likely that changes in *enforcement* practices have dramatically narrowed the gap in delinquency between girls and boys. Specifically, behaviors that were once categorized as status offenses (noncriminal offenses like "runaway" and "person in need of supervision") and domestic violence have been increasingly relabeled as violent or criminal offenses, potentially accounting for increases in girls' arrest trends.

A recent study of juvenile robbery illustrates the trend of relabeling status offenses as more serious offenses. Chesney-Lind and Paramore (2001) examined police files that focused on robbery incidents resulting in arrest from two time periods (1991 and 1997). They found that the proportion of robbery arrests involving girls within this time period more than tripled. However, a closer analysis of the data revealed that the problem of juvenile robbery in the City and County of Honolulu was largely characterized by slightly older youth bullying and "hijacking" younger youth for small amounts of cash and occasionally jewelry. In effect, less serious offenses appeared to be swept up into the system, perhaps as a result of changes in school policy and parental attitudes.

Girls' case files in various states highlight a similar trend of relabeling domestic violence as more serious offenses. Mayer (1994) reviewed over 2,000 cases of girls referred to Maryland's juvenile justice system for "person-to-person" offenses. Although virtually all of these offenses (97.9%) involved "assault," about half of them were "family centered" and involved such actions as "a girl hitting her mother and her mother subsequently pressing charges" (p. 2). Similarly, Acoca (1999) examined nearly 1,000 girls' files from four California counties and found that a majority of their offenses involved assault. However, a close reading of these girls' case files revealed that most of their assault charges were the result of "nonserious, mutual combat situations with parents" (p. 8). Case descriptions ranged from self-defense (e.g., "father lunged at [daughter] while she was calling the police about a domestic dispute. [Daughter] hit him") to trivial arguments (e.g., girl arrested "for throwing cookies at her mother").

In sum, these research findings call into question the actual reported prevalence of female juvenile delinquency. The use of arrest data alone fails to paint an accurate picture of girls' violence and aggression. Research using self-report data and case files challenges the notion that girls are getting more violent. Nevertheless, recent statistics have identified the rising use of detention with girls and the disproportionate representation of girls of color in detention. These issues will be discussed next.

Rising Detentions and Racialized Justice

Recent research has illustrated the increasing use of detention for girls and the different ways that girls of color are processed through the juvenile justice system. National data indicate that between 1989 and 1998, detentions involving girls increased by 56% compared to a 20% increase in boys' detentions (Harms, 2002). The increase is most likely related to recent legislation, which has made it easier for girls to be detained for status offenses such as running away (Sherman, 2000). Further, research has illustrated that detention centers are being utilized with girls for a significantly longer duration than with boys, with 60% of the girls detained for more than 7 days, compared to only 6% of the boys (Shorter, Schaffner, Schick, & Frappier, 1996). Quite possibly, this may be related to the lack of gender-specific programs nationally.

Another critical issue is different ways girls of color are processed through the juvenile justice system compared to Caucasian girls. For example, Miller (1994) found that Caucasian girls were significantly more likely to be recommended for a treatment as opposed to a "detention oriented" placement than either African American or Latina girls. In fact, 75% of the Caucasian girls were recommended for a treatment-oriented facility compared to 34.6% of the Latinas and only 20% of the African American girls. Other evidence of this pattern is reported by Bartollas (1993) in his study of youth confined in institutional placements in a midwestern state. His research sampled female adolescents in both public and private facilities. The "state" sample (representing the girls in public facilities) was 61% black, whereas the private sample was 100% white. Little difference, however, was found in the offense patterns of the two groups of girls. These findings suggest the development of a two-track juvenile justice system—one track for girls of color and another for white girls.

CHARACTERISTICS OF GIRLS' AGGRESSION

What distinguishes girls' aggression from boys' aggression? Some studies have found very few observable differences between the two genders in manifestations of aggressive behaviors such as conduct problems (Tiet, Wasserman, Loeber, McReynolds, & Miller, 2001) and bullying (Gropper & Froschl, 2000). Nonetheless, much attention has been given in the research literature to the unique aspects of girls' aggression. This section will discuss two of these aspects—relational aggression and the effects of trauma.

Relational Aggression

Research has suggested the presence of gender-specific forms of aggression in childhood and adolescence. These studies support the hypothesis that boys' aggression tends to be more "overt," such as hitting or pushing others or threatening to beat others up, whereas girls' aggression tends to be more "relational," or focused on damaging another child's friendships or feelings of inclusion by the peer group (Crick & Grotpeter, 1995; Lagerspetz & Bjorkqvist, 1994). Bjorkqvist and Niemela (1992), for example, found that when types of verbal aggression (e.g., gossiping, spreading rumors) were included in their overall measurement of aggression, only about 5% of the variance was explained by gender. Research has suggested that relationally aggressive youth experience more difficulties with social adjustment, as they are significantly more disliked and lonelier than nonrelationally aggressive peers (Crick & Grotpeter, 1995). These findings suggests that girls' aggression is rooted in significant relationships within childhood and adolescence and, unlike boys', is less likely to focus on typical male behaviors such as physical dominance and intimidation. It also highlights the historically patriarchal context of the conceptualization of aggression and the need to acknowledge its gender-specific manifestations.

The Effects of Trauma

Research suggests that trauma, particularly in the form of physical and sexual abuse, has a profound impact on the development of female juvenile delinquency. Artz (1998), for example, found that violent girls in Canada reported significantly greater rates of victimization and abuse than their nonviolent counterparts and reported great fear of sexual assault, especially from their boyfriends. Artz found that 20% of violent girls reported that they were physically abused at home (compared to 10% of violent boys and 6.3% of nonviolent girls) and that roughly one out of four violent girls had been sexually abused compared to one in 10 of nonviolent girls.

Similarly, studies of girls on the streets are showing high rates of both sexual and physical abuse. For example, McCormack, Janus, and Burgess (1986) found that 73% of female runaways in a runaway shelter in Toronto had been sexually abused, versus 38% of the male runaways. They also found that sexually abused female runaways were more likely than their nonabused counterparts to engage in delinquent or criminal activities such as substance abuse, petty theft, and prostitution. No such pattern was found among the male runaways.

Finally, studies of youth entering the juvenile justice system in Florida have compared the "constellations of problems" presented by girls and boys entering detention and have found that female youth are more likely than male youth to have abuse histories (Dembo, Sue, Borden, & Manning, 1995; Dembo, Williams, & Schmeidler, 1993). Dembo et al. (1995) concluded, "Girls' problem behavior commonly relates to an abusive and traumatizing home life, whereas boys' law violating behavior reflects their involvement in a delinquent life style" (p. 21).

RISK FACTORS FOR FEMALE JUVENILE DELINQUENCY

The Office of Juvenile Justice and Delinquency Prevention (OJJDP, 1998) has outlined several risk factors that predispose girls to delinquency. These factors are consistent with recently published literature focused on predicting risk of violence in other populations (Roberts & Rock, 2002). OJJDP states that, although every female juvenile offender is unique, she is likely to have some of the following risk factors:

- Poverty
- Residence in a community with a high crime rate
- Ethnic minority status
- History of poor academic performance
- Victimization by physical/sexual/emotional abuse and/or exploitation
- Drug and/or alcohol abuse
- Lack of hope for the future

Again, the effects of trauma, in the form of physical, sexual, and/or emotional abuse, are particularly significant for girls. In most cases, girls were victims themselves before they became offenders (Girls Incorporated, 1996; OJJDP, 1998).

PRACTICE PRINCIPLES

The early practice principles related to female juvenile delinquency reflected what Belknap et al. (1997) described as an "add-women-and-stir" approach. In other words, early delinquency treatment strategies typically reflected boys' social and behavioral issues, and it was assumed that girls would benefit from these strategies (Belknap et al., 1997; Chesney-Lind & Shelden, 1998). However, the unique etiology of female juvenile delinquency coupled with the relative lack of success of girls in the treatment setting suggests that practice principles for girls should be different from those for boys. Chesney-Lind and Okamoto (2001) described three characteristics in the treatment literature that reflect the unique practice principles in working with girls: (a) Girls are much more emotional than boys in the treatment setting, (b) girls have distinctly different needs from boys in the areas of life skills training and education, and (c) girls elicit unique countertransference reactions from practitioners. This section will discuss these characteristics and focus on their implications to working with girls. Promising programs related to girls will also be discussed.

Addressing Emotionality

One of the distinguishing characteristics of female youth receiving juvenile justice services is their level of emotionality compared to boys. Observations by both youth practitioners and girls themselves suggest that girls' complex issues, and the emotional way in which they are expressed, can be very challenging to address in the therapeutic setting (Alder & Hunter, 1999; Baines & Alder, 1996). Typically, girls' emotionality is framed negatively. For example, youth-serving practitioners in Baines and Alder's study frequently referred to young women as "hysterical." They stated that female youth engage in attention-seeking behaviors, such as threats of self-mutilation and property destruction, which are extremely demanding and time consuming.

Although the majority of the research focuses on the negative aspects of female emotionality, Chesney-Lind and Freitas (1999) described some of the positive functions of this personality trait. In their study of youth workers' perceptions and feelings in working with girls, they found that some workers felt that the emotional nature of girls made it easier for them to engage with the population and establish rapport. Along these lines, girls' emotionality may also promote verbal introspection and insight, which is often lacking for many boys in treatment. As one practitioner in their study stated, "Girls strike up relationships more easily with each other, compared to boys, [and at a] quicker rate" (p. 11).

This literature suggests that gender-specific programming should be structured to meet the emotional and relational needs of girls. Chesney-Lind and Okamoto (2001) discussed the need for lower staff-to-client ratios so that relationship building between therapeutic staff and female youth could be promoted. Literature has suggested that the practitioner-client relationship significantly affects the level of client engagement in treatment (Prinz & Miller, 1996) and degree of therapeutic change of the client (Drisko, 2002). Appropriately addressing the relational and emotional aspects in working with girls through relationship building might subsequently affect these other critical areas of treatment.

Life Skills Training/Education

Much of the gender-specific treatment literature has highlighted the unique psycho-educational needs of girls. OJJDP (1998) has identified several specific program components related to prosocial education (e.g., decision making, problem solving, anger and stress management, assertive communication) that are essential to the development of comprehensive, gender-specific services. An example of the application of these components is the Staff-Secured Detention Program for Female Juvenile Offenders in Philadelphia, Pennsylvania (Juvenile Justice Evaluation Center, n.d.; OJJDP, 1998). This program uses one of the few evidence-based residential treatment models (the teaching family model) and has adapted the staff training and teaching components of the program to reflect the specific psychoeducational needs of girls. Staff are trained in dealing with issues such as female hygiene, sexual acting out, pregnancy prevention, and eating disorders. The teaching components focus on life skills such as effective communication, independent living, problem solving, and anger management. One study found that girls in teaching family model group homes had significantly lower rates of criminal offenses than girls in comparison group homes during treatment (Kirigin, Braukmann, Atwater, & Wolf, 1982). This finding suggests that teaching gender-specific life skills in a supportive and nurturing environment can positively affect girls' lives.

Countertransference

A developing body of practice literature has focused on the impact of youth-serving practitioners' feelings and perceptions on their work with girls. Research focused on these practitioners has suggested that they experience unique countertransference feelings when working with girls. The term *countertransference* refers to practitioners' emotional (and subsequent behavioral) reactions to a client's behavior (Brandell, 1992). Research on the countertransference feelings of youth-serving practitioners suggests that these feelings influence practice with high-risk youth, including juvenile offenders, at

the system, agency, and practitioner levels (Kersten, 1990; Okamoto, 2001; Okamoto & Chesney-Lind, 2000).

The emotional reactions to working with girls have been suggested in the literature by the questions such as "Are girls more difficult to work with?" (Baines & Alder, 1996) or "Are girls more difficult to handle?" (Kersten, 1990). Some research has suggested that girls are not necessarily harder to work with but are just "different" (Alder & Hunter, 1999; Chesney-Lind & Freitas, 1999). Nonetheless, research and anecdotal evidence suggest that female youth evoke strong feelings in the practitioners that work with them. Kersten (1990), for example, described how workers in juvenile justice institutions enforced policies more strictly with female inmates and resorted to punitive measures, such as "time-out" procedures, more frequently with girls.

Countertransference reactions are particularly significant in the therapeutic relationship between male youth-serving practitioners and their female youth clients. In particular, male youth-serving practitioners have identified fears related to liability/litigation (i.e., being sued by a client or client's parents) when working with girls (Okamoto, 2002). Preliminary theory development of the male practitioner/female youth client relationship suggests that these fears are related to girls' past sexual abuse trauma from men and the subsequent rage related to this trauma directed toward male authority figures (Okamoto, 2002). Male youth-serving practitioners manage these feelings in a variety of ways, such as managing proximity and physical touch and maintaining appropriate boundaries with female youth clients (Baines & Alder, 1996; Okamoto, in press). These studies also highlight the importance of gender-specific training and supervision of youth-serving practitioners related to self-awareness, boundary setting, and managing feelings and perceptions in working with female youth (St. Germaine & Kessell, 1989).

Promising Programs

Establishing promising programs for girls is a challenge on many different levels. In general, there are issues in treatment research regarding the ability to move established, "evidence-based," or "efficacious" youth programs from the research setting to the real-world setting. These issues have been discussed in the literature as "transportability" issues (Schoenwald & Hoagwood, 2001) and call into question the actual clinical utility of efficacious programs for youth, including those that supposedly affect girls. Even if issues related to "transportability" of efficacious youth programs are ignored, the actual findings of many of these programs have not been examined specifically with girls in mind. Thus, identifying promising programs specifically for girls appears to be somewhat "speculative" in nature.

Nevertheless, Chesney-Lind and Okamoto (2001) described several types of programs thought to be effective in working with girls—cognitive-behavioral interventions, including behavioral parent training, and dialectical behavior therapy (DBT; Linehan, 1993). The efficacy of DBT with female juvenile offenders was supported by one study (Trupin, Stewart, Boetsy, & Beach, 1999). The program, originally developed for use with adult women with borderline personality disorder, focuses on cognitive-behavioral strategies but also incorporates methods to deal with the emotional dysregulation that is indicative of borderline personality disorder. The therapist acts as "both coach and consultant to the patient, and actively works to cultivate a positive interpersonal and collaborative relationship throughout the course of treatment" (Trupin et al., 1999, p. 5). According to their findings, female juvenile offenders receiving the treatment demonstrated a significant reduction of behavior problems (e.g., aggression, parasuicidal behavior, and class disruption). The findings of this study support the importance in addressing the emotional and relational aspect in working with girls discussed earlier in this chapter.

Most recently, LeCroy and Daley (2001) have evaluated a comprehensive, school-based prevention program for adolescent girls called the *Go Grrrls* Program. The 12-session psychoeducational program focuses on gender-specific issues such as media messages about girls and women, self-esteem, friendships and dating, and problem-solving strategies. The program is delivered to about 8 to 10 girls after school and uses discussion, role play, and games to deliver the content of the program. Preliminary data indicate that the program had a significant positive effect on attitudes toward body image, assertiveness, self-efficacy, and self-esteem compared to a control group. Though this program does not focus exclusively on delinquent girls, it reflects many of the aspects of gender-specific programming described by OJJDP (1998) and is one of the few programs that has been developed for and evaluated with a focus on girls.

CONCLUSION

Are girls closing the gender gap in violence? On the basis of the examination of the research presented in this chapter, the answer to this question would most likely be "no." Nevertheless, there are many girls in the juvenile justice and mental health service system that are in need of effective services. Of the 26 programs identified as "potentially promising" by the OJJDP, only *two* were tailored to the specific needs of girls (Girls Incorporated, 1996). Most likely, the lack of gender-specific services relates to the increasing use of detention for girls. Though this chapter discussed several recent innovations in gender-specific programming, more research needs to be directed toward identifying and elucidating effective practice principles when working with girls. These principles might incorporate methods in addressing relational aggression, female emotionality, gender-specific life-skills training, and counter-transference when working with girls. Future research should also attempt to create links between service providers, academicians, and researchers in order to advance the field of gender-specific research and to ultimately meet the unique needs of delinquent girls (OJJDP, 1998).

REFERENCES

Acoca, L. (1999). Investing in girls: A 21st century challenge. *Juvenile Justice, 6*(1), 3–13.

Alder, C., & Hunter, N. (1999). *"Not worse, just different"? Working with young women in the juvenile justice system.* Melbourne, Australia: University of Melbourne, Criminology Department.

Artz, S. (1998). *Sex, power, and the violent school girl.* Toronto: Trifolium.

Baines, M., & Alder, C. (1996). Are girls more difficult to work with? Youth workers' perspectives in juvenile justice and related areas. *Crime and Delinquency, 42*, 467–485.

Bartollas, C. (1993). Little girls grown up: The perils of institutionalization. In C. Culliver (Ed.), *Female criminality: The state of the art* (pp. 469–482). New York: Garland.

Belknap, J., Dunn, M., & Holsinger, K. (1997). *Moving toward juvenile justice and youth-serving systems that address the distinct experience of the adolescent female.* Cincinnati, OH: Gender Specific Services Work Group.

Bjorkqvist, K., & Niemela, P. (1992). New trends in the study of female aggression. In K. Bjorkqvist & P. Niemela (Eds.), *Of mice and women: Aspects of female aggression.* San Diego: Academic Press.

Brandell, J. R. (Ed.). (1992). *Countertransference in psychotherapy with children and adolescents.* Northvale, NJ: Jason Aronson.

Brener, N. D., Simon, T. R., Krug, E. G., & Lowry, R. (1999). Recent trends in violence-related behaviors among high school students in the United States. *Journal of the American Medical Association, 282*, 330–446.

Budnick, K. J., & Shields-Fletcher, E. (1998). *What about girls?* (OJJDP Publication No. 84).

Washington, DC: U.S. Department of Justice.

Centers for Disease Control. (2000). *Youth risk behavior surveillance—United States, 1999.* Atlanta: Author.

Chesney-Lind, M., & Freitas, K. (1999). *Working with girls: Exploring practitioner issues, experiences and feelings* (Rep. No. 403). Honolulu: University of Hawai'i at Manoa, Social Science Research Institute.

Chesney-Lind, M., & Okamoto, S. K. (2001). Gender matters: Patterns in girls' delinquency and gender responsive programming. *Journal of Forensic Psychology Practice, 1*(3), 1–28.

Chesney-Lind, M., & Paramore, V. (2001). Are girls getting more violent? Exploring juvenile robbery trends. *Journal of Contemporary Criminal Justice, 17*(2), 142–166.

Chesney-Lind, M., & Shelden, R. G. (1998). *Girls, delinquency, and juvenile justice* (2nd ed.). Belmont, CA: Wadsworth.

Crick, N. R., & Grotpeter, J. K. (1995). Relational aggression, gender, and social-psychological adjustment. *Child Development, 66*, 710–722.

Dembo, R., Sue, S. C., Borden, P., & Manning, D. (1995, August). *Gender differences in service needs among youths entering a juvenile assessment center: A replication study.* Paper presented at the annual meeting of the Society of Social Problems, Washington, DC.

Dembo, R., Williams, L., & Schmeidler, J. (1993). Gender differences in mental health service needs among youths entering a juvenile detention center. *Journal of Prison and Jail Health, 12*, 73–101.

Drisko, J. W. (2002, January). *Common factors in psychotherapy effectiveness: A neglected dimension in social work research.* Paper

presented at the annual meeting of the Society for Social Work and Research, San Diego, CA.

Federal Bureau of Investigation. (2001). *Crime in the United States 2000.* Washington, DC: Government Printing Office.

Girls Incorporated. (1996). *Prevention and parity: Girls in juvenile justice.* Indianapolis: Girls Incorporated National Resource Center.

Gropper, N., & Froschl, M. (2000). The role of gender in young children's teasing and bullying behavior. *Equity and Excellence in Education, 33*(1), 48–56.

Harms, P. (2002). *Detention in delinquency cases, 1989–1998* (OJJDP Fact Sheet No. 1). Washington, DC: U.S. Department of Justice.

Huizinga, D. (1997). *Over-time changes in delinquency and drug use: The 1970's to the 1990's.* Unpublished report, Office of Juvenile Justice and Delinquency Prevention.

Juvenile Justice Evaluation Center. (n.d.). Staff-Secured Detention Program for Female Juvenile Offenders. Retrieved September 20, 2002, from www.jrsa.org/jjec/programs/gender/.

Kersten, J. (1990). A gender specific look at patterns of violence in juvenile institutions: Or are girls really "more difficult to handle"? *International Journal of the Sociology of Law, 18*, 473–493.

Kirigin, K. A., Braukmann, C. J., Atwater, J. D., & Wolf, M. M. (1982). An evaluation of Teaching-Family (Achievement Place) group homes for juvenile offenders. *Journal of Applied Behavior Analysis, 15*, 1–16.

Lagerspetz, K. M. J., & Bjorkqvist, K. (1994). Indirect aggression in boys and girls. In L. R. Huesmann (Ed.), *Aggressive behavior: Current perspectives* (pp. 131–150). New York: Plenum.

LeCroy, C. W., & Daley, J. (2001). *Empowering adolescent girls: Examining the present and building skills for the future with the Go Grrrls Program.* New York: W. W. Norton.

Linehan, M. M. (1993). *Cognitive-behavioral treatment of personality disorder.* New York: Guilford.

Males, M., & Shorter, A. (2001). *To cage and serve.* Unpublished manuscript.

Mayer, J. (1994, July). *Girls in the Maryland juvenile justice system: Findings of the Female Population Taskforce.* Presentation to the Gender Specific Services Training Group, Minneapolis, MN.

McCormack, A., Janus, M. D., & Burgess, A. W. (1986). Runaway youths and sexual victimization: Gender differences in an adolescent runaway population. *Child Abuse and Neglect, 10,* 387–395.

Miller, J. (1994). Race, gender and juvenile justice: An examination of disposition decision-making for delinquent girls. In M. D. Schwartz & D. Milovanovic (Eds.), *The intersection of race, gender and class in criminology* (pp. 219–246). New York: Garland.

Office of Juvenile Justice and Delinquency Prevention. (1998, October). Guiding principles for promising female programming. Retrieved March 9, 2000, from www.ojjdp.ncjrs.org/ pubs/ principles/contents.html.

Okamoto, S. K. (2001). Interagency collaboration with high-risk gang youth. *Child and Adolescent Social Work Journal, 18*(1), 5–19.

Okamoto, S. K. (2002). The challenges of male practitioners working with female youth clients. *Child and Youth Care Forum, 31*(4), 256-268.

Okamoto, S. K. (In press). The function of professional boundaries in the therapeutic relationship between male practitioners and female youth clients. *Child and Adolescent Social Work Journal.*

Okamoto, S. K., & Chesney-Lind, M. (2000). The relationship between gender and practitioners' fears in working with high-risk youth. *Child and Youth Care Forum, 29,* 373–383.

Poe-Yamagata, E. & Butts, J. A. (1996). *Female offenders in the juvenile justice system.* Pittsburgh, PA: National Center for Juvenile Justice.

Prinz, R. J., & Miller, G. E. (1996). Parental engagement in interventions for children at risk for conduct disorder. In R. D. Peters & R. J. McMahon (Eds.), *Preventing childhood disorders, substance abuse, and delinquency* (pp. 161–183). Thousand Oaks, CA: Sage.

Roberts, A. R., & Rock, M. (2002). An overview of forensic social work and risk assessments with the dually diagnosed. In A. R. Roberts & G. J. Greene (Eds.), *Social workers' desk reference* (pp. 661–668). New York: Oxford University Press.

Schoenwald, S. K., & Hoagwood, K. (2001). Effectiveness, transportability, and dissemination of interventions: What matters when? *Psychiatric Services, 52,* 1190–1197.

Sherman, F. (2000). What's in a name? Runaway girls pose challenges for the justice system. *Women, Girls and Criminal Justice, 1*(2), 19–20, 26.

Shorter, A. D., Schaffner, L., Schick, S., & Frappier, N. S. (1996). *Out of sight, out of mind: The plight of girls in the San Francisco juvenile justice system.* San Francisco: Center for Juvenile and Criminal Justice.

St. Germaine, E. A., & Kessell, M. J. (1989). Professional boundary setting for male youth workers with female adolescent clients. *Child and Youth Care Quarterly, 18,* 259–271.

Tiet, Q. Q., Wasserman, G. A., Loeber, R., McReynolds, L. S., & Miller, L. S. (2001). Developmental and sex differences in types of conduct problems. *Journal of Child and Family Studies, 10*(2), 181–197.

Trupin, E. W., Stewart, D., Boesky, L., & Beach, B. (1999). *Effectiveness of a dialectical behavior therapy program for incarcerated female juvenile offenders.* Unpublished manuscript.

18

Optimizing Adjudication Decisions Based on a National Evaluation of Juvenile Boot Camps

Gaylene S. Armstrong
Doris Layton MacKenzie
David B. Wilson

A Day in the Life of a Juvenile Boot Camp Cadet

The arrival of a new squad of "cadets," or recently adjudicated juvenile delinquents, at a typical juvenile boot camp in the United States sets in motion an intense reception and integration process. The rude awakening that often welcomes the juvenile delinquents into the facility attempts to induce them into the regimented lifestyle that will surround them for the subsequent months of their sentence. The reception and integration process, or intake, can be a powerful experience for the cadets. From the day that these youth enter the camps, their reality changes: A door is now referred to as a "portal," other residents are referred to as "cadets" in their "squad," supervising correctional officers are referred to by military titles such as "captain" or "major." In some boot camps, cadets must even refer to themselves as "this inmate" rather than using first-person pronouns. In part, the integration process into the boot camp attempts to symbolically strip away the juvenile delinquent's previous street identity and attitude in preparation for new positive attitudes and behaviors.

Boot camp environments are meant to be tough in dealing with the cadets; a stern, no-excuses attitude tends to prevail throughout. Immediately upon arrival, the boot camp staff members force the cadets to take responsibility for the actions that resulted in their boot camp placement. During intake, staff members fully disseminate program expectations and standards to the cadets, familiarize them with military lingo, and list off the schedule of their daily activities. The cadet's presentation of himself or herself also changes during the intake process. Cadets are provided military-style clothing and boots. As part of the boot camp's standards, cadets' clothing must be neatly ironed and boots must be polished to sheen. Further, male cadets receive military-style haircuts. Along with these strict standards for behavior and presentation of self comes the threat of discipline for misbehavior or failure to comply with rules.

Although there is large variation between boot camp programs in their focus on military elements such as drill and ceremony, and physical training versus rehabilitative programs, one aspect is certain—early to bed, early to rise, and not one unaccounted minute will pass during each day. In their new surroundings, cadets learn to adhere to a daily schedule with the military precision evident throughout the boot camp. The following example reflects a typical boot camp schedule:

0500   First Call, Hygiene Call, Barrack Details

0530–0535	Reveille Formation, Raise Flag
0535–0630	Physical Training, Kitchen Patrol (KP)
0630–0755	Breakfast, Details, Hygiene
0755–0800	Inspection Formation
0800–1200	Rotation Through Academics, Training Activity, or Detail Team
1200–1255	Lunch Cadet Debriefing
1255–1300	Accountability Formation
1300–1700	Rotation Through Academics, Training Activity, or Detail Team
1700–1705	Retreat Formation, Lower Flag
1705–1800	Supper
1800–1845	Squad Debriefing, D-Tours, KP, Personal Time
1845–1945	Classes, Snack, Showers
1945–2045	Study Hall
2045–2100	Squad Preparation for Next Day's Training

2100   Lights Out, TAPS

Philosophies behind boot camp programs vary greatly; however, one goal that acts as a common denominator for any juvenile correctional program is that the juvenile delinquents desist from delinquent behavior. In addition to programming, various aspects inherent to the boot camps attempt to facilitate desistance. The assignment of juveniles to squads of cadets forces them to work together to accomplish tasks and learn to depend on others in a positive manner. The structure of daily activities attempts to instill a sense of responsibility. The presentation of self according to a military standard, in both dress and behavior, attempts to instill pride and respect for oneself as well as respect for others and authority figures. To varying extents, boot camp programs also attempt to encourage individual change through rehabilitative programs, including anger management, substance abuse treatment, and counseling.

ARE BOOT CAMPS THE OPTIMAL RESPONSE FOR ADJUDICATED JUVENILES?

Despite their continuing popularity, correctional boot camps for juveniles remain controversial. The debate involves questions about the impact of the camps on the adjustment and behavior of juveniles while they are in residence and after they are released. According to advocates, the atmosphere of the camps is conducive to positive growth and change (Clark & Aziz, 1996; MacKenzie & Hebert, 1996). In contrast, critics argue that many of the components of the camps are in direct opposition to the type of relationships and supportive conditions that are needed for quality therapeutic programming (Andrews, Zinger, et al., 1990; Gendreau, Little, & Groggin, 1996; Morash & Rucker, 1990; Sechrest, 1989). Research on the recidivism of releasees from correctional boot camps has not been particular helpful in settling the controversy over the camps. Neither adult nor juvenile boot camps appear to be effective in reducing recidivism. In general, no differences are found in recidivism when boot camp releasees are compared to comparison samples who served other sentences or who were confined in another type of juvenile facility (MacKenzie, 1997; MacKenzie, Brame, McDowall, & Souryal, 1995).

This chapter will recount recent research by MacKenzie, Wilson, Armstrong, and Gover (2001) that examines the experiences of 2,668 juveniles confined in 26 boot camps and compares these experiences to those of 1,848 juveniles residing in 22 traditional facilities. We examine whether juveniles in the boot camps experience more anxiety and depression in comparison to those in traditional facilities and whether these experiences are related to perceptions of the environment. In addition, we compare the changes juveniles make during residency in the facilities. Specifically, we were interested in changes in stress (anxiety, depression), and impulsivity, social bonds, and antisocial attitudes. The latter characteristics have been found to be associated with criminal activity and are therefore reasonable targets for intermediate change during the time juveniles reside in the facilities. We also measured juveniles' perceptions of the environment in order to examine whether they perceived the environments of the two types of facilities differently and, if so, whether these perceptions were related to the type of changes they made while in the facilities.

Critics of boot camps propose that some juveniles will experience more difficulties than others in the boot camps due to the confrontational nature of the interactions between the juveniles and the staff (Morash & Rucker, 1990). The boot camps are proposed to be particularly stressful for girls and for juveniles who have experienced a past history of family violence. To examine this proposal, we compared the impact of the boot camps on girls and boys and investigated whether there were differences in the impact of the boot camps on those who had experienced family violence.

For a subset of respondents, data were collected at two points in time, enabling us to examine change in anxiety and depression as well as social bonds, impulsivity, and social attitudes during the time juveniles were in the two different types of facilities. These characteristics are theoretically associated with

criminal behavior. Increased social bonds and positive social attitudes and, conversely, decreased impulsivity are anticipated to be associated with reductions in later criminal activity. Thus, facilities that have an impact on these characteristics of juveniles may be successful in reducing recidivism.

We begin by reviewing the research literature establishing the importance of understanding the environments of facilities and the effects of different environments on the residents. Following this, we review the literature on juveniles' adjustment in facilities, the changes juveniles are hypothesized to make during their time in residential facilities, and the association of these changes with future criminal activities.

The Perceived Environment

The impact of the prison environment on inmates' adjustment and behavior both while they are in the facility and when they are released has been well established in the research literature (Ajdukovic, 1990; Goffman, 1961; Johnson & Toch, 1982; Moos, 1968; Wright, 1985, 1991; Wright & Goodstein, 1989; Zamble & Porporino, 1988). Facilities possess unique characteristics that "impinge upon and shape individual behavior" (Wright & Goodstein, 1989, p. 266).

Few people who have visited correctional boot camps doubt that the environments of these facilities are radically different from the environment of traditional facilities (Lutze, 1998; MacKenzie & Hebert, 1996; MacKenzie & Parent, 1992; Styve, MacKenzie, Gover, & Mitchell, 2000). Juveniles in these facilities are awakened early each day to follow a rigorous daily schedule of physical training, drill and ceremony, and school. They are required to follow the orders of the correctional staff. Orders are often presented in a confrontational manner, modeled after basic training in the military. Summary punishments such as push-ups are frequently used to sanction misbehavior. In comparison to traditional juvenile facilities, boot camps appear to be more physically and emotionally demanding for the residents. In fact, research on adult boot camps suggests that inmates in the boot camps will voluntarily drop out even if this means they will have to serve a longer term in the prison than they would if they completed the boot camp (MacKenzie & Souryal, 1995).

Perceptions of inmates in the different types of facilities would be expected to reflect the differences in environments. Continuing controversy exists about the appropriateness of the camps for managing and treating juvenile delinquents. Advocates of the boot camps argue that the focus on strict control and military structure provides a safer environment conducive to positive change (Steinhart, 1993; Zachariah, 1996). From their perspective, the intense physical activity and healthy atmosphere of the camps provide an advantageous backdrop for treatment and education (Clark & Aziz, 1996; Cowles & Castellano, 1996).

Critics disagree with this perspective (Morash & Rucker, 1990) and claim that the confrontational nature of the camps is diametrically opposed to the constructive, interpersonally supportive environment necessary for positive change to occur (Andrews, Zinger, et al., 1990). According to critics, juveniles in the boot camps, when compared to youth in traditional facilities, will perceive their facilities as less caring and less therapeutic, and, in general, as preparing them less for reentry into the community. Furthermore, juveniles may be worried about their safety while they are in a boot camp facility. Given the hypothesized negative environmental characteristics, youth in boot camps would be expected to experience much more stress than youth in traditional facilities.

Morash and Rucker (1990) are particularly concerned that the boot camps will have a detrimental impact on girls and on both girls and boys who have experienced abuse. The confrontational nature of the interactions between staff and inmates is expected to be particularly problematic for these youth. For those who have a history of abuse and for girls who have dependency issues, these interactions will be reminiscent of the difficulties they faced in an abusive relationship; as a result, the environment will be particularly stressful for them and counter-therapeutic.

The environments of the facilities would also be expected to have an impact on the types of changes inmates make during their time in the different facilities. For example, research demonstrates that treatment programs with particular characteristics are effective in changing antisocial attitudes. An environment that emphasizes therapeutic programming instead of physical activity would be expected to have a greater impact on such attitudes (Andrews & Bonta, 1998; Goodstein & Wright, 1989).

In summary, we proposed that juveniles in the boot camps and the traditional facilities would perceive the environment of their institutions differently and that the characteristics of the environments would have an impact on the level of stress experienced by the residents and on the changes they made in social bonds, antisocial attitudes, and impulsivity.

Stress

One concern regarding the boot camps is whether the environment creates dysfunctional stress for the participants. Some levels of stress may actually be beneficial. For example, critical life events may create stress, and this stress may result in changing the trajectories of the lives of those involved in criminal activities (Sampson & Laub, 1993). Instead of continuing in the previous path (e.g., a criminal path), youth may change and make a commitment to family, school, or employment. The stress created by the critical life event in such cases has a functional and beneficial impact. In contrast, some stress is so severe that an individual's level of functioning is compromised. In such cases, the stress is considered dysfunctional.

The adjustment of inmates to the environments of correctional facilities has been the topic of numerous research studies (e.g., Goodstein & Wright, 1989). A concern has been that institutions such as prisons create a total environment that may severely limit inmates' development and create dysfunctional stress, particularly in youth (Goffman, 1961; Johnson & Toch, 1982; Moos, 1968; Wright, 1991). Critics of boot camps fear that the demanding nature of the boot camp environment will be beyond the coping ability of youth and, as a result, will be detrimental to them.

In contrast, advocates argue that boot camp is a healthy environment that creates the type of stress that will lead youth to reevaluate their lives and make changes. Some level of stress may be effective in bringing about change. For example, Zamble and Porporino (1990) found that adult inmates experienced stress when they first entered prison. This was also the time inmates were most willing to enter programs designed to help them make changes in their lives. Zamble and Porporino proposed that the stress associated with entry to prison might be instrumental in getting inmates to reevaluate their lives and take steps toward positive change. This proposal is similar to the type of critical life event Sampson and Laub (1993) considered conducive to bringing about changes in life trajectories. That is, entering a new situation like a residential facility may be the type of event that leads to changes in the trajectory of the lives of some juveniles. As a result of this event, the juveniles may become more prosocial and begin to build ties and bonds with conventional social institutions.

Frequently, boot camp staff refer to the early period in boot camps as a period when they "break down" youth before they begin the "build-up period." The question is whether the breakdown period creates functional or dysfunctional stress: That is, do the youth in the boot camp experience the type of anxiety that will result in a reevaluation of their lives and a decision to make changes, or is the stress so severe that they become depressed, anxious, and unable to adequately function in the new environment? Critics would suggest that the stress in boot camps is so severe as to be dysfunctional; advocates of the camps argue that it is the type of stress that leads to positive changes.

Changing Youth

If institutional programs are going to have an impact on the future criminal activities and adjustment of youth, the programs must change the youth in some way. These intermediate changes can be thought to be signals of the impacts the facilities will have on the future criminal activities of the youth. This research examines adjustment and short-term change in boot camp facilities and compares these to the changes juveniles in traditional facilities make. Three correlates of criminal activity are social bonds, impulsivity, and antisocial attitudes.

These characteristics are theoretically and empirically associated with criminal activity and other antisocial behavior. We begin by reviewing evidence that these characteristics are associated with criminal behavior and that changes in the characteristics are associated with changes in criminal activity.

Increasing Social Bonds

Evidence exists that increases in social bonds are associated with declines in criminal activity. According to Sampson and Laub (1993), informal social controls form a structure of interpersonal bonds linking individuals to social institutions such as work, family, and school. These ties or bonds are important in that they create obligations and restraints that impose significant costs for translating criminal propensities into action. Although Sampson and Laub acknowledged that there is continuity in individual antisocial behavior, they argued, unlike the continuity theorists (e.g., Gottfredson & Hirschi, 1990), that such continuity does not preclude large changes in individuals' offending patterns. In a reanalysis of the Glueck and Glueck data, Sampson and Laub (1993) found support for the proposal that childhood antisocial behavior and deviance can be modified over the life course by adult social bonds. Job stability and marital attachment were significant predictors of adult crime even when childhood delinquency and crime in young adulthood were statistically controlled.

Further evidence that criminal propensity can be modified comes from research by Horney, Osgood, and Marshall (1995) that examined the self-reported criminal activities of offenders. The authors found that life circumstances indicative of changes in social bonds and commitment to conformity influenced offending behavior even over relatively short time periods.

Similar to the findings from research with adults, increased social bonds have been found to be associated with declines in the criminal activities of juveniles (Jang, 1999; Simons, Johnson, Conger, & Elder, 1998). For example, Simons et al. (1998) found that stronger ties to family and school and decreased affiliation with deviant peers lowered the probability that youth who had behavior problems during childhood would graduate to delinquency during adolescence.

In summary, the research on social bonds demonstrates that increased social bonds are associated with decreased criminal activity. The research does not demonstrate how or why bonds change. Sampson and Laub (1993) proposed that bonds may change as a function of critical life events that lead individuals to reevaluate their life and begin to make positive changes. Theoretically, such a critical life event could occur for juveniles who enter a residential facility. If the experience of being in the facility or the programs provided in the facility increase the attitudes of commitment to conformity or ties the juveniles have to social institutions like family, work, and school, then theoretically the future criminal activities of these youth may decrease. The major characteristics of boot camps do not suggest that these programs will incorporate elements that would increase ties or commitments to conventional activities outside the facility. Restrictions on visitation may limit contact with the outside, and the environment of the camps is very different from the environment of work or school outside the camps. The traditional facilities may be much more likely to strengthen these ties or attitudes. Theoretically, a critical life event such as entering an institution could initiate changes in ties or attitudes. If either type of facility did have an impact on attitudes or times, we would anticipate that this would be a hopeful sign that such changes would be associated with a reduction in future criminal activities for the participants.

Impulsivity and Control

The connection between impulsivity and criminal activities is well established. According to Gottfredson and Hirschi's (1990) general theory of crime, antisocial acts are committed by people with low self-control. Impulsivity is one of the major characteristics of such individuals. Theorists interested in individual differences in temperament and personality have also emphasized the need to consider differences in impulsivity. For example, in her psychosocial control theory, Mak (1990, 1991) emphasized the importance for understanding criminal activity and delinquency of individual differences in thinking through consequences, a preference for immediate gratification, poor planning, and a

lack of patience. These impulsive characteristics are similar to the temperament and personality characteristics Glueck and Glueck (1950) linked to persistent and serious delinquent behavior.

Numerous key criminological studies have shown that impulsivity is a strong correlate of delinquent and criminal behavior (Caspi et al., 1994; Farrington, 1998; Glueck & Glueck, 1950; Loeber, Farrington, Stouthamer-Loeber, Moffitt, & Caspi, 1998; White et al., 1994). In comparison to nondelinquents, delinquents show markedly higher levels of impulsivity. These results hold despite differences in whether impulsivity is measured by self-reports, teachers, independent raters, staff psychologists, or parents. Stronger impulsivity is related to increases in official measures of offending and delinquency, self-reported criminal activities, and childhood behavior problems as reported by teachers, mothers, and peers. The association between impulsivity and crime is stronger than associations between crime and intelligence or socioeconomic status (White et al., 1994).

Controversy exists regarding whether an individual's impulsivity can be changed during the life course. Gottfredson and Hirschi (1990) asserted that "people who lack self-control will tend to be impulsive" (p. 90) and that variation in self-control is a latent trait that provides the primary explanation for individual differences in involvement in antisocial behavior throughout the life course. In contrast, others believe that individuals can change during their life course. For example, the life course perspective views life course trajectories as a sequence of events and transitions that either accentuate or redirect behavioral tendencies (Elder, 1992; Simons et al., 1998). From this perspective, characteristics such as antisocial behavior and impulsivity are associated with criminal activity, but a trajectory may change as a result of life circumstances or critical life events. In their study, Simons et al. (1998) found evidence that the correlation between childhood and adolescent deviant behavior reflects a developmental process, as proposed by those with a life course perspective, rather than the latent antisocial trait proposed by Gottfredson and Hirschi (1990).

A critical life event that may change a juvenile's life trajectory is institutionalization in a juvenile facility. Impulsivity is a particular target for change in boot camps. The rigorous structure in the camps and the strict requirements for military bearing are designed, in part, to get youth to think before they act. We anticipate that this is one characteristic of juveniles that would change as a result of the boot camp experience. The traditional facilities are not expected to affect a youth's impulsivity.

Antisocial Attitudes

According to correctional theorists, treatment programs that are effective in reducing recidivism have certain clearly defined characteristics (Andrews, Bonta, & Hoge, 1990; Andrews & Kiessling, 1980; Glaser, 1974; Gendreau & Ross, 1979, 1987; Palmer, 1974). These authors argue that "appropriate" treatment delivers services to higher-risk cases, uses styles and modes of treatment that are capable of influencing criminogenic "needs," and is matched to the learning styles of offenders. Criminogenic needs are defined as those that are dynamic or changeable as opposed to static (not changeable) and that are directly related to the criminal behavior of the offender. Meta-analyses examining the effectiveness of treatment programs have supported the proposed importance of these "appropriate" treatment characteristics (Andrews, Zinger, et al., 1990; Lipsey, 1992). Procriminal or antisocial attitudes have consistently shown significant associations with criminal behavior for adults (Andrews & Bonta, 1998; Bonta, 1990) and youthful offenders (Shields & Ball, 1990; Shields & Whitehall, 1994). The evidence showing the association between procriminal or antisocial attitudes and criminal behavior makes these prime criminogenic needs and therefore targets for change in correctional treatment.

Summary

In summary, there is strong empirical evidence that social bonds, antisocial attitudes, and impulsivity are associated with criminal activity. Recent research supports the proposal that these characteristics do change during the life course. The question is how this change can be initiated. Life course theorists propose that critical life events may bring about change in adolescence or adulthood. One such critical life event, at least for some adolescents, may be incarceration in a juvenile correctional facility. Differences in the environments and programming in correctional boot camps and traditional facilities led us to predict that the impacts of these facilities on the youth who spend time there would be different. Given the environment and programming in the boot camps, we anticipated that the camps might reduce the impulsivity of the youth who reside there. On the other hand, we anticipated that the traditional facilities might be more apt to change the social bonds and antisocial attitudes of the youth who reside there. For correctional facilities to have an impact on the future offending behavior of youth, these are the changes we would hope to observe during residency in a juvenile facility.

METHOD

In April 1997, all juvenile boot camps in operation in the United States, excluding Hawaii and Alaska, were identified for inclusion in this study. At that time, 50 privately and publicly funded secure residential boot camps were identified. All facilities were contacted and asked to participate, and 27 agreed. The 23 programs that did not participate did so for various reasons, including parental consent issues, staffing and resource limitations, and impending program closure. Thus, the 27 boot camps agreeing to participate in this project represented 54% (27 out of 50) of the residential juvenile boot camps operating in 1997 and unfortunately cannot be considered a random sample of the population of facilities.[1]

For each boot camp agreeing to participate, a comparison facility was sought to allow the contrast of youths' experiences in a boot camp with youths' experiences in traditional juvenile correctional facilities. Comparison facilities were selected by identifying those secure residential facilities where the juveniles would have been confined if the boot camp programs had not been in operation. This method of selection ensured that the residents at the comparison facilities were as similar as possible to the boot camp residents. With this definition of a comparison facility in mind, the facility administrator at each boot camp or an individual from within the state's juvenile justice department recommended the most appropriate comparison facility. Comparison facilities were then contacted and asked to join the research project. All comparison facilities identified agreed to participate in the research project. Although there were 26 boot camp facilities included in the study, there were only 22 comparison facilities. There were two reasons for this discrepancy. First, two boot camps did not have a viable comparison facility within the state. Second, in two states, the same non-boot camp facility was identified as the most appropriate comparison for two different boot camps. In these instances, one facility served as the comparison for each of the two boot camps.

Participants

A full census of all juveniles at each facility on two occasions was sought. A total of 4,516 juveniles were surveyed, 2,668 from the boot camp facilities and 1,848 from the traditional facilities. The overall response rate for this survey was high and represented 85% of the juvenile population at the surveyed facilities. A common reason for nonparticipation was a juvenile's overriding need to be somewhere else at time of survey administration, such as a court hearing or medical visit outside of the facility. A small number of youths started the survey but chose to not complete it. A total of 2,473 were surveyed at the Time 1 administration and 2,030 at the Time 2 administration. The first administration of the survey was designed to include juveniles shortly after their entry into the boot camp program. The second administration of the survey was designed to include juveniles just prior to release from the boot camp. The time interval between the two survey administrations in the comparison facilities was matched to the time interval between administrations for the corresponding boot camp. The interval between Time 1 and Time 2 administrations ranged from 3 months to 8 months, with a median of 4 months. The Time 2 administration included 530 juveniles, 264 in boot camps and 266 in traditional facilities, who also were surveyed at Time 1. This subsample of the data is the major focus of this chapter.

Juvenile Survey

The survey questionnaire for the youths included 266 questions. Thirteen questions were open-ended (primarily demographics items), with the remaining questions based on either a 4- or 5-point Likert response scale or a yes-no dichotomous response format. Overall, there was a high completion rate of over 85% of the population. Surveys were administered to groups of 15 to 20 youths in classroom-type settings in accordance with prevailing ethical principles. A videotaped presentation of the survey was shown on a large television providing instructions and the survey questions to ensure uniform administration and provide assistance to juveniles with reading difficulties.[2]

Administrator Interview and Institutional Records

A structured interview was conducted with the facility administrator or administrators to obtain information about the facilities where the juveniles resided. Some items in the interview survey required information from institutional records (e.g., hours of treatment per week) that was obtained by the administrator after the completion of the interview. Researchers placed follow-up telephone calls within 2 weeks of the site visit to obtain outstanding information.

The interview included 264 items and provided information on a variety of factors, including the size of the facility (the average number of juveniles who usually resided in the facility), how selective the facility could be about who entered the facility (Selectivity Index), the seriousness of the delinquency history of the juveniles who were admitted to the facility (Seriousness Index), the number of hours the juveniles participated in treatment in a 1-week period, the contact juveniles had with the outside (Contact With Outside Index), the staff-to-juvenile inmate ratio, the juveniles' average length of stay, and whether someone at the facility collected or obtained information on the juveniles who were released, including information on rearrest for delinquent or criminal activities, return to school, residence with family, and reinstitutionalization.[3]

Measures

Individual-Level Measures

Five individual-level composite measures were the primary focus of the analyses below: depression, anxiety, commitment to conventional behavior (social bonds), dysfunctional impulsivity, and antisocial attitudes. Additional individual-level variables included in this study were age, gender, ethnicity, the number of self-reported nonviolent arrests, the number of self-reported violent arrests, an indicator of history of family violence and child abuse, indicators of alcohol and drug abuse, peer criminality, criminal history, perceptions of the environment, and amount of time the youth had been in the facility at the time of the survey. The construction of these measures is discussed below.

Depression. Five Likert-type items were taken from the Jesness Inventory (Jesness, 1962) to measure depression. These items were intended to measure state characteristics of depressed mood rather than trait characteristics of depression. The five items were summed and averaged. As such, the range for the Depression Index was from 1 to 5, with a mean of 3. The internal consistency of these items was high (α = .77). High scores on this scale measured more depression. This scale was thought to indicate severe and, as previously described, potentially dysfunctional stress.

Anxiety. Six dichotomous (yes-no) items assessing state anxiety were combined to create the anxiety measure. The internal consistency was adequate (α = 0.71). These six items were drawn from Spielberger, Gorsuch, and Lushene (1970). The goal of these items was to examine differences in the stress and anxiety levels of the youths. High scores on this scale indicated increased anxiety. In comparison to the Depression Scale, this scale was designed to reflect low to moderate stress.

Dysfunctional Impulsivity. Four dichotomous (yes-no) questions composed the Dysfunctional Impulsivity Scale. The items of this scale focused on the cognitive aspects of impulsivity (i.e., thinking before acting or speaking). The internal consistency of this scale was adequate

(α = 0.66). High scores on this scale reflected high impulsivity.

Social Attitudes. The Social Attitudes Scale was a composite of 35 true-false items from the Antisocial Subscale of the Jesness Inventory (Jesness, 1962).[4] This scale measures attitudes of the juveniles toward conventional aspects of society such as authority figures. The internal consistency of this scale was adequate (α = 0.65). Previous research with this scale has demonstrated that it measures short-term change in confined youthful offenders and that this change is associated with recidivism.

Commitment to Conventional Behavior. A measure of commitment to conventional behavior was constructed from three Likert-type items assessing the importance of education, work, and spending time with family. The internal consistency for this scale was low (α = 0.56). High scores on this scale indicate commitment to conventional behaviors or bonds. Originally, three separate scales had been developed to indicate commitments to family, school and work; however, factor analyses (eigenvalues, scree plot) indicated that the items formed one scale.

History of Family Violence. An index of the degree of physical and sexual abuse and neglect within the family of origin was constructed from nine Likert-type items. All but two of the items dealt with physical abuse directed either at the youth or at other family members. This scale had good reliability (α = 0.85).

Alcohol and Drug Abuse. Scales indicating alcohol and drug abuse were created from 10 dichotomous (yes-no) items. These items dealt with lifetime substance use. For purposes of the present study, a composite Alcohol and Drug Abuse Scale was constructed from these items and showed good reliability (α = 0.77). High scores indicated higher levels of substance abuse.

Peer Criminality. Peer criminality was measured using four 5-point items asking the youth if their closest group of friends prior to arrest were in trouble with the law, incarcerated, involved with gangs, and users of drugs and alcohol. The internal reliability coefficient for this measure was adequate ($\alpha = 0.71$).

Criminal History. An index of criminal history was constructed as a composite of age at first arrest and the natural log of the number of previous commitments, number of prior nonviolent offenses, and number of prior violent offenses. The internal consistency of this scale was adequate ($\alpha = 0.69$).

Perception of Facility Environment. The youth were asked a series of 129 Likert-type questions (with five response choices from *strongly agree* to *strongly disagree*) designed to measure their perception of the facility environment. These items were rationally constructed to represent 13 major dimensions of the juvenile facility: control, resident danger, danger from staff, environmental danger, activity, care, risks to residents, quality of life, structure, justice, freedom, programs, and emphasis on the individual. Scales were scored so that higher scores reflected higher perceptions in the direction of the name of the scale (e.g., more control, more danger, more activity). These facets have been previously identified in the literature as important elements of juvenile residential facilities (MacKenzie, Styve, & Gover, 1998). At the individual level (e.g., $n = 2,473$ individuals at Time 1), these 13 measures were correlated, with the absolute value of the correlations ranging from .05 to .63.

Facility-Level Measures

At the facility level ($N = 48$ facilities), the 13 facility-level measures were, on average, highly correlated, with a mean absolute correlation of .55 and a range from .05 to .90. Because of these high correlations and the limited degrees of freedom for facility-level analyses ($df = 48$), we performed an exploratory factor analysis with the 129 items that made up the 13 measures. Examination of the eigenvalues suggested either a one- or a three-factor solution, and both solutions produced interpretable results. The three-factor solution produced factors we judged to measure the following aspects: (a) the therapeutic warmth of the environment (e.g., "A counselor

is available for me to talk to if I need one"), (b) the general level of hostility (e.g., "Residents have to defend themselves against other residents in this institution"), and (c) the degree of freedom and choice available to the youths (e.g., "Residents choose the type of work they do here").

In comparison to the three-factor solution, the single-factor solution appears to represent how positively, in general, the youth perceive the environment (staff care, staff are fair, learning useful skills, program is helpful, help in staying focused on future goals, etc.). When the individual-level factor scores for the three-factor solution are aggregated up to the facility, the degree of facility-level hostility is negatively correlated with the facility-level therapeutic warmth ($r = -.88$). Thus, although these factors appear distinct on the basis of the youths' perceptions, facilities that tend to be high on therapeutic warmth are, not surprisingly, low in perceived hostility and dangerousness.

The goal of this study was to assess the impact of the facility environment on the changes juveniles made while in the facilities using the five indicator variables (Depression, Anxiety, Dysfunctional Impulsivity, Social Attitudes, Commitment to Conventional Behavior) discussed above, not the impact of an individual's perception of the environment on his change while residing in the facility. Thus, we averaged the perception of the environment measures across individuals to create a facility-level rating on each of the three dimensions using the three-factor solution for the environmental measures.

Other facility-level measures, obtained from administrator interviews and institutional records, that were used in the analyses below included an index of the seriousness of the criminal histories of the youths admitted to the facilities, the admission selectivity of the facility, and the facility size. The index of seriousness was constructed as a composite score of dichotomous questions (yes-no) that determined whether the facility accepted specific categories of juvenile delinquents (e.g., juveniles with a history of violence or juvenile convicted of arson). The index of selectivity was constructed as a composite of dichotomous questions (yes-no) that demonstrated the stringency of admissions of the juveniles (did facility personnel interview juveniles prior to entry, did juveniles have to pass physical, mental, and medical examinations in order to enter). The index demonstrated the extent of input by the individual facilities in choosing the juveniles that were admitted into their facilities. High scores on the Seriousness and the Selectivity Indices indicated, respectively, that the facility accepted juveniles with more serious criminal histories and that facility personnel could be more selective about which juveniles were permitted to enter the facility. These variables provide a mechanism for examining the variability across and within the boot camps and traditional facilities.

RESULTS

Juvenile Characteristics

Comparisons of the boot camps and traditional facilities on the individual characteristics of the juveniles for the total sample and the pretest-posttest sample are shown in Tables 18.1 through 18.3. The demographic characteristics between these two facility types were comparable. The boot camps tended to have a higher percentage of girls, although boys dominated both facility types. The traditional facilities had a population that was more criminally involved, on average, than the boot camps, with a substantially higher mean number of prior nonviolent and violent arrests. Furthermore, at the time of the survey, the typical youth in the traditional facility had resided in the facility roughly twice as long as the typical boot camp youth. This reflects the generally longer lengths of stay in these facilities relative to the boot camps and the method of determining when to conduct the surveys at each facility.

The two samples were highly similar on the psychosocial indices. A statistically significant but small difference was observed in the history of family violence, with higher levels of previous violence reported by the youths in the traditional facilities. A small difference was also observed on the Commitment to Conventional Behavior Index, with higher levels reported by the youth in the boot camps. Although the traditional facilities

were selected because they were facilities to which the boot camp youths would have been sent in the absence of the boot camp, the general impression from these data is that the traditional facilities also serve youth who are more seriously delinquent, on average, than the youth admitted to the boot camps. It appears that whereas all of the boot camp youth may have been appropriate for the comparison facility, not all of the youth at the comparison facility may have been appropriate for the boot camp facilities.

Juveniles' Perception of the Facility Environment

The juveniles' perception of the environment differed between the two facility types (Table 18.4). Surprisingly, the boot camps were perceived, on average, as more therapeutic and less hostile than the traditional facilities. Consistent with expectation, the youths perceived the boot camps as more restrictive of personal freedom and choice than the traditional facilities. These findings are consistent with the qualitative observations made within the facilities by the research staff. Within the typical juvenile boot camp, the increased structure does not appear to be associated with an increase in hostility or perceived danger from staff (an element of this factor). The greater selectivity of the boot camps in admissions criteria (see below) may also contribute to a safer overall environment if the more troubled and potentially violent youth are not allowed admission. These differences remained after statistically adjusting for measured characteristics of the youth: that is, the characteristics presented in Tables 18.1 through 18.3. Thus, the evidence suggests that the observed differences represent actual differences in the environments and not just differential perceptions of comparable environments. We cannot determine from these data, however, whether the differences are produced by the structural, organizational, programmatic, and staffing aspects of a facility or by the juveniles themselves. That is, a facility with a higher proportion of violent offenders may be genuinely more dangerous, despite staffing and organization aspects. It is likely that both the characteristics of a program and the juveniles served contribute to the environmental conditions.

Table 18.1   Demographics by Facility Type

Variable by Facility Type	Full Sample		Pre/Post Sample[a]	
	%	N	%	N
Gender (% Male)				
Boot camps	92	2,390	93	264
Traditional facilities	96	1,578	98	265
Race (% Caucasian)				
Boot camps	33	2,382	43	264
Traditional facilities	31	1,566	35	266
Race (% African American)				
Boot camps	36	2,382	30	264
Traditional facilities	33	1,566	24	266
Race (% Hispanic)				
Boot camps	19	2,382	16	264
Traditional facilities	20	1,566	17	266
Race (% Other)				
Boot camps	12	2,382	11	264
Traditional facilities	16	1,566	24	266

NOTE: None of the above differences between boot camps and traditional facilities were statistically significant under a population-averaged (facility) logistic regression model.

Facility Characteristics

On average, the boot camps were smaller and were more selective about the entrants than the traditional facilities (Table 18.5). Traditional facilities permitted juveniles with more serious criminal histories to enter the program and were generally less selective about whom they admitted. The typical length of stay in the traditional facilities was nearly double that of the boot camps. Only 46% of the boot camps and 32% of the traditional facilities had any follow-up information on the releasees, including whether the youth were returned to the same facility sometime after being released.

Initial Levels of Anxiety and Depression

The first hypothesis to be addressed was whether boot camp youths had higher initial levels of depression and anxiety. Although some individuals are generally more anxious or depressed than other individuals, depression and anxiety are not static, and an individual's level of each will rise and fall depending on life stressors

and environmental circumstances. The transition into an institutional setting, whether it is a traditional juvenile delinquency facility or a boot camp, is stressful and may lead to increased depression and anxiety for some youths. The boot camp, with its highly structured militaristic style and reputation, may be a more stressful environment, at least initially, for juveniles.

To examine this issue, we selected all survey respondents who completed the survey during their first month of residency. This resulted in a sample of 774 juveniles from boot camps and 274 juveniles from comparison facilities. The mean levels of both depression and anxiety were highly similar between the boot camp facilities and the traditional facilities for this sample (3.1 and 3.0 respectively for depression on a 5-point scale and 1.5 and 1.4 respectively for anxiety on a 1-point scale) (Table 18.6). Not surprisingly, a simple test of this difference using a nested analysis of variance (Kirk, 1982; StataCorp, 1999) showed that the slightly higher values for the boot camp facilities were not statistically significant.

A history of family violence is a risk factor for affective disorders, such as depression and anxiety. It was hypothesized not only that a

Table 18.2   Individual Characteristics by Facility Type and Sample

Variable by Facility Type	Full Sample			Pre/Post Sample[a]		
	Mean	SD	N	Mean	SD	N
Age						
Boot camps	16.0	1.2	2,383	15.9	1.2	264
Traditional facilities	16.3	1.3	1,570	16.0	1.5	266
Number of Prior Nonviolent Arrests[b]						
Boot camps	6.6*	8.3	1,407	6.1*	4.7	260
Traditional facilities	9.3	10.0	982	10.5	11.8	261
Number of Prior Violent Arrests[b]						
Boot camps	1.6*	2.7	1,409	1.6	2.0	260
Traditional facilities	2.8	4.5	1,000	2.7	3.9	260
Number of Months Resided in Facility[b]						
Boot camps	2.7*	3.3	2,310	1.6	3.0*	257
Traditional facilities	6.0	7.9	1,500	5.3	7.1	260
Family Violence[b]						
Boot camps	1.6*	.6	2,362	1.5	.6	262
Traditional facilities	1.7	.7	1,545	1.7	.8	264
Alcohol Abuse Scale						
Boot camps	1.3	.3	2,370	1.3	.3	262
Traditional facilities	1.3	.3	1,556	1.4	.3	264
Drug Abuse Scale						
Boot camps	1.4	.3	2,374	1.4	.3	262
Traditional facilities	1.5	.3	1,553	1.5	.3	264
Peer Criminality						
Boot camps	3.3	1.0	2,319	3.3	1.0	260
Traditional facilities	3.4	1.0	1,516	3.4	1.0	262

a Values for the pre/post sample represent the first measurement for the 530 youths who were measured on two occasions.
b Analysis of variance performed on logged values.
* The difference between groups is statistically significant at $p < .05$.
NOTE: Mean difference tested using a nested analysis of variance, with facilities nested within facility type.

history of family violence would be related to the initial level of anxiety and depression of the juveniles but also that it would interact with facility type. We presumed that the more aggressive "in-your-face" atmosphere of the boot camps would be more traumatic for juveniles with a history of family violence and would therefore lead to a higher level of anxiety and depression. We tested this hypothesis for both anxiety and depression using a random-effects regression model estimated via maximum likelihood. Two regression analyses were estimated, one for depression and one for anxiety, each regressed on both individual-level and facility-level variables (see Table 18.7). These analyses were restricted to boys because there

were only four girls in traditional facilities for this subsample of the data set. Both analyses showed a statistically significant relationship between a history of family violence and level of anxiety and depression. As expected, facilities that were perceived, on average, as more hostile had higher levels of both anxiety and depression. Contrary to expectation, however, the interactions of facility type and facility-level hostility with history of family violence were not statistically significant. On the basis of these data, a history of family violence does not appear to interact with the type of facility (boot camp or traditional) or with the degree of perceived hostility. The regression analyses did show that youths perceiving the facility as more

Table 18.3   Psychosocial Measures by Facility Type and Sample

	Full Sample			Pre/Post Sample[a]		
Variable by Facility Type	Mean	SD	N	Mean	SD	N
Depression						
Boot camps	3.0	1.0	2,355	3.2	1.0	260
Traditional facilities	3.1	1.0	1,529	3.2	1.0	263
Anxiety						
Boot camps	1.4	.3	2,338	1.5	.3	257
Traditional facilities	1.4	.3	1,520	1.4	.3	261
Dysfunctional Impulsivity						
Boot camps	1.6	.3	2,326	1.7	.3	261
Traditional facilities	1.6	.3	1,506	1.6	.3	260
Antisocial Attitudes						
Boot camps	1.5	.1	2,320	1.5	.1	261
Traditional facilities	1.5	.1	1,490	1.5	.1	259
Commitment to Convention Behavior						
Boot camps	4.4*	.7	2,332	4.3	.7	257
Traditional facilities	4.2	.9	1,524	4.2	.9	259

a Values for the pre/post sample represent the first measurement for the 530 adolescents who were measured on two occasions.
* The difference between groups is statistically significant at $p < .05$.
NOTE: Mean difference tested using a nested analysis of variance, with facilities nested within facility type.

hostile and having less freedom and choice relative to their peers in the same facility were more likely to be anxious and depressed. It may well be that anxious and depressed youths are more likely to perceive their environment negatively.

Changes in Anxiety, Depression, Social Bonds, Dysfunctional Impulsivity, and Prosocial Attitudes

A main question of this study is whether the boot camp and traditional facilities produce positive changes in correlates of delinquency. That is, during youths' stay in a facility, do they become less impulsive, increase their bonds to conventional society, decrease their antisocial attitudes, and become less anxious and less depressed? To address this issue, we examined a subsample of the study that was measured on two occasions, ranging from 1 to 6 months between occasions, 4 months on average. For this sample (264 boot camp respondents and 266 traditional facility respondents), a maximum likelihood estimated random effects

regression model (Bryk & Raudenbush, 1992; Stata Corp, 1999) was used to examine change in the five outcome variables.

Shown in Table 18.7 are the mean Time 1, Time 2, and difference scores for the five outcome variables by type of facility for the pretest-posttest sample. The boot camps observed larger average changes in the desired direction for all five outcomes. In standardized mean difference effect size units, this difference was largest for prosocial attitudes, depression, and dysfunctional impulsivity. The observed effect sizes for the traditional facilities were all less than 0.10 in absolute values; these are small effects by most standards. The magnitude of change for the boot camp facilities range from a small effect for bonds to conventional behavior to a modest effect for prosocial attitudes.

For purposes of analysis, residualized change scores were used, as is common practice in the analysis of change (see Campbell & Kenny, 1999). Table 18.8 presents the regression coefficients for four regression models applied to each of the five outcomes. Model 1 simply tests whether the boot camp facilities differ in the

Table 18.4 Perception of Facility Environment Measures by Facility Type and Sample

Variable by Facility Type	Full Sample			Pre/Post Sample[a]		
	Mean	*SD*	*N*	*Mean*	*SD*	*N*
Therapeutic Environment						
Boot camps	3.8*	.7	2,341	3.7*	.7	263
Traditional facilities	3.3	.7	1,508	3.5	.7	262
Hostile Environment						
Boot camps	2.5*	.8	2,343	2.5*	.7	263
Traditional facilities	2.8	.8	1,506	2.7	.8	262
Freedom and Choice						
Boot camps	2.3*	.7	2,329	2.0	.7	263
Traditional facilities	2.7	.8	1,486	2.7	.7	262
Positive Environment Composite						
Boot camps	0.07*	.36	2,388			
Traditional facilities	−0.12	.36	1,568			

a Values for the pre/post sample represent the first measurement for the 530 adolescents who were measured on two occasions.
* The difference between groups (boot camps vs. traditional facilities) is statistically significant at $p < .05$.
NOTE: Mean difference tested using a nested analysis of variance, with facilities nested within facility type.

amount of change from the traditional facilities. For depression and social attitudes, boot camps observed greater change in the desired direction. Model 2 tests whether there is a relationship between the overall facility rating and amount of change. As with facility type, this effect was significant for the depression and social attitudes regression models. This is not surprising, for facility type and the overall facility rating are correlated. Models 1 and 2 do not control for known individual differences between the boot camps and traditional facilities. It is not surprising that the regression effects for anxiety, social bonds, and dysfunctional impulsivity were statistically nonsignificant, for the amount of change observed on these variables was small.

Model 3, which incorporated individual-level covariates, including age, race, history of alcohol and drug abuse, criminal history, and history of family violence and child abuse, provides a more realistic test of the relationship between facility characteristics and individual change. The coefficients for these individual characteristics were assumed to be fixed: that is, constant across facilities. From Table 18.2, we know that the pretest-posttest sample differed across facility type in the average months in the facility. Therefore, this variable was also included in the analysis to control for any linear relationship between amount of change and time in facility at first measurement. At the facility level, coefficients were estimated for the facility type, the composite indicator of the facility environment (facility mean across the individual perceptions by the youths), the selectivity of the facility, and the seriousness of the offenses of the juveniles admitted. Several interaction effects were also tested. We theorized that the facility environment might moderate the relationship between race, criminal history, and history of family violence, and amount of change on the outcome variables. We also theorized that the facility type might moderate the relationship between history of family violence and amount of change.

Facility type was statistically nonsignificant across all models with the full complement of covariates. For both change in depression and change in social attitudes (a predictor of delinquency), however, the overall rating of the facility environment was related to change in the desired direction. It appears that the boot camp versus traditional facilities distinction is far less relevant then how positively the youth perceive an environment to be. Recall that in these models, it is the facility means of the youths' perceptions that is used. Presumably, the composite of all of the youths' perceptions of the environment

Table 18.5   Facility-Level Descriptive Statistics

Variable by Facility Type	Mean	SD	N
Size*			
Boot camps	70.9	64.4	25
Traditional facilities	150.5	156.7	22
Selectivity Index*			
Boot camps	.6	.2	26
Traditional facilities	.3	.2	22
Seriousness Index*			
Boot camps	1.0	0.4	26
Traditional facilities	1.5	0.4	22
Level of Contact Permitted With the Outside			
Boot camps	.7	.2	22
Traditional facilities	.8	.1	21
Average Length of Stay*			
Boot camps	4.5	2.3	25
Traditional facilities	8.3	4.0	22
Follow-up Information on Youth			
Boot camps	46%		25
Traditional facilities	32%		22

*$p < .01$.

Table 18.6   Levels of Depression and Anxiety for Respondents Surveyed During First Month of Stay

Variable	Mean	SD	N
Depression			
Boot camps	3.1	1.0	765
Traditional facilities	3.0	1.0	267
Anxiety			
Boot camps	1.5	.3	760
Traditional facilities	1.4	.3	266

produced an index of the facility environment that is relatively independent of each individual youth's perceptions, although it might be affected by the composition of youths completing the survey.

A hypothesis of these analyses was that facility type would interact with history of family violence. As expected, youths with histories of family violence changed less in social attitudes, on average, than youth without histories of family violence. Also, as expected, this relationship interacted with facility type and was stronger for boot camp facilities. That is, there is only a slight relationship between history of family violence and change in social attitudes for the traditional facilities. The boot camp environment appears detrimental (or at least less therapeutic), on the basis of these data, for youths with a history of family violence. This pattern of effect, albeit statistically nonsignificant, was consistent across all five regression models. Thus, youths with a history of family violence exhibit less positive change overall, yet fare better relative to their peers in traditional facilities.

An unexpected finding was the relationship between race and change in social attitudes and the interaction of this effect with the overall rating of the facility environment. On average, African Americans exhibited less positive change in social attitudes. Furthermore, a plot of the regression function shows that the relationship between the overall rating of the facility environment and change in social attitudes was not evident for African Americans. The average amount of change in social attitudes was roughly equal across facilities rated differentially on overall rating. These two regression coefficients were statistically significant at the rather liberal level of $p < .10$ and were not hypothesized effects. Therefore, these findings need replication for any confidence to be placed in them.

A final finding in this regression model worth noting is the positive relationship between the time in facility at first measurement and the amount of positive change in social attitudes. This coefficient suggests that larger changes in social attitudes tend to occur in later periods of a youth's stay in these facilities.

The amount of reduction in depressed mood was related, as expected, to a history of family violence. The higher the level of prior family violence, the less decrease in depressed mood between administrations of the survey instrument. Also as expected, the youths in environments judged positively were more likely to have decreased in depressed mood between survey administrations, as were youths in facilities that were highly selective of the youth admitted.

Table 18.7 Random Effects Regression Analyses (HLM) of Residualized Change for the Five Outcome Variables on the Pre/Post Sample, Excluding Girls[a]

Variable	Anxiety	Depression?	Social Bonds	Dysfunctional Impulsivity	Social Attitudes[a]
Model 1					
Facility Type[b]	0.027	0.313***	−0.093	0.060*	−0.044**
Model 2					
Positive Environment	−0.038	−0.358***	0.060	−0.037	0.043***
Model 3					
Individual Level					
Age	0.005	−0.040	−0.005	0.003	−0.007
Race (African Am.)	−0.056*	0.150	−0.041	−0.002	−0.025*
by + Environment	0.048	0.290	0.126	0.051	−0.059*
Drug/Alcohol Abuse	0.023	0.064	−0.046	0.008	−0.002
Criminal History	0.001	−0.058	0.022	0.010	−0.001
by + Environment	0.062*	0.060	−0.089	0.000	−0.016
History Fam. Violence	0.164	0.899*	−0.510	0.095	−0.163**
by Facility Type	−0.098	−0.326	0.228	−0.034	0.089**
by + Environment	−0.077	−0.311	0.263	−0.038	0.052
Time in Facility	0.002	0.013	−0.012*	−0.001	0.003**
Facility Level					
Facility Type[c]	0.004	0.157	−0.062	−0.061	−0.008
+ Environment	−0.044	−0.261**	−0.080	−0.048	0.048**
Seriousness	0.043*	−0.226	0.072	0.069	−0.037
Selectivity	0.117	−0.236*	0.151	−0.131	0.006
Intercept	−0.258	0.381	0.279	−0.020	0.221***

p < .10; ** *p* < .05; ***p* < .01.
a Regression model statistically significant, *p* < .05.
b Sample sizes were (individuals/facilities): anxiety (446/41), depression (450/41), social bonds (444/41), dysfunctional impulsivity (443/41), social attitudes (453/41).
c Facility type coded as 1 for boot camp and 2 for traditional facility.

The regression models for anxiety, social bonds, and dysfunctional impulsivity were statistically nonsignificant. This may be due in part to the small amount of change observed on these three outcomes (see Table 18.7). In particular, a ceiling effect was observed for the measure of social bonds at Time 1, leaving little room for improvement in the scores of the youths. Although there were a few significant regression coefficients across these models, little confidence can be placed in these findings given a nonsignificant overall regression model.

DISCUSSION

Boot camps for delinquent juveniles are a modern alternative to traditional detention and treatment facilities, although the notion that strict discipline and physical exercise will "straighten out" wayward youth has a long history. The debate surrounding boot camps has focused on the potential stressfulness of the environment and the plausibility that the confrontation and militaristic style will be harmful to the juveniles, particularly those with a history of abuse. This study contributes to the debate by examining the environment of boot camps relative to traditional facilities as perceived by the youths in the facilities, the initial stress levels of the youths in the two types of facilities, and the intermediate changes of the youths on variables associated with future offending behavior.

Contrary to the expectation of the critics of boot camps, the juveniles perceived the boot

Table 18.8   Change in Anxiety, Depression, Bonds to Conventional Behavior, Dysfunctional Impulsivity, and Social Attitudes for the Pre/Post Sample

Variable	Time 1		Time 2		Difference		N	Effect Size
	Mean	SD	Mean	SD	Mean	SD		
Anxiety								
Boot camps	1.49	0.34	1.42	0.30	−0.06	0.36	254	−0.16
Traditional facilities	1.42	0.32	1.40	0.30	−0.02	0.32	255	−0.05
Depression								
Boot camps	3.17	0.99	2.86	1.05	−0.30	1.16	256	−0.25
Traditional facilities	3.34	0.98	3.16	0.99	−0.09	1.06	258	−0.07
Bonds to Conventional Behavior								
Boot camps	4.34	0.71	4.35	0.79	0.02	0.84	253	0.02
Traditional facilities	4.20	0.87	4.12	0.83	−0.09	0.87	251	−0.08
Dysfunctional Impulsivity								
Boot camps	1.66	0.35	1.61	0.35	−0.05	0.38	255	−0.12
Traditional facilities	1.64	0.34	1.65	0.33	0.02	0.35	249	0.05
Prosocial Attitudes								
Boot camps	1.49	0.13	1.53	0.15	0.05	0.14	252	0.36
Traditional facilities	1.48	0.14	1.49	0.15	0.01	0.14	248	0.07

NOTE: None of the mean gain score differences between facility types were statistically significant ($p \leq .01$) using a nested ANOVA model. Effect size computed as the difference score divided by the pooled within facility standard deviation at Time 1.

camp environments more favorably than the traditional facilities. These differences in perceptions remained after accounting for measured differences in the characteristics of the youths across the two facility types. Not only did the youths in the boot camps generally feel safer, but they perceived the environment to be more therapeutic or helpful. Thus, the fears that the boot camps, in general, would be hostile, negative environments appear not to have been realized. Though the boot camps were more structured and placed more constraints on the freedom of the juveniles, the implementation of the boot camp model for juveniles does not appear to produce environments that are perceived by juveniles as negative, relative to existing alternatives. On the basis of observational information gained through site visits to all of the surveyed facilities, we believe that this finding reflects the positive atmosphere of

many but not all of the boot camps. Most boot camps have strict rules and discipline for disobedience; however, despite this, or because of this, close and caring relationships seem to form between youth and staff.

A concern regarding boot camps is that the militaristic environment may contrast so sharply with the past home and community experiences of the juveniles that the camps will produce harmfully high levels of stress, resulting in high levels of depression and anxiety. It was not possible with the available data to determine if anxiety and depression among the youths in this study were at dysfunctional levels. A contrast between the initial levels of anxiety and depression between traditional and boot camp facilities, however, showed that youths in boot camps do not appear to have higher levels of anxiety and depression than comparable youths in traditional facilities. Considering the positive perception of the boot camp environment, this finding is not surprising, although it is counter to the expectation of many. Initial levels of depression and anxiety were related, however, to a history of family violence or abuse. Contrary to expectation, this relationship was not mediated by the facility type.

We hypothesized that the structured, disciplined nature of boot camps would increase the effectiveness of these facilities at reducing impulsivity among juveniles relative to traditional facilities. Furthermore, we anticipated that traditional facilities would be more effective at modifying a youth's social bonds and antisocial attitudes. These predictions were not confirmed. The raw differences in the mean change from pretest to posttest favored the boot camps for all three of these intermediate outcomes. These differences were substantially attenuated, and statistically nonsignificant, once the facility environment variable was included in the model, as well as characteristics of the individuals and other facility features. Thus, it appears that any differences in the effects of boot camps relative to traditional facilities on these variables can be explained by how positively the youths perceived the environment.

There is concern that the boot camp environment may be detrimental to youth with abuse histories (e.g., Morash & Rucker, 1990). This study provides some support for this view. For the antisocial attitude measure, youth with abuse histories exhibited substantially less change in the desired direction. Furthermore, this effect was twice as large for boot camps as for traditional facilities. That is, there was a statistically significant interaction between facility type and abuse history for antisocial attitudes, suggesting that boot camps may be ineffective and potentially detrimental to persons with a history of family violence.

An unexpected finding that deserves additional research was an interaction between the perceptions of the facility environment and race/ethnicity (African American vs. other). For African Americans, there was virtually no relationship between the characteristics of the facility environment, as measured by our single factor, and change in social attitudes, whereas non-African Americans exhibited greater change in the desired direction as the environment became more positive. As the result of an exploratory analysis, the finding may represent sampling error. However, if it is confirmed by additional research, it points to the need to examine the effect of environmental conditions on juvenile adjustment and change separately for African Americans relative to Caucasians and other racial/ethnic groups.

Almost anyone who visits a juvenile correctional boot camp recognizes the large difference between the environment of the camps and the environment of more traditional juvenile facilities. The question is whether this is a positive atmosphere conducive to positive growth and change or whether it is detrimental to juveniles and is in opposition to a high-quality therapeutic environment. Our findings suggest that, at least from the perspective of the juveniles residing in the facilities, the boot camps are a more positive environment than traditional facilities. Boot camp residents perceive their environments as less hostile and more therapeutic than juveniles in traditional facilities. Furthermore, according to their self-reports, they are no more (or less) anxious or depressed even during the early period in boot camps, when adjustment is hypothesized to be the most difficult. The boot camps also appear to have a more positive impact on the juveniles with regard to antisocial attitudes and depression; however, this effect appears to be related to the more positive atmosphere, not whether a facility is a boot camp. The only problematic impact of the boot camps was for juveniles with a history of abuse and family violence. These youth did not do as well in the boot camps as they did in the more traditional facilities.

Several selection bias effects are obvious in our data. First, juveniles sent to boot camps may differ from those sent to traditional facilities. Juveniles sent to boot camps may be those who would not otherwise be incarcerated, or they may be adjudicated for less serious crimes and sent to the boot camp because it requires a shorter period of confinement.

When we compared the characteristics of the two samples, and given our knowledge of these facilities, it appears that the boot camp youth were appropriate for the traditional facilities but that not all those in the traditional facilities would be appropriate for the boot camps. To control for this in our multivariate analysis, we included measured characteristics of the youth. Our analyses are unlikely to completely control for all selection bias. Though we cannot rule out all selection bias, our examination of the data leads us to believe that this is not a major threat to our conclusions.

Prior research examining boot camp facilities has not demonstrated any differences in recidivism when those released from boot camps are compared to those released from traditional facilities (MacKenzie, 1997). One possible reason for this finding is that the two types of facilities being compared in the prior studies were similar in environmental characteristics. Our results suggest that whether a facility is called a boot camp is less important than the characteristics of the facility environment. Facilities perceived as having more positive environments will be more apt to have an impact on social attitudes, and in past research these attitudes have been found to be associated with recidivism. Despite a generally more positive assessment of the boot camp environment by the youth, both boot camp and traditional facilities varied greatly on these measures.

Overall, we found only small changes during their time in the facilities in the characteristics of these juveniles that are related to delinquent behavior. This is disappointing.

This finding may reflect deficiencies in the scales or the short period of time between the pre- and post-measures of change. However, if this change truly reflects the very limited change these juveniles make during their time in the facilities, it is worrisome because the characteristics we measured have been linked to criminal behavior. This suggests to us that these facilities will have a very limited impact on the future delinquent and criminal activities of these youth. Disappointingly, few of these facilities had any information about the juveniles who left their care. Few even knew if the juveniles returned to the same facility, and fewer still had any information about whether the juveniles had recidivated, returned to a community school, or found employment. We wonder how staff and administrators who view their mission as the rehabilitation of juveniles can plan and improve programs if they do not know what happens to the youth once they leave the facility.

NOTES

1. Though 27 boot camps agreed to participate, one of the sites was very distinct from all other programs due to its 3-year length and transition from a boot camp into a detention program. As a result of these anomalies, this program was excluded from analyses here.

2. A copy of this survey is available from the authors.

3. A copy of this survey is available from the authors.

4. We label this scale Social Attitudes because a high score on it reflects more positive social attitudes, or conversely less antisocial attitudes.

REFERENCES

Ajdukovic, D. (1990). Psychosocial climate in correctional institutions: Which attributes describe it? *Environment and Behavior, 22,* 420–432.

Andrews, D. A., & Bonta, J. (1998). *The psychology of criminal conduct.* Cincinnati, OH: Anderson.

Andrews, D. A., Bonta, J., & Hoge, R. D. (1990). Classification for effective rehabilitation: Rediscovering psychology. *Criminal Justice and Behavior, 17,* 19–52.

Andrews, D. A., & Kiessling, J. J. (1980). Program structure and effective correctional practices: A summary of the CaVIC research. In R. R. Ross & P. Gendreau (Eds.), *Effective correctional treatment* (pp. 439–463). Toronto: Butterworth.

Andrews, D. A., Zinger, I., Hoge, R. D., Bonta, J., Gendreau, P., & Cullen, F. T. (1990). Does correctional treatment work? A clinically relevant and psychologically informed meta-analysis. *Criminology, 28,* 369–404.

Bonta, J. (1990, June). *Antisocial attitudes and recidivism.*

Paper presented at the annual meeting of the Canadian Psychological Association, Ottawa, Ontario.

Bryk, A. S., & Raudenbush, S. W. (1992). *Hierarchical linear models: Applications and data analysis methods.* Newbury Park, CA: Sage.

Campbell, D. T., & Kenny, D. A. (1999). *A primer on regression artifacts.* New York: Guilford.

Caspi, A., Moffitt, T. E., Silva, P. A., Stouthamer-Loeber, M., Krueger, R. F., & Schmutte, P. S. (1994). Are some people crime-prone? Replications of the personality-crime relationships across countries, genders, races and methods. *Criminology, 32,* 163–195.

Clark, C. L., & Aziz, D. W. (1996). Shock incarceration in New York State: Philosophy, results, and limitations. In D. L. MacKenzie & E. E. Hebert (Eds.), *Correctional boot camps: A tough intermediate sanction.* Washington, DC: National Institute of Justice.

Cowles, E. L., & Castellano, T. C. (1996). Substance abuse programming in adult correctional boot camps: A national overview. In D. L. MacKenzie & E. E. Hebert (Eds.), *Correctional boot camps: A tough intermediate sanction.* Washington, DC: National Institute of Justice.

Elder, G. H., Jr. (1992). The life course. In E. F. Borgatta & M. L. Borgatta (Eds.), *The encyclopedia of sociology.* New York: Macmillan.

Farrington, D. P. (1998). Individual differences and offending. In M. Tonry (Ed.), *The handbook of crime and punishment.* New York: Oxford University Press.

Gendreau, P., Little, T., & Groggin, C. (1996). A meta-analysis of the predictors of adult offender recidivism: What works! *Criminology, 34,* 575–607.

Gendreau P., & Ross, R. R. (1979). Effective correctional

treatment: Bibliotherapy for cynics. *Crime and Delinquency, 25,* 463–489.

Gendreau, P., & Ross, R. R. (1987). Revivification of rehabilitation: Evidence from the 1980s. *Justice Quarterly, 4,* 349–408.

Glaser, D. (1974). Remedies for the key deficiency in criminal justice evaluation research. *Journal of Research in Crime and Delinquency, 10,* 144–154.

Glueck, S., & Glueck, E. T. (1950). *Unraveling juvenile delinquency.* Cambridge, MA: Harvard University Press.

Goffman, E. (1961). *Asylums: Essays on the social situation of mental patients and other inmates.* Garden City, NY: Doubleday.

Goodstein, L., & Wright, K. (1989). Adjustment to prison. In L. Goodstein & D. L. MacKenzie (Eds.), *The American prison* (pp. 253–270). New York: Plenum.

Gottfredson, M. R., & Hirschi, T. (1990). *A general theory of crime.* Stanford, CA: Stanford University Press.

Horney, J. D., Osgood, W., & Marshall, I. H. (1995). Criminal careers in the short-term: Intra-individual variability in crime and its relationship to local life circumstances. *American Sociological Review, 60,* 655–673.

Jang, S. J. (1999). Age-varying effects of family, school, and peers on delinquency: A multilevel modeling test of interactional theory. *Criminology, 37,* 643–686.

Jesness, C. F. (1962). *Jesness Inventory.* North Tonawanda, NY: MultiHealth Systems.

Johnson, R., & Toch, H. (Eds.). (1982). *The pains of imprisonment.* Beverly Hills, CA: Sage.

Kirk, R. E. (1982). *Experimental design* (2nd ed.). Belmont, CA: Wadsworth.

Lipsey, M. (1992). Juvenile delinquency treatment: A meta-analytic inquiry into the

variability of effects. In T. Cook (Ed.), *Meta-analysis for explanation: A casebook.* New York: Russell Sage.

Loeber, R., Farrington, D., Stouthamer-Loeber, M., Moffitt, T. E., & Caspi, A. (1998). The development of male offending: Key findings from the first decade of the Pittsburgh Youth Study. *Studies on Crime and Crime Prevention, 7,* 141–171.

Lutze, F. (1998). Do boot camp prisons possess a more rehabilitative environment than traditional prison? A survey of inmates. *Justice Quarterly, 15,* 547–563.

MacKenzie, D. L. (1997). Criminal justice and crime prevention. In L. W. Sherman, D. C. Gottfredson, D. MacKenzie, J. Eck, P. Reuter, & S. Bushway (Eds.), *Preventing crime: What works, what doesn't, what's promising.* Washington, DC: Office of Juvenile Justice and Delinquency Prevention.

MacKenzie, D. L., Brame, R., McDowall, D., & Souryal, C. (1995). Boot camp prisons and recidivism in eight states. *Criminology, 33,* 401–430.

MacKenzie, D. L., & Hebert, E. E. (Eds.). (1996). *Correctional boot camps: A tough intermediate sanction.* Washington, DC: National Institute of Justice.

MacKenzie, D. L., & Parent, D. (1992). Boot camp prisons for young offenders. In J. M. Byrne, A. J. Lurigio, & J. Petersilia (Eds.), *Smart sentencing: The emergence of intermediate sanctions.* Newbury Park, CA: Sage.

MacKenzie, D. L., & Souryal, C. (1995). Inmate attitude change during incarceration: A comparison of boot camp with traditional prison. *Justice Quarterly, 12,* 325–354.

MacKenzie, D. L., Styve, G. J., & Gover, A. R. (1998). Performance-based standards for juvenile corrections.

Corrections Management Quarterly, 2, 28–35.

MacKenzie, D. L., Wilson, D. B., Armstrong, G. S., & Gover, A. R. (2001). The impact of boot camps and traditional institutions on juvenile residents. *Journal of Research in Crime and Delinquency, 38,* 279–313.

Mak, A. S. (1990). Testing a psychological control theory of delinquency. *Criminal Justice and Behavior, 17,* 215–230.

Mak, A. S. (1991). Psychosocial control characteristics of delinquents and nondelinquents. *Criminal Justice and Behavior, 18,* 287–303.

Moos, R. H. (1968). The assessment of the social climates of correctional institutions. *Journal of Research in Crime and Delinquency, 5,* 173–188.

Morash, M., & Rucker, L. (1990). A critical look at the idea of boot camp as a correctional reform. *Crime and Delinquency, 36,* 204–222.

Palmer, T. (1974, March). The Youth Authority's community treatment project. *Federal Probation, 38,* 3–14.

Sampson, R. J., & Laub, J. H. (1993). *Crime in the making.* Cambridge, MA: Harvard University Press.

Sechrest, D. D. (1989). Prison "boot camps" do not measure up. *Federal Probation, 53,* 15–20.

Shields, I. W., & Ball, M. (1990, June). *Neutralization in a population of incarcerated young offenders.* Paper presented at the annual meeting of the Canadian Psychological Association, Ottawa, Ontario.

Shields, I. W., & Whitehall, G. C. (1994). Neutralizations and delinquency among teenagers. *Criminal Justice and Behavior, 21,* 223–235.

Simons, R. L., Johnson, C., Conger, R. D., & Elder, G., Jr. (1998). A test of latent trait versus life-course perspectives on the stability of adolescent antisocial behavior. *Criminology, 36,* 217–244.

Spielberger, C. D., Gorsuch, R. L., & Lushene, R. E. (1970). *Manual for the State-Trait Anxiety Inventory.* Palo Alto, CA: Consulting Psychologists Press.

StataCorp. (1999). *Stata Statistical Software: Release 6.0.* College Station, TX: Stata Corporation.

Steinhart, D. (1993, January/February). Juvenile boot camps: Clinton may rev up an old drill. *Youth Today,* pp. 15–16.

Styve, G. J., MacKenzie, D. L., Gover, A. R., & Mitchell, O. (2000). Perceived conditions of confinement: A national evaluation of juvenile boot camps and traditional facilities. *Law and Human Behavior, 24,* 297–308.

White, J. L., Moffitt, T. E., Caspi, A., Bartusch, D. J., Needles, D. J., & Stouthamer-Loeber, M. (1994). Measuring impulsivity and examining its relationship to delinquency. *Journal of Abnormal Psychology, 103,*192–205.

Wright, K. N. (1985). Developing the Prison Environment Inventory. *Journal of Research in Crime and Delinquency, 22,* 257–277.

Wright, K. N. (1991). A study of individual, environmental, and interactive effects in explaining adjustment to prison. *Justice Quarterly, 8,* 217–242.

Wright, K., & Goodstein, L. (1989). Correctional environments. In L. Goodstein & D. L. MacKenzie (Eds.), *The American prison* (pp. 253–270). New York: Plenum.

Zachariah, J. K. (1996). An overview of boot camp goals, components, and results. In D. L. MacKenzie & E. E. Hebert (Eds.), *Correctional boot camps: A tough intermediate sanction.* Washington, DC: National Institute of Justice.

Zamble, E., & Porporino, F. J. (1988). *Coping, behavior, and adaptation in prison inmates.* New York: Springer-Verlag.

PART V

CRITICAL ISSUES IN THE CRIMINAL COURT SYSTEM

19

SENTENCING OPTIONS AND THE SENTENCING PROCESS

CASSIA C. SPOHN

The most difficult task facing a criminal court judge is sentencing a defendant. . . . Judges are aware of the horrible conditions awaiting defendants sentenced to penal institutions, just as they are cognizant of the general ineffectiveness of the probation sentence. Compassionate judges are aware of the inability of either form of sentence to adequately prevent the defendant from returning to society even more embittered and committed to a continued life of crime.

Paul Wice[1]

In February of 2001, Colorado District Court Judge David Lass sentenced Nathan Hall, a former lift attendant at the Vail ski resort, to 90 days in county jail for the ski slope collision that killed Alan Cobb in 1997. Hall, who was 18 years old at the time of the incident, was skiing out of control and at a high rate of speed when he collided with Cobb on an intermediate ski run. He was convicted of criminally negligent homicide.

Judge Lass also sentenced Hall to 3 years of intensive supervision probation following the 90-day jail sentence. In addition, he ordered Hall to perform community service and make financial restitution to Cobb's family for funeral and travel expenses and for psychological counseling. In announcing the sentence, Judge Lass said that although Hall didn't deserve a prison sentence, he did need to be punished for his role in Cobb's death. "He is still a young man, and he has done some immature and irresponsible things," Lass said. "Probably some of the things that have been said by others in this courtroom may be more important than my decision or anything I can say."[2]

As Nathan Hall's case illustrates, an individual who has been convicted of a crime faces a number of different sentence alternatives, depending on the seriousness of the crime and the willingness of the judge to experiment with

---

Spohn, C. C. (2002). Sentencing Options and the Sentencing Process. In Spohn, C. C. (Ed.), *How Do Judges Decide?* (pp. 33-77). Thousand Oaks, CA: Sage. Used with permission.

alternative sentences. The judge has more discretion and thus more opportunities to tailor the sentence to fit the individual offender if the crime is a misdemeanor or a less serious felony. In this type of case, the judge might impose a fine, order the offender to perform community service, place the offender on probation, or impose some other alternative to incarceration. The judge's options are more limited if the crime is serious or if the offender has a lengthy prior criminal record. In this situation, a jail or prison sentence is likely; the only question is how long the offender will serve.

In this chapter, we discuss the options available to the judge at sentencing. We begin with the death penalty, which is the ultimate sanction that society can impose on the guilty. We then discuss incarceration and the various alternatives to incarceration. We conclude the chapter by explaining that the sentences imposed on (and served by) offenders actually result from a "collaborative exercise" that involves legislators and criminal justice officials other than the judge.

The Judges' Options at Sentencing

The Death Penalty

In the United States, 38 states and the federal government have statutes that authorize the death penalty.[3] Imposed for a variety of offenses in the past, including armed robbery and rape,[4] the death penalty today is imposed almost exclusively for first-degree murder. It is, however, a penalty that is rarely applied. In 1999, nearly 15,000 persons were arrested for murder and nonnegligent manslaughter,[5] but only 272 persons were sentenced to death.[6]

A sentence of death does not necessarily mean that the offender will be executed. The offender's conviction or sentence might be overturned by a higher court, the sentence might be commuted to life in prison by the governor, or the offender might die in prison. Of the 6,365 prisoners under the sentence of death from 1977 to 1999, 598 (9.4 percent) were executed and 2,240 (35.2 percent) received some other type of disposition; the remaining 3,527 inmates were still incarcerated. The average amount of time these offenders had been on death row was 7 years and 7 months.[7]

Current death penalty statutes have a number of common features.[8] Most are *guided-discretion statutes* that allow the death penalty to be imposed only if at least one statutorily defined aggravating circumstance is present. Although the aggravating circumstances vary among jurisdictions, the list typically includes such crimes as murder for hire, murder of more than one person, murder of a police officer, murder involving torture, or murder during the commission of another crime, such as armed robbery or sexual assault.

Most jurisdictions also require a bifurcated trial in capital cases. The first stage involves the determination of the defendant's guilt or innocence. If the defendant is convicted of a capital crime—that is, a crime for which the death penalty is an option—a separate sentencing proceeding is held. At this stage, evidence regarding the aggravating and mitigating circumstances of the case is presented. The jury and/or the judge weigh the evidence and decide whether the defendant should be sentenced to death or should receive a lesser sentence of life without parole, life, or a specified term of years. Finally, most death penalty statutes also provide for automatic review of the conviction and death sentence by the state's highest court. If either the conviction or the sentence is overturned, the case can be sent back to the trial court for retrial or resentencing.[9]

The role played by the judge in the capital sentencing process varies from state to state. In some jurisdictions, such as Texas and Louisiana, the jury decides whether the defendant should be sentenced to death or not. In these jurisdictions, the jury's decision cannot be overturned by the trial court judge. In other jurisdictions, such as Alabama and Florida, the jury recommends the sentence, but the judge can overrule the jury. Under Florida law, for example, the jury in a capital case makes an advisory sentence recommendation to the court, but the judge is required to independently weigh the aggravating and mitigating circumstances and determine whether the offender should be sentenced to life imprisonment or death.[10] In some states, such as Nebraska, the death penalty decision is made either by the trial judge or by a panel of judges.

Although, as noted above, the death penalty is used infrequently, its importance as a sentencing alternative cannot be ignored. As we explain in detail in Chapter 5, questions have been raised about the fairness of the death penalty process. Critics charge that the death penalty is imposed in an arbitrary and capricious manner and that there is compelling evidence of racial discrimination in the capital sentencing process.

Incarceration

Unlike the death penalty, which can be imposed only for first-degree murder and a handful of other offenses, a jail or prison sentence is an option in most criminal cases. This includes misdemeanors as well as felonies. The Texas Penal Code, for example, categorizes misdemeanors as Class A, Class B, or Class C; Class A misdemeanors are the most serious, Class C the least serious. Although offenders convicted of Class C misdemeanors cannot be sentenced to jail, those convicted of Class B offenses can be confined in jail for up to 180 days, and those found guilty of Class A offenses can be sentenced to jail for as long as a year.[11] In most jurisdictions, sentences of less than a year are served in a local jail, and those of a year or more are served in a state prison. All offenders who are tried in U.S. District Courts and receive a prison sentence are incarcerated in a federal prison; there are no federal jails.

For some offenses, a prison sentence is not simply an option, it is required. All jurisdictions in the United States now have laws that prescribe mandatory minimum terms of incarceration for selected crimes. For example, 41 states have mandatory sentences for repeat or habitual offenders and for crimes involving possession of a deadly weapon. Most states also require minimum prison sentences for certain types of drug offenses: trafficking, selling drugs to minors, or selling drugs within 1,000 feet of a school.[12] At the federal level, more than 100 crimes are subject to laws requiring 2- to 20-year minimum sentences. Mandatory minimum provisions can be circumvented by prosecutors who refuse to charge offenses that trigger a minimum sentence[13] or by judges who either refuse to convict or ignore the statute and impose less than the mandatory minimum sentence,[14] but these "tough-on-crime" laws limit the options available to the judge at sentencing. They generally require the judge to sentence the offender to prison for a specified period of time.

Judges' options are also limited by the type of sentencing system used in their jurisdiction (see Box 19.1). Some state laws require judges to impose indeterminate sentences.[15] In these states, the legislature specifies a minimum and a maximum sentence for a particular offense or category of offenses. In sentencing an offender, the judge either imposes a minimum and a maximum sentence from within this range or, alternatively, determines only the maximum sentence that the offender can serve. Assume, for example, that the sentence range for armed robbery is 5 to 20 years. In the first scenario, the judge would determine both the minimum and the maximum sentence. He or she might sentence the offender to 5 to 10 years, 10 to 15 years, 5 to 20 years, or any other range of years within the statutory minimums and maximums. In the second scenario, the judge would determine only the maximum penalty; the minimum penalty would be automatically applied. In other words, regardless of whether the maximum penalty was 10 years, 15 years, or 20 years, the minimum would always be 5 years. In either case, the actual amount of time the offender will serve is determined by the parole board on the basis of its judgment as to whether the offender has been rehabilitated or has simply served enough time. An armed robber sentenced to an indeterminate sentence of 5 to 10 years, then, will serve at least 5 years but no more than 10 years, depending on the parole board's assessment of his case.

In other states, judges impose determinate sentences, which are fixed-term sentences that may be reduced if the offender behaves while incarcerated (i.e., through "good-time credits").[16] The offender's date of release is based on the sentence imposed minus any good-time credits. The parole board may supervise offenders who have been released from prison, but it does not determine when offenders will be released. In states that have adopted this type of system, the legislature provides a presumptive range of confinement for various categories of offenses. Although some offenses are nonprobationable, for most crimes, the judge has the discretion to determine whether the offender will be incarcerated and if so, for how long. If, for example, the

---

Box 19.1   Sentencing Systems in State and Federal Jurisdictions

Indeterminate Sentence

The legislature specifies a minimum and a maximum sentence for each offense or class of offenses. The judge imposes either a minimum and a maximum term of years or the maximum term only. The parole board decides when the offender will be released from prison.

Determinate Sentence

The legislature provides a presumptive range of confinement for each offense or class of offenses. The judge imposes a fixed term of years within this range. The offender serves this sentence minus time off for good behavior.

Mandatory Sentence

The legislature requires a mandatory minimum prison sentence for habitual offenders and/or offenders convicted of certain crimes. Examples include use of a weapon during the commission of a crime, drug trafficking, and selling drugs to minors.

Sentencing Guidelines

A legislatively authorized sentencing commission establishes presumptive sentencing guidelines. The guidelines are typically based on the seriousness of the offense and the offender's prior record. Judges are required to follow the guidelines or explain in writing why they did not.

---

presumptive range of confinement for robbery is 4 to 15 years and the statute does not specify that offenders convicted of robbery must be sentenced to prison, the judge could impose either a probation sentence or a prison sentence of anywhere from 4 to 15 years. If the sentence was 10 years in prison, the offender would serve that time minus credit for good behavior.

In still other states and at the federal level, judges' incarceration options are limited by presumptive sentencing guidelines.[17] In jurisdictions that use this model, a sentencing commission develops guidelines based on the seriousness of the offense and the offender's prior criminal record, which judges are required to use in determining the appropriate sentence. Judges are allowed to depart from the guidelines and impose a more severe or less severe sentence than the guidelines require, but they must provide a written justification for doing so. The judge's decision to depart either upward or downward can be appealed to a state or federal appellate court.

Felony Sentences in 1996

Recent data on the types of sentences imposed on felony offenders reveal that most of them are incarcerated; less than one third receive "straight probation" (i.e., probation only, not jail or prison followed by probation).[18] As shown in Exhibit 19.1, 38 percent of all offenders convicted of felonies in state courts in 1996 were sentenced to prison, 31 percent were sentenced to jail, and 31 percent were placed on probation. The proportion of federal offenders sentenced to prison in 1996 was even higher. Of all federal offenders, 80 percent were sentenced to prison, 64 percent received a sentence of a year or more, and 16 percent received a sentence of less than a year. As one would expect, the likelihood of a prison sentence was highest for offenders convicted of violent crimes. This was especially true of federal offenders: 85 percent of all federal offenders convicted of violent crimes were sentenced to prison for more than a year. Federal offenders convicted of drug offenses also faced high odds of imprisonment. In fact, the incarceration rates for drug offenders and violent offenders convicted in U.S. District Courts were nearly identical.

Exhibit 19.1 also displays the average lengths of prison, jail, and probation sentences. The typical state felony offender received a prison sentence of just over 5 years, whereas the typical federal offender received a sentence of 6.5 years. Again as one would expect, offenders convicted of violent offenses received substantially longer average prison sentences than those convicted of property crimes or drug offenses. The sentences imposed on state property offenders averaged longer than those imposed on federal property offenders. In contrast, the average sentence imposed on federal drug offenders (89 months) was substantially longer than the average sentence imposed on state drug offenders (51 months).

It is important to note that the amount of prison time the offender receives at sentencing is not the amount of time that he or she will actually serve in prison. There are two reasons for this. As noted earlier, many states have indeterminate sentencing systems in which the parole board decides when the offender is fit to be released. In these states, the amount of time served equals the amount imposed at sentencing only in the relatively rare cases in which the defendant is never paroled. In addition, in most states, prison inmates can earn early release for good behavior and automatic good-time credits are awarded in a number of states. Both of these factors reduce the amount of time the offender will serve in prison.

As shown in Exhibit 19.1, state prisoners serve only 45 percent of the sentence imposed. The estimated time to be served is slightly higher for violent offenses and somewhat lower for property and drug offenses. In contrast, the expected time to be served for federal inmates is a uniform 85 percent. By law, all federal inmates—with the exception of those sentenced to life imprisonment, who must serve the sentence in full—must serve a minimum of 85 percent of

the sentence imposed. This means that offenders sentenced in federal court receive sentences averaging 16 months longer than the ones imposed by state courts but they will serve 3 years and 3 months longer than state prison inmates. The differences for drug offenses are even larger: Federal drug offenders will serve 4.5 years longer than state drug offenders.

Incarceration, then, is a widely prescribed and frequently imposed sentencing option. In the United States, in fact, it is now the "option of choice" for offenders convicted of felonies in both state and federal courts.

Probation

The primary alternative to incarceration is probation. A straight probation sentence does not entail confinement in jail or prison.[19] Rather, the judge imposes a set of conditions agreed to by the offender, who is then released into the community. The court retains control over the offender while he or she is on probation; if the conditions of probation are violated, the judge can modify the conditions or revoke probation and sentence the offender to jail or prison.

Probation was developed primarily as a means of diverting juvenile offenders and adults convicted of misdemeanors and nonviolent felonies from jail or prison. However, as shown in Exhibit 19.1, just under a third of all state felony defendants and 20 percent of all federal felony defendants were sentenced to probation in 1996. Although straight probation sentences were imposed more frequently when the offender was convicted of a nonviolent crime, a significant minority (21 percent) of offenders convicted of violent felonies in state courts were placed on probation. This included 28 percent of all offenders convicted of aggravated assault, 21 percent of those convicted of sexual assault, 13 percent of those convicted of robbery, and 5 percent of those convicted of murder and manslaughter.[20] As these figures indicate, probation is not reserved exclusively for nonviolent offenders.

Although the proliferation of statutes requiring mandatory minimum prison sentences has reduced the number of offenses for which probation is an option, judges retain wide discretion in deciding between prison and probation for

Exhibit 19.1 Sentences Imposed on Felony Offenders in State and Federal Courts, 1996

Type of Sentence	Offenders Convicted in State Courts			Offenders Convicted in Federal Courts		
	Prison	Jail	Probation	Prison ≥ 1 year	Prison < 1 year	Probation
All offenses (percentage)	38	31	31	64	16	20
Violent offenses[a]	57	22	21	85	7	8
Property offenses[b]	34	28	38	29	29	42
Drug offenses[c]	35	37	28	84	7	8
Average sentence length (months)						
All offenses	62	6	41	78	7	39
Violent offenses	105	7	48	107	8	42
Property offenses	49	6	40	35	7	39
Drug offenses	51	6	42	89	9	42
Average estimated time served in prison: months/percentage						
All offenses	28/45%			67/85%		
Violent offenses	53/51%			91/85%		
Property offenses	21/42%			30/85%		
Drug offenses	21/41%			76/85%		

SOURCE: Adapted from *Felony Sentences in the United States, 1996* (Table 3, Table 7, and Table 11), by U.S. Department of Justice, Bureau of Justice Statistics, 1999. Washington, DC: Author.
a Includes murder, sexual assault, robbery, aggravated assault, and other violent crimes.
b Includes burglary, larceny, motor vehicle theft, forgery, fraud, and embezzlement.
c Includes drug trafficking and drug possession.

offenses that are not subject to mandatory minimums. Most state statutes allow judges to impose probation unless they believe that (a) the offender is likely to commit additional crimes if released, (b) the offender is in need of treatment that can be provided more effectively in jail or prison, or (c) probation would be inappropriate given the seriousness of the offender's crime. Many statutes also specify the conditions that "shall be accorded weight in favor of withholding sentence of imprisonment."[21] The Nebraska statute, for example includes such aspects as the offender's motivation or intent, the role played by the offender in the crime, provocation on the part of the victim, the offender's criminal history, and the burden that imprisonment would place on the offender's dependents. Other state statutes similarly specify that probation is the preferred sentence unless the crime is

serious and the offender is likely to commit additional crimes if not locked up.

The Conditions of Probation

All offenders placed on probation are expected to meet regularly with their probation officer and to obey all laws. But these are minimum requirements: Most probationers face additional restrictions. A recent national survey of adults on probation found that 36 percent of all probationers had three or four additional conditions attached to the sentence; 46 percent had five or more conditions.[22] As shown in Exhibit 19.2, which lists the conditions imposed on probationers (felony and misdemeanor) in 1995, most offenders were also required to pay court costs, supervision fees, or fines. About a third were required to pay restitution to the victim. About 40 percent of the probationers were required to enroll in some form of substance abuse treatment. Consistent with this, about a third were required to submit to drug testing, and 8.4 percent were required to refrain from using alcohol or drugs. Forty percent of the offenders were also required to maintain employment or enroll in educational or vocational training programs. Relatively few offenders, on the other hand, were subject to requirements designed to monitor or restrict their movement (e.g., curfews, house arrest, electronic monitoring) or orders to stay away from bars or other places of business.

Offenders who violate the conditions of their probation may be called before the court to explain the circumstances of the violation. At the probation violation hearing, the judge can continue or extend the offender's probation, with or without additional conditions. If the violations are serious enough or if the offender repeatedly violates the conditions of probation, the judge may schedule a hearing to determine whether the offender's probation should be revoked and a different sentence alternative imposed, generally incarceration in jail or prison. In 1995, 18 percent of the adult probationers surveyed by the Bureau of Justice Statistics had at least one disciplinary hearing during the course of their probation. Most were brought back to court for "technical violations": for example, failure to meet regularly with the probation officer, failure to pay fines or restitution, or failure to attend or complete a drug treatment program. More than a third, however, violated probation by being arrested for new offenses.[23] Although most of these hearings resulted in a continuation of the offender's probation, just under a third led to revocation of probation and incarceration of the offenders.[24]

Revocation of probation is more likely if the offender is arrested for a new offense, particularly a new felony offense, than if the offender is cited for a technical violation. If, for example, an offender on probation for a felony drug offense tests positive for illegal drugs, the judge might continue his or her probation with a stipulation that the individual submit to more frequent random drug tests and complete a drug treatment program. If, on the other hand, the offender is arrested for possession of cocaine,

Exhibit 19.2	Conditions of Sentences of Adult Probationers, 1995
Condition of Sentence	*Percentage*
Fees, fines, court costs	84.2
Supervision fees	61.0
Fines	55.8
Court costs	54.5
Substance abuse treatment	41.0
Alcohol	29.2
Drug	23.0
Employment and training	40.3
Maintain employment	34.7
Education/training	15.0
Alcohol/drug restrictions	38.2
Alcohol	32.5
Drug	8.1
Restitution to victim	30.3
Community service	25.7
Restrictions	21.1
No contact with victim	10.4
Driving restrictions	5.3
Other treatment	17.9
Sex offenders program	2.5
Psychiatric/psychological counseling	7.1
Other counseling	9.2
Confinement/monitoring	10.1
Boot camp	0.5
House arrest/electronic monitoring	3.7
Curfew	0.9
Restriction to movement	4.2

SOURCE: Adapted from *Characteristics of Adults on Probation, 1995 (Table 8)*, by U.S. Department of Justice, Bureau of Statistics, 1997. Washington, DC: Author.

the judge would probably revoke his or her probation and sentence that individual to prison. Offenders who repeatedly commit technical violations also risk eventual revocation of probation and incarceration.

Intensive Supervision Probation

We noted above that probation was developed as an alternative to incarceration for juvenile offenders and for first-time adult offenders convicted of misdemeanors. Skyrocketing increases in prison and jail populations coupled with concerns about the costs of incarcerating nonviolent offenders led to increasingly large numbers of felony offenders being placed on probation. Some policymakers and researchers questioned the wisdom of this, citing high recidivism rates for felony probationers. One of the most influential of these studies was an analysis of recidivism among adult felony offenders placed on probation in Los Angeles and Alameda Counties (California).[25] The authors of this study reported that recidivism rates among these probationers were high: 65 percent were arrested and 51 percent were charged and convicted during the 40-month follow-up period. Moreover, 18 percent of the probationers were convicted of serious violent crimes. These results led the researchers to conclude that "felony probation has been a high-risk gamble."[26]

Findings such as this led to a search for "alternative sanctions that punish but do not involve incarceration."[27] One of these alternative, or intermediate, sanctions is intensive supervision probation (ISP), which, as the name suggests, involves closer scrutiny of offenders and more restrictions on their behavior. Candidates for ISP include offenders, typically those convicted of felonies, deemed too serious or dangerous for routine probation but not so serious or dangerous that they must be incarcerated. An unemployed car thief whose crimes are becoming more frequent or escalating in seriousness might be eligible for intensive supervision probation. A serial rapist, on the other hand, would be viewed as too dangerous to qualify for the program, and a college student convicted of possession of a small amount of marijuana would be perceived as a better candidate for routine probation than intensive supervision probation.

Offenders placed on intensive supervision probation are required to meet frequently with their probation officers. Some programs require as many as five face-to-face contacts per week, whereas others require less frequent face-to-face meetings but include routine curfew checks or daily phone calls.[28] In addition, these probationers are routinely ordered to perform community service, maintain employment and/or attend school or job training programs, complete substance abuse treatment programs, submit to random drug tests, and abide by curfews and other restrictions. Offenders in ISP programs may also be subject to house arrest and electronic monitoring.

Intensive supervision probation, then, is a penalty with "more punitive bite than simple probation"[29] that is designed to reform and rehabilitate offenders in the community while minimizing the risk that such offenders pose to the community.

Although the concept of ISP has almost universal support among policymakers and practitioners, questions regarding its effectiveness and its application have surfaced. One issue is whether ISP is more effective than routine probation in preventing or reducing reoffending. Most evaluations conclude that it is not.[30] In fact, a multisite experimental evaluation in which offenders were randomly assigned to ISP or to other sentence alternatives found that offenders sentenced to ISP not only had recidivism rates comparable to other offenders but also had *higher* rates of technical violations and probation revocation.[31] The authors of this study attributed this higher rate of "failure" in part to the nature or structure of the ISP program. In other words, because they were subject to more restrictive conditions as well as heightened surveillance, ISP participants faced a greater risk of failure than did offenders on routine probation.[32] Others suggest that the results of this evaluation indicate that ISP is succeeding, not failing. They suggest that if ISP is seen as a means of incapacitating offenders in the community, then high rates of technical violations and probation revocation signal that probation officers "are watching their clients closely enough to make the commission of new crimes quite difficult."[33]

A second issue is *net widening*, which in this case, refers to the use of ISP for probation-bound rather than prison-bound offenders. An important goal of ISP is to divert nonserious, nonviolent offenders from the prison system. Thus, offenders selected for participation in ISP should be those who otherwise would have gone to prison. But critics charge that ISP is not used in this way. They charge that judges and probation officials use ISP for offenders who otherwise would have received regular probation. Hence, the term net widening.

According to Morris and Tonry, the "mechanism that serves to widen the net is obvious."[34]

Judges who in the past have given the benefit of the doubt to some offenders, sentencing them to probation rather than prison despite reservations about doing so, view more restrictive alternatives such as ISP as more appropriate sentences for these offenders. As Morris and Tonry note, if judges have given "a probationary sentence to the offender for whom imprisonment seemed too severe and probation too lenient, the newly available intermediate punishment will be just what's wanted."[35]

Net widening can be avoided, or at least minimized, by the use of *back-end* rather than *front-end* programs. A front-end program is one in which the sentencing judge makes the decision to sentence the offender to prison, to regular probation, or to ISP probation. This allows the judge to widen the net by using ISP in place of regular probation rather than as a substitute for imprisonment. In contrast, a back-end program selects ISP participants from those already sentenced to prison. In New Jersey, for example, an offender who is sentenced to prison and meets the eligibility criteria for the program can apply for resentencing to ISP. Applicants are carefully screened using risk-assessment tools, and those who qualify are released from prison and placed on ISP for 18 months.[36] This approach seems less susceptible to, but not immune from, net widening. Judges who previously might have placed, albeit reluctantly, medium-risk offenders on regular probation might decide to sentence them to prison instead, anticipating that they will be released into the ISP program.[37] Both front-end and back-end programs, then, may present "almost irresistible temptations to judges and corrections officials to use them for offenders other than those for whom the program was created."[38]

As we will see, questions about effectiveness and net widening have been raised about many other intermediate sanctions developed to reduce prison overcrowding and supervise offenders in the community.

Other Intermediate Sanctions

It should be clear from the information presented thus far that incarceration (in jail or in prison) and probation (routine or intensive supervision) are the principal sentences imposed on criminal offenders in the United States.

Despite their widespread use, however, they are not the only sentencing options. Indeed, a variety of *intermediate sanctions* are available: boot camps, house arrest and electronic monitoring, day reporting centers, community service, restitution, and fines. Like intensive supervision probation, these alternative punishments are intended to fill the void between routine probation and protracted imprisonment. They are designed to free up prison beds and "to provide a continuum of sanctions that satisfies the just deserts concern for proportionality in punishment."[39]

A number of scholars have urged judges to reduce their reliance on prison and probation and to incorporate intermediate sanctions into their philosophy of sentencing.[40] Although they recognize that the alternatives developed thus far will not *replace* prison or probation as the primary punishments imposed on criminal offenders, these scholars have called on judges to sentence fewer nonviolent offenders to prison, to shorten the prison sentences they do impose, and to develop a "comprehensive punishment system" incorporating a wide range of intermediate punishments.[41] As Morris and Tonry contend,

> At present, too many criminals are in prison and too few are the subjects of enforced controls in the community. We are both too lenient and too severe; too lenient with many on probation who should be subject to tighter controls in the community, and too severe with many in prison and jail who would present no serious threat to community safety if they were under control in the community.[42]

Morris and Tonry argue that a just sentencing system is one which includes a wide range of punishments and not just a choice between prison and probation. They further suggest that widespread use of these intermediate sanctions will make it possible "to move appreciable numbers who otherwise would be sentenced to prison into community-based punishments, having a roughly equivalent punitive bite but serving both the community and the criminal better than the prison term."[43]

Austin and Irwin[44] are less sanguine about the prospects for significant reductions in imprisonment through the use of intermediate sanctions. They argue that "well-intentioned alternatives have had marginal impact on reducing prison populations."[45] They attribute this both to the net-widening phenomenon discussed earlier and to the fact that policymakers and the public "are increasingly disenchanted with probation and other forms of community sanctions."[46] Austin and Irwin also contend that even widespread use of intermediate punishments, which primarily target offenders who would otherwise serve only short prison terms, would not significantly reduce the United States prison population, which is increasingly composed of offenders serving long prison sentences. According to these authors, "The single most direct solution that would have an immediate, dramatic impact on prison crowding and would not affect public safety is to shorten prison terms."[47]

In the following sections, we briefly describe four intermediate sanctions: boot camps, house arrest and electronic monitoring, community service, and monetary penalties.

Boot Camps

Correctional boot camp, or shock incarceration, programs target young, nonviolent felony offenders who do not have extensive prior criminal records.[48] Modeled on the military boot camp, the correctional boot camp emphasizes strict discipline, military drill and ceremony, and hard labor and physical training. Most also provide substance abuse counseling and educational and vocational training. Offenders selected for these programs, which typically last 3 to 6 months, are separated from the general prison population. They live in barracks-style housing, address the guards by military titles, and are required to stand at attention and obey all orders. Although the structure and focus of the programs vary widely, their primary goal is to divert young offenders from "a life outside the law using the same tactics successfully employed by the military to turn civilians into soldiers."[49]

Prison boot camps began in Georgia and Oklahoma in 1983. Today, these programs are found in 29 states. Most serve young adult males, but programs that target women and juvenile offenders are becoming increasingly popular.[50] Most programs limit eligibility to offenders who are under 30, but some have a higher upper-age limit. Georgia, which has one of the largest programs, admits offenders aged 17 to 35. Massachusetts takes offenders up to 40 years old. About a third of the programs restrict eligibility to first offenders; the others admit first-time felony offenders or those without prior prison sentences.

Despite their popularity, prison boot camps are controversial. The primary point of contention is the military model on which the programs are based. Critics charge that the harsh discipline characteristic of prison boot camps is not an effective way to change offenders. They also contend that too much time is devoted to military training and hard physical labor and too little to therapeutic treatment, education, and vocational training. Supporters counter that the military atmosphere encourages reform and rehabilitation. They argue that "the stress created in boot camp may shake up the inmates and make them ready to change and take advantage of the treatment and aftercare programs offered."[51]

The controversy surrounding boot camps prompted the National Institute of Justice to fund a number of single- and multisite evaluations of their effectiveness. One of the earliest studies found that boot camp participants in Louisiana did not have lower recidivism rates than either offenders who served time in traditional prisons or offenders who were placed on probation.[52] The results of an evaluation of programs in eight states were similarly disappointing.[53] This study found that although boot camp participants had more positive attitudes than regular prison inmates while they were incarcerated, they did not adjust more positively than traditional inmates to community supervision. Moreover, the recidivism rates of the boot camp participants were generally similar to those of comparable offenders who served their time in state prisons or local jails. In the three jurisdictions in which boot camp participants did have significantly lower recidivism rates, the programs emphasized education and counseling rather than military training. They also required 6 months of intensive supervision of offenders who were released into the community. These results led the author of this study to conclude that "the core elements of boot camp programs—military-style discipline, hard labor, and physical training—by themselves did not reduce offender recidivism."[54]

House Arrest and Electronic Monitoring

In September of 2000, baseball All-Star Darryl Strawberry pleaded guilty to driving under the influence of medication and leaving the scene of an accident. Strawberry, who was serving an 18-month probation sentence for drug and solicitation charges, was on his way to a meeting with his probation officer when he blacked out and rear-ended a car stopped at a red light. He apparently had taken a prescription sleeping medication prior to getting behind the wheel of his car. Florida Circuit Court Judge Florence Foster sentenced him to 2 years' house arrest. He was ordered to remain in his home except for specifically permitted outings.

House arrest, with or without electronic monitoring, is an alternative sanction used primarily for nonviolent offenders such as Darryl Strawberry. Offenders placed on house arrest, which is also referred to as *home confinement,* are ordered to remain at home for designated periods of time. They are allowed to leave only at specified times and for specific purposes: to obtain food or medical services, to meet with a probation officer, and sometimes, to go to school or work. Both back-end and front-end programs exist. In some jurisdictions, such as Oklahoma, offenders are released early from jail or prison sentences if they agree to participate in a home confinement program. In other jurisdictions, such as Florida, house arrest is a front-end program in which offenders who otherwise would be sentenced to jail or prison are instead confined to their homes. Both types of programs "aim simultaneously to offer a community sentence that is seen as burdensome and intrusive . . . and to reduce pressure on overcrowded prisons and jails."[55]

Offenders placed on house arrest are often subject to electronic monitoring, which is a means of ensuring that they are at home when they are supposed to be. The most popular system today is one in which the offender is fitted with a wrist or ankle bracelet that he or she wears 24 hours a day. The bracelet serves as a transmitter. It emits a constant radio signal to a home monitoring unit, which is attached to the offender's home phone. The monitoring unit informs the monitoring center when offenders enter and leave home; it also sends a message if they tamper with or attempt to remove the bracelet. The monitoring officer is informed when the offender deviates from the preapproved schedule: if, for example, he or she returns from school at 9 p.m. but was expected at 4 p.m. Ideally, the officer will respond immediately to a violation notice. The offender can be terminated from the program for tampering with the device, for repeated unauthorized absences from home, and for a variety of other violations.

The use of electronic monitoring began in 1983 in New Mexico and Florida. Although there were only 95 offenders on electronic monitoring in 1986, by the early 1990s, there were more than 400 programs involving 12,000 offenders. During the next decade, the number of offenders placed on electronic monitoring continued to grow. By 1998, approximately 95,000 offenders, including pretrial detainees, juveniles, and adult probationers and parolees, were being monitored electronically.[56] Among adults on probation in 1997, more than 15,000 were being monitored.[57]

Like other intermediate sanctions, house arrest with electronic monitoring is designed to be a cost-effective alternative to jail or prison that does not jeopardize public safety. The results of studies designed to evaluate its effectiveness in meeting these objectives are inconsistent. There have been no multisite studies in which offenders are randomly assigned to house arrest, regular probation, or incarceration, and the studies that have been done have been criticized for "shoddy or weak research designs."[58] Some studies report recidivism rates of less than 10 percent for offenders on electronic monitoring, whereas others report rates of 40 percent or higher.[59] As the authors of a comprehensive review of the research concluded, "we know very little about either home confinement or electronic monitoring."[60]

Community Service Orders

An underused but potentially important intermediate sanction is the community service order. Rather than being sentenced to jail or placed on routine probation, offenders are ordered to perform a certain number of hours of unpaid work at schools, hospitals, parks, and other public and private nonprofit agencies.

Thus, an accountant convicted of fraud might be sentenced to provide advice to poor taxpayers, a baseball player convicted of a drug offense might be required to lecture to junior high school students on the dangers of using drugs, and traffic offenders might be required to pick up trash along highways or in public parks. As Morris and Tonry observe, the list of community service projects "is limited only by the imagination of the sentencing judge and the availability of some supervision to ensure that the offender fulfills the terms of the sentence."[61]

Use of community service orders in the United States dates from 1966, when judges in Alameda County, California, began to sentence low-income women found guilty of traffic offenses to unpaid labor in the community. The concept spread rapidly in the United States and elsewhere; but today, community service orders are used much more frequently in the United Kingdom and Europe than in the United States and for a greater variety of offenses. In fact, laws passed in England, Wales, Scotland, and the Netherlands specifically authorize the use of community service orders as an alternative to short-term incarceration.

The most comprehensive community service program in the United States is New York City's Community Service Sentencing Project (CSSP), established in 1979.[62] CSSP is one of the few community service programs in the United States to be designed as an alternative to jail or prison. It places repeat misdemeanor offenders on work crews in their home neighborhoods and supervises them as they complete 10 to 15 days of unpaid labor. Each year, about 1,000 adult offenders are ordered to remove graffiti, clean sidewalks, paint buildings, or perform other types of community service work in low-income neighborhoods. An early evaluation of the program revealed that it was effective. Offenders were diverted from prison, rearrest rates of CSSP participants were similar to those of a matched group of offenders who served short jail sentences, and the program was cost-effective.[63] A more recent evaluation found that 73 percent of the participants successfully completed the program and that very few were arrested for new offenses while they were participating in the program.[64]

With the exception of the program in New York City, which targets misdemeanants only, community service is not widely used as an alternative punishment in the United States. Here, the primary use of community service is as a condition of probation or as a punishment for minor traffic offenses. Some judges are reluctant to use this alternative as a stand-alone sanction for misdemeanors and less serious felonies. This may reflect a belief that requiring an offender to perform community service is not adequate punishment for most crimes. Judges may believe, in other words, that removing graffiti or cleaning up parks for 200 hours is not as onerous as even a short stint in jail. Those who advocate wider use of intermediate sanctions disagree. They contend that community service should not be regarded as merely a "slap on the wrist." Rather, it is a constructive and burdensome penalty "that is inexpensive to administer, that produces public value, and that can to a degree be scaled to the seriousness of crimes."[65] Advocates of restorative justice agree. They suggest that community service can be used "to restore victims to wholeness, to hold offenders accountable," and to repair "the damage to the community fabric resulting from crime."[66]

Monetary Penalties

Monetary penalties, such as fines, fees, and restitution to the victim, are frequently imposed on offenders convicted of misdemeanors and felonies in American courts. Every year, millions of offenders convicted of traffic offenses and less serious misdemeanors are ordered to pay fines that are to some degree calibrated to the seriousness of the crime. Offenders placed on probation are required to pay fees for probation supervision, substance abuse treatment, urinalysis, and use of electronic monitoring equipment. And offenders who steal or damage the property of others or cause physical or emotional injuries are ordered to pay restitution to the victims, often as a condition of probation.

Although these types of monetary penalties are common, their use as stand-alone sentences is not. Fines, for example, are seldom used as alternatives to incarceration for offenders convicted of less serious felonies; they are often used only in conjunction with probation for more serious misdemeanors. Although the fine is "unambiguously punitive,"[67] studies reveal that judges do not regard it as "a meaningful alternative to incarceration or probation."[68] The use of fines also poses dilemmas for the courts. If the amount of a fine is commensurate with the seriousness of an offense, with no consideration given to the ability of the offender to pay the fine, then fines will be relatively more burdensome to the poor than to the rich. Low-income offenders will have a more difficult time coming up with money to pay the fines—indeed, they may resort to more crime to raise the money needed—and those who are unable to pay the fines may be jailed for nonpayment. Wealthier offenders, then, get double benefits: The fines they pay are less punitive punishments, and they evade jail because they have money to pay the fines.

To avoid these potential problems, a number of jurisdictions use what is referred to as a *day fine.* Imported from Europe and Latin America, the day fine is calibrated both to the seriousness of the offense and to the offender's ability to pay. Rather than requiring all offenders convicted of shoplifting to pay $1,000, for example, judges determine how many *punishment units* each offender deserves. Typically, each punishment unit is equal to 1 day's pay or some fraction of a day's pay. Two offenders convicted of shoplifting might each be ordered to pay five punishment units, equal to 5 days' pay. If one offender made $50 per day and the other earned $500 per day, the fines paid by the two offenders would differ. The first would pay $250, the second $2,500.

The day fine has obvious advantages over the traditional fine. Because it can be tailored more precisely to an offender's ability to pay, the day fine is more equitable. It is also more likely to be paid in full, which means that offenders are less likely to be called back to court or sentenced to jail for nonpayment. But there are disadvantages as well. Courts that use the day fine must define a unit of punishment in terms other than jail time or standard dollars to be paid, establish the range of units to be imposed on offenders convicted of various offenses, and devise a means of translating these units into dollars. Implementing a system of day fines, in other words, involves "changing—or, at least accommodating—existing habits, customs, and laws."[69]

Whereas fines, whether traditional or day fines, and fees go into state and federal government coffers, restitution is money paid by the offender to the victim or to a victim compensation fund.[70] In requiring restitution, courts attempt to calculate the costs of the crime to the victim and order the offender to repay those costs. Thus, victims might be compensated for financial losses, for medical expenses, or for the costs of traveling to and from the courthouse for criminal proceedings. Because not all offenders, particularly those who are incarcerated, are equally able to pay restitution, most jurisdictions require all offenders to pay fees to victim compensation funds. Victims can apply to these funds and obtain money to recover their losses. These funds, however, are "notoriously inadequate."[71] They do not cover all crime victims and they seldom pay the full costs of the victimization.

The Future of Intermediate Sanctions

More than a decade ago, Attorney General Richard Thornburgh called for the development of a "portfolio of intermediate punishments." As he pointed out,

> We . . . know that there are many for whom incarceration is not appropriate. But is simple probation sufficient? Particularly when probation officers are carrying caseloads far beyond what is manageable? We need to fill the gap between simple probation and prison. We need intermediate steps, intermediate punishments.[72]

Others have made similar claims and predicted that the repertoire of punishments advocated by the attorney general would be developed and implemented (see Box 19.2). In 1990, for example, Morris and Tonry made the following prediction:

> Concerns for justice and fairness in sentencing will lead, in time, probably within 25 years, to the creation in most American states of comprehensive systems of structured sentencing discretion that encompass a continuum of punishments from probation to imprisonment, with many intermediate punishments ranged between.[73]

Box 19.2   Innovation (or Bias) in the Courtroom?

In 2000, Dayne Brink appeared before Douglas County (Nebraska) Judge Lyn White for violating his probation in a domestic violence case. While on probation, Brink allegedly violated the protection order obtained by his former wife. He was arrested for driving while intoxicated and carrying a concealed weapon and failed to attend a mandatory batterers' intervention program.

Judge White sentenced him to the maximum possible sentence: 180 days in jail and a $1,000 fine. She also ordered Brink to read Shakespeare's tragic play *Othello,* in which a jealous husband kills his wife. When questions were raised about the propriety of the sentence, Judge White stated that requiring Brink to read *Othello* was appropriate because it was "rationally related to the defendant's crime."

Brink appealed his sentence, arguing that it was excessive and that Judge White was biased against him. District Court Judge Joseph Troia vacated the sentence and sent the case back to County Court for resentencing by a different judge. In the decision, Troia wrote, "Although the sentence fell within the statutory parameters, the court finds that the trial court's actions and comments could cause a reasonable person to question the impartiality of the judge."

SOURCE: "Judge Takes Self Off Cases," *Omaha World Herald,* 2001, February 13, pp. A1-2.

As we have seen, Morris and Tonry may have been overly optimistic about the future of intermediate sanctions. As we enter the 21st century, prison, jail, and probation sentences remain the dominant forms of punishment imposed on criminal offenders, especially those convicted of felonies. In fact, state and federal prison statistics suggest that the trend has been the *opposite* of what Thornburgh called for and Morris and Tonry predicted. As shown in Exhibit 19.3, the rate of incarceration, which was relatively stable from 1925 through 1975, skyrocketed from 1980 to 1998. Despite stable or falling crime rates, the number of offenders incarcerated in state and federal prisons ballooned from 329,821 in 1980 to 1,284,894 in 1999, an increase of nearly 400 percent. At the turn of the century, 6.3 million people, 3.1 percent of all adult residents, were on probation, in jail or prison, or on parole in the United States.[74] Rather than replacing incarceration and probation with "a portfolio of intermediate punishments," the United States "has been engaged in an unprecedented imprisonment binge"[75] during the past two decades.

The question, of course, is why this is so. Austin and Irwin suggest that it can be attributed in large part to the public's fear of drugs and crime, which led to a movement for more punitive sentences for felony offenders in general and for drug offenders in particular.[76] Tonry argues that it also reflects a view on the part of judges and policymakers that "only imprisonment counts" as punishment. He suggests that this emphasis on absolute severity "frustrates efforts to devise intermediate sanctions for the psychological (not to mention political) reason that few other sanctions seem commensurable with a multiyear prison sentence."[77] If, in other words, the philosophy of just deserts demands that felony offenders be punished harshly and only incarceration is regarded as harsh punishment, then nonincarceration sentences will be viewed as inappropriately lenient. Unless these attitudes change, it seems unlikely that intermediate sanctions will play anything more than a supporting role in sentencing policies and practices in the United States.

SENTENCING AS A COLLABORATIVE EXERCISE

Seated on a raised bench, clothed in black robes, and wielding a gavel that he uses to maintain order in the courtroom, the judge pronounces sentence on the defendant, who has been found

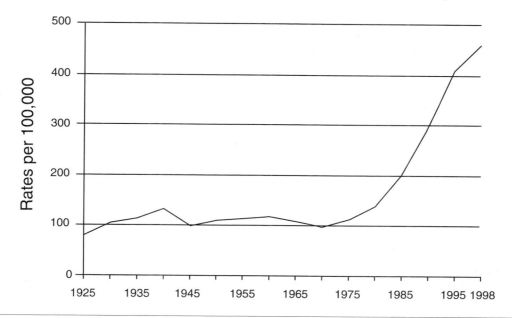

Exhibit 19.3   Incarceration Rates, 1925-1998

SOURCE: Adapted from Figure 1.1, James Austin and John Irwin, 2001. *It's About Time: America's Imprisonment Binge,* 3rd Ed. Belmont, CA: Wadsworth.

guilty of robbery and sexual assault. "After considering all of the facts and circumstances in the case," he states, "I have decided to impose the maximum sentence: 15 years in prison. I believe that this is the appropriate sentence given the heinousness of the crime and the fact that the defendant has a prior conviction for aggravated assault."

Scenes such as this are played out everyday in courtrooms throughout the United States. Media accounts of these sentencing proceedings report, *"The judge* sentenced the defendant to prison," *"The judge* gave the defendant 3 years probation and ordered him to pay a $2,000 fine," or *"The judge* sentenced the defendant to death." Although accurate, these accounts are misleading. The judge clearly plays an important role in the sentencing process—in many cases, the lead role—but others play supporting roles. The sentences defendants eventually receive are produced by a collaborative exercise involving legislators, prosecutors, jurors, probation officials, trial court judges, corrections officials, and possibly appellate court judges.

In the following sections, we discuss "the many participants and decisions that together constitute 'sentencing.'"[78] These decisions are summarized in Box 19.3. We first describe how decisions made by Congress and state legislatures structure the sentencing process. We then discuss the role played by criminal justice officials other than the judge. We assume that a crime has been reported, a suspect has been arrested, and the case has been forwarded to the prosecuting attorney.

Legislative Restrictions on Judicial Discretion

Congress and state legislatures make the basic policy decisions regarding sentencing. As explained above, these bodies decide whether the sentences imposed by judges will be indeterminate, determinate, mandatory, or structured by sentencing guidelines. They decide whether and under what conditions the death penalty can be imposed, the circumstances under which probation is the preferred sentence, and to some extent, the types of intermediate sanctions that are available. They also decide, for each offense or class of offenses, what the penalty range will be.

Box 19.3   Sentencing as a Collaborative Exercise

Legislature

- Makes the basic policy decisions regarding sentencing
- Decides whether and how to use capital punishment
- Determines the types of sentences that are to be imposed
- Specifies penalties for the offense or classes of the offenses

Prosecuting Attorney

- Decides whether to file charges and what charges to file
- Engages in plea bargaining on the charges and sentence
- Recommends a sentence

Jury

- Decides whether to acquit or convict
- Decides whether to convict of original or lesser included charge
- Recommends or (rarely) determines sentence

Probation Officer

- Conducts the presentence investigation
- Recommends a sentence to the judge
- Supervises offenders on probation

Judge

- Has final responsibility for determining the sentence

Corrections Officials

- Determine where prison sentence will be served
- Reduce sentence by awarding good-time credits
- Determine date of release on parole

Appellate Court Judges

- Decide whether to overturn conviction and/or sentence

These decisions obviously constrain the discretion of judges. If, as is the case in 12 states, the state legislature has decided not to enact a statute authorizing capital punishment, a judge in that state cannot sentence an offender to death. Even if he or she believes that capital punishment is an appropriate penalty and is convinced that the offender deserves to die, the judge must abide by the legislature's decision. If the legislature has decided that sentences are to be indeterminate, the judge cannot disregard the law and impose a fixed term of years. Conversely, in states with determinate sentencing, the judge cannot impose a minimum and maximum sentence and order corrections officials to determine when the offender is fit to be released. And in states with mandatory minimum sentences or sentencing guidelines, there is an expectation that judges will comply or explain their reasons for noncompliance.

The laws passed by legislative bodies generally place tighter restrictions on judges' discretion in more serious cases and fewer restrictions in less serious cases. This is particularly true with respect to the decision to sentence the offender to prison or not. Offenses regarded as particularly serious or offenders deemed especially dangerous may be nonprobationable. The Illinois Criminal Code, for example, states that a variety of options (probation, imprisonment, an order directing the offender to repair the damage caused by the crime, a fine, and/or an order directing the offender to make restitution to the victim) are appropriate for "all felonies or misdemeanors." The statute also lists 15 offenses for which "[a] period of probation, a term of periodic imprisonment, or conditional discharge shall not be imposed." These nonprobationable offenses include the following: first-degree murder, any Class X felony (the most serious classification), possession of more than 5 grams of cocaine, residential burglary, aggravated battery of a senior citizen, a forcible felony related to the activities of an organized gang, carjacking, and a second or subsequent conviction for a hate crime.[79]

Judges can circumvent, or at least attempt to circumvent, some of these legislative restrictions in several ways. For example, if a defendant charged with a nonprobationable offense is being tried by a judge, he or she can find the defendant guilty of a lesser included offense that will not trigger the mandatory sentence. Similarly, judges can impose sentences that fall outside the penalty range or guidelines enacted by the legislature and then state, either from the bench or in writing, why they decided to depart. If the defendant is convicted of a collateral charge, such as use of a weapon, which carries a mandatory sentence enhancement, the judge can reduce the sentence for the main charge so the overall sentence is no longer than it would have been if the defendant had been convicted only of the main charge. Generally, however, as U.S. District Court Judge Irving R. Kaufman wrote in 1960, "The judge must take this legislative guide and apply it to the particular circumstances of the case before him. He must work within the legislative formula, even if he does not agree with it."[80]

The Role Played by the Prosecutor

All decision makers in the American criminal justice system have a significant amount of unchecked discretionary power, but the prosecutor stands apart from the rest. The prosecutor decides who will be charged, what charge will be filed, who will be offered a plea bargain, and the type of bargain that will be offered. The prosecutor may also recommend the offender's sentence. As Supreme Court Justice Jackson noted in 1940, "the prosecutor has more control over life, liberty, and reputation than any other person in America."[81]

None of the discretionary decisions made by the prosecutor is more critical than the initial decision to prosecute or not, which has been characterized as "the gateway to justice."[82] Prosecutors have wide discretion at this stage in the process. There are no legislative or judicial guidelines on charging, and a decision not to file charges is ordinarily immune from review. As the Supreme Court noted in *Bordenkircher v. Hayes*,

> So long as the prosecutor has probable cause to believe that the accused committed an offense defined by statute, the decision whether or not to prosecute, and what charge to file or bring before a grand jury generally rests entirely in his discretion.[83]

The discretion exercised by the prosecutor at charging has obvious implications for sentencing. If the prosecutor decides not to file charges, either because he or she believes that the defendant is innocent or that the defendant is guilty but a conviction is unlikely, the case is closed; the defendant will not be convicted or sentenced. If the prosecutor does file charges, the number and seriousness of the charges filed may affect the severity of the sentence imposed by the judge if the defendant is convicted. The death penalty is the most striking example of this. If the prosecutor decides not to request the death penalty, the death penalty will not be an option at sentencing. Conversely, if the prosecutor does seek the death penalty, the defendant may in fact be sentenced to death. The sentences imposed in noncapital cases can similarly depend on the charges filed. Suppose, for example, that the police in Chicago arrest a woman with 6 grams of cocaine in her purse, which, as we have seen, is a nonprobationable offense in Illinois. The prosecutor could charge her with possession of some lesser amount, thereby avoiding the mandatory prison sentence. Sentence severity may also be affected by the prosecutor's decision to charge the defendant as a habitual offender, to add weapons charges that enhance the sentence for the main offense, or to charge the defendant with multiple counts of the same offense.

Plea Bargaining and Its Effect on Sentence Outcomes

Another way the prosecutor affects sentencing is through plea bargaining.[84] The predominance of jury trials in television dramas such as *Law and Order* and *The Practice* not withstanding, most convictions result from guilty pleas. In 1996, for instance, 91 percent of all felony convictions in state courts were the result of guilty pleas.[85] The rate was even higher—94 percent—for defendants convicted of felonies in federal courts.[86] Although the actual number of guilty pleas that result from plea bargains is unknown, most experts would argue that some type of negotiation between the prosecutor and the defendant (or the defense attorney) occurs prior to the entry of the guilty plea.

There are several different forms of plea bargaining, and they can be used alone or in combination with one another. Consider the following scenario: A 19-year-old man holds a gun to the head of a convenience store clerk, demands money, and escapes on foot with a bag containing more than $1,000. He is spotted and arrested by the police 30 minutes later. The prosecutor reviewing the case files one count of aggravated robbery, punishable by 3 to 10 years in prison, and one count of use of a weapon during the commission of a felony, which adds an additional 2 years to the sentence. The plea negotiations in this case might center on the charges filed. If the defendant agrees to plead guilty, the prosecutor might reduce the aggravated robbery charge to a less serious charge of robbery or dismiss the weapons charge. Both types of charge reductions would reduce the potential sentence the defendant faces.

The plea negotiations might also revolve around the sentence. In exchange for a guilty plea, the prosecutor might agree to a sentence of 3 years on the aggravated robbery and 2 years on the weapons charge, with the sentences to be served concurrently rather than consecutively. Alternatively, the prosecutor might agree "to stand mute at sentencing" or agree to a "sentence lid." In the first instance, he or she would not recommend any particular sentence or challenge the defense attorney's presentation of mitigating evidence and recommendation for leniency. In other words, the prosecutor would say nothing about the sentence the defendant should receive. In the second instance, the prosecutor would recommend that the judge impose a sentence that would not exceed the sentence lid. In this case, for example, he or she might state that the sentence should be no greater than 3 years in prison.

Both charge reductions and sentence agreements limit the judge's options at sentencing. Judges have little recourse if the prosecutor decides to reduce the number or the severity of the charges in exchange for a guilty plea. Because the charging decision "generally rests entirely in his [the prosecutor's] discretion,"[87] the judge ordinarily cannot refuse to accept the plea to a reduced charge and force the defendant to go to trial. In *United States v. Ammidown*, for example, the U.S. Court of Appeals ruled that a judge "had exceeded his discretion"[88] by rejecting a plea agreement because he believed that the "public interest" required the defendant to be tried on a greater charge. Although the justices stated that the trial court should not "serve merely as a rubber stamp for the prosecutor's decision,"[89] they ruled that judges cannot reject agreements reached between the prosecution and defense unless they determine that the prosecutors have abused their discretion. Moreover, the court said,

> The question is not what the judge would do if he were the prosecuting attorney, but whether he can say that the action of the prosecuting attorney is such a departure from sound prosecutorial principle as to mark it an abuse of prosecutorial discretion.[90]

Sentence agreements also reduce the judge's discretion, even though the prosecutor does not have any official authority to impose sentence. If, for example, the prosecution and the defense negotiate an agreement whereby the defendant agrees to plead guilty and the state agrees that a probation sentence is the appropriate disposition, the judge must either (a) accept the plea agreement and place the defendant on probation or (b) reject the agreement and allow the defendant to withdraw his guilty plea.[91] As the Supreme Court stated in the case of *Santobello v. New York*, "When a plea rests in any significant degree on a promise or agreement of the prosecutor, so that it can be said to be part of the inducement or consideration, such promise must be fulfilled."[92] Moreover, judges face organizational pressure to approve plea agreements. Like other members of the courtroom workgroup, they view the guilty plea as an efficient and effective method of case disposition. As a result, they are unlikely to reject sentence agreements that make a high rate of guilty pleas possible.

In summary, charging and plea-bargaining decisions made by prosecuting attorneys have significant effects on sentence outcomes. The charges filed by the prosecutor determine the limits of the defendant's legal liability, and concessions made during plea negotiations may reduce the sentence.

The Jury's Role in Sentencing

As we have seen, most criminal convictions result from guilty pleas. Very few defendants exercise their right to a trial, and many of those who do are tried by judges, not juries. In 1993, for example, 90 percent of all offenders convicted of felonies in Cook County (Chicago), Illinois, pled guilty; 9 percent were tried by a judge, and less than 1 percent were tried by a jury.[93] Jury trials comprised a somewhat larger proportion of all dispositions in U.S. District Courts in 1998: 94 percent of these federal defendants pled guilty, 5 percent had a jury trial, and 1 percent had a bench trial.[94]

Although the rarity of jury trials limits the jury's overall influence on sentencing, decisions made by the jury do affect sentence outcomes in cases that are tried before juries. The jury decides whether to convict the defendant or not. The jury may also decide whether to convict the defendant for the offense charged or for a lesser included offense. In a murder case, for example, the jury might have the option of finding the defendant guilty of first-degree murder, second-degree murder, or manslaughter. Like the prosecutor's charging decisions, these conviction decisions affect the sentences that will be imposed. Defendants charged with first-degree murder but convicted of manslaughter will be sentenced more leniently than if they were convicted of the more serious charge.

Most jury trials result in convictions. For example, in 1996, in the 75 largest counties in the United States, 84 percent of all defendants tried by juries were convicted.[95] A jury's decision to acquit the defendant usually means that the state has failed to prove its case beyond a reasonable doubt. Sometimes, however, the jury votes to acquit despite overwhelming evidence that the defendant is guilty. In this case, the jury ignores, or nullifies, the law.

Jury nullification, which has its roots in English common law, occurs when a juror believes that the evidence presented at trial establishes the defendant's guilt, but nonetheless votes to acquit. The juror's decision may be motivated either by a belief that the law under which the defendant is being prosecuted is unfair or by an objection to the application of the law in a particular case. In the first instance, a juror might refuse to convict a defendant charged in U.S. District Court with possession of more than 5 grams of crack cocaine because

he or she believes that the long prison sentence mandated by the law is unfair. In the second instance, a juror might vote to acquit a defendant charged with petty theft and also charged as a habitual criminal and facing a mandatory life sentence—not because the juror believes the law itself is unfair but because he or she believes that this particular defendant does not deserve life in prison.[96] Similarly, an African American juror might heed Paul Butler's call for "racially based jury nullification" (see Box 19.4) and refuse to convict African Americans charged with drug offenses and less serious property crimes.[97]

Jurors clearly have the power to nullify the law and vote their conscience.[98] If the jury votes to acquit, the double jeopardy clause of the Fifth Amendment prohibits reversal of the jury's decision. The jury's decision to acquit, even in the face of overwhelming evidence of guilt, is final and cannot be reversed by the trial judge or an appellate court. In most jurisdictions, however, jurors do not have to be told that they have the right to nullify the law.[99]

The jury's effect on sentence outcomes is not limited to its power to acquit the defendant of all charges or convict the defendant of lesser included charges. In a number of jurisdictions, the jury either determines the sentence or makes a sentence recommendation to the judge. As explained earlier, juries play a prominent role in sentencing in capital cases: In 6 states, the jury makes a sentencing recommendation to the judge; in 6 states, the jury decides the sentence but the judge can alter it; and in 18 states, the jury makes the final sentencing decision. The jury's role is more circumscribed in noncapital cases. Texas is the only state in which the jury determines the final sentence, and there are only 3 additional states (Arkansas, Missouri, and Virginia) in which the jury determines the original sentence but the sentence can be altered by the trial judge.[100] This reluctance to entrust the sentencing function to the jury in noncapital criminal cases may reflect legislators' fears that doing so would lead to inconsistent or wildly disparate sentences for similar offenders. Unlike the capital sentencing process, which is tightly constrained by death penalty statutes, the noncapital sentencing process is more discretionary.

Box 19.4   Racially Based Jury Nullification

In a provocative essay published in the *Yale Law Journal* shortly after O.J. Simpson's acquittal on murder charges, Paul Butler, an African American professor at George Washington University Law School, called for "racially based jury nullification." Arguing that there are far too many young black men in prison, Butler urged African American jurors to refuse to convict black defendants charged with nonviolent crimes, regardless of the strength of evidence arrayed against them. According to Butler, the "black community is better off when some nonviolent law-breakers remain in the community rather than go to prison" (p. 677).

Butler asserted that black jurors have a moral responsibility "to emancipate some guilty black outlaws" (p. 679). He argued that the criminal justice system is racially biased and suggested that white racism, which "creates and sustains the criminal breeding ground which produces the black criminal" is the underlying cause of much of the crime committed by blacks. Because, in other works, the operation of the criminal law does not reflect or advance the interests of black people, jurors can—and should—"opt out of American criminal law" (p. 714). They should do this by generally refusing to convict blacks charged with nonviolent victimless crimes such as possession of drugs and, depending on the circumstances, sometimes refusing to convict blacks charged with property offenses and more serious drug trafficking offenses. According to Butler, "The race of a black defendant is sometimes a legally and morally appropriate factor for jurors to consider in reaching a verdict of not guilty" (p. 679).

SOURCE: "Racially Based Jury Nullification: Black Power in the Criminal Justice System," by Paul Butler, 1995. *Yale Law Journal, 105.*

The Probation Department and the Presentence Investigation

Prior to sentencing an offender, the judge may order a presentence investigation. In most jurisdictions, the probation department conducts the investigation into the circumstances of the offense and the background of the offender and reports its findings and sentencing recommendation to the judge. Probation officers may interview the offenders, the offenders' families, other persons familiar with the offenders, and the victims. They will also gather information on offenders' prior criminal records, education, employment history, and family circumstances.[101]

The purpose of the presentence investigation is to provide the judge with more detailed information about the crime and the offender. Because most offenders plead guilty, the judge may have relatively little information about the case. In fact, he or she may know nothing more than the minimal facts necessary to support acceptance of the guilty plea. The presentence investigation report typically describes the nature of the offense, the role played by the offender in the offense, the offender's past and current circumstances, and the effect of the crime on the victim and the community. The report may also include the officer's assessment of the offender's remorse (or lack thereof) and risk of future criminal offending. In some jurisdictions, the assessment of the offender's likelihood of recidivism is based on the officer's subjective impressions; in others, it is based on a risk-screening instrument. On the basis of the information gathered, the officer indicates whether the offender is a good candidate for probation or for some other alternative to incarceration.

Studies reveal that judges do follow the sentencing recommendations in presentence reports.[102] One study, for example, found 95 percent agreement between the probation officer's recommendation and the judge's sentence when the officer recommended probation; the rate of agreement was 88 percent when the officer recommended against probation.[103] Researchers disagree, however, about the meaning of these findings. Some argue that probation officers play a key role in the sentencing process and conclude that judges "lean heavily" on their professional advice.[104] Others contend that the probation officer's recommendation typically matches the recommendation of the prosecutor. They assert that the probation officer is "largely superfluous"[105] and plays little more than a "ceremonial" role in the sentencing process.[106]

Rosecrance, who examined the presentence investigation process in California, is even more critical. Based on interviews with probation officers in two California counties, he concluded that "probation presentence reports do not influence judicial sentencing significantly but serve to maintain the myth that criminal courts dispense individualized justice."[107] He found that probation officers based their recommendations primarily on the seriousness of the offense and the offender's prior record. Early in the process, they used these two factors to "type" the defendant—that is, to place the defendant into a dispositional category: diversion, probation, prison, or some other outcome. After determining the appropriate disposition in the case, the probation officer then gathered other information that could "legitimate" this decision. As Rosecrance noted, "Probation officers do not regard defendant typings as tentative hypotheses to be disproved through inquiry but rather as firm conclusions to be justified in the body of the report."[108]

Rosecrance contends that the primary purpose of the presentence report is to give the appearance of individualized justice. As he states, "Although judges do not have the time or the inclination to consider individual variables thoroughly, the performance of a presentence investigation perpetuates the myth of individualized sentences."[109] This suggests that the effect of the presentence report on sentencing is illusory rather than real.

Corrections Officials and Sentencing

The casual observer of the American criminal justice system probably assumes that the sentencing process ends when the judge pronounces sentence from the bench and the sentence imposed is the sentence that will be served. As we have seen, these assumptions are not necessarily correct. The sentence that the offender serves may be "refined" by corrections officials, who play a number of important roles in the sentencing process.

We already have discussed the role played by probation officers, who conduct the presentence investigation and supervise offenders placed on probation. Corrections officials also make a number of decisions regarding offenders sentenced to prison. In many states, they decide where the offender will serve his or her sentence: in a maximum-, medium-, or minimum-security institution or in a correctional facility in the community. In making this decision, they use the information contained in the presentence investigation report, as well as diagnostic and risk-assessment tools that they administer after the offender arrives in the prison system.

In addition, corrections officials can award or withhold good-time credits, which reduce the amount of time the offender will serve. Good-time credits are one way for corrections officials to control inmates. Those who abide by the rules will be awarded the maximum time off, whereas those who violate the rules may have some or all of their credits taken away. Offenders in a few states do not earn good-time credits. In other states, only those convicted of less serious offenses are eligible for time off for good behavior. The rate at which offenders earn good time also varies. In a number of states, good time accrues at a rate of 1 day off for every day served. In other states, offenders can earn a specified number of days off for each month served. Offenders in Maine and Nevada, for example, can earn up to 10 days per month, and those in Massachusetts earn from 2.5 to 12.5 days per month.[110]

In a majority of the states, the parole board also plays an important role in the sentencing process. In states with indeterminate sentencing, the judge sets the maximum (and sometimes the minimum) sentence the offender can serve, but the parole board determines the actual date of release. In states with determinate sentencing or sentencing guidelines, the parole board may or may not have the authority to release offenders prior to the expiration of their sentences. Congress, for instance, abolished discretionary parole release with the passage of the Sentencing Reform Act of 1984.[111] Federal offenders serve the entire sentence minus a maximum of 15 percent off for good behavior.

California's determinate sentencing statute also abolished parole release. Pennsylvania and Utah, on the other hand, adopted sentencing guidelines but retained discretionary release on parole.[112]

Although some offenders serve the maximum term, most are released on parole after serving a portion of the sentence. Offenders serving time in states with "truth-in-sentencing" laws, which attempt to reduce the discrepancy between the sentence imposed and the time served, will serve a larger proportion of their sentence than those incarcerated in states without such laws. As explained earlier, the average state inmate serves 45 percent of his or her sentence, whereas the typical federal inmate serves 85 percent. During the 1990s, the number of adults on parole grew by 32.7 percent. There were 531,407 parolees in 1990 and 704,964 in 1998.[113]

The decision to release the offender prior to the expiration of the sentence is made by the parole board. Today, most parole boards are not directly linked to or controlled by the staff of the correctional institution. They are either independent bodies whose members are appointed for fixed terms or consolidated boards that include corrections officials, representatives of social service agencies, and private citizens. The members of the parole board make individualized release decisions, taking into consideration such factors as the seriousness of the offense, the offender's past criminal history, the offender's behavior while incarcerated, and, in the case of a violent crime, the wishes of the victim or the victim's survivors. In many jurisdictions, the board uses written parole guidelines that structure the release decision.[114]

Like probationers, parolees are released into the community and are required to abide by a set of conditions designed to ensure that they do not return to a life of crime. They may be required, for example, to meet regularly with their parole officers, maintain employment, enroll in educational or vocational training, continue substance abuse treatment, and abide by curfews and other restrictions on their behavior. Offenders who are rearrested or who otherwise violate the conditions of their release may be returned to prison to serve the balance of the sentence.

The Role of Appellate Courts

Appellate courts can alter the outcomes of criminal cases by overturning offenders' convictions or sentences. Although offenders do not have a constitutional right to appeal their convictions, every jurisdiction has created a statutory right to appeal to one higher court. The purpose of this is to ensure that proper procedures were followed by police, the prosecutor, the judge, and the jury. A state court defendant who believes that his conviction was obtained improperly can appeal his conviction to the intermediate appellate court or, in states that do not have a two-tiered appellate court system, to the state supreme court. Similarly, a defendant who has been tried and convicted in U.S. District Court can appeal his conviction to the U.S. Court of Appeals. If the appellate court sustains the appeal and rules that procedures were violated, the court will overturn the conviction and send the case back to the trial court. The case then may be retried or dismissed. If the appellate court rules against the offender, he or she can appeal to the next highest court, but that court does not have to hear the appeal.

The ability of appellate courts to alter sentences imposed by trial court judges is more limited. The U.S. Supreme Court has ruled as follows:

> [That] review by an appellate court of the final judgement in a criminal case . . . is not a necessary element of due process of law, and that the right of appeal may be accorded by the state to the accused upon such conditions as the state deems proper.[115]

Although all states with death penalty statutes provide for automatic appellate review of death sentences, only half of the states permit appellate review of noncapital sentences that fall within statutory limits.[116] The standards for review vary. In some states, appellate courts are authorized to modify sentences deemed "excessive," but in other states only sentences determined to be "manifestly excessive," "clearly erroneous," or "an abuse of discretion" can be altered.[117] A defendant sentenced under the federal sentencing guidelines can appeal a sentence that is more severe than the guidelines permit. Federal law also allows the government to appeal a sentence that is more lenient than provided for in the guidelines. If an offender appeals his or her sentence and the appeal is sustained, the sentence must be corrected. An appellate court decision to vacate the sentence does not mean, however, that the offender will escape punishment. As the Supreme Court stated in 1974, "The Constitution does not require that sentencing should be a game in which a wrong move by the judge means immunity for the prisoner."[118] Thus, the case will be sent back to the trial court for resentencing.

The Sentencing *Process*

The sentence imposed on an offender who has been found guilty of a crime is the result of a collaborative exercise that involves legislators and criminal justice officials other than the judge. It results from a sentencing *process,* not from a single sentencing *decision.* Legislators determine the sentencing policies that judges are to apply. Prosecutors and jurors decide whether the offender will face punishment. Prosecutors and occasionally jurors also determine the charges on which the punishment will be based. Prosecutors, jurors, and probation officers may provide sentencing recommendations to the judge, and corrections officials supervise and monitor offenders and determine the amount of time that those who are incarcerated actually serve. Appellate court judges can overturn convictions and vacate sentences.

Although the decisions made by legislators and other criminal justice officials obviously limit the judge's options and constrain his or her discretion, the ultimate responsibility for determining the sentence does in fact rest with the judge. Sentencing guidelines and mandatory minimum sentencing provisions restrict but do not eliminate judicial discretion. Charge reductions by the prosecutor and convictions by the jury on lesser included charges may reduce the offender's potential sentence, and sentence recommendations may influence the judge's decision, but they do not dictate what the sentence will be. The judge determines who goes to prison and who does not and how long the

sentence will be. As Judge Irving R. Kaufman wrote, more than 40 years ago:

> If the hundreds of American judges who sit on criminal cases were polled as to what was the most trying facet of their jobs, the vast majority would almost certainly answer, 'sentencing.' In no other judicial function is the judge more alone; no other act of his carries greater potentialities for good or evil than the determination of how society will treat its transgressors.[119]

Some judges, particularly those at the federal level, might argue that the sentencing reforms enacted during the past 25 years have reduced their position from a leading role to a supporting role in the sentencing process, but most of them would agree that sentencing is, as Judge Kaufman put it, "the judge's problem."

Notes

1. Wice, Judges and Lawyers: *The Human Side of Justice,* p.273.
2. "Skier Receives 90-Day Term In Fatal Crash; Hall Remorseful, Plans Appeal," Steve Lipsher, Denver Post, February 1, 2001, p. A-01.
3. The following states do not have capital punishment statutes: Alaska, Hawaii, Iowa, Maine, Massachusetts, Michigan, Minnesota, North Dakota, Rhode Island, Vermont, West Virginia, and Wisconsin.
4. In 1977, the Supreme Court ruled that the death penalty was a disproportionately severe penalty for the rape of an adult woman [Coker v. Georgia, 433 U.S. 584 (1977)].
5. Federal Bureau of Investigation, Uniform Crime Reports, Crime in the United States, 1999, Table 29.
6. U.S. Department of Justice, Bureau of Justice Statistics (BJS), Capital Punishment 1999, p. 1.
7. Ibid., pp. 11-12.
8. In 1972, the U.S. Supreme Court struck down federal and state laws that permitted wide discretion in the imposition of the death penalty. In the case of Furman v. Georgia [408 U.S. 153 (1972)], the Court characterized death penalty decisions under these laws as "arbitrary and capricious" and ruled that they constituted cruel and unusual punishment in violation of the Eighth Amendment. In the wake of this decision, most states enacted new death penalty statutes. These statutes were of two types: those that required the death penalty for all offenders convicted of specified crimes and those that allowed the death penalty to be imposed depending on the presence of

certain aggravating circumstances. In 1976, the Court struck down the mandatory death sentence statutes [Woodson v. North Carolina, 428 U.S. 280 (1976); Roberts v. Louisiana, 428 U.S. 325 (1976)], but upheld the guided discretion statutes [Gregg v. Georgia, 428 U.S. 153 (1972); Jurek v. Texas, 428 U.S. 262 (1976); Proffitt v. Florida, 428 U.S. 242 (1976)].

9. U.S. Department of Justice, BJS, Capital Punishment 1999, pp. 2-4.

10. Florida Statutes, 2000, Title XLVII, chap. 921.141.

11. Texas Statutes, Revised 2000, Title 3, § 12.03 to 12.23.

12. U.S. Department of Justice, BJS, National Assessment of Structured Sentencing, pp. 21-23. See also, Tonry, Sentencing Matters, chap. 5.

13. See, for example, Shichor and Sechrest, *Three Strikes and You're Out: Vengeance as Public Policy*; and *U.S. Sentencing Commission, Mandatory Minimum Penalties in the Federal Criminal Justice System*.

14. Heumann and Loftin, "Mandatory Sentencing and the Abolition of Plea Bargaining: The Michigan Felony Firearms Statute."

15. The Bureau of Justice Assistance conducted a national survey of sentencing practices in the United States in 1993 and 1994. They classified 29 states and the District of Columbia as having "primarily" indeterminate sentencing. (A few states use indeterminate sentences for some classes of offenses and determinate sentences for others.) See U.S. Department of Justice, Bureau of Justice Assistance, National Assessment of Structured Sentencing, pp. 19-21.

16. Ibid. The Bureau of Justice Assistance survey revealed that five states—Arizona, California, Illinois, Indiana, and Maine—used determinate sentencing.

17. Ibid. The Bureau of Justice Assistance categorized 16 states as having sentencing guidelines. Of these, 10 had presumptive guidelines and 6 had advisory or voluntary guidelines.

18. U.S. Department of Justice, BJS, Felony Sentences in the United States, 1996.

19. This is in contrast to a "mixed" or "split" sentence, which involves a term of incarceration followed by period of probation.

20. U.S. Department of Justice, BJS, Felony Sentences in the United States, 1996, Table 4.

21. Nebraska Statutes, 29-2260.

22. U.S. Department of Justice, BJS, Characteristics of Adults on Probation, 1995, p. 7.

23. Ibid., Table 12.

24. Ibid., Table 13.

25. Petersilia, Turner, Kahan, and Peterson, Granting Felons Probation: Public Risks and Alternatives.

26. Ibid., p. 45.

27. Petersilia, Expanding Options for Criminal Sentencing, p. 4.

28. Morris and Tonry, Between Prison and Probation: Intermediate Punishments in a Rational Sentencing System; Petersilia, Expanding Options for Criminal Sentencing.

29. von Hirsch and Ashworth, Principled Sentencing, p. 325.

30. See Petersilia and Turner, "Intensive Probation and Parole" and U.S. General Accounting Office, Intermediate Sanctions: Their Impacts on Prison Crowding, Costs, and Recidivism Are Still Unclear.

31. Petersilia and Turner, "Intensive Probation and Parole."

32. For a discussion of the "piling up" of intermediate sanctions and its effect on failure rates, see Blomberg and Lucken, "Intermediate Punishment and the Piling Up of Sanctions."

33. Petersilia, Expanding Options for Criminal Sentencing, pp. 30-31.

34. Morris and Tonry, Between Prison and Probation, p. 225.

35. Ibid.

36. Clear and Hardyman, "The New Intensive Supervision Movement."

37. Ibid., pp. 182-183.

38. Tonry, Sentencing Matters, p. 101.

39. Ibid.

40. See, for example, Austin and Irwin, It's About Time: America's Imprisonment Binge; Beckett and Sasson, The Politics of Injustice; Mauer, Race to Incarcerate; Morris and Tonry, Between Prison and Probation; Tonry, Sentencing Matters.

41. Morris and Tonry, Between Prison and Probation, chap. 2.

42. Ibid., p. 3.

43. Ibid., p. 32.

44. Austin and Irwin, It's About Time.

45. Ibid., p. 245.

46. Ibid., p. 245.

47. Ibid., p. 246.

48. For a detailed description of correctional boot camps, see Correctional Boot Camps: A Tough Intermediate Sanction, edited by MacKenzie and Hebert.

49. Ibid., p. vii.

50. U.S. Department of Justice, BJS, State Court Organization 1998, Table 47.

51. MacKenzie and Hebert, Correctional Boot Camps, p. xi. MacKenzie and Parent, "Shock Incarceration and Prison Crowding in Louisiana."

52. MacKenzie and Brame, "Shock Incarceration and Positive Adjustment During Community Supervision"; MacKenzie, Brame, McDowall, and Souryal, "Boot Camp Prison and Recidivism in Eight States."

53. MacKenzie, Brame, McDowall, and Souryal, "Boot Camp Prison and Recidivism," p. 351.

54. Morris and Tonry, Between Prison and Probation, p. 213.

55. National Law Enforcement Corrections Technology Center, Keeping Track of Electronic Monitoring, p. 1.

56. U.S. Department of Justice, BJS, Bureau of Justice Statistics, Correctional Populations in the United States, 1997.

57. Renzema, "Home Confinement Programs: Development, Implementation, and Impact," p. 49.

58. Lilly, Ball, Curry and McMullen, "Electronic Monitoring of the Drunk Driver: A Seven-Year Study of the Home Confinement Alternative."

59. Baumer and Mendelsohn, "Electronically Monitored Home Confinement: Does It Work?" p. 66.

60. Morris and Tonry, Between Prison and Probation, p. 152.

61. A detailed discussion of the program can be found in McDonald, Punishment Without Walls: Community Service Sentences in New York City.

62. Ibid.

63. Caputo, "Community Service for Repeat Misdemeanor Offenders in New York City."

64. Tonry, Sentencing Matters, p. 121.

65. Hahn, Emerging Criminal Justice: Three Pillars for a Proactive Justice System, p. 147.

66. McDonald, Greene, and Worzella, "Day Fines in American Courts: The Staten Island and Milwaukee Experiments," p. 1.

67. Cole, Mahoney, Thornton, and Hanson, as cited in Morris and Tonry, Between Prison and Probation, p. 127. Regarding Fines as a Criminal Sanction.

68. McDonald, Greene, and Worzella, Day Fines in American Courts, p. 5.

69. For a detailed discussion of the philosophical justification of restitution as a goal of punishment, see Abel and Marsh, Punishment and Restitution.

70. Clear, Harm in American Penology: Offenders, Victims, and Their Communities, p. 129.

71. Thornburgh, Opening Remarks.

72. Morris and Tonry, Between Prison and Probation, p. 19.

73. U.S. Department of Justice, BJS, Prisoners in 1999.

74. Austin and Irwin, It's About Time, p. 1.

75. Ibid., pp. 4-7.

76. Tonry, Sentencing Matters, p. 128.

77. Blumstein, Cohen, Martin, and Tonry, Research on Sentencing: The Search for Reform, Vol. I, p. 41.

78. Illinois Compiled Statutes Annotated, 730 ILSC § 5-5-3 (1996).

79. Kaufman, "Sentencing: The Judge's Problem," at http://www.theatlanticmonthly/unbound/flashbks/death/kaufman.htm

80. Davis, Discretionary: A Preliminary Inquiry, p. 190.

81. Kerstetter, "Gateway to Justice: Police and Prosecutorial Response to Sexual Assaults Against Women," p. 182.

82. Bordenkircher v. Hayes, 434 U.S. 357, 364 (1978).

83. For a detailed discussion of the guilty plea process, see Nardulli, Eisenstein, and Flemming, The Tenor of Justice: Criminal Courts and the Guilty Plea Process.

84. U.S. Department of Justice, BJS, State Court Sentences of Convicted Felons, 1996, Table 4.2.

85. U.S. Department of Justice, BJS, Federal Criminal Case Processing, 1998, Table 4.

86. Bordenkircher v. Hayes, 434 U.S. 357, 364 (1978). United States v. Ammidown, 162 U.S. App. D.C. 28, 497 F.2d 615 (1973).

87. Ibid.

88. Ibid. However, it should be noted that other appellate courts have adopted a less restrictive standard. In United States v. Bean [564 F.2d 700 (5th Cir. 1977)], the Court of Appeals for the 5th Circuit ruled that "A decision that a plea bargain will result in the defendant's receiving too light a sentence under the circumstances of the case is a sound reason for the judge's refusing to accept the agreement."

89. Rule 11 of the Federal Rules of Criminal Procedure, which applies to cases adjudicated in federal court, states the following:

If the court accepts the plea agreement, the court shall inform the defendant that it will embody in the judgment and sentence the disposition provided for in the plea agreement. . . . If the court rejects the plea agreement, the court shall, on the record, inform the parties of this fact, advise the defendant personally in open court . . . that the court is not bound by the plea agreement, afford the defendant the opportunity to then withdraw his plea, and advise the defendant that is he persists in this guilty plea. . . . The disposition of the case may be less favorable to the defendant than that contemplated by the plea agreement.

90. Santobello v. New York, 404 U.S. 257 (1971).

91. Spohn and DeLone, "When Does Race Matter? An Analysis of the Conditions Under Which Race Affects Sentence Severity."

92. U.S. Department of Justice, BJS, Compendium of Federal Justice Statistics, 1998, Table 4.2.

93. U.S. Department of Justice, BJS, Felony Defendants in Large Urban Counties, 1996, p. 26.

94. Dodge and Harris, "Calling a Strike a Ball: Jury Nullification and 'Three Strikes' Cases."

95. Butler, "Racially Based Jury Nullification: Black Power in the Criminal Justice System."

96. Scheflin and Van Dyke, "Merciful Juries: The Resilience of Jury Nullification."

97. See, for example, United States v. Dougherty, 473 F.2d 1113 (D.C. Cir., 1972).

98. U.S. Department of Justice, BJS, State Court Organization 1998, Table 46.

99. Clear, Clear, and Burrell, Offender Assessment and Evaluation.

100. Carter and Wilkins, "Some Factors in Sentencing Policy"; Hagan, "The Social and Legal Construction of Criminal Justice: A Study of the Presentence Process"; Hagan, Hewitt, and Alwin,

"Ceremonial Justice: Crime and Punishment in a Loosely Coupled System"; Kingsnorth and Rizzo, "Decision-Making in the Criminal Courts: Continuities and Discontinuities"; Rosecrance, "Maintaining the Myth of Individualized Justice: Probation Presentence Reports"; Walsh, "The Role of the Probation Officer in the Sentencing Process."

101. Carter and Wilkins, "Some Factors in Sentencing Policy."

102. Walsh, "The Role of the Probation Officer in the Sentencing Process," p. 363.

103. Kingsnorth and Rizzo, "Decision-Making in the Criminal Courts: Continuities and Discontinuities."

104. Hagan, Hewitt, and Alwin, "Ceremonial Justice: Crime and Punishment in a Loosely Coupled System."

105. Rosecrance, "Maintaining the Myth of Individualized Justice," p. 369.

106. Ibid., p. 380.

107. Ibid., p. 381.

108. U.S. Department of Justice, BJS, State Court Organization 1998, Table 50.

109. 18 U.S.C. §§ 3551-3626 and 28 U.S.C. §§ 991-998.

110. U.S. Department of Justice, BJS, State Court Organization 1998, Table 50.

111. U.S. Department of Justice, BJS, Probation and Parole in the United States, 1998, p. 1.

112. For a discussion of the development and implementation of parole guidelines, see Bottomley, "Parole in Transition: A Comparative Study of Origins, Developments, and Prospects for the 1990s," and Gottfredson, Wilkins, and Hoffman, Guidelines for Parole and Sentencing.

113. Murphy v. Com. of Massachusetts, 177 U.S. 155 (1900). U.S. Department of Justice, BJS, State Court Organization, Table 45.

114. Miller, Dawson, Dix, and Parnas, Prosecution and Adjudication, p. 1153.

115. Bozza v. United States, 330 U.S. 160 (1947).

116. Kaufman, "Sentencing: the Judge's Problem," at http://www.theatlanticmonthly/unbound/flashbks/death/kaufman.htm

117. Miller, Dawson, Dix, and Parnas, Prosecution and Adjudication, p. 1153.

118. Bozza v. United States, 330 U.S. 160 (1947).

119. Kaufman, "Sentencing: the Judge's Problem," at http://www.theatlanticmonthly/unbound/flashbks/death/kaufman.htm

20

DEBATING DEATH

Critical Issues in Capital Punishment

BEAU BRESLIN AND DAVID R. KARP

Deserves death! I daresay he does. Many that live deserve death. And some die that deserve life. Can you give that to them? Then be not too eager to deal out death in the name of justice.

Gandalf (J. R. R. Tolkien, *The Two Towers*)

In April 1999, John Howley, a lawyer and graduate of the college where we teach, gave one of us an unexpected phone call. He was writing a clemency petition to Virginia Governor James Gilmore to spare the life of Death Row inmate Calvin Swann. Working on the case after hours and for free, he needed some help. Were there some students who could do some research for the case, he asked? Eight students volunteered and, for 10 days, worked day and night to help construct the petition before Swann's scheduled execution on May 12, 1999. Their efforts were successful. Four hours before execution, Gilmore granted Calvin Swann clemency.

For these students, the Swann case helped illustrate many of the controversial issues in capital punishment. Swann is a poor, black man from the rural South. He is mentally ill. He had an inexperienced court-appointed lawyer defending him. In this chapter, we explore the larger issues surrounding Swann's case, such as poverty, race, geography, and mental illness. We also examine the question of constitutionality, the effectiveness of the death penalty as a deterrent, the execution of the innocent, public opinion, and the impact of the death penalty on victims' families. We do not pretend to cover these issues exhaustively; instead, we try to introduce you to the major topics in the current debate about the death penalty.

IS CAPITAL PUNISHMENT CONSTITUTIONAL? THE LANDMARK CASES OF *FURMAN* AND *GREGG*

In the post–World War II era, many Western nations abolished capital punishment. West Germany banned it in 1949, the United Kingdom in 1973, Canada in 1976, France in

1981, and Italy in 1994 (Bedau, 1997). Many people believed that the United States would follow the path that has now been taken by all other Western nations. Although the United States has made use of capital punishment since the nation's founding, by 1968 executions had effectively ended (see Figure 20.1). Surprisingly, the hiatus lasted only 9 years, and executions began again with some frequency. Moreover, the number of Death Row inmates has increased dramatically to 3,593 in 2000 (see Figure 20.2). In 1972, the Supreme Court ruled capital punishment unconstitutional. Yet the landmark decision in *Furman v. Georgia* (1972) was reversed 4 years later in another famous Supreme Court case, *Gregg v. Georgia* (1976). We had the death penalty, then we didn't, and then we had it again. Here's what happened.

The Court's decision in *Furman v. Georgia* (1972) marks not only the beginning of the modern era of capital jurisprudence but also the first (and only) time the U.S. Supreme Court has ruled that the practice offends the Constitution. In the case, a deeply divided Court opined that the application of capital punishment under a Georgia criminal statute transgressed the Eighth Amendment's prohibition against cruel and unusual punishment. Two justices believed that the death penalty was unconstitutional—period. Nothing would change their views on the matter. Another two justices thought the death penalty *in the way it was currently administered* was unconstitutional. But they held open the possibility that practices could be changed to make it constitutional. A fifth justice, though uncertain whether capital punishment was always or just currently unconstitutional, joined the four to make the majority opinion the law of the land. Capital punishment in America was officially over.

Two justices—William Brennan and Thurgood Marshall—argued that capital punishment violated the Constitution in all circumstances. Their position centered on a shared belief that capital punishment was degrading to human dignity, that it was imposed arbitrarily, that its popularity was waning, and that it was no more effective in preventing crime than less severe punishments. Justice Brennan wrote that "in comparison to other punishments today, the deliberate extinguishment of human life by the state is uniquely degrading to human dignity"

(*Furman v. Georgia,* 1972, p. 291). Quoting a statement by the court in an earlier opinion, he wrote in *Furman* that capital punishment violated the "evolving standards of decency that mark the progress of a maturing society" *(Trop v. Dulles,* 1958, p. 101).

The court, however, was divided. Justice Lewis Powell did not see evidence of "evolving standards of decency." In his view, the public strongly supported capital punishment. "In a democracy, the first indicator of the public's attitude must always be found in the legislative judgments of the people's chosen representatives" (*Furman v. Georgia,* 1972, pp. 436–437). The fact that 40 states, the District of Columbia, and the federal government still permitted executions seemed to undermine any contention that the American people were repelled by the practice. The dissenting justices thought Brennan and Marshall were being disingenuous by relying on abstract notions of human dignity and standards of decency to overturn Georgia's death penalty statute.

Indeed, public opinion about the death penalty has fluctuated and has by no means steadily declined. Figure 20.3 shows how support has generally increased in the last decades. Figure 20.4, however, shows that support varies across segments of our society. Men support it more than women, whites more than people of color, and Republicans more than Democrats or independents.

The dissenting justices in the *Furman* case were not just concerned about normative standards. They were also worried about judicial activism. The dissenters argued that an independent judiciary—a judiciary that is in no way accountable to the populace—should avoid grounding its decisions on the private moral convictions of individual justices. Judicial restraint, Justice Powell wrote, demands that the judiciary defer to legislative judgment in cases where the Constitution is silent. Capital punishment is not outlawed by the constitutional text, so Powell concluded that "the sweeping judicial action undertaken [by the majority] reflects a basic lack of faith and confidence in the democratic process" (*Furman v. Georgia,* 1972, pp. 464–465).

Yet Furman prevailed over the dissenting voices of these four jurists. The key to understanding how the Court's interpretation of

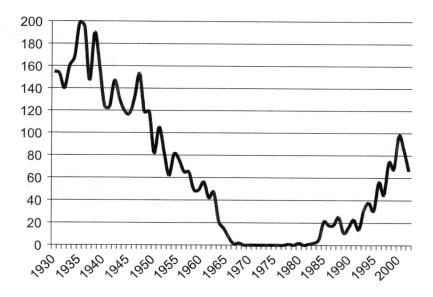

Figure 20.1   Executions Since 1930 in the United States

SOURCE: Snell (2001).

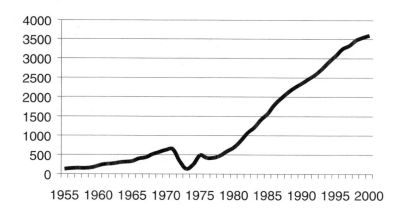

Figure 20.2   Numbers of Death Row Inmates Since 1930

SOURCE: Snell (2001).

the Eighth Amendment led to the halting of executions is to recognize that three justices— William Douglas, Potter Stewart, and Byron White—shared deep concerns about the administration of death sentences. Although they were unwilling to admit that capital punishment in all circumstances violated the Constitution, White and Stewart both agreed with Brennan that the punishment of death was unique in its finality and that it was imposed with unconstitutional randomness. Justice Stewart even equated its

application with the randomness of being struck by lightning. They insisted that as long as states allowed it to be so "wantonly and freakishly imposed," the death penalty would breach the Eighth Amendment's protection against cruel and unusual punishment. Death was different, they concluded, and thus it should be administered with the utmost fairness and caution.

Where Stewart and White disagreed with Brennan was over the possibility that the system

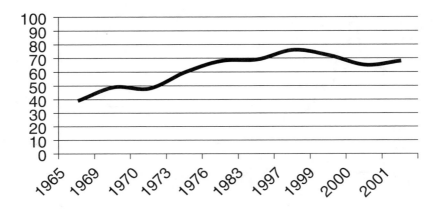

Figure 20.3   Percentage of Americans Who Support the Death Penalty

SOURCE: Maguire and Pastore (2001a).

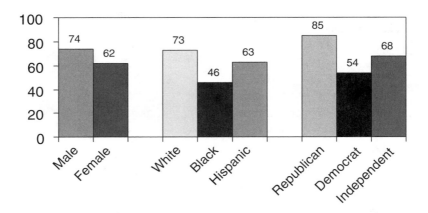

Figure 20.4   Percentages of Various U.S. Groups Supporting the Death Penalty

SOURCE: Maguire and Pastore (2001a).

could be fixed. Justices Stewart and White argued that if certain procedures were in place, capital punishment would be a constitutionally acceptable form of punishment. At a minimum, those procedures had to better protect defendants facing execution. The irrevocability of death, they said, demanded mechanisms that minimize the possibility of arbitrary death sentences. A higher degree of due process—"super" due process, as it has come to be known—was imperative when the state sought to end a life.

The result was a series of revised statutes emerging from various statehouses.[1] State lawmakers began crafting new legislation that embraced a higher form of procedural protection.

Laws across the country now mandated that capital trials be conducted in a bifurcated fashion, with the first phase of the two-part trial devoted solely to determining the guilt of the defendant and the second phase—the penalty phase—used to determine the appropriateness of a sentence of death. In addition, state laws further required that any capital jury weigh "aggravating circumstances" against "mitigating circumstances" when considering whether to impose death. Aggravating circumstances are those defined as contributing to the general depravity of the criminal or the crime itself, and they are typically specified by law. The aim of requiring aggravating circumstances is to narrow the field of those eligible for death to only the most barbaric few. A first-degree murder conviction, therefore, no longer carries a possible death sentence *unless* the offender was, during the course of committing the murder, *also* engaged in another felony (burglary, rape, etc.). Similarly, a jury can now consider an offender's prior conviction for capital crimes, or the fact that he or she murdered a police officer, as testament to the seriousness of his or her actions.

The new state laws also permitted juries to hear mitigating circumstances—circumstances that ultimately *diminish* the offender's overall culpability for the capital crime. The list of mitigating circumstances most often includes personal characteristics (mental retardation, childhood abuse, etc.) that may soften or temper the jury's inclination toward death. The point of including mitigating circumstances in the capital trial is, once again, to safeguard the defendant and give him or her every reasonable chance of avoiding execution. Finally, the new laws also stipulated that all capital convictions should be automatically reviewed by the state's highest court.

By a 7-2 margin, with only Justices Brennan and Marshall dissenting, the Court upheld the new regulations just 4 years after *Furman*. The case of *Gregg v. Georgia* (1976) successfully tested the specifics of Georgia's modified death penalty statute against the protective scope of the Eighth Amendment. The key players in the *Furman* decision—Justices Stewart and White—were now convinced that the modified law acknowledged the gravity of the death sentence. They were persuaded that Georgia's revised statute provided for a heightened level of constitutional protection, and thus they joined the *Furman* dissenters (as well as John Paul Stevens, who had been appointed to the High Court a year earlier) in validating the practice of capital punishment. The country was back in the business of executing capital criminals.

SUPER DUE PROCESS: HAS IT GUARANTEED FAIR TRIALS?

Since *Gregg v. Georgia,* any problems associated with the trial process should have been solved. Yet a recent study by James Liebman et al. (2000) found that plenty still remain. The researchers examined all death verdicts from 1973 to 1995, the period following *Furman*. They discovered that, because of prejudicial error during trial, state and federal courts eventually vacate 68% of all death sentences. "In other words, courts found serious reversible error in nearly 7 of every 10 of the thousands of capital sentences that were fully reviewed during that period." Of those, only 18% result in a second death sentence at retrial; 82% result in either a lesser sentence or an outright acquittal. Liebman et al.'s study notes that the two most common errors in death penalty cases are (a) "egregiously incompetent defense lawyers who didn't even look for—*and demonstrably missed*—important evidence that the defendant was innocent or did not deserve to die; and (b) police or prosecutors who *did* discover that kind of evidence but *suppressed* it, again keeping it from the jury" (emphasis in original).

Current Supreme Court Justices Ruth Bader Ginsburg and Sandra Day O'Connor, in separate speeches over the past few years, have highlighted constitutional concerns about adequate legal representation of capital defendants (O'Connor, 2001). "People who are well represented at trial do not get the death penalty," notes Ginsburg. "I have yet to see a death case among the dozens coming to the Supreme Court on eve-of-execution stay applications in which the defendant was well represented at trial" (Ginsburg, 2001).

Further, it has become more—not less—difficult to remedy errors at trial. The Anti-Terrorism and Effective Death Penalty Act (AEDPA), a congressional statute enacted in 1996, amends the federal judicial code to establish severe limitations for habeas corpus actions by state prisoners. Habeas appeals in federal courts have been widely used to review state court death penalty decisions. Previously, Death Row inmates were permitted to raise a variety of concerns on appeal, but now the AEDPA specifies that the federal courts must approve successive appeals. The law reduces the number of capital appeals and the window of opportunity for filing them. In short, it speeds up executions by making it increasingly difficult for capital offenders to turn to the federal judiciary for relief. Now that the number and scope of potential appeals are statutorily restricted, it is not likely that we will see much remedy for the high numbers of errors in capital trials.

EXECUTING THE INNOCENT

While campaigning for the presidency, Texas Governor George W. Bush stated that "as far as I'm concerned, there has not been one innocent person executed since I've been the governor" (Yang, 2000). This is a confident statement for a governor who oversaw more executions than any other governor in the nation, presiding over approximately one every 2 weeks while he was in office. Another Republican governor, George Ryan of Illinois, was less confident. In 2000, even though he was a supporter of the death penalty in principle, Ryan declared a moratorium on all Illinois executions. "I now favor a moratorium, because I have grave concerns about our state's shameful record of convicting innocent people and putting them on death row," Governor Ryan said. "And, I believe, many Illinois residents now feel that same deep reservation. I cannot support a system, which, in its administration, has proven to be so fraught with error and has come so close to the ultimate nightmare, the state's taking of innocent life" (Ryan, 2000).

Defense attorney Barry Scheck and his colleagues offer a chilling account of the "ultimate nightmare" (Scheck, Neufeld, & Dwyer, 2001, pp. 273–288). For ex-marine Kirk Bloodsworth, March 22, 1985, was a tough day. That was the day he was sentenced to die. He was 24 years old. Judge J. William Hinkel had this to say: "The most terrible of all crimes, murder, rape, sodomy . . . was committed upon the most helpless of all our citizens, a trusting little girl. The torture that she endured and the horror that was visited upon her is beyond my mere words to describe or express. . . . Therefore, it is the judgment of this Court that the Defendant be committed to the Division of Correction for the purpose of carrying out the sentence of death" (p. 287). June 28, 1993, was a better day. Eight years after sentencing, now 32, Kirk was released from prison, not to heaven or hell, but back to ordinary life in the state of Maryland as a free citizen. It was not that the crime committed was any less terrible. It was just that Kirk didn't commit it.

On a warm summer day in Baltimore, two boys were busy catching fish in a small pond. A 9-year-old girl named Dawn Hamilton came wandering by in search of her cousin Lisa. The boys hadn't seen her. At the same time, a blond, curly-haired man with a mustache joined them. Hearing that Lisa was lost, he offered to help Dawn look for her. As they walked along a wooded path, a woman in her yard saw them and heard them calling for Lisa. Five hours later, Dawn Hamilton was found in the woods. She had been raped and murdered.

Based on descriptions given by the woman and the two boys who had seen the man with Dawn, a composite sketch was created and published in the Baltimore *Evening Sun*. Among the 286 tips called in to the police, one person thought Kirk Bloodsworth, a man living in Cambridge, Maryland, looked a lot like the sketch. Baltimore detectives brought him in for questioning, and Kirk had no alibi. He couldn't remember where he had been on the day of the crime; it had been his day off. However, he did say that he hadn't been to the suburban development where the crime occurred; he didn't even know where it was. The detectives then showed a Polaroid picture of Kirk along with other photos in a line-up to the two boys who had been fishing. Eight-year-old Jackie didn't think any of the photos looked like the man he had seen. Ten-year-old Chris thought Kirk did look like the man but that his hair was the wrong color; it was too red. Nevertheless, it was enough for the detectives to get an arrest warrant. The two boys and the woman in the yard watched the news that night, closely watching a handcuffed Kirk as he was escorted to jail. Later, in a police line-up all three witnesses were asked to identify the man they had seen near the pond. Although uncertain, each eventually chose Kirk. But was Kirk the man they had seen at the pond, or was his image familiar because they had seen him on TV?

It was this eyewitness testimony that convicted Kirk and got him a death sentence. But he got lucky. As part of a project at the Cardozo School of Law in New York City,[2] defense attorneys Barry Scheck and Peter Neufeld were able to use a new DNA test on semen found on Dawn Hamilton's underwear. The results were clear—it was not Kirk Bloodsworth's. DNA testing may help combat this problem, freeing innocent men like Kirk. But not every case has evidence amenable to DNA testing. It is not a panacea. Nevertheless, DNA testing is now demonstrating that eyewitness testimony is unreliable and that the court system is imperfect. Innocent people are wrongfully convicted. Even before DNA testing, Hugo Bedau and Michael Radelet (1987) documented 463 cases of wrongful convictions in capital cases over the last century. As we wrote the first draft of this chapter, we received an e-mail announcement that on April 8, 2002, Ray Krone was released from prison in Arizona. He is the 100th inmate to be released from Death Row since the death penalty was reinstated in 1973. It was bad enough that Krone, like Kirk Bloodsworth, was imprisoned for years for a crime he did not commit. Worse, he could have been killed for it.

Not everyone is as lucky as Kirk Bloodsworth and Ray Krone, if we can even call them lucky. Radelet and Bedau (1998) described the tragic case of Jesse Tafero, who was executed in Florida in 1990. He and Sonia Jacobs were both sentenced to death for a murder they had allegedly committed together. When Tafero was executed, the electric chair malfunctioned and his head caught on fire before he died. Two years later, Jacobs was cleared of the crime. The U.S. Court of Appeals concluded that she and Tafero were convicted because the prosecution suppressed evidence of their innocence and made use of a witness who lied about their role in the crime. (The witness lied because he was the real killer.) No Florida official has admitted error or expressed remorse for executing the wrong man. Just like Governor Ryan, we are worried about errors made in the court system. We wonder how many innocent people have been executed, how many innocent prisoners are on Death Row, and how many innocent people will be sentenced to death in the years to come.

DEATH ROW: HOME FOR AFRICAN AMERICANS, THE POOR, AND SOUTHERNERS

When we examine Death Row across America, it becomes apparent that capital punishment is applied unevenly. Three disparities stand out. Blacks are disproportionately represented on Death Row. So are poor people. And Death Rows are crowded in some states but empty in others.

As Table 20.1 illustrates, 38 states and the federal government allow for the death penalty, and 12 have abolished it. And as the table also reveals, the rate of executions varies widely among those that do have the death penalty. Texas executed 239 people from 1977 to 2000, whereas many states executed just a few or none at all.[3] Pro–capital punishment states are overwhelmingly concentrated in the South, the Midwest, and the far West. Indeed, the entire southern half of the country—from Georgia to Arizona, and from Virginia to Florida—retains the option of executing capital defendants. Though we might expect more uniformity in the federal system, geography matters once again. More than 50% of all capital cases come from less than 6% of the federal districts. And 10% of all cases originated in one single district—the Eastern District of Virginia (U.S. Department of Justice, 2000).

Eleven years after capital punishment was officially reinstated following the Supreme Court's decision in *Gregg,* Georgia found itself once again at the center of a constitutional firestorm. This time, the problem was race. Warren McCleskey, a middle-aged Georgia man, was convicted and sentenced to death for the 1978 murder of an Atlanta police officer. On appeal, McCleskey's attorneys argued that the Georgia death penalty statute was being carried out with discriminatory malice. Their claim rested on the findings of a study conducted by Baldus, Pulanski, and Woodworth (1990).

The Baldus team examined more than 2,000 murders in the state of Georgia from 1973 to 1978. Like previous researchers of race and capital punishment, they wanted to know if black defendants were more likely to get death sentences than whites. It seemed likely because Death Row is disproportionately black. Forty-three percent of all Death Row inmates today are black, even though blacks constitute only 12% of the population (Maguire & Pastore, 2001c). Such disproportionality can be explained by criminality or racism. If blacks disproportionately commit capital crimes, they will be disproportionately represented on Death Row. Alternately, it could be a racist criminal justice system that is more content to send a black person to death than a white.

Table 20.1   Geographic Application of the Death Penalty

Death Penalty Jurisdictions	Executions 1977–2000	No Death Penalty
Alabama	23	Alaska
Arizona	22	District of Columbia
Arkansas	23	Hawaii
California	8	Iowa
Colorado	1	Maine
Connecticut	0	Massachusetts
Delaware	11	Michigan
Florida	50	Minnesota
Georgia	23	North Dakota
Idaho	1	Rhode Island
Illinois	12	Vermont
Indiana	7	West Virginia
Kansas	0	Wisconsin
Kentucky	2	
Louisiana	26	
Maryland	3	
Mississippi	4	
Missouri	46	
Montana	2	
Nebraska	3	
Nevada	8	
New Hampshire	0	
New Jersey	0	
New Mexico	0	
New York	0	
North Carolina	16	
Ohio	1	
Oklahoma	30	
Oregon	2	
Pennsylvania	3	
South Carolina	25	
South Dakota	0	
Tennessee	1	
Texas	239	
Utah	6	
Virginia	81	
Washington	3	
Wyoming	1	
Federal system	0	

SOURCE: Snell (2001).

Baldus et al. did not find the Georgia court system to be overtly racist. "Post-*Furman* . . . there is no evidence of race-of-defendant discrimination statewide" (p. 185). However, they discovered a more subtle form of racism that reveals itself when one is looking at the race of the defendant *and* the race of the victim. A defendant's likelihood of receiving a death sentence was 4.3 times greater if the *victim* was white. That is to say, a defendant who had murdered a white person was more than 430% more likely to receive a death sentence than one who had murdered a black person.[4] Because McCleskey—a black man—was convicted of murdering a white police officer, he had a greater chance of getting a death sentence than if he had been white and the police officer had been black.

The Supreme Court, however, was not convinced. A majority of justices did not find *purposeful* discrimination on the part of the Georgia legislators who crafted the death penalty statute, nor was there any hint that the prosecutors, jurors, or judge in the *McCleskey* case were motivated by racial prejudice (*McCleskey v. Kemp,* 1987). According to Justice Powell, who penned the majority opinion for the Court, a violation of the Fourteenth Amendment's Equal Protection Clause requires that the defendant show discriminatory intent by some actor involved in the process. This McCleskey could not do. Like so many criminal justice systems around the country, the framework in Georgia allows for discretion in the individual prosecutor's decision to seek the death penalty. Moreover, it allows for discretion in the jury's choice of punishment. The Supreme Court concluded that any system that offers wide prosecutorial and jury discretion is bound to reveal some degree of bias. But the nature and scale of prejudice uncovered by the Baldus study did not transgress the Equal Protection Clause of the Fourteenth Amendment.

Bolstered by the decision in *McCleskey*, proponents of the death penalty now maintain that racial disparities in capital punishment mirror similar problems in the entire criminal justice system. Race is a factor in many trials, say supporters, and collectively we are trying to correct these injustices. These same individuals argue that the way to remedy the problem of racial prejudice is not by throwing out the system altogether and letting criminals go free but by working more effectively within the statutory rules set by state and federal legislatures. No one suggests, in other words, that we eliminate life sentences simply because there may be some racial bias in noncapital cases. Hence, it does not make sense to abandon capital punishment because some findings hint at prejudice. The answer, claims Ernest Van den Haag (1997), may be as simple as executing more whites.

McCleskey is the most important Supreme Court case addressing race and capital punishment. But the decision did not end the controversy about race—far from it. On May 9, 2002, Governor Parris Glendening imposed a moratorium on all executions in Maryland pending the results of a study of racial bias in capital cases currently underway at the University of Maryland. And David Baldus and his team of social scientists have continued their research. They recently published findings of both racial and class disparities in the Philadelphia area, and the results complement their earlier research in Georgia (Baldus, Woodworth, Zuckerman, Weiner, & Broffitt, 1998). According to the Philadelphia study, there is "substantial" evidence to suggest that the race of the defendant still has some bearing on the outcome of a capital trial. And there is more evidence that the race of the victim is an important factor. From these findings, the Baldus group infers that the discretionary power of juries contributes meaningfully to the disproportionate presence of black defendants on Death Row.

The Philadelphia study also explores the issue of class in a rigorous and systematic way. The Baldus group found that economic position of the defendant correlates with capital sentencing. Socioeconomic status of the victim is also a factor. The likelihood of a defendant getting the death penalty increases as his or her status decreases and as the socioeconomic status of the victim increases. The richer the defendant or the poorer the victim, the less likely it is that a prosecutor will seek the death penalty and a jury will impose it.

Most capital offenders are so poor that they must rely on the aid of court-appointed lawyers—lawyers who do not have the time or resources to represent their clients well. It is thus not surprising that, as Justice William O. Douglas once remarked, there are simply no rich people on Death Row. He wrote in *Furman*: "One searches our chronicles in vain for the execution of any member of the affluent strata

of this society. The Leopolds and Loebs are given prison terms, not sentenced to death" (*Furman v. Georgia,* 1972, pp. 251–252). Justice Ginsburg has similarly voiced concern over the overwhelming population of indigents on Death Row. She and Justice O'Connor have both publicly noted that the quality of representation is so poor when the state appoints defense counsel in death penalty cases that many convictions come to the U.S. Supreme Court under considerable suspicion (Ginsburg, 2001; O'Connor, 2001). Like race or place, they both insist, socioeconomic status should not be the deciding factor that separates life and death.

EXECUTING THE MENTALLY ILL AND THE MENTALLY RETARDED

Diagnosed as a severe schizophrenic, Calvin Swann had been in and out of mental institutions for most of his adult life; and when he was not in state-run mental hospitals, he was most likely in prison. He would speak to objects that were not there, rant about nothing in particular, and periodically engage in self-mutilation. His last crime—the one for which he faced death—involved the murder of Conway Richter. Because of a door left ajar, Swann entered Richter's home demanding money. When Richter refused, Swann fired a single shotgun blast into his chest, killing him instantly. A year later, a Virginia jury sentenced Calvin Swann to death. Ultimately, however, the governor commuted his sentence because of Swann's mental illness.

By some estimates, as many as 16% of all prisoners are plagued by some mental incapacity, and the percentage is even higher on Death Row (Ditton, 1999). Death Row inmates suffer from a variety of mental health diseases, such as schizophrenia and severe depression. Although distinct, another significant portion of Death Row inmates are considered mentally retarded, a condition defined by "significantly subaverage intellectual functioning" combined with limitations in adaptive skill areas like communication, self-care, and socialization (American Association on Mental Retardation, 2002). Together, the presence of so many on Death Row with some degree of mental shortcoming raises concerns about the moral culpability of these individuals and their capacity to understand their fate.

The judiciary's attitude toward the mentally ill has been generally protective. Insanity, for example, is ordinarily viewed as a legitimate defense against criminal action. Concerning capital punishment more specifically, the Supreme Court has decided that the Eighth Amendment prohibits the execution of the insane. Writing for the majority in *Ford v. Wainwright* (1986), Justice Marshall noted that in every stage of history the execution of the insane has been widely condemned. From our earliest forebears to the present, Americans have viewed the state-sponsored killing of those who lack "the capacity to come to grips with their own conscience" as offensive (*Ford v. Wainwright,* 1986, p. 409). It does not comply with our commitment to the dignity of the individual, Marshall wrote, and neither does it conform to the high standard of decency we have set for society itself. Justice Marshall then concluded by adding that there is no retributive value served by executing the insane. Subjecting the insane to death, he said, "provides no example to others and thus contributes nothing to whatever deterrence value is intended to be served by capital punishment" (*Ford v. Wainwright,* 1986, p. 407).

Just 3 years after *Ford v. Wainwright*, however, the United States Supreme Court addressed the issue of mental retardation and capital punishment with ambiguous results. The case was *Penry v. Lynaugh* (1989), and it involved a challenge to the Texas death penalty statute by a moderately retarded man under the Eighth Amendment's "cruel and unusual punishment" provision. John Paul Penry's attorneys argued that their client's mental capacity (equivalent to that of a six-and-a-half-year-old at the time of the 1980 trial), his severe antisocial behavior, and his low IQ (between 50 and 63) meant that he had neither the competency to stand trial nor the capacity to understand that he might eventually be executed. Defense attorneys further argued that the retarded lack the moral culpability required to justify the imposition of the death penalty. They insisted that there is a growing national consensus against executing the mentally retarded. Accordingly,

they said, executing Penry would constitute a transgression of the Eighth Amendment.

The Supreme Court did not agree. By a narrow 5-4 margin, the majority rejected the argument that there exists a national consensus against executing the mentally retarded. Despite pleas by the American Association on Mental Retardation and other interest groups hoping to save Penry's life, Justice O'Connor wrote that there was insufficient evidence to suggest any national movement, particularly when one considers that only two states have laws that prohibit the execution of the mentally retarded (since 1989, 16 more states have outlawed the practice). In her estimation, a "severely" or "profoundly" retarded individual might warrant the court's protection against execution, but Penry was neither severely nor profoundly retarded. He had been defined as only "moderately retarded" and was thus found competent to stand trial by a jury of his peers.

Justice O'Connor's opinion did, however, indicate that mental retardation is a powerful mitigating factor to be considered by a jury during sentencing. "In sum," she wrote, "mental retardation is a factor that may well lessen a defendant's culpability for a capital offense. But we cannot conclude today that the Eighth Amendment precludes the execution of any mentally retarded person of Penry's ability convicted of a capital offense simply by virtue of his or her mental retardation alone" (*Penry v. Lynaugh,* 1989, p. 340). Giving hope to abolitionists, her opinion also left open the possibility that the Court might reconsider the question of a national consensus against executing the mentally retarded sometime in the future.

The opportunity to revisit that issue arose in 2002. The case was *Atkins v. Virginia,* and it involved a Virginia man who, by most accounts, has an IQ of 60. His case garnered national attention because it was accompanied by a movement among many state legislatures to exclude the mentally retarded from capital sentences. Since 1989, when only two states exempted retarded persons from capital punishment, 16 additional states have now outlawed the practice. That was enough for the Supreme Court to reverse its earlier conclusion that a national consensus had not yet developed. In an opinion written by Justice Stevens, the majority of the Court characterized the movement away from executing the mentally retarded as a "dramatic shift in the legislative landscape that provides powerful evidence that today our society views mentally retarded offenders as categorically less culpable than the average criminal." Stevens further noted that it was "not so much the number of these states that is significant, but the consistency of the direction of change," particularly when one considers the recent popularity of anticrime legislation around the country.

The case is likely to have a wide impact. Some estimate that as many as 200 or more Death Row inmates will be released into the general prison population as a result of the Atkins decision. At the center of the Court's opinion is a concern that the protections afforded criminal defendants do not work as well when applied to the mentally retarded. Stevens cited their "diminished capacities to understand and process information, to communicate, to abstract from mistakes and learn from experience, to engage in logical reasoning, to control impulses, and to understand others' reactions" as evidence that they "face a special risk of wrongful execution." Over a scathing dissent by Justice Scalia, Stevens concluded that the intellectual deficiencies of mentally retarded persons do not exempt them from criminal punishment but certainly "diminish their personal culpability."

DETERRENCE: DOES THE DEATH PENALTY MAKE CRIMINALS THINK TWICE?

One of the most compelling arguments in favor of the death penalty is that of deterrence, a term that has two meanings—"special deterrence" and "general deterrence." *Special deterrence* refers to an intervention that prevents an offender from committing crimes in the future. It is hard to argue against the effectiveness of the death penalty on this point. And many juries vote in favor of death penalty for this reason alone—they want to make sure that the offender can never again threaten public safety. Then again, a sentence of life without parole is also an effective special deterrent. But jurors sometimes believe that a "life sentence" does not really

mean "life." As one capital juror noted, "We all had decided if we were absolutely sure that he would never have gotten out of prison we wouldn't have given him the death penalty. But we were not sure of that" (Lifton & Mitchell, 2000, p. 156). Recently, there has been a movement to tighten the rules so that "life without parole" means exactly what it says. As this occurs, juries may believe that such a life sentence offers the same deterrent benefit and may increasingly choose this option.

General deterrence refers to the broadcast message of the death penalty: would-be killers will think twice if they know they could be executed. The well-publicized execution of one is a lesson for others. Many Americans believe that the death penalty is a general deterrent. In a 2001 national poll, respondents were asked, "Do you feel that executing people who commit murder deters others from committing murder, or do you think that such executions don't have that much effect?" Forty-two percent of Americans believed that executions are a deterrent (Maguire & Pastore, 2001b). Are they correct?

The deterrence argument is not an argument about morality or justice; it is an argument about crime control and prevention. It suggests that the death penalty can be a rational strategy to lower the homicide rate. In this way, it moves the debate from ethics to science, offering a hypothesis that can be verified through empirical research. So let's take a close look at the research on general deterrence. Does the death penalty reduce murder in America?

In 1980, a prominent criminologist named Thorsten Sellin (1980) published a study comparing homicide rates in states that had the death penalty and those that did not. If the death penalty were a general deterrent, we would expect lower homicide rates in states that impose death. That is not what Sellin found, however. A more recent analysis demonstrates, in fact, that states with the death penalty have a much higher homicide rate than states that have abolished the death penalty (Kappeler, Blumberg, & Potter, 2000). Another way Sellin looked at deterrence was to look at homicide trends over time. If the death penalty was a deterrent, then states that abolished the death penalty should show an increase in homicides after abolition, and states that reinstated the death penalty after a period of abolition should see a decline after reinstatement. Again, Sellin found no effect. The trends continued independently of changes in sentencing.

Perhaps the studies were too general. Perhaps if we selected one arena for close inspection, we would find a deterrent effect. For instance, in states that have capital punishment, most people (presumably criminals too) know that killing a police officer is an offense that makes one highly eligible for the death penalty. Moreover, most people know that the police will work long and hard to solve such a crime. Perhaps, then, capital punishment deters the killing of police officers because the would-be killers know that they are likely to get caught and to get a death sentence. Does capital punishment offer a second "bulletproof vest" to police officers, increasing their safety by deterring would-be "cop killers"? Bailey and Peterson (1994) conducted a study to examine this question for the period from 1976 to 1989. Over this period in the United States, 1,204 police officers were killed and 120 offenders were executed, 12 of them for killing police officers. A close examination of the relationship between homicides and executions nationally yielded findings that were consistent with those from other studies of deterrence. The death penalty did not reduce homicides of police officers—the murder rates were not different between states that had the death penalty and those that did not. The death penalty has not provided extra protection to police officers as was hoped.

Some scholars have examined the possibility that execution not only is ineffective as a deterrent but may even *increase* homicide rates. This is called the "brutalization hypothesis." This theory suggests that some citizens learn that they live in a "brutal" society in which killing is normative, even endorsed by the state. As such, when citizens learn of executions, such as with highly publicized cases, some may be more inclined to commit homicides themselves. Murder rates, according to the brutalization hypothesis, will increase soon after an execution. Three recent studies examined this, finding evidence supporting the brutalization hypothesis. Cochran, Chamlin, and Seth (1994) looked at homicide rates in Oklahoma 1 year

prior to and 1 year following the execution of Charles Troy Coleman, the state's first execution in 25 years. Ernie Thomson replicated their study twice, examining the effect of Arizona's execution of Donald Eugene Harding (Thomson, 1997) and of California's execution of Robert Alton Harris (Thomson, 1999). In all three studies, the researchers noticed that homicide rates increased after the executions. Although more research is necessary to fully support the brutalization hypothesis, the findings are troubling. What if the death penalty not only fails to deter murder but actually encourages it?

The evidence clearly suggests that the death penalty is not a deterrent. Each time there has been a serious evaluation of the research, the reviewer has drawn this same conclusion. This was true for Thorsten Sellin in 1967, for Franklin Zimring and Gordon Hawkins in 1986, and for William Bailey and Ruth Peterson in 1997, who wrote, "The available evidence remains 'clear and abundant' that, as practiced in the United States, capital punishment is no more effective than imprisonment in deterring murder" (p. 155).

But intuitively we often feel it should—the evidence doesn't seem to conform to our gut feelings, which tell us that the more severe the penalty, the less likely someone will be to break the rule. Of course, this assumes a basic rationality on the part of a killer. He or she must pause and quietly reflect on the possibility of punishment. Because many murders are not premeditated, it is unlikely even this assumption can be held. Nevertheless, severity of punishment is important for deterrence. No one would argue that homicide should go unpunished or that murderers should pay a fine as we do with traffic tickets. Severity of punishment is calibrated to the severity of the offense, partly for reasons of justice and partly for deterrence. Yet fine-grained distinctions in severity may be irrelevant for deterrence. Would a thief be deterred if he knew that he might get a 24-month sentence but not be deterred if the sentence was for 18 months? Would a killer be deterred if he knew he would get the death penalty but not deterred if he would get life without parole? Both are harsh sentences. The deterrent value of severity, if it exists, is not likely to be different between two types of severe sentences but only between the mild and the severe.

Finally, for deterrence, severity may not be as important as certainty and swiftness. The threat of punishment works only when the potential offender believes that he or she will definitely be caught, that the punishment will definitely be applied, and that this will happen pretty quickly. In Cesare Beccaria's classic essay on criminology, published in 1764, he wrote, "One of the greatest curbs on crimes is not the cruelty of punishments, but their infallibility. . . . The certainty of punishment, even if it be moderate, will always make a stronger impression than the fear of another which is more terrible but combined with the hope of impunity" (Beccaria, 1764/1985, p. 85). In our country, the rarity and uncertainty and fallibility of the death penalty combine to undermine any possible deterrent benefit that can be gained by its severity.

THE MORAL DEBATE: RETRIBUTIVE VERSUS RESTORATIVE VISIONS OF JUSTICE

Thus far we have examined the instrumental logic of the death penalty—whether the public supports it, whether the Constitution affirms it, whether it deters others from committing murder, whether innocent defendants are executed, and whether it is applied unfairly by class, race, or place. For many people, however, all of these issues are secondary. The real question is moral—is capital punishment the right thing to do, even if it has some unfortunate side effects?

In a recent poll, American citizens who supported the death penalty were asked to explain their reasons (Maguire & Pastore, 2001d). Ten percent gave a simple reason—it saves money. Although the cost-saving rationale makes intuitive sense, they happen to be wrong about this. States spend much more money on capital cases than they would have by incarcerating the defendants for life. One study found that Florida spent six times as much money in capital cases as they would have spent by treating them as noncapital cases. Another found that California would save $90 million per year if it abolished the death penalty (Kappeler et al., 2000). The point, however, is that cost savings—even if erroneous—are not the driving reason for most

supporters of the death penalty. Neither is deterrence. Only 16% of the survey respondents offered this as a reason for their support. Earlier we reported that 42% of Americans believed that the death penalty was a deterrent, but even so, this was *not* the major explanation for their support.

For those who support the death penalty, the reason given most frequently (by 60% of those polled) is retribution. They believe the offender deserves it. They believe death is the just response. They believe it is called for in the Bible. They believe it is the way to honor victims. They believe execution most clearly signals our society's repulsion for the crime. According to legal philosopher Ernest Van den Haag (1997):

> They believe that everyone who can understand the nature and effects of his acts is responsible for them, and should be blamed and punished, if he could know that what he did was wrong. Human beings are human because they can be held responsible, as animals cannot be. In that Kantian sense the death penalty is a symbolic affirmation of the humanity of both victim and murderer. (p. 454)

The retributive perspective is the dominant theme of our society's justice system. It is based on the belief that accountability should be measured on a scale of proportional punishment—a light sentence for a minor crime and the "ultimate" sentence for the worst crimes. Just as it would be inappropriate to execute a burglar, it would be inappropriate *not* to execute a murderer. The scales of "Lady Justice" are brought back to equilibrium by applying a punishment that "fits the crime." The focus of justice is not practical or instrumental; it is moral. We do what is necessary to affirm society's standards, even if it is costly, even if we make occasional errors.

Recently, another punishment philosophy has gained national attention. It is called "restorative justice" and provides a different moral perspective on the problem of accountability (Braithwaite, 2002; Zehr, 1990). As in retributive justice, a focus is placed on balancing the scales of justice. Through his or her crime, the offender has incurred a debt that must be paid back to victims and to society. But in restorative justice, unlike retributive justice, this debt is not paid through *passive* submission to punishment. The offender does not receive a proportional amount of harm to equate with the harm caused. Accountability is defined as an *active* responsibility on the part of the offender to make amends for what he or she has done. This is relatively simple in minor crimes. If an offender has broken a window, justice requires that he or she fix it or pay for it. But how can murderers be held accountable when they cannot bring their victims back to life?

It is important in both perspectives of justice that offenders are prevented from profiting from their crimes—from getting away with it. The difference is in how they are held accountable. From a restorative perspective, a prison sentence is a strange way to "pay one's debt to society" because the victim suffers once as a result of the crime and then suffers again because he or she is a taxpayer who must pay for the offender's incarceration. Meanwhile, the offender does nothing to compensate the victim. Of course, incarceration serves a very important other purpose—incapacitation. We impose prison sentences not only for moral reasons but also for the instrumental purpose of preventing the offender from inflicting more harm upon victims and society. But restorative advocates believe offenders should engage in a variety of tasks that make amends. Earning money to pay restitution is the most basic task. Victims shouldn't pay for their offender's incarceration; offenders should be paying victims back for the harm they caused. Their labor should also be used to pay for victims' services—programs that help people recover from the traumatic experience of victimization.

Advocates of restorative justice believe that the death penalty is not a good way to hold murderers accountable. Instead, they argue that justice is best served when offenders actively work to make amends in whatever way they can, for the rest of their lives. By focusing on the offender's obligations, restorative justice prioritizes the needs of victims. What victims researchers have discovered is that victims have many needs, that much could be done to help in their healing process, and surprisingly, that offenders themselves can sometimes play a positive role in the healing process.

EXECUTING TIMOTHY McVEIGH: DID EXECUTION BRING RELIEF TO VICTIMS' FAMILIES?

Timothy McVeigh was executed on June 11, 2001, for his bombing of the Alfred P. Murrah Federal Building in Oklahoma City in 1995. It was an unusual event because the nation learned about how differently victims' families reacted to the death penalty. We expected comments like those from Gloria Buck, the niece of a bombing victim:

> I know some people didn't get closure, didn't get anything from this, but I got a lot from it. Timothy McVeigh, he's done. I really wouldn't want to be where he is right now. I thought about that right when they announced he was dead. I feel like I can finally start walking away from this. (Tuchman, 2001)

For Buck, McVeigh's execution marked the end of one chapter in her healing process. Many victims are told that the death penalty is necessary so that they can achieve a feeling of closure, move on with their lives, and overcome the feeling of trauma associated with violent victimization. Kathleen Treanor, the mother and daughter-in-law of bombing victims, however, noted that closure is not so easily achieved:

> I don't think anything can bring me any peace or anything from this. I'll always face the loss of my daughter. I'll never get over that. When I die and they lay me in my grave is when I'll have closure. That's when I'll stop grieving for my daughter. (Tuchman, 2001)

Bud Welch, whose daughter was killed in the bombing, opposed the execution. He had this to say:

> At first I was in absolute pain. All I wanted was to see those two people fried. I was smoking three packs of cigarettes a day and drinking heavily. . . . I knew that the death penalty wasn't going to bring her back, and I realized that it was about revenge and hate. And the reason Julie and 167 others were dead was because of the very same

thing: revenge and hate. . . . Once I turned loose of that revenge and hate, the feel-good was tremendous. After that, I was able to get things sorted out. I started getting a handle on my drinking and cutting back on smoking. I was able to start reconciling things within myself. (Zehr, 2001, pp. 60–62)

For Welch, the execution did not aid in his healing; what helped was overcoming his own impulses toward "revenge and hate."

Most family members of homicide victims experience post-traumatic stress disorder, and it typically takes years before the worst symptoms disappear (Freedy, Resnick, Kilpatrick, Dansky, & Tidwell, 1994)—symptoms such as repeatedly reliving the trauma in painful recollections or nightmares, emotional numbing, estrangement from others, insomnia, difficulty concentrating or remembering, and guilt about surviving when others did not (Goldenson, 1984). Unfortunately, the criminal justice process often gets in the way of victim recovery. Part of the problem is that victims do not receive the services they need and expect. Often they are not provided good information about how criminal cases are proceeding or how they can participate (Day & Weddington, 1996). Capital cases are particularly onerous because the proceedings last for years. We have two unanswered questions. First, given that 68% of initial death penalty sentences are overturned at a later date (Liebman et al., 2000), does this "roller-coaster" ride for family members prolong their suffering? Second, do family members for whom the offender received a life sentence at the outset recover more quickly than those in death penalty cases? No one yet has the answers to these questions.

Recently, restorative justice practitioners have begun programs that enable homicide victims' family members to meet with offenders. As we are now discovering, many victims want to meet with offenders in order to tell them of the pain they caused, to learn from the offender why he or she committed the crime, and to see if the offender has taken responsibility. The results from these meetings and in other cases where victims of severe violence meet with their offenders are striking because of the healing they seem to provide. According to Mark Umbreit and Betty Vos (2000):

> The overall effects of the mediation session reported by victims included the following: they had finally been heard, the offender now no longer exercised control over them, they could see the offender as a person rather than a monster, they felt more trusting in their relationships with others, they felt less fear, they were no longer preoccupied with the offender, they felt peace, they would not feel suicidal again, and they had no more anger. (p. 66)

In capital cases, Umbreit and Vos found that the mediation sessions also have an impact on victim attitudes toward the death sentence. In one case, they found that "the granddaughter and the sister [of the victim] both reported that initially they had wanted to execute him themselves; now they were more ambivalent" (p. 80). In other words, part of what they hope to personally achieve through the execution might be found in victim-offender mediation. Of course, much more research is necessary to explore this topic. However, we speculate that although some families are relieved by the execution, other families are not well served by it, in part because it denies them the chance for mediation. The lesson of McVeigh's execution is that the death penalty cannot always be imposed on behalf of victims or their surviving relatives. In fact, a national organization called Murder Victims' Families for Reconciliation uses the motto "Not in Our Name" in its opposition to the death penalty.[5]

CONCLUSION

To capture an accurate sense of the frustration felt by both supporters and opponents of capital punishment, consider the career of the late Harry A. Blackmun, former Associate Justice of the U.S. Supreme Court. One of the first cases Blackmun heard as a member of the High Court was *Furman v. Georgia*. His opinion, although short in comparison to those penned by Justices Brennan, Douglas, and Powell, was no less emotionally charged. He wrote of his "distaste, antipathy, and abhorrence, for the death penalty," insisting that "cases such as these provide an excruciating agony of the spirit." "For me," he remarked, "[capital punishment] violates childhood's training and life's experiences, and is not compatible with the philosophical convictions I have been able to develop. It is antagonistic to any sense of 'reverence for life'" (*Furman v. Georgia,* 1972, pp. 405–406).

Yet despite his confessed aversion, Blackmun ultimately disagreed with the opinion shared by the majority of his brethren that the practice of state-sponsored executions violated the Eighth Amendment to the Constitution. He voted in favor of upholding Georgia's death penalty statute and, had he been able to secure one more vote, would have supported the execution of William Furman. In the end, Justice Blackmun's moral opposition to the capital punishment was exceeded by his faith in the democratic process. He argued that the duty of a judge was to interpret the Constitution with dispassion and that personal moral belief should never creep into the Court's decisions. The forum for eliminating capital punishment was therefore to be found in America's legislatures—not in its courts.

Twenty years later, Blackmun's faith in the democratic process had all but eroded. He wrote in *Callins v. Collins* (1994):

> From this day forward, I no longer shall tinker with the machinery of death. For more than 20 years I have endeavored—indeed, I have struggled—along with a majority of the Court, to develop procedural and substantive rules that would lend more than the mere appearance of fairness to the death penalty endeavor. Rather than continue to coddle the Court's delusion that the desired level of fairness has been achieved and

the need for regulation eviscerated, I feel morally and intellectually obligated simply to concede that the death penalty experiment has failed. (p. 1145)

In Blackmun's mind, the same issues that perpetually keep capital punishment in the headlines—issues ranging from the race and socioeconomic status of the offender, to the question of executing the innocent and the possibility of closure for the victims' families—are the ones that are the most difficult to address. But just the same, these are the issues that most require our attention. Death is different, he was quick to remind us, and thus we must approach the ultimate punishment with the degree of humility and respect it demands. Until we do that, Blackmun believed that our only rational response was to hope that more capital defendants experience the same walk away from Death Row that Calvin Swann once enjoyed.

NOTES

1. In all, death penalty statutes were revised in 35 states.

2. For more information about Scheck and Neufeld's Innocence Project, check out www.innocenceproject.org.

3. Timothy McVeigh was the first person to be executed in the federal system in the post-*Furman* era. He was executed in 2001.

4. Another report from the same year yielded similar findings (U.S. General Accounting Office, 1990). The GAO report was a systematic analysis of all the studies of racial discrimination in capital punishment conducted during the 1970s and 1980s.

5. To learn more about this organization, go to www.mvfr.org. To learn about victims' families who support the death penalty, explore www.murdervictims.com.

REFERENCES

American Association on Mental Retardation. (2002). Fact sheet: Frequently asked questions about mental retardation. Retrieved October 2002 from www.aamr.org/Policies/faq_mental_retardation.shtml

Atkins v. Virginia, No. 008452, 260 Va. 375, 534 S. E. 2d 312 (2002).

Bailey, W. C., & Peterson, R. D. (1994). Murder, capital punishment, and deterrence: A review of the evidence and an examination of police killings. *Journal of Social Issues, 50,* 53–74.

Bailey, W. C., & Peterson, R. D. (1997). Murder, capital punishment, and deterrence: A

review of the literature. In H. A. Bedau (Ed.), *The death penalty in America: Current controversies* (pp. 135–161). New York: Oxford University Press.

Baldus, D., Pulaski, C., & Woodworth, G. (1990). *Equal justice and the death penalty: A legal and empirical analysis.* Boston: Northeastern University Press.

Baldus, D., Woodworth, G., Zuckerman, D., Weiner, N. A., & Broffitt, B. (1998). Symposium: Racial discrimination and the death penalty in the post-Furman era: An empirical and legal overview, with recent findings from Philadelphia. *Cornell Law Review, 83,* 1638.

Beccaria, C. (1985). *Essay on crimes and punishments* (H. Paolucci, Trans.). New York: Macmillan. (Original work published 1764)

Bedau, H. A. (Ed.). (1997). *The death penalty in America: Current controversies.* New York: Oxford University Press.

Bedau, H. A., & Radelet, M. (1987). Miscarriages of justice in potentially capital cases. *Stanford Law Review, 40,* 21–179.

Braithwaite, J. (2002). *Restorative justice and responsive regulation.* New York: Oxford University Press.

Callins v. Collins, 510 U.S. 1141 (1994).

Cochran, J. K., Chamlin, M. B., & Seth, M. (1994). Deterrence or brutalization? An impact assessment of Oklahoma's return to capital punishment. *Criminology, 32,* 107–134.

Day, L. E., & Weddington, M. M. (1996). *Grief and justice in families of homicide victims: Initial results from a study of the impact of the criminal justice system.* Paper presented at the annual meeting of the American Society of Criminology, Chicago.

Ditton, P. M. (1999). *Mental health and treatment of inmates and probationers.* Washington, DC: Bureau of Justice Statistics.

Ford v. Wainwright, 477 U.S. 399 (1986).

Freedy, J., Resnick, H., Kilpatrick, D., Dansky, B., & Tidwell, R. (1994). The psychological adjustment of recent crime victims in the criminal justice system. *Journal of Interpersonal Violence, 9,* 450–468.

Furman v. Georgia, 408 U.S. 238 (1972).

Ginsburg, R. B. (2001, April). *In pursuit of the public good: Lawyers who care.* University of the District of Columbia, David A. Clarke School of Law, Joseph L. Rauh Lecture, Washington, DC.

Goldenson, R.M. (1984). *Longman dictionary of psychology and psychiatry.* New York: Longman.

Gregg v. Georgia, 428 U.S. 153 (1976).

Kappeler, V. E., Blumberg, M., & Potter, G. W. (2000). *The mythology of crime and criminal justice.* Prospect Heights, IL: Waveland.

Liebman, J. S., Fagan, J., Gelman, A., West, V., Davies, G., & Kiss, A. (2000). A broken system, Part II: Why there is so much error in capital cases, and what can be done about it. Retrieved October 2002 from Columbia University Law School Web site: www.law.columbia.edu/brokensystem2/index2.html.

Lifton, R. J., & Mitchell, G. (2000). *Who owns death? Capital punishment, the American conscience, and the end of executions.* New York: HarperCollins.

Maguire, K., & Pastore, A. L. (Eds.). (2001a). Attitudes toward the death penalty, United States, selected years 1965–2001 (Table 2.61). In Sourcebook of criminal justice statistics. Retrieved April 5, 2002, from www.albany.edu/sourcebook/.

Maguire, K., & Pastore, A. L. (Eds.). (2001b). Attitudes toward the deterrent effect of the death penalty, by demographic characteristics, United States, 2001 (Table 2.67). In Sourcebook of criminal justice statistics. Retrieved April 5, 2002, from http://www.albany.edu/sourcebook/.

Maguire, K., & Pastore, A. L. (Eds.). (2001c). Prisoners under sentence of death by race, ethnicity, and jurisdiction, on Oct. 1, 2001 (Table 6.80). In Sourcebook of criminal justice statistics. Retrieved April 5, 2002, from www.albany. edu/sourcebook/.

Maguire, K., & Pastore, A. L. (Eds.). (2001d). Reported reasons for favoring the death penalty for persons convicted of murder, United States, 2001 (Table 2.66). In Sourcebook of criminal justice statistics. Retrieved April 5, 2002, from www.albany.edu/sourcebook/.

McCleskey v. Kemp, 481 U.S. 279 (1987).

O'Connor, S. D. (2001, October). Remarks at the annual meeting of the Nebraska State Bar Association, Lincoln, NE.

Penry v. Lynaugh, 492 U.S. 302 (1989).

Radelet, M. L., & Bedau, H. A. (1998). The execution of the innocent. *Law and Contemporary Problems, 61,* 105–124.

Ryan, G. H. (2000). Governor Ryan declares moratorium on executions, will appoint commission to review capital punishment system. Illinois Government News Network press release. Retrieved April 2002 from State of Illinois Web site: www.state.il.us/gov/press/00/Jan/morat.htm.

Scheck, B., Neufeld, P., & Dwyer, J. (2001). *Actual innocence: When justice goes wrong and how to make it right.* New York: Penguin.

Sellin, T. (1967). *Capital punishment.* New York: Harper & Row.

Sellin, T. (1980). *The penalty of death.* Thousand Oaks, CA: Sage.

Snell, T. L. (2001). *Capital punishment 2000.* Washington, DC: Bureau of Justice Statistics.

Thomson, E. (1997). Deterrence versus brutalization: The case of Arizona. *Homicide Studies, 1,* 110–128.

Thomson, E. (1999). Effects of an execution on homicides in California. *Homicide Studies, 3,* 129–150.

Trop v. Dulles, 356 U.S. 86, 101 (1958).

Tuchman, G. (2001). Witnessing McVeigh's last moments: Survivors, victims' families differ on death penalty but support viewing execution. Retrieved October 2002 from CNN Web site: www.cnn.com/SPECIALS/2001/okc/stories/tuchman.families.html.

Umbreit, M., & Vos, B. (2000). Homicide survivors meet the offender prior to execution. *Homicide Studies, 4,* 63–87.

U.S. Department of Justice. (2000). *The federal death penalty system: A statistical survey (1988–2000).* Washington, DC: Author.

U.S. General Accounting Office. (1990). *Death penalty sentencing: Research indicates pattern of racial disparities.* Washington, DC: Author.

Van den Haag, E. (1997). The death penalty once more. In H. A. Bedau (Ed.), *The death penalty in America: Current controversies* (pp. 445–456). New York: Oxford University Press.

Yang, C. M. (2000). Bush defends death penalty: Texas governor denies deficiencies as execution looms. Retrieved October 2002 from ABC News Web site: http://abcnews.go.com/sections/politics/DailyNews/Bush_Death_Penalty000621.html.

Zehr, H. (1990). *Change lenses.* Scottsdale, PA: Herald.

Zehr, H. (2001). *Transcending: Reflections of crime victims.* Intercourse, PA: Good Books.

Zimring, F. E., & Hawkins, G. (1986). *Capital punishment and the American agenda.* New York: Cambridge University Press.

21

CRITICAL PROBATION FAILURES AND PROMISING INNOVATIONS

MICHAEL JACOBSON

CHARLES LINDNER

In 1991, in response to looming reductions in financial resources, the New York City Department of Probation undertook a massive organizational restructuring initiative. Because of rapidly declining revenues, then Mayor David Dinkins was ordering substantial budget reductions in every city agency. Because probation officer caseloads were already approaching 200, the executive staff of the Probation Department made the decision to radically alter the way it delivered probation supervision services as a way to absorb substantial budget reductions. The staff felt strongly that simply increasing caseloads even further would simply render probation supervision completely ineffective and decided that redesigning probation "from scratch" would be a better way to cope with less funding.

This article will begin by providing some historical context of how probation has been viewed by policy makers and analysts during the 20th century. It will then detail the ways in which the New York City Department of Probation attempted to "remake" itself and will describe the major programs now operating in the department. Finally, it will raise a number of questions about whether this restructuring was effective in public safety or management terms.

A SHORT HISTORY OF PROBATION

Only a small number of criminal justice concepts are widely shared by academics, practitioners, and the public. One concept most commonly agreed upon is the general ineffectiveness of probation services. Academics repeatedly cite research studies showing high rates of recidivism (Gordon, 1991; Petersilia & Turner, 1996; Walker, 2001), practitioners talk of frequent violations by probationers, and the public generally perceives probation as merely a "slap on the wrist." Criticism of probation not only dates from the earliest years of its creation but has been consistently expressed throughout its history and continues today. Moreover, many of the most vocal and astringent criticisms have been expressed by the prominent scholars of the day.

Rothman's (1980) study of early probation is replete with critiques of systematic flaws, including excessive caseloads, inadequate services, token supervision, insufficient salaries, untrained probation officers, and an overall underfunding of probation agencies. He concluded that the system failed both early and uniformly, stating:

> Of the facts there is no dispute. Investigatory committees of all types, whether composed of legislators, social workers, district attorneys, blue-ribbon grand jurors, or concerned citizens, returned a similar verdict: probation was implemented in a most superficial, routine, and careless fashion, as "a more or less hit-or-miss affair," a "blundering ahead." (p. 83)

Probation agencies developed throughout the nation over time, with the sentence of probation steadily increasing in popularity. But despite the increased use of the probation sentence, criticism continued unabated, causing Lindner (1993) to refer to this phenomenon as "Probation's First 100 Years: Growth Through Failure." Illustrative of this phenomenon was Sheldon Glueck's criticism of probation services in 1933, which were not unlike that of Rothman, cited above. Glueck stated that "in too many places, probation work is superficially conducted instead of intensively cultivated" (p. 184). Glueck also noted that probation was characterized by poorly trained officers, poor supervision and record keeping, and a lack of research.

Similar thoughts were expressed 4 years later by Pauline V. Young (1937) of the University of Southern California and coeditor of *Social Work Technique* in her book *Social Treatment in Probation and Delinquency.* Dr. Young stated that she agreed with Glueck "when he maintains that probation, without certain standards, becomes a 'farcical game of chance.' That is, what probation proposes to be and what it has so far succeeded in becoming are frequently very different matters"(p. 14).

As the years progressed and the number of offenders under supervision continued to grow, so did the criticisms expressed by academicians and practitioners. Paul Tappan, a highly respected law professor, noted in 1960 that there was a conflict between the idealistic methods and goals of casework theory and the poor resources given to probation agencies. Nor did Charles Silberman (1978) find the juvenile probation system any better: "After 75 years of talk about substituting rehabilitation for punishment,

it is difficult to find the rehabilitation, and all too easy to find the punishment" (p. 453).

Equally damaging to the image of probation were the numerous instances of extremely negative feedback received from investigating commissions and major research studies. As early as 1927, the New York State Crime Commission concluded that probation officer responses to questions of frequency of home visits "abounded in evasions" and caused the commission to publicly doubt that visits were being made even on a monthly basis.

In 1939, the U.S. Attorney General found that "only a minority of the probation departments conduct any regular training programs for their personnel" (p. 108). The report went on to say that "because of the disorderliness and inadequacy of case recording, it is impossible in many instances to ascertain either the quality or the amount of supervision given by a probation unit" (p. 306).

Among the many negative studies of probation, two published in the 1950s are especially significant in that they documented a ubiquitous weakness of probation services: infrequent and inadequate client contacts. England's (1955) study of a federal probation office found serious weaknesses in both offender supervision and client services. Of a sample of 490 adult probationers who had completed supervision without revocation and were discharged before December 1, 1944, there was an average of only one officer-client contact once every two and a half months. Moreover, only about 25% of the 490 persons in the sample were the beneficiaries of officer-specific services and/or treatment, with the services being primarily advice or referrals. Similarly, Diana's (1955) study of a state probation office established that of a group of 280 delinquents placed under supervision in a single year, for an average time of sixteen and a half months, most were visited only once, and, in turn, most visited the probation office on only one occasion during their period of supervision.

In 1967, the President's Commission on Law Enforcement and Administration of Justice produced the *Task Force Report on Corrections,* one of the most comprehensive and prestigious studies of the major components in the field of criminal justice. The topic of probation was included in the *Task Force Report.* Evaluating probation on a national basis, the report expressed numerous and significant criticisms of probation:

> Probation services have been characteristically poorly staffed and often poorly administered. Despite that, the success of those placed on probation, as measured by not having probation revoked, has been surprisingly high. (p. 28)

> [Probation is] badly undermanned in general by staff who are too often under-trained and almost always poorly paid. Probation agencies only occasionally mount the type of imaginative programs that fulfill their potential for rehabilitation. (p. 29)

> More manpower is needed for probation services than is now available. (p. 29)

> Over the country, then, probation services to misdemeanants are sparse and spotty. Some exceptions are . . . in large metropolitan areas. . . . Even here caseloads are too high to permit adequate pre-sentence investigations and meaningful supervision of probationers. (p. 76)

In 1985, a Rand Foundation study (Petersilia, 1985) reported that "during the 40 month period following being placed on probation, 65% of the total sample were rearrested and 53% had official charges filed against them" and concluded that "felony probation" presented a serious threat to public safety (p. 3). Lauen (1998) studied the effectiveness of a number of probation and parole programs and reported that "the evidence that probation and parole are effective correctional treatments is weak. . . .They might have a marginal effect on some offenders for short periods of time" (p. 33). And one of the largest studies of the effectiveness of probation supervision (Langan & Cunniff, 1992) studied approximately 79,000 felons, or one fourth of the 306,000 felons sentenced to probation in 1986. The study found that while the felons were still on probation, 43% were rearrested for a felony. Half of the arrests were for violent crimes, including murder, rape, robbery, and aggravated assault. Significantly, "Probationers under intensive supervision programs were arrested more frequently than those under less scrutiny" (p. 8).

Any discussion of evaluations of probation supervision effectiveness must mention Martinson (1974), whose negative review of 231 correctional programs placed a damper on the effectiveness of treatment in general. Noting that "with few and isolated exceptions, the rehabilitative efforts that have been reported so far have had no appreciable effect on recidivism" (p. 22), Martinson's work had a chilling effect on all correctional treatment programs. The extensive coverage of his work in professional criminal justice journals and textbooks, although referring to correction programs in general, had a significant effect on probation treatment programs as well. Although a few years later Martinson (1979) wrote that, under certain conditions, many correctional programs had in fact reduced recidivism, his original conclusions on the overall failure of rehabilitative programs were the most often cited. Martinson's original study was greatly influential and, as James Q. Wilson (1980) pointed out several years later, laid the groundwork for increased interest in deterrence, incapacitation, and retribution as features of the criminal justice system. Certainly, Martinson's original study contributed to the current crime control model of longer and harsher prison sentences that we see in the United States today.

Though space limitations prevent a more complete cataloging of criticisms of probation supervision effectiveness, a small number of more recent statements by prominent academicians are especially relevant. Van den Haag (1975) was of the opinion that probation supervision, in most cases, is only a formality. Conrad (1985) stated that "in the present circumstances the survival of the idea of probation as a service is in jeopardy" (p. 421). Morris and Tonry (1990) similarly castigated ordinary probation, stating that in many cities it had "degenerated into ineffectiveness under the pressure of excessive caseloads and inadequate resources" (p. 6). Byrne (1988), citing the continually increasing numbers of offenders under probation supervision and the resultant excessively large caseloads borne by officers, argued that the crowding of probation "poses a more immediate threat to the criminal justice process and to community protection" than does prison overcrowding (p. 1). Walker (2001) wrote:

In short, most offenders placed on probation "fail." . . . Many experts have serious doubts about whether traditional probation programs provide any kind of meaningful treatment. The level of supervision has always been very minimal, involving a meeting with a probation officer (PO) once a month. The PO fills out the required reports, and that is that. (p. 218)

Champion (2002) found that "standard probation is considered by many critics to be the most ineffective probationary form." He attributed limitations on probation effectiveness to high caseloads, inadequate staffing, overworked officers, varying clientele, and often excessive and unnecessary paperwork. As a result, he concluded that "standard probation supervision in many jurisdictions is essentially no supervision at all" (172). Many would agree with Rosecrance's (1986) assessment of the problems faced by probation:

Judicial support for probation services has eroded, public support has diminished, legislative backing has wavered. Probation officers themselves question the efficacy and purposefulness of their actions, while probationers seriously doubt that any good will come from their contacts with probation officials. (p. 25)

Despite the incessant criticism of probation for the lack of client/probation officer contacts throughout the history of probation, it does not appear that the situation has been corrected. In a survey conducted by the National Center on Institutions and Alternatives in 1997, "Over 100 probation officers reported that they saw at-risk offenders face to face less than five times per year" (Wooten & Hoelter, 1998, p. 30).

Critiques of juvenile probation services are generally equally negative. Lundman (1993), for example, found that most juvenile probation officers are plagued by inordinately large caseloads and excessive paperwork, resulting in an ability to "devote sustained attention and effort to only some of their probationers and even then, only for short periods of time. For many probationers, regular probation necessarily consists of little more than an occasional contact with an over-worked officer" (121–122).

COMPETITION FOR SCARCE RESOURCES

Though the criticisms of probation work expressed above may in many instances be valid, it should be recognized that probation supervision today is more difficult than ever before. Originally, probation was intended to serve, and did serve, basically a misdemeanant population—typically nonviolent offenders. Rarely would the number of convicted felons under probation supervision exceed 10% of the total probationer population. In New York State, for example, over a 14-year period ending on September 30, 1921, the number of convicted felons on probation amounted to approximately 9% of the total probationer population (New York State Probation Commission, 1923). Rothman (1980) similarly found that "in a state like New York, a little over 90 percent of probationers in 1914 were misdemeanants and only 10 percent felons" (p. 108).

Over the years, with the insatiable demands of institutional facilities for more space, the probation population nationwide was transformed into a felony population. This was at least in part because many of the felons placed under probation supervision would have been incarcerated in past years. New York City probation is illustrative of the dramatic shift away from misdemeanor populations toward a felony population: Currently, approximately 75% of the probationer population are convicted felons. Not only does a felony population present increased supervision difficulties for probation, but studies reflect high numbers of special-needs offenders among felony offenders, such as substance abusers, an especially difficult population to supervise. The New York City Department of Probation now estimates that about 50% of all new admissions are drug users (J. Alpern, Assistant Commissioner of the New York City Department of Probation, personal communication, March 2002).

Many probation officers, not only in New York City but nationwide, perceive that over the years caseloads have become more difficult to supervise. In a nationwide study of probationer/parole personnel, it was reported that "at least three-fourths of the respondents believe that the supervision needs of offenders are greater now than in the past. Thus, not only are the numbers larger, the offenders are also a more difficult group to manage" (Guynes, 1988, p. 8). In addition to these concerns, many officers in New York City find fieldwork stressful because they often make their visits in high-crime, drug-ridden areas where dangerous weapons proliferate (Lindner & Koehler, 1992). This is especially true in many probation departments, including the New York City Department of Probation, where policy does not permit the arming of probation officers except for those assigned to special units.

As is true of probation agencies throughout the nation, inadequate funding contributes to the many problems faced by the New York City Department of Probation. As with all city agencies, there is a competition for law enforcement funding, with priorities going to police, courts, and institutional corrections. As noted by Clear and Cole (2000):

> Corrections must vie for funding, not only with other criminal justice agencies, but also with agencies dealing with education, transportation, social welfare, and so on. . . . Understandably, corrections does not always get the funding it needs; people may want garbage collected regularly more than they want quality correctional work performed. . . . It is not easy to win large budgets to help people who have broken the law. (p. 17)

The inadequate funding of probation on a nationwide basis was reported by Petersilia (1985), who found that the funding of probation had not kept up with the expansion of probation caseloads. She found that 25¢ of every dollar spent on criminal justice goes to correction, with only 3¢ of that quarter spent on probation, and that whereas most criminal justice agencies on a nationwide basis received increased funding over the past 10 years, only probation received fiscal reductions.

Probation has long been the Cinderella of the criminal justice system: overworked, inadequately funded, low in status, and subjected to frequent criticism. Barlow (2000) noted that "probation officers are among the lowest-paid criminal justice professionals" (p. 646). He cited a number of states where the average salary of probation officers is especially low,

including Arkansas ($19,058), Indiana ($21,576), Tennessee ($21,528), Illinois ($26,000), Nebraska ($26,572), and South Dakota ($26,800). Officers in other states do not earn much higher salaries. Citing low salaries, inadequate staffing, unrealistic caseloads, and an overall lack of support for probation services, Barlow (2000) painted a dismal picture of probation work:

> By the early 1990's probation was clearly in a state of crisis, with little sign of improvement on the horizon. Underfunded and overwhelmed–and seemingly unappreciated by politicians and the public alike–the probation officer could see little reason to cheer. (p. 647)

Not only does inadequate funding contribute to excessive caseloads and limitations on available services, but the low salaries of officers contribute to the failure to attract high-quality staff and result in excessive staff turnover. It is generally recognized that smaller caseloads are not necessarily a guarantee of success. Experiments conducted during the 1960s and 1970s to find the optimal caseload size found "that caseload reduction alone did not significantly reduce recidivism in adult probationers" (Clear & Cole, 2000, p. 188). Nevertheless, caseloads that today can range from 250 to 1,000 are certain guarantees of failure.

The underfunding of probation is not new but has occurred from the earliest years of probation's history. In 1991, the realities of probation's inadequate budgets were noted by Kaplan, Skolnick, and Feeley:

> Probation agencies all over the country have been asked to supervise more criminals, and more serious criminals, yet their budgets are being cut. The justification for the budget cuts has been that probation agencies serve less serious offenders and can therefore manage with less funding. The greatest irony is that most of the cuts were made to free more resources for prison construction and operation. The result has been a staggering increase in some probation agencies' caseloads. In many jurisdictions, "low-risk" probationers have virtually no supervision; they only have to mail in a postcard at specified intervals. (p. 533)

In 1967, the *Task Force Report on Corrections* positively cited New York State's partial funding of local probation departments. It noted that

> in New York State a 50 percent reimbursement is made to county probation departments for operating expenditures, thus enabling the State to set standards for education and compensation of staff and for administration and record-keeping. The National Survey found that this subsidy promoted better-qualified staff, more adequate salaries, lower staff turnover, and smaller caseloads. (President's Commission, 1967, p. 80)

In recent years, the state has cut back on this funding. In the early 1980s, New York State reimbursed almost 50% of the New York City Department of Probation's budget. In 2001, that percentage has been reduced to 31%. Thus, the state had systematically defunded probation services in New York City and in all the state's counties over the last 20 years. Though New York City essentially replaced the withdrawn state funds with local tax levy monies, there has been little money left to bolster probation services. The state has withdrawn funding for probation over the years, yet it has continued to entirely fund the city's Intensive Supervision Program (ISP). The reason that this program continues to be 100% state funded despite drastically reduced state funding for probation is that it is designed to divert convicted felons from state prison. The state clearly believes it is in its financial interest to fully fund the ISP because the program's goals are to divert offenders from more expensive prison sentences.

Additionally, there is little federal money available for probation departments to tap. Unlike police or correction agencies, which received billions of dollars through the federal 1994 Crime Act, no funding at all was allowed for probation agencies. Thus, if cities and states reduce probation budgets, there is essentially no chance that funding can be obtained from the federal government. In fact, even with New York City making up for the state's reduced funding for probation, supervision

caseloads in New York City have grown from 148 in 1990 to 221 in 2000 (City of New York, 1991, 2000). Though probation officer caseloads were rising during the 1980s and 1990s in New York City, the same was not true for child protective workers or for New York City school teachers. From 1989 to 2001, child protective workers' caseloads actually decreased significantly from 16.3 to 12.4 cases per worker, and the average high school class size stayed relatively constant, increasing only from 30.4 students to 31 students (City of New York, 1989, 2001).

Not only has the New York City Department of Probation long experienced inadequate funding, but it appears that the situation will be exacerbated rather than improved. In fact, in the mayor's preliminary budget for fiscal year 2003, the Probation Department is slated to lose almost $13 million in total funds, or about 16% of its total budget (City of New York, 2002).

EFFORTS TO DO MORE WITH LESS

Plagued by inadequate funding, excessive caseloads, high staff turnovers, a negative public image, and other severe limitations impeding effective services, the New York City Department of Probation has, in recent years, nevertheless sought to improve probation work with a number of innovative techniques. Some of the most significant are discussed below.

The New York City Department of Probation is the second largest in the country. At any time, the department is responsible for the supervision of more than 60,000 adults and 4,000 juveniles. In city fiscal year 1999, the agency prepared approximately 53,000 presentence investigations (City of New York, 2001). These investigations are primarily used to provide the court with an overview of the crime, a social history study of the defendant, and a recommendation for disposition so as to assist the court in sentencing. In addition, if the defendant receives a probation sentence, the report serves to assist the supervising officer in formulating case management plans for supervision of the probationer.

RESTRUCTURING THE AGENCY

In the early 1990s, the Department of Probation, together with a number of other New York City agencies, was ordered to downsize so as to reduce operating costs. The city was experiencing a severe downturn in revenues, and Mayor Dinkins ordered all his agencies to cut their budgets. This was especially difficult for an agency like Probation, which was already underfunded and crippled by excessive workloads. Funding simply did not match steadily increasing caseloads. Moreover, the disproportionate numbers of felony cases were a further drain on agency resources and heightened the risk of public condemnation should a probationer receive negative media exposure for committing a particular heinous violent crime.

In 1991, the Probation Department and the city's Office of Management and Budget (OMB) agreed that Probation's planned budget reduction of about $3 million would be deferred if the agency undertook a systematic restructuring of its adult supervision services. Because neither OMB nor Probation wanted the budget reduction to simply result in huge caseload increases, the agency was given the opportunity to completely redesign supervision services from scratch. The rationale was that if resources were going to be even further diminished, a process that questioned and potentially changed a number of assumptions about supervision might well result in a more effective supervision system. The agency then embarked on a large-scale restructuring of its adult supervision function and used a business process redesign methodology, receiving technical assistance from the Bell Atlantic telephone company.

Recognizing that inadequate staffing prevented all probationers from receiving the same degree of supervision, in terms of either surveillance or treatment services, it was decided that probation services would be entirely restructured through the use of new classification systems, use of structured group work for high-risk cases, and use of the latest technology. The redesign itself was put together over a year-and-a-half period by a team of probation officers, supervisors, and clerical staff.

The goal of the restructured system was to identify and supervise most closely those probationers whom the agency believed to be at the highest risk for committing violent crime. In fact, the overall goal of the restructured system was to reduce violent recidivism. The new risk instrument or classification system was designed to flag those probationers who required intensive and invasive controls, unlike the many low-risk probationers, who were not likely to commit a serious crime and who could function with only token supervision. Similarly, treatment/rehabilitative services were not needed for all probationers and, being always in short supply, were not to be wasted when there was neither need nor probationer motivation for change. In effect, the probation classification system was based on a triage model, ensuring that intensive surveillance and intervention would be used for those who most needed these services. Probationers unlikely to commit another serious crime, or not in need of treatment, would receive minimum levels of supervision.

Classification systems are not a new concept in probation. For many years, in the absence of formal classification systems, probation officers unofficially and informally classified offenders on their caseloads according to levels of risk and services required. Subsequently, formal classification systems were developed, often based on risk and service needs.

THE RESTRUCTURED NEW YORK CITY DEPARTMENT OF PROBATION

The New York City system was designated the Adult Supervision Restructuring (ASR) system and referred to as " New York's Bold Experiment" by the New York Times (DOP Web Page available at: http://www.nyc.gov/ html/prob/html/accompl.html,accessed 9/29/01). ASR appears to be quite methodical and is based on extensive research. Its goals are as follows:

1. Public protection is increased by provision of the lion's share of surveillance and controls to higher-risk probationers. Accordingly, high-risk probationers are supervised in smaller

caseloads. These probationers participate in structured cognitive training groups.

2. Treatment services are delivered on a need basis. Unlike early probation, the belief that a probation sentence, in and of itself, is an indication of a need for treatment has been discarded.

3. Consistent with a behavioral modification model, probationers may be moved to a more or less intrusive level of supervision on the basis of their behavior and the time remaining of their probation sentence. This provides a degree of motivation for good behavior.

The reengineering process of the agency has been completed and is now in operation throughout the city of New York. At the heart of the strategy is a computerized system, known as the Adult Restructuring Tracking System (ARTS), "which tracks offenders as they move through the various stages of the probation system. The computerization gives probation officers and management instant access to up-to-date information on the probation population" and is useful in trends and studies of agency effectiveness (NYC DOP Web Site available at: http://www.nyc.gov/html/ prob/html/ asr.html, accessed 9/28/01).

Upon being sentenced to probation, all offenders are assigned to an Intake and Assessment Unit for classification. Those classified as nonviolent will be assigned to a minimal supervision unit. Probationers classified as demonstrating a high potential for violence undergo a secondary assessment that includes a structured interview, a mental health screening, and a determination as to the suitability for group counseling.

Enforcement Track

On the basis of the secondary assessment, high-risk probationers are placed in one of four specialized units (designated by different colors) of an "Enforcement Track."

The Blue Unit

Probationers assigned to the Blue Unit are males, ages 16 to 20, who have been classified as violence prone. This unit is characterized by small caseloads and frequent probation officer-client contacts to enhance community protection. An important component of supervision of this category of probationers is mandatory group counseling, based on a cognitive skills development model, "which was designed to change the way in which [probationers] think and make decisions" (NYC DOP Web Site). The counseling groups are limited to no more than 15 probationers and meet twice weekly for 90-minute sessions over 19 weeks. Areas covered include anger management, interpersonal relations, self-care, and impulse control. The content and structure of these small group work sessions were developed by a team of social work professors from the Hunter School of Social Work in New York City.

Probationers who successfully complete the counseling program and who do not commit serious behavioral infractions are usually transferred to the Green Unit after about 8 months. This is in effect a reward, consistent with behavioral modification models, as the Green Unit is less intrusive and restrictive. Those failing to successfully complete the program are transferred to the Amber Unit.

The Amber Unit

The Amber Unit provides supervision of high-risk probationers who failed to complete the group sessions while in the Blue Unit or who were believed to be inappropriate for group counseling. As in the Blue Unit, caseloads are small, with frequent contacts between probation officer and probationer. Although there is no group counseling, referrals are made to community service agencies when appropriate. Referrals include employment counseling, mental health services, substance abuse programs, and assistance for domestic abuse problems. Probationers remain in this track for approximately 1 year, and upon graduation from this unit they are promoted to the Green Unit.

The Green Unit

This unit supervises probationers who have graduated from the Blue or Amber Units, an ISP, or an electronic monitoring program. Officers in this unit are "relapse prevention specialists" who attempt to build on the progress already made by probationers. When indicated, referrals are made to community service agencies. Caseloads are somewhat larger than those of the Blue or Amber Units, and the focus of offender supervision is the identification of "realistic goals and taking the necessary steps to realize them. Probationers usually remain in the Green Unit for one year before graduating to either the Special Conditions or the Reporting Tracks" (NYC DOP Web Site).

The Red Unit

This unit supervises probationers transferred from other units because of a pending allegation of violation of probation. Upon transfer to this unit, the alleged violation is referred to court for calendaring. In the interests of public safety, probationers under supervision in this unit are subject to frequent contacts with the supervising probation officer. The officers in this unit prepare presentence investigations of offenders under their supervision and additionally "monitor the status of incarcerated offenders with Violation of Probation charges pending. Probationers remain in the Red Unit until the Violation of Probation has been resolved" (NYC DOP Web Site).

The Special Conditions Track

In addition to the four Enforcement Tracks cited above, there is a Special Conditions Track, which is for the supervision of probationers who (a) are not considered violence prone but are subject to court-ordered special conditions or (b) are classified as violence prone and, though graduated from the Green Unit, have not yet completed their court-imposed special conditions. Court-imposed special conditions, for example, include the payment of a fine more than $1,000, orders of restitution and/or community service, or referrals for treatment services, including substance abuse, mental health, or domestic violence prevention programs.

Probation officers in this track are responsible for making appropriate referrals and monitoring compliance. As a general rule, probationers remain in this track until the special conditions have been implemented or successfully carried out for a period of 12 months. Upon successful completion of the conditions of probation, they are transferred to the Reporting Track.

The Reporting Track

Probationers assigned to the Reporting Track receive minimal supervision. Contact with the agency is through automated reporting kiosks in lieu of face-to-face contacts, and unless there is known probationer misconduct, there is no mandated officer-offender contact. However, probationers are seen when they request appointments with their officers. The Reporting Track is based on technology that resembles automated bank teller machines. The kiosks are capable of identifying probationers through the use of hand geometry technology.

The Probation Department does not have any trend data that can show the impact of these group work interventions on rearrest or recidivism rates.

THE STATE OF TECHNOLOGY IN PROBATION

In a 1985 article entitled "Probation and the Hi-Technology Revolution," Moran and Lindner noted that "the hi-technology revolution has had far less impact in the field of probation" than in other professions (p. 25). Though that may have been true in the past, many probation agencies are now incorporating technological innovations into their practices. Dunworth (2001) noted that law enforcement, which would include probation departments, "lags behind many other elements of society" (p. 52). Nevertheless, he asserted that there was an "inevitability about that adoption. In the end, law enforcement will not have a choice. The IT revolution will have to be embraced" (p. 52).

Pagers

Ogden and Horrocks (2001) reported on the assignment of pagers to supervisees in the U.S. Probation and Pretrial Office in the District of Utah. As supervisees are required to have their pagers on at all time and to respond immediately to a message, the pager has been found to be a reliable means of communication. In those unusual cases where the supervisee is without a telephone, it is expected that he or she will go to a public telephone or neighbor to call the probation/pretrial officer. The use of pagers allows for contact with the supervisee at any time and enables the officer to communicate immediate instructions and directions. It has proved to be effective in reducing wasted efforts in attempting field visit contact, which is especially significant in rural areas. Pagers can also serve as an alternative or enhancement to electronic monitoring, as the supervisee can be reached at any time and in any place.

Computer Applications

The Utah office has also improved the efficiency of its officers through the installation of a digital dictation system, which allows chronological case recordings to be made from any telephone and at any time. This system gives greater flexibility to the officer as to the time and place of the recording and enables records to be kept always current in that they can be transcribed and entered into the computer with less than a 2-day turnaround.

Unquestionably, enhanced technological application will improve case management systems in community-based corrections. Illustrative is the federal judiciary's Probation and Pretrial Services Automated Case Tracking-Electronic Case Management System (PACTS), which "allows officers to electronically collect pertinent case-related information to produce statistical and workload reports. The case management portion (ECM) helps officers collect, manipulate, and recall case-management-specific information" (Cadigan, 2001, p. 25). It is believed that this comprehensive case management system will make data more accessible to a larger number of users, enable them to more easily use the information in their professional duties, and allow for more efficient and effective investigations and supervision. Anticipated benefits are reductions in both paper waste and the inefficiency inherent in a paper-controlled world, the elimination of data redundancy, increased validation of data, and other improvements in efficient agency function.

Obviously, there will be continued growth in the use of computers in community-based corrections agencies. It is anticipated that virtually all agencies will have computers available to their officers, together with notebook computers, which are especially helpful in fieldwork. The computer not only is invaluable for inputting information into agency files but will be increasingly used by officers in the investigative process. Cadigan (1998) found that probation officers have used the Internet

> to obtain information regarding street and prison gangs, militia groups, and "hate groups." Other officers have used the Internet to obtain information on the newest suggested techniques for defeating drug testing. One officer used the Internet to detect web pages developed by a sex offender with a special condition prohibiting him from using the Internet. (p. 16)

Telecommuting

Telecommuting is another technological innovation currently in use by the U.S. Probation Offices in the Middle District of Florida. It enables an officer to work at home or other telework facility for at least one regularly scheduled workday a week. Long used in private industry, telecommuting is believed to reduce agency cost by eliminating and/or consolidating unused office space. This benefit can be especially important in offices where space is at a premium. In addition, telecommuting may also increase officer job satisfaction, reduce time and expenses related to travel to a work site, and contribute to increased productivity.

Greek (2001), however, warned that in many instances telecommuting does not prove to be effective in reducing overhead cost or productivity. Some employees miss the social interaction with fellow employees, the ability to immediately share ideas and/or concerns with coworkers. Equally important, some employees, lacking a strong sense of discipline, may be easily diverted from their employment duties by home or family matters. Greek cautioned that telecommuting is not right for everyone. Probation "officers must be organized, self-motivated, disciplined, positive, and capable of working with little direct supervision" (p. 52).

Automated Reporting Kiosks

The use of kiosks as an aid to supervision techniques appears to be at the forefront of the technology revolution. A number of community-based correctional agencies already employ kiosks, and others have indicated plans to implement similar strategies (Ogden & Horrocks, 2001). The New York City Department of Probation was one of the first community-based corrections agencies to use automated reporting kiosks in the supervision of low-risk offenders.

Probation's restructuring process resulted in the use of kiosks to keep track of the over 15,000 low-risk probationers in New York City. The kiosk system was designed to replace a system of questionnaires that required low-risk probationers to drop off a monthly form as well as make a mandatory officer contact once every 6 months. Because the agency had no sure way of knowing who exactly was dropping off these questionnaires to probation offices and because thousands of questionnaires were backlogged before being entered into the agency's database, it was decided the system would be totally scrapped. The designers of the new restructured system felt that improved technology could far better identify and keep track of these probationers and that requiring a 6-month contact served no substantive purpose whatsoever.

The kiosks themselves resemble bank ATM machines. They have hand geometry technology that requires probationers to take an image of their handprint that is matched against an original handprint scan. There is a screen where probationers read the questions they must answer and a keyboard where they enter responses. There is also an audio component that reads the questions to probationers along with the written questions.

Reporting to the kiosk is made monthly, though the technology allows for more or less frequent reporting. Those reporting are low-risk probationers and those without special court conditions and those with a higher risk of violent recidivism who have already successfully completed a minimum of 18 months of more intensive supervision (J. Alpern, personal communication, March 2001). The time needed for reporting to the kiosk is about 3 minutes. The kiosk can be programmed in several languages.

To start the machine, the probationer enters an identification number, followed by the match of his or her handprint. Once the probationer's identity is verified, he or she is asked a number of questions, including whether he or she has been arrested since the last check-in. If the probationer's identity is not verified or if the answers indicate that the probationer is giving answers that seem to contradict prior responses or violate the conditions of probation, the computer sends a silent signal to the kiosk attendant. The attendant will then summon a probation officer to immediately talk to the probationer.

Preparation for use of the kiosk begins with the collection of information regarding a new probationer. This is done in a one-on-one meeting between the probationer and the officer. The information collected includes the probationer's name and other identifying data as well as a variety of case-specific criminal justice data such as conviction charge and prior convictions. In addition, a picture of the probationer is taken along with an image of his or her handprint. All this information is stored in the kiosk database. The kiosks themselves are located in all the city's five boroughs and are in the probation supervision offices. All probationer information is stored on the agency's database and is updated each time a probationer reports to a kiosk. In fact, the kiosk system essentially makes the probationers themselves the prime data enterers in the agency. The kiosks collect, and send to the agency's mainframe computer, all data collected on probationer activities, including whether they are still working, in school, or in drug treatment and whether they have been rearrested.

When the probationer is finished reporting, the kiosk will automatically generate the next appointment according to a predetermined schedule set by the reporting-track probation officer. The probationer will be given a receipt with the next date, and the agency's computer system will note the next date the probationer is to report.

A failure of the probationer to report to the kiosk automatically generates letters to the residence of the probationer. The series of letters sent get more threatening in tone, with the final letter saying that a violation will be filed and a warrant immediately issued if the probationer does not immediately report. According to the agency, the nonreporting rate for reporting-track probationers is between 11% and 12%. This compares extremely favorably to the nonreporting rate for probationers under the questionnaire system, who had an overall non-reporting rate of over 40% (J. Alpern, personal communication, 2001). Similarly, rearrests cause the probation kiosk staff to be notified, so they may in turn call the probationer for an interview if he or she is not being held in jail.

It is important to keep in mind that the purpose of the kiosk system is not to replace direct contact between probationers and the probation officer. Rather, it is simply designed to keep track of the thousands of probationers who used to just submit a monthly questionnaire to an agency clerk. As a purely monitoring effort, it is far more efficient than the use of questionnaires. The technology allows the positive identification of the reporting probationer, the immediate collection of data, total flexibility of creating reporting schedules, and the automatic generation of letters in the event of nonreporting. The fact that, according to the agency, nonreporting rates have been cut by almost 75% compared to the questionnaire system is a testament to the narrowly defined role that kiosks are designed to play in New York City's probation system.

The Adult Services Restructuring project is now going through additional refinements. The current Commissioner of the Probation Department, Martin Horn, is planning to collapse the different tracks described above into only two basic tracks: high-risk and reporting. Within the high risk track will be those classified as high risk as the current risk instrument calls for and there will be an additional group that does not fall into this category through their risk classification but have other risks such as domestic violence or mental illness. The Red track will be eliminated and individual officers will be responsible for their own violation cases. All other cases will fall into the reporting track. The Commissioner estimates that eventually about 75% of all cases will be in the reporting track.

Finally, the Department will begin to experiment with and install telephone integrated voice response systems in its field offices to gradually replace the kiosks. This phone system, which will be integrated with the Department's computer system, is capable of verifying identification through voice recognition and is able to collect the same kinds of information as the kiosk. The probationer will still use a keypad, attached to the phone, to respond to the same kinds of questions that appear on the kiosks. One of the reasons the Commissioner is pursuing this technology is that with the great majority of cases eventually in the reporting track, many more of these phone systems can be installed in probation offices than can kiosks. Additionally, it has the potential to eventually allow the probationers to call from their home phones and to serve as a curfew checking system (Horn, 2002).

INTENSIVE PROBATION SUPERVISION

Intensive probation supervision is an intermediate punishment in that it is between probation and incarceration: more punitive and invasive than probation but less restrictive than incarceration. Intensive probation has become increasingly popular for the supervision of high-risk offenders.

Champion (2001) noted that "after 25 years, we have yet to devise a standard definition of what is meant by intensive supervised probation" (p. 456). Abadinsky (2000) similarly concluded that "intensive supervision actually takes many shapes" (p. 411). And Morris and Tonry (1990) stated that although there are many intensive probation supervision programs throughout the nation,

> their diversity is such that the term has almost ceased to have useful meaning. Their common feature, however, is that more control is to be exerted over the offender than that described as probation in that jurisdiction and that often these extra control mechanisms involve restrictions on liberty of movement, coercion into treatment programs, employment obligations, or all three. (p. 180)

Though designs vary, a number of characteristics are common to most programs. These include smaller caseload size (often a maximum of 25), supervision of higher-risk offenders, more frequent client-officer and collateral contacts, mandatory referrals for treatment, education, or vocational training, more intrusive conditions of probation, and aggressive elements of surveillance and control.

Some ISPs use a two-person team in which one officer is responsible for overall supervision while the other handles offender surveillance. Some see team supervision as a means of addressing the role conflict inherent in probation supervision, tension between the need to control and monitor as well as to perform casework. As Morris and Tonry note:

> . . . . Many jurisdictions are experimenting with team supervision of those on intensive probation, duties being divided between two probation officers in relation to those often conflicting purposes. (Morris & Tonry, 1990, p. 184)

One strength of juvenile ISPs is the cost savings when they serve as an alternative to incarceration (Champion, 2001). In addition, if the selection of offenders placed in intensive supervision units primarily consists of those who otherwise would have been incarcerated, then these programs serve to reduce the crowding of correctional facilities. Illustrative is the claim of probation administrators and evaluators in Georgia "that the numbers of people diverted from prison to intensive supervision probation have eliminated a need to build two new prisons" (Morris & Tonry, 1990, p. 233).

In addition, ISPs are consistent with principles of caseload classification, as they ensure that high-risk probationers will receive intrusive supervision because of smaller caseloads, the allocation of more resources, increased controls, and more frequent officer-offender contacts. Hopefully, the increased levels of supervision intensity will heighten public safety. Finally, ISPs may help to improve the tarnished image of probation among the public, academicians, and criminal justice practitioners. Though many consider ISPs to be little more than what probation originally was meant to be, and should be, the heightened surveillance, punitiveness, and offender accountability are more akin to the public's expectations of probation.

For an agency faced with a limited budget, the utilization of intensive probation caseloads is limited by its increased costs compared to those of traditional probation. In Georgia, the basic probation cost per offender was $1.01 per day, in contrast to intensive probation, which was $4.07 per day. In Illinois, the cost of basic probation was about $2.75 compared to the $10.95 cost of intensive supervision (Abadinsky, 2000, p. 418).

The effectiveness of intensive probation supervision is extensively and passionately debated. Early studies were highly positive in terms of recidivism and the reduction of prison admissions. However, in 1989, the U.S. General Accounting Office examined 18 studies of ISPs and concluded that ISPs do not reduce recidivism. The report noted that "as implemented, the programs provide significant community protection while the offender is in the program but offer little to assist the offender in reducing future criminality" (McCarthy, McCarthy, & Leone, 2001, p. 175).

In 1993, the National Institute of Justice evaluated 14 ISPs in nine states, including over 2,000 offenders. McCarthy et al. (2001) stated that the programs "generally failed to reduce prison crowding, save money, or reduce recidivism" (p. 176). In summary:

> Although the levels of contact varied considerably, the programs succeeded in providing enhanced surveillance. Surveillance alone, however, seemed to have little impact on recidivism. At the end of one year, 37 percent of ISP participants and 33 percent of controls were arrested, with the difference largely attributable to enhanced controls on the ISP offenders. (p. 176)

In reviewing the same National Institute of Justice study, DiMascio (1997) concluded that "ISP had not lowered the cost of correctional services and did not significantly reduce recidivism" (p. 317).

In 1998 Joan Petersilia wrote that ISPs in general are seldom used for prison diversion, do not reduce the number of prison-bound defendants, and increase the number of technical violations, thus driving up total system costs. Petersilia (1998) did find, however, that programs that provided treatment and additional services obtained some reductions in recidivism, especially among high-risk or drug offenders.

New York City, like most other large probation departments around the country, has a large ISP. In fiscal year 2000, the department supervised almost 2,300 probationers in intensive supervision. Of those, about 1,600 were adults and 700 were supervised in juvenile ISPs. The number of adults on ISP has doubled since 1990, when the Probation Department supervised about 800 probationers (City of New York, 1991, 2000). Clearly, this is one area of probation that the state is willing to continue to invest in as a way to try to control prison costs. Probation officer caseloads are 25 to 1, or about one tenth what they are in regular adult probation and a quarter of the general juvenile supervision caseload. Like all ISPs, these programs are designed to keep adults from being sent to prison or juveniles being placed in secure state-run facilities. Also, like most other ISPs, there are no reliable data showing the diversion savings from either the adult or juvenile programs.

Criticism of ISPs continues to be strong. Major criticisms are that they foster net widening; serve to increase rates of technical violations; fail to reduce the number of prison beds; do not reduce recidivism rates; are expensive when compared to traditional probation rather than incarceration; and drain agency resources from traditional caseloads (Abadinsky, 2000; Champion, 2002; Latessa & Allen, 1999). It should be noted, however, that there is considerable variation among ISPs, so various programs show widely diverse indicators of effectiveness. Research was conducted by the New York State Division of Parole in 1980 on intensive supervision (Abadinsky, 2000). But although the New York City Department of Probation is one of the largest in the nation and has both juvenile and adult supervision programs, it was unable to provide any recent, comprehensive study.

OPERATION NEIGHBORHOOD SHIELD

Operation Neighborhood Shield was initiated by the New York City Department of Probation in October 1997 and was modeled after Boston's Operation Night Light, which subsequently became Operation Tracker. The essence of these programs is to form a partnership between probation and the police to enhance the supervision of certain groups of high-risk probationers. It is hoped that this will be achieved by eliminating some of the blind spots of communication between the police and probation through the two agencies' collaboration in the supervision of certain probationers (Champion, 2002). Probation-police partnership programs have become expeditiously popular and include Maryland's "HotSpots" initiative; Redmond, Washington's SMART Partnership; and Operation Spotlight, created by the National Center on Institutions and Alternatives. Like all of these programs, Operation Spotlight "focuses investigative and supervision services of police and probation systems on the at-risk offenders already in the community" (Wooten & Hoelter, 1998, pp. 30.). Illustrative is Redmond's SMART Partnership, where "police officers received training and actually began direct monitoring of felons who were under Department of Corrections supervision" (Morgan & Marrs, 1998, p. 171). The New York City Probation program titled Operation Neighborhood Shield focuses on the apprehension of probation violators with outstanding warrants and on probation absconders. The teams also conduct nighttime home visits on probationers identified as domestic violence offenders, sex offenders, child abusers, gang members and drug offenders. (NYC DOP Web Site available at: http://www. nyc. gov/html/prob/html/adult_serv.html, accessed 10/22/02). The team supervision approach allows probation officers to provide the police with identification of probationers, together with court-ordered conditions of probation. The agency reports that "in fiscal year 2001, Nightwatch teams executed 1147 warrants and made over 35,000 field checks (Office of the Mayor, Office of Operations 2002). All participating probationers must submit to random drug testing and those who test positive must get drug treatment. In the counties of New York City in which Neighborhood Shield is in existence, specially designated judges monitor the cases of probationers in the program.

In addition to the police-probation monitoring of high-risk probationers, Operation Neighborhood Shield is designed to connect probationers and their families with the services they need to help an offender make a transition into school or work. Referrals to these services are made through collaboration with a number of community-based agencies.

Offender accountability is an important component of the program. Probationers, usually working in teams, perform community service for nonprofit organizations in the neighborhood as part of their sentence. It is believed that community service is especially beneficial to neighborhoods that have endured excessive rates of crime. Examples of community service performed under program auspices include graffiti removal, cleanups of abandoned lots, and painting of churches and community centers.

A second major goal of Operation Neighborhood Shield is to have probation relate more closely to the community that it serves. On the basis of many of the concepts of problem-oriented policing, community advisory boards have been created, and regular meetings are held between residents and probation to discuss neighborhood concerns. The Department of Probation provides assistance to the community in addressing these concerns. The rationale for this component of the program is the belief that probation should not solely function out of a centrally located court building but that probation officers should know and understand the community and be able to work with the probationers in their own neighborhood. Although this approach to probation services is not new, having been tried many times before by different probation departments, the agency's commitment seems stronger in terms of resources and financial aid than in the past.

Interestingly, though many view these community support programs as arising primarily from problem-oriented policing and community-oriented policing, they are also a reflection of concepts underlying the probation officer's role in the early 1900s. Influenced by the philosophy of the Progressive Movement, the early probation officer was seen as a friend to the probationer (Rothman, 1980). Moreover, "the probation officer was not to lose sight of the duty of upgrading the community" (Rothman, 1980, p. 65). These early concepts of the probation officer's role as a provider of support for both the offender and the community arose from the custom of "friendly visitation," in which

middle- and upper-class women visited the homes of the poor to give advice and support. In addition, the Settlement House Movement sought to improve conditions of the poor and the neighborhoods in which they lived (Lindner, 1992; Lindner & Savarese, 1984). These probation officer role definitions were totally consistent with the ideology of the Progressive Movement.

It should also be remembered that many of the principles underlying probation-police partnership programs were advocated by Morris and Tonry in 1990. These authors suggested a closer liaison between probation and the police to provide greater surveillance of offenders. They noted "that unlike probation officers, the police are on duty 24 hours a day and in particular at the times when the probation office is not open—which are also the favored times both for crimes and for breaches of conditions of probation" (p. 185).

Walker (2001) noted that Operation Night Light and related programs are seen by critics as "often more punishment than rehabilitation oriented." He argued that "this trend reflects the extent to which recent developments in corrections have been completely dominated by the conservative crime control agenda" (p. 222). Most important, he referred to the "Boston programs as aggressive enforcement efforts," with the involvement of probation and parole being irrelevant "to the traditional rehabilitation goals of probation or parole" (p. 227). There is little question that current police-probation and parole partnerships have a far stricter emphasis on enforcement and control than traditional probation. Though these programs have become exceedingly popular, there is no reason that probation cannot also perform its function of assisting reintegration through treatment, job training and placement, and other services. Simply turning probation and parole into enforcement arms of the police will not serve the long-term interests of probationers or public safety.

Nova Ancora

One of the unique programs run by the department is Nova Ancora ("new anchor"). Essentially a job training and employment program, it was initiated in the 1980s with a

$100,000 grant from the Ford Foundation. It was suspended in 1990 because of budgetary cutbacks but was reactivated in 1994. The mission of the program is to reduce recidivism by providing life skills preparation, vocational training, and employment for probationers (Gelormino & Weiss, 1997, pp. 1–4). Nova Ancora recruits businesses and trade unions to provide on-the-job training and subsequent employment for probationers. Over the years, probationers have found employment in positions ranging from entry-level jobs to relatively skilled labor, including employment in computer technician, construction work, clerical, and food catering services positions.

Employment of probationers is often a key force in preventing recidivism. However, it is not easy for many to obtain satisfactory employment on their own. As long ago as 1969, Glaser, in his study of the prison and parole systems, clearly spoke to the issue of satisfactory employment in relation to parole success:

> It seems reasonable to believe that employment is usually a major factor making possible an integrated "style of life" which includes non-recidivism, successful marriage, and satisfaction in other social relationships. (p. 221)

> Available statistics indicate that young workers are particularly subject to unemployment and also are more prone than older workers to engage in crime when blocked in efforts to procure a legitimate income (or when inordinately expensive in their tastes). (p. 275)

The statement by the President's Commission (1967) in their *Task Force Report on Corrections* is still true today:

> It is difficult for probationers . . . to find jobs. They are frequently poor, undereducated, and members of minority groups. They may have personal disabilities—behavior disorders, mental retardation, poor physical health, overwhelming family problems. And they have in any case the stigma of a criminal record to overcome. (p. 32)

In a memorable speech entitled "Prisons Without Fences," former Chief Justice Warren Burger (1983) expressed similar concerns regarding the failure of our correctional institutions to adequately prepare inmates for their return to the free community. He cautioned:

> It is predictable that a person confined in a penal institution for two, five or ten years, and then released, yet still unable to read, write, spell or do simple arithmetic and not trained in any marketable vocational skill, will be very vulnerable to returning to a life of crime. And very often the return to crime begins within weeks after release. What job opportunities are there for an unskilled, functional illiterate who has a criminal record? (p. 5)

In addition, some probationers are barred from certain types of employment because of their inability to obtain necessary licenses or by restrictions in their probation or parole conditions of behavior. Many are handicapped by the stigmatization inherent in the incarceration experience, together with negative public attitudes about former inmates. Moreover, as noted by Davidoff-Kroop (1983), probationers and parolees "who are often undereducated and with few skills learn that finding work is problematical and frustrating" (p. 1). She indicated that the employment problem is more severe for parolees than for probationers because of the separation from the community resulting from incarceration.

As a general rule, many probationers and parolees lack vocational skills necessary for certain types of work, have never held a real job, and have been totally dependent upon daily "shape-up" work. Finally, many simply do not have the basic reading or mathematical skills to be able to function in even the least demanding jobs. Officers have obtained positions for probationers and parolees as messengers or grocery delivery men, only to find that the probationers were discharged shortly afterward because they could not read the customers' addresses. In 1996, a program evaluation compared the rearrest rates of "successful" participants (defined as those who remained in the program for 3 months or more) with those of a control group of probationers who did not participate in the program. It was found that the "successful" participants had a 50% lower rearrest rate (Gelormino & Weiss, 1997). Similarly, Grogger (1989), in a study in California, found a close association between unemployment/underemployment and recidivism. Nor has the thinking changed in recent years. According to Cromwell, Del Carmen, and Alario (2002),

> It is not unreasonable to suggest that meaningful employment is the most important issue for most probationers and parolees. Not only does employment provide financial support for the offender and his or her family, but it is also crucial for establishing and maintaining self-esteem and personal dignity—qualities that are seen by most authorities as essential to successful reintegration into the community. (p. 119)

Petersilia and Turner (1996) succinctly summed up the importance of adequate work opportunities for former inmates returning to the community. In noting the high rates of recidivism of offenders, they reported:

> Experts debate the reasons for such high recidivism rates, but all agree that the lack of adequate job training and work opportunities is a critical factor. Offenders often have few marketable skills and training and, as a result, have a difficult time securing legitimate employment. With no legitimate income, many resort to crime. (p. 1)

Because of the strong relationship between unemployment and recidivism, it is essential that community-based correctional agencies provide employment placement services for their clientele. Not only should officers assigned to this unit provide job referrals, but a proactive function would include job development. This would require the solicitation of civil service agencies as well as private firms, cooperation with labor unions, the development of funding grants to promote employment opportunities, collaboration with state labor departments, and referrals to private employment agencies. Equally important, supervisees should have the opportunity to enhance their job-related skills so that their worth in the employment market is enhanced. Training sessions, for example, would be given in basic reading and writing, communication skills, computer technology, likely areas of employment, and résumé writing.

In addition, training/counseling sessions should also focus on skills in seeking and maintaining a job and advancing in an employment setting. Individual counseling related to employability should be used in conjunction with this training to help the offender prepare for rejection, relate to individual employment concerns, and respond to sensitive issues such as how much should be revealed of one's criminal record on an employment application. Too frequently, designers of training in employment skills are not sensitive to offender needs for the most basic employment skills. In reality, many are not familiar with the extensive advertisements for employment in the *New York Times* and other major newspapers, the role of employment agencies, and the free employment referrals offered by many community-based social service and offender advocacy organizations.

Equally important, the training seminars must cover expected behavior that is commonplace to better educated persons and/or persons with work experience but surprisingly unfamiliar to many offenders. This would include, for example, the use of an alarm clock to ensure timely reporting for work, acceptable dress, appropriate language, and overall behavioral expectations. Ideally, employment preparation should be initiated in prisons, where there is always a need for activities to fill free time. If they are to succeed, these prison programs must be readily available, meaningful, intensive, and structured so that inmates can regularly attend, rather than when convenient to the officers. Programs in name only are doomed to fail, and high rates of continued recidivism can be expected to continue.

The New York City Department of Probation's Nova Ancora program is valuable in that it is consistent with attempts to reduce recidivism through employability by providing probationers with both job training and employment referral services. The department might want to explore a significant expansion of the program, in terms of both the number of participants and enriched programming. Despite the realities of anticipated budget reductions, an active solicitation of funding through private grants might prove successful.

The department reports that in fiscal year 2001, 1403 probationers were enrolled in vocational and life skills training through the Nova Ancora program (City of New York, Office of the Mayor, Office of Operations, 2002). Currently, the program conducts 2-day pre-employment workshops and support groups for newly employed probationers.

DOMESTIC VIOLENCE CASES

Like many agencies, including the police and the courts, the New York City Department of Probation is now giving special attention to domestic violence cases. This is understandable because of statutory changes, the demands of victim support organizations, media attention, and the realization that domestic violence victims have usually been incessantly and often barbarously brutalized. Moreover, it is well recognized that in most situations of domestic violence, not only is the spouse victimized, but the children are as well, either physically or emotionally.

An increased public awareness of both the seriousness and extensiveness of violence inherent in spousal battering has brought about numerous and meaningful changes in society's administration of domestic violence occurrences. Illustrative are the proliferation of batterer intervention programs (Hanson, 2002), more stringent laws, proactive policing, a more rapid and receptive court response, specialized training of criminal justice personnel, and demonstrably increased concerns for the victim throughout the criminal justice system (Roberts & Kurst-Swanger, 2002). Along with a heightened awareness of the ravages of domestic violence, a number of technological innovations now afford greater protection for the victims of batterers. These include police and corrections computerized offender data, including data on the possible existence of an order of protection, and improved communication through car faxes, cell phones, pagers, and video surveillance. Control of the offender has been increased through court-mandated electronic monitoring. Many victims of battering receive added protection by being provided with "a home electronic monitor (e.g., panic alarm or pendant), also known as the abused woman's active emergency response pendant, which can deter the batterer from violating his restraining order" (Roberts & Kurst-Swanger, 2002, p. 117).

Further assistance to a number of victims of domestic violence will be provided through the New York State Probation Domestic Violence Intervention Program. Funded by the U.S. Office of Justice Programs, several agencies, including the New York State Division of Probation and Correctional Alternatives, will work collaboratively to "assist probation departments in developing local policies and protocols and ensuring that local probation department work with local victim service programs and other criminal justice agencies to provide a consistent response to domestic violence" (Roberts & Kurst-Swanger, 2002, p. 139).

The New York City Department of Probation has adopted a number of programs designed not only to augment victim protection but to enable the offender to handle his anger without resorting to violence. A number of strategies have been undertaken to reduce domestic violence:

1. All probation officers receive specialized training designed to enhance their supervision of domestic violence cases. This training includes explanations of the causation and manifestations of domestic violence, a review of related statutes and case law, skills training for improved supervision, treatment strategies, and appropriate community resources.

2. Specially trained officers have been assigned to the domestic violence parts of appropriate courts to prepare customized presentence reports on domestic violence cases. These reports provide assistance to the judiciary in passing sentence in that they provide not only a history of the social factors of the family life, including a study of the nature and patterns of domestic violence, but also a suggested sentence recommendation. In addition, they provide assistance to the court in imposing special conditions intended to prevent further episodes of violence and offer recommendations for therapeutic services.

3. The agency is involved in a collaborative pilot program with the Special Domestic Violence Court of the Kings County Courts and the Kings County District Attorney's Office to provide innovative protection for victims of domestic violence. Using state-of-the-art technology, a system designated as Juris Monitor provides an early warning inside of the home of the victim at the approach of a probationer who was ordered out of the marital home.

Probationers convicted of domestic violence are often required to wear an electronic ankle bracelet as a condition of probation. Accordingly, if the probationer is within 500 feet of the victim's home, the ankle bracelet will set off a monitoring alarm within the home. Immediate notification will also be made to both the local police precinct and the Department of Probation. As the program is new, no statistics on its effectiveness could be supplied by the agency. (NYC DOP Web Site available at: http://www.nyc.gov/html/prob/html/accompl.html, accessed on 9/28/01)

It should be noted that, in the mayor's preliminary budget for fiscal year 2003, the department is slated to lose almost all of its specialized domestic violence initiatives (City of New York, Office of Management and Budget 2002).

MONITORING PROBATION CURFEWS

The monitoring of youth curfews has long been a problem for probation agencies. Unable to track the many probationers with court-imposed conditions of probation, probation officers were forced to rely on parental complaints of curfew violations or reports obtained from the police based on a new arrest. This method of monitoring curfews was haphazard and usually ineffective because many parents are inclined to be protective of their child.

The department has improved this aspect of supervision through updated technology. Probationers with curfew limitations are, for the most part, juveniles and young adults. Those enrolled in the monitoring program have a voiceprint made for the system and are provided with a pager that can be used only with the monitoring system. "When the pager sounds, the offender has a limited period of time to call the system from an authorized location. Utilizing caller ID, the system reads the location of the incoming call and confirms the probationer's identity through a voiceprint analysis" (NYC DOP Web Site available at: http://www.nyc. gov/html/prob/html/accompl.html accessed on 9/28/01)]

The system is programmed to provide immediate notification to the Probation Department if there is no response, if someone other than the probationer responds, or if the response is made from an unauthorized location. Because an assigned surveillance probation team responds only to ostensible curfew violations, a single squad can monitor a large amount of probationers. Unlike earlier efforts to audit probationers' compliance with curfew restrictions, today's technologically based system makes monitoring more reliable and inexpensive and enables the offender to understand that curfew restrictions are strictly controlled.

AFTERCARE SUPERVISION

The department's creation of a juvenile aftercare program is consistent with research that has indicated that "the rehabilitation of serious, chronic juvenile offenders does not end with their release from secure confinement" (Wiebush, McNulty, & Le, 2000, p. 1). The Office of Juvenile Justice and Delinquency Prevention (OJJDP), for example, has developed an Intensive Aftercare Program (IAP) that "seeks to reduce recidivism among high-risk juvenile parolees by providing a continuum of supervision and services during institutionalization and after release" (Wiebush et al., p. 1). Illustrative are Intensive Community Based Aftercare Programs at various locations, including Clark County (Las Vegas); Denver, Arapaho, Douglas, and Jefferson Counties (metropolitan Denver), Colorado; Essex (Newark) and Camden Counties, New Jersey; and the City of Norfolk, Virginia. These programs combine intensive supervision in the community, characterized by small caseloads and weekday, weekend, and evening supervision, with active and intensive services and supports through a variety of community resources. Their importance is an emphasis on transition and postinstitutional correctional programming and supervision "in part because research findings indicate that gains made by juvenile offenders in correctional facilities quickly evaporate following release" (Altschuler & Armstrong, 1998, p. 1).

The department currently operates Linden House, a community-based center located in Kings County, which provides intensive supervision and social services for youths discharged from secure placements and residing in their own homes while awaiting court action. Linden

House is operated under contract with the New York State Office for Children and Family Services.

This program is basically a carryover from the department's long-running Alternative to Detention Program, which permits selected juveniles to reside at home, in lieu of secure detention, pending court disposition of their cases. As a result, many juveniles are spared the trauma and possible physical abuse of detention confinement. In addition, the city's exceedingly expensive costs of juvenile detention are reduced, and attorneys have greater access to their young clients for preparation of a legal defense (Lindner, 1981).

Children confined in detention are usually high-risk offenders who are alleged to have committed a serious act of delinquency. As a result, selection for participation in the program is based on risk considerations as to whether the juvenile will keep his or her court appointment and/or might commit an act of delinquency while awaiting court disposition. Youths in the program are closely supervised, generally have curfew restrictions, and report to the center every school day. Daytime programming includes counseling, educational classes, and cultural and recreational activities (Lindner, 1981). The department reports that in fiscal year 2000, 104 youngsters were served by the program (City of New York, 2001). Linden House was honored in 1998 by the Office for Children and Family Services as the State's Aftercare Site of the Year (NYC DOP Web Site available at: http://www.nyc.gov/html/prob/html/accompl. html, accessed on 9/28/01). Though honored in 1998, this program will be closed in 2002. The state did not include any funding for the program in its latest budget. Though anecdotally many success stories are attributed to Linden House, the department did not keep any recidivism or other data on which any measures of success or failure could be made.

CONCLUSION

That the New York City Department of Probation looks and operates differently in some very fundamental ways today as opposed to a decade ago is unarguable. The institution of structured group interventions for the highest-risk probationers, the use of automated kiosks and other technology, and the recent implementation of a large police-probation partnership are all fairly new ways of "doing business." Other programs, such as the department's jobs program and ISP, have been in existence far longer. Many of these newer programs would not have ever come into existence were it not for the threat of budget reductions. On their face, they seem a reasonable and innovative response to a budget crisis. At least in the case of the kiosks, there seems to be some very preliminary evidence that, compared to the prior system of questionnaires, there has been some gain in effectiveness.

There are, however, far more questions about both the restructuring and the basic operations of the New York City Department of Probation that remain unanswered. Like most probation departments, the New York City Department has a paucity of data and analysis that might make a public safety or management argument that its reforms have been successful. There are essentially no available data on how successful the group work has been compared to individual counseling in terms of recidivism or rearrests. The same is true for ISP and the juvenile aftercare program. This lack of data and analysis is not unusual for underfunded public organizations such as the City Probation Department. When a probation agency is as starved for resources and has incredibly high and growing caseloads, it is a difficult administrative decision to allocate resources to collecting and analyzing data. The immediate relevance of data collection and analysis seems to pale in comparison to directing resources to core functions like supervision and monitoring. The problem, of course, is that without these data, probation agencies cannot make any arguments that what they do is effective or is better than some other past practice. In the absence of data, it is all too easy for policy to be made on the basis of anecdotal evidence or no evidence at all.

Especially in times of fiscal crises, this situation becomes particularly problematic for agencies with little evaluative data. New York City is now facing its worst budget crisis since at least the early 1970's and possibly ever. If the New York City Probation Department cannot make a solid analytical case that what it does is effective, benefits public safety and saves money, the agency will doubtless be subject to huge budget reductions from which it may be difficult to recover. Agencies like probation departments that do not have the high public profile or public support of police or fire agencies must be able to rely on data and analysis to make a case for increased funds or at least to protect current funding. In the absence of this data, they are the most likely agencies to be targeted for huge budget reductions.

The irony of this situation is that much of what the New York City Department of Probation did over the last several years may well be effective in terms of recidivism or preventing violations or simply in terms of keeping track of its probationers. There is simply not enough information to make any judgment one way or another as to the combined impact of all these organizational changes. New York City

Probation may well have an interesting success story to tell that can guide other probation departments and public agencies facing resource shortages and overwhelming demands on limited funds. This probation department and others could contribute greatly to the state of knowledge about what works and what doesn't by making the admittedly difficult decision to invest funds and effort in evaluation of what they do. There are universities across the country that would be eager to partner with probation departments to help evaluate their programs. Probation agencies should take advantage of that resource. This is not simply an academic issue. Without the data and rigorous evaluation necessary for probation as a field to make a case that there are cost-effective programs that can also protect and enhance public safety, it will be all too easy for policy makers to keep probation in the dire financial situation it has found itself in for decades.

REFERENCES

Abadinsky, H. (2000). *Probation and parole: Theory and practice* (7th ed.). Upper Saddle River, NJ: Prentice Hall.

Altschuler, D. M., & Armstrong, T. L. (1998, July). Intensive juvenile aftercare as a public safety approach. *Corrections Today, 60*(4), 118-121.

Barlow, H. D. (2000). *Criminal justice in America.* Upper Saddle River, NJ: Prentice Hall.

Burger, W. F. (1983). Factories with fences. *Pace Law Review, 4*(1), 5.

Byrne, J. M. (1988). *Probation.* Washington, DC: U.S. Department of Justice .

Cadigan, T. P. (1998). Officers are making good use of the Internet. *News and Views, Administrative Office of the United States Courts, Federal Corrections and Supervision Division,* 23.

Cadigan, T. P. (2001, September). PACTS:ECM. *Federal Probation, 65,* 25–30.

Champion, D. J. (2001). *The juvenile justice system: Delinquency, processing, and the law* (3rd ed.). Upper Saddle River, NJ: Prentice Hall.

Champion, D. J. (2002). *Probation, parole and community corrections* (4th ed.). Upper Saddle River, NJ: Prentice Hall.

City of New York, Office of the Mayor, Office of Operations. (1989). *Mayor's management report.* New York: Author.

City of New York, Office of the Mayor, Office of Operations. (1991). *Mayor's management report.* New York: Author.

City of New York, Office of the Mayor, Office of Operations. (2000). *Mayor's management report.* New York: Author.

City of New York, Office of the Mayor, Office of Operations. (2001). *Mayor's management report.* New York Author.

City of New York, Office of the Mayor, Office of Operations. (2002). Mayor's management report. Preliminary Fiscal 2002.

City of New York, Office of the Mayor, Office of Management and Budget. (2002). *Preliminary budget for the city of New York.* New York: Author.

Clear, T. R., & Cole, G. F. (2000). *American corrections* (5th ed.). Belmont, CA: Wadsworth.

Conrad, J. P. (1985). The penal dilemma and its emerging solution. *Crime and Delinquency, 31,* 411–422.

Cromwell, P. F., Del Carmen, R. V., & Alario, L. F. (2002). *Community-based corrections* (5th ed.). Belmont, CA: Wadsworth/Thomson Learning.

Davidoff-Kroop, J. (1983). *An initial assessment of the Division of Parole's employment services.* Albany: New York State Division of Parole.

Diana, L. (1955). Is casework necessary? *Forum, 34*(1), 1–8.

DiMascio, W. M. (1997). *Seeking justice: Crime and punishment in America.* New York: Edna McConnell Clark Foundation.

Dunworth, T. (2001, September). Criminal justice and the IT revolution. *Federal Probation, 65,* 52–63.

England, R. (1955, September). A study of post-probation recidivism among five hundred federal offenders. *Federal Probation, 19,* 10–16.

Gelormino, L., & Weiss, E. J. (1997). Nova Ancora model program. *Corrections Today*, vol. 59.

Glaser, D. (1969). *The effectiveness of a prison and parole system* (abridged ed.). Indianapolis: Bobbs-Merrill.

Glueck, S. (1933). The significance and promise of probation. In Glueck, S. (Ed.), *Probation and criminal justice.* New York: Macmillan.

Gordon, D. (1991). *The justice juggernaut: Fighting street crime, controlling citizens.* New Brunswick, NJ: Rutgers University Press.

Greek, C. E. (2001, June). The cutting edge: A survey of technological innovation. *Federal Probation, 65,* 51–53.

Grogger, J. (1989). *Employment and crime.* Sacramento: California Bureau of Criminal Statistics and Special Services.

Guynes, R. (1988). *Difficult clients, large caseloads plague probation, parole agencies.* Washington, DC: Government Printing Office.

Hanson, B. (2002). Interventions for batterers: Program approaches, program tensions. In A. R. Roberts (Ed.), *Handbook of domestic violence intervention strategies: Policies, programs, and legal remedies.* New York: Oxford University Press.

Horn, M. Personal communication, 10/24/2002.

Kaplan, J., Skolnick, J., & Feeley, M. M. (1991). *Criminal justice: Introductory cases and materials* (5th ed.). Westbury, NY: Foundation.

Langan, P. A., & Cunniff, M. A. (1992). *Recidivism of felons on probation, 1986–89.* Washington, DC: Government Printing Office.

Latessa, E. J., & Allen, H. E. (1999). *Corrections in the community.* Cincinnati, OH: Anderson.

Lauen, R. J. (1998). *Community managed corrections: And other solutions to America's prison crisis.* Laurel, MD: American Correctional Association.

Lindner, C. (1981). The utilization of day/evening centers as a alternative to secure detention of juveniles. *Journal of Probation and Parole,* (13).

Lindner, C. (1992). The probation field visit and office report in New York State: Yesterday, today, and tomorrow. *Criminal Justice Review, 17*(1).

Lindner, C. (1993). Probation's first 100 years: Growth through failure. *Journal of Probation and Parole.*

Lindner, C., & Koehler, R. J. (1992). Probation officer victimization: An emerging concern. *Journal of Criminal Justice, 20,* 53–62.

Lindner, C., & Savarese, M. (1984). The evolution of probation: University settlement and the beginning of statutory probation in New York City. *Federal Probation, 48*(3), 3–12.

Lundman, R. J. (1993). *Prevention and control of juvenile delinquency.* New York: Oxford University Press.

Martinson, R. (1974). What works? Questions and answers about prison reform. *Public Interest, 35,* 22.

Martinson, R. (1979). New findings, new views: A note of caution regarding sentencing and reform. *Hofstra Law Review, 7,* 243–258.

McCarthy, B. R., McCarthy, Jr., B. J., & Leone, M. C. (2001). *Community based corrections* (4th ed.). Belmont, CA: Wadsworth.

Moran, T. K., & Lindner, C. (1985). Probation and the hi-technology revolution: Is a reconceptualization of the traditional probation officer role inevitable? *Criminal Justice Review, 10*(1).

Morgan, T., & Marrs, S. D. (1998). Redmond, Washington's SMART Partnership for police

and community corrections. In J. Petersilia (Ed.), *Community corrections: Probation, parole, and intermediate sanctions.* New York: Oxford University Press.

Morris, N., & Tonry, M. (1990). *Between prison and probation.* New York: Oxford University Press.

New York State Probation Commission. (1923). *Sixteenth annual report.* Albany, NY: J. B. Lyon.

New York City Department of Probation. Retrieved from http://www.nyc.gov/html/prob/home.html

New York State Crime Commission. (1927). *Report of the Crime Commission* (New York State Legislative Document No. 94). Albany, NY: J. B. Lyon.

Ogden, T. G., & Horrocks, C. (2001, September). Pagers, digital audio, and kiosk: Officer assistants. *Federal Probation, 65,* 35–41.

Petersilia, J. (1985). *Probation and felony offenders* (National Institute of Justice Research in Brief). Washington, DC: Government Printing Office.

Petersilia, J. (1998). *A decade of experimenting with intermediate sanctions: What have we learned?* Washington, DC: U.S. Department of Justice, National Institute of Justice.

Petersilia, J., & Turner, S. (1996). Work release in Washington: Effects on recidivism and corrections costs. *Prison Journal, 76*(2), 138.

President's Commission on Law Enforcement and Administration of Justice. (1967). *Task Force Report on Corrections.* Washington, DC: Government Printing Office.

Roberts, A. R., & Kurst-Swanger, K. (2002). Police responses to battered women: Past, present, and future. In A. R. Roberts (Ed.), *The handbook of domestic violence intervention strategies: Policies, programs, and legal remedies.*

New York: Oxford University Press.

Rosecrance, J. (1986). Probation supervision: Mission Impossible. *Federal Probation, 50*(1), 25–31.

Rothman, D. J. (1980). *Conscience and convenience: The asylum and its alternatives in Progressive America.* Boston: Little, Brown.

Silberman, C. E. (1978). *Criminal violence, criminal justice.* New York: Random House.

Tappan, P. W. (1960). *Crime, justice and correction.* New York: McGraw-Hill.

U.S. Attorney General. (1939). *Survey of release procedures: Vol. 2. Probation.* Washington, DC: Government Printing Office.

Van den Haag, E. (1975). *Punishing criminals.* New York: Basic Books.

Walker, S. (2001). *Sense and nonsense about crime and drugs* (5th ed.). Belmont, CA: Wadsworth.

Wiebush, R. G., McNulty, B., & Le, T. (2000, July). *Implementation of the Intensive Community-Based Aftercare Program.* Washington, DC: U.S. Department of Justice, Office of Justice Programs, Office of Juvenile Justice and Delinquency Prevention.

Wilson, J. Q. (1980). "What works" revisited: New findings on criminal rehabilitation. *Public Interest, 61,* 3–17.

Wooten, H. B., & Hoelter, H. J. (1998). Operation Spotlight: The community probation-community police team process. *Federal Probation, 62*(2), 30–35.

Young, P. V. (1937). *Social treatment in probation and delinquency.* New York: McGraw-Hill.

22

PACTS^ECM

Probation and Pretrial Services Automated Case Tracking-Electronic Case Management System

TIMOTHY P. CADIGAN

On April 1, 2001, the federal judiciary began implementing the Probation and Pretrial Services Automated Case Tracking-Electronic Case Management System (PACTS^ECM). The result of years of planning, requirements definition, design, development, and testing, this implementation will position the federal probation and pretrial services system to utilize the technological tools of an advanced case management system on a daily basis. This article looks at the many implications and issues arising from a task of this magnitude and explores what the future can hold once this technological base is established. Areas of discussion include the application itself, design and development issues, implementation issues, potential benefits, business process change issues, and an exploration of future potential.

THE PACTS^ECM APPLICATION

The PACTS^ECM system is both a case tracking and a case management tool. The case tracking component (PACTS) allows officers to electronically collect pertinent case-related information to produce statistical and workload reports. The case management portion (ECM) helps officers collect, manipulate, and recall case-management-specific information.

This promotes more efficient and effective defendant/offender supervision and investigations for the district. Overall, the PACTS^ECM makes information more easily accessible to an expanded number of users and allows those users to manipulate the information in a manner more consistent with the professional activities they perform.

Cadigan, T. P. (2001, September). PACTS: Probation and Pretrial Services Automated Case Tracking-Electronic Case Management System. *Federal Probation, 65*(2), 25-30.

PACTS^{ECM} is a "total" information system. It includes functionality for: 1) electronic generation, storage, and retrieval of all investigation and supervision case information; 2) electronic retrieval for judiciary personnel of vital case information, including the presentence report, pretrial services report, and chronological records; 3) integrated access to the criminal component of the Case Management-Electronic Case Files (CM-ECF) project; and 4) electronic imaging of defendants/offenders—their tattoos, homes, vehicles, or other appropriate images.

The project team has worked closely with automation staff, data quality analysts, officers, supervisors, and administrative support staff from many districts to ensure that users' needs are addressed and that operational requirements are reflected in the data structure and user interface of the new Informix-based system. The intended audience for the PACTS^{ECM} application is the entire staff of probation and pretrial services offices. Probation and pretrial services operations involve approximately 7800 authorized positions in 509 locations. There are 93 district headquarters probation offices, 56 of which are combined probation and pretrial services offices and 37 of which have separate pretrial services offices.

PACTS^{ECM} is a browser-enabled application that is accessed through the federal judiciary's Intranet. It replaces its predecessor, PACTS Unify. However, it has been enhanced in two significant ways. The first is by expanding and redesigning the data structures in the database and the second is by using contemporary software tools and web technology. The enhanced database structures allow multiple IDs to be stored for each client. It also permits maintenance and search of historical sentence and historical address information. The software tools make it possible for PACTS^{ECM} to have graphical navigational tools such as drop-down lists and tabbed dialog boxes, display digital images, and link to resources outside the database. For example, the application links directly to Mapquest.com to provide officers with point-and-click access to directions to the defendant/offender's home.

The major features included as part of the first version of the software are a utility to make data conversion easier for data managers, a defendant/offender module, a treatment module, a pretrial services module, and a probation module. A number of standard reports and forms are available and the application provides for the required statistical extractions. Functionality will be added with Versions 2–4 of the software in generally the following order:

1. Automated Chronological records (Chronos);

2. Drug detection event tracking;

3. Completion of all forms and other "canned" reports;

4. Probation/Pretrial Services Case Plans and Reviews;

5. Fine and restitution tracking;

6. On-Line Case Assignment;

7. PS-2 Pretrial Services Interview Work-sheet;

8. Electronic Monitoring;

9. Pre-sentence Report Disclosure Tracking; and

10. Interfaces to other databases including the FBI's National Crime Information Center (NCIC) 2000.

DESIGN AND DEVELOPMENT

In-house development had been the normal mode for software development projects in the judiciary virtually throughout its history. At the time PACTS^{ECM} was ready for development, the judiciary had recently been using off-the-shelf (COTS) software, with modifications for accounting and personnel applications, but all case-related systems had been produced internally, including PACTS-Unix, used in most probation and pretrial services offices. The proposal for development of the PACTS^{ECM} application combined the strengths of both the in-house and out-sourced development strategies previously used. The approach provided the necessary resources to complete the project in a timely manner and reduce the impact of other judiciary automation efforts on the timely completion of PACTS^{ECM}. The judiciary was able to take advantage of substantial short-and long-term cost-saving opportunities, and the AO could effectively respond to requests from court users for enhanced automated functionality to manage the judiciary's vital information resources.

By using both in-house and outsourced talent, the PACTS[ECM] project team combined institutional and technical knowledge unique to the judiciary with a body of expert technical skills and knowledge in Informix and other state-of-the-market programming tools using the judiciary's Informix contract and other government agency contracts as needed. Combining resources in this manner allowed managers to reliably and more flexibly schedule highly skilled technical staff on the project— i.e., place the right people with the right skills on the right tasks at the right times.

This development approach was attractive because it made use of the considerable expertise and experience in the AO and in court units, including a cadre of in-house development personnel who are well-trained and productive, using fourth-generation languages (4GLs). In addition, federal pretrial services and probation offices (as distinct from state and local jurisdictions) offered considerable institutional knowledge and experience that no contractor could approximate, let alone duplicate. Finally, hourly labor costs of in-house personnel were lower than the most inexpensive contractor resources. Therefore, we used the contractor labor (which was considerably more expensive than the in-house labor) sparingly and only when necessary.

PACTS[ECM] IMPLEMENTATION

The PACTS[ECM] system is being deployed in a test wave of 14 courts, beginning on April 1, 2001. Recurring waves of 6 to 8 courts are scheduled to start at two-month intervals beginning February 1, 2002. Each wave will cover a nine-month implementation period. Prior to the start of the first wave in February 2002, changes in implementation will be made as appropriate based on the experiences of the test wave.

PACTS[ECM] implementation occurs when 1) applicable district staff are trained to use the PACTS[ECM] application; 2) technical tasks concerned with hardware and software installation and operations are successfully completed; and 3) the legacy database is converted to PACTS [ECM]. *The PACTS[ECM] Implementation Kit* assists the district by providing guidance,

checklists, activities, suggested actions, and examples of documents, and provides the district with references to resource materials available through the J-Net.

The kit is divided into three sections: "getting started," "operations," and "systems," based on the nature of the activities covered and the intended audience. The "getting started" section is of interest to all participants. It lays the foundation for implementation. The "operations" section focuses on activities leading up to district staff being able to use the system. The target audience for this section includes chief probation and pretrial services officers, deputies, supervisors, data quality analysts (DQA), training specialists, and any officers assigned to assist in implementation. The "systems" section, which provides guidance on hardware, software, and database issues, is of most interest to the district's systems manager and systems staff. However, the "operations" and "systems" areas overlap. Decisions made by operations personnel will affect the work systems personnel must do to set up and support the system. Similarly, the systems staff expertise with supporting automated systems will be useful to the operations staff as they make key decisions or perform implementation activities. The district PACTS[ECM] project manager and the systems manager work closely together to ensure the successful implementation of the system.

IMPLEMENTATION TASKS

Perhaps the most useful tool within *The PACTS[ECM] Implementation Kit* is the PACTS[ECM] Implementation Project Plan. The project plan is a Microsoft Project file that can be used by district management as a quick reference to the tasks that must be accomplished in order to successfully implement PACTS[ECM]. Tasks can be checked off as they are completed, thus showing what has been accomplished and what is left to be done. The chief probation and pretrial services officers should meet with their district PACTS[ECM] project manager on a weekly basis to review the status of the project plan. The project plan also contains recommended start and end dates for each task, the anticipated duration of each task, and a timeline for the tasks. At the beginning of the implementation period, the district's PACTS^{ECM} project manager will have received a copy of the project plan customized with dates appropriate to that district's start date. Each district has a PACTS^{ECM} Implementation Coordinator from the Systems Deployment and Support Division (SDSD) within the Administrative Office assigned to support it in the implementation effort. The PACTS^{ECM} Implementation Coordinator works with the project manager to track progress according to the customized plan. The district's PACTS^{ECM} project manager may also wish to use the customized plan to manage the project using the Microsoft Project software.

Data Conversion—A Critical Cross-Functional Activity

Before a district can begin using the new PACTS^{ECM} as its tracking and case management system, the data stored on the old PACTS-Unify system must be transferred, or "converted," to the new PACTS^{ECM} system. The physical transfer of the data is a largely technical task performed by the district's systems staff as the last step before beginning live operations on the new system. However, a great deal of preparation needs to be accomplished early in implementation to ensure a smooth conversion of data. The most time-consuming task for most districts will be "cleaning" the data stored in PACTS. This task can begin as early as possible in implementation and is a collaborative effort between operations and systems personnel. Data-conversion software necessary for performing this task is supplied to the districts.

Training

Training for PACTS^{ECM} is comprised of two primary components: application training for end users and technical training for technical staff who must support the application. The end-user training includes a train-the-trainer segment, as the majority of end-user training will be conducted in each district by district personnel who participated in the application training course defined here.

The PACTS^{ECM} Application training course is designed to provide the necessary understanding and skills for the end user to successfully apply the newly-developed PACTS^{ECM} software. Informational and introductory sessions will explain the enhanced functionality. Participants will be guided through the browser-based menu and on-line help links and develop an understanding of how the application applies to probation and pretrial services. Participants will also be prepared to deliver training to in-court personnel. The target audience includes data entry clerks, data quality analysts (DQA), and training specialists who will be responsible for training the remainder of the office staff. The course teaches participants to:

- Identify differences between PACTS^{ECM} and the former PACTS Unify system;
- Confidently docket events on Pretrial and Probation cases;
- Create and modify client records and related events for both pretrial services and probation;
- Learn to generate reports and utilize on-line forms; and
- Incorporate PACTS^{ECM} training materials into the court's training plan.

The class is delivered in two distinct components designed to accommodate both separate and combined pretrial services and probation.

The technical training is comprised of several classes: 1) Database Administration, 2) Systems Administration, and 3) Informix SQL. All technical training classes are provided in San Antonio, Texas at the judiciary's information technology training center.

This Database Administration course is designed to provide probation and pretrial systems staff with the technical information required for implementing and operating PACTS^{ECM}. The course includes overviews of the application (modules, contents, navigation, enter data, query data, etc.), physical hardware/ software architecture, and logical application architecture of DB schema. It identifies tables that will require local population and maintenance and review procedures for managing these tables. It discusses linking to resources outside the DB, implementation of login security algorithm, and maintenance of the NT and report servers and software. Finally, it presents security issues and the relationship of WordPerfect templates and Crystal Reports templates to report server software.

The systems administration course is intended for Informix Dynamic Server and Informix Dynamic Server system administrators. Participants learn the skills necessary to successfully administer one or more database servers: configure and initialize a database server instance, configure and test client connectivity, configure and manage memory and disk usage, plan and implement system maintenance tasks, and configure the server for optimal OLTP or decision support.

Finally, the Informix Structured Query Language (SQL) course covers the Data Manipulation Language (DML) portion of SQL. Participants learn to create SELECT, INSERT, UPDATE, DELETE, LOAD, and UNLOAD statements, simple and complex joins, and subqueries. In addition, the course covers the basic configuration of an Informix instance, logical and physical log maintenance, archiving and restoring, and troubleshooting of basic configuration problems.

BENEFITS OF THE PACTS[ECM] SYSTEM

The PACTS[ECM] system offers both intangible and quantifiable benefits to the end user or to the public at large. Intangible benefits are those benefits that are real, but difficult or impossible to quantify accurately or precisely. The quantifiable benefits have been assigned cash values. First, as probation officers and pretrial services officers use PACTS[ECM] as a tool in their daily duties, paper waste should be reduced. The intention is to move to an environment in which the workstation becomes the usual medium for disseminating information, with a paper copy printed only on demand. However, it is difficult to predict human behavior: one manager may demand a paper copy of virtually everything, while another will be content with electronic dissemination of documents. Thus, we make no attempt to project savings in paper. Second, and probably more important, but even more difficult to quantify, the accuracy and effectiveness of services should be increased by a benefit that, for lack of a better name, can be called data quality. PACTS[ECM] will increase data quality in two ways:

1. *Elimination of data redundancy.* This has two aspects:

- The PACTS[ECM] database, using a fully relational database management system, will eliminate to as large an extent as possible redundancies of data.
- The forms producing capability of PACTS[ECM] will integrate discrete data elements with form templates, thus eliminating the need to re-key data into multiple sources.

2. *Increased validation of data.* This will mainly be accomplished through use of standard tables for the various codes, and through cross-validation of user inputs based on the business rules (e.g., detention hearing date cannot be earlier than initial hearing date).

Third, increased efficiencies in productivity of probation and pretrial services officers will free them from their paper-intensive world to dedicate their energies to conducting more thorough and complete investigations, implementing better supervision practices, ensuring community safety, and improving enforcement of pretrial release and sentence conditions imposed by judicial officers.

Fourth, the PACTS[ECM] information system will place the judiciary in a better position to respond to the grievances of victims, and to coordinate and share information with other law enforcement agencies. Although neither of these uses is part of the charter of the PACTS[ECM] project, both are benefits to the public at large that will accrue. They are not quantifiable, but they are real. The presence of a coordinated, validated, up-to-date information system from which details about federal probation and pretrial services defendants/offenders and their offenses can be quickly and accurately retrieved will increase efficiency, accuracy, and timeliness over the current manual methods.

Quantifiable benefits of the PACTS[ECM] system fall into two general categories: increases in efficiency specifically related to forms production; and increases in general efficiency. Tables 1 and 2 illustrate how even very modest cost avoidance associated with forms production and increases in general productivity can produce dramatic results when multiplied across the entire user community. To demonstrate the possible efficiencies that could be achieved with PACTS^ECM, the project team traveled to the Western District of Texas probation and pretrial services offices to conduct testing comparing current methodologies of document production and PACTS^ECM methodologies of electronic forms development. Participating in the testing were staff from the AO's Systems Deployment and Support Division, Applications Maintenance and Development Division, and Federal Corrections and Supervision Division, and the probation and pretrial services offices from the Western District of Texas. Separate testing was conducted in each office.

The group agreed to test five forms for purposes of this analysis. Those forms are the initial case supervision plan (ICSP), travel permit, Form 14-A Request for Arrest Record, Flash Notice Request, and Form 7A Conditions of Supervision. These forms were selected by the group because of their frequency of use and because they ranged in complexity from a simple one-page form to the more elaborate multipage case plan. The goal of the testing was to develop a base of knowledge to generalize to all forms without performing testing on all forms.

That testing demonstrated a wide range of average efficiencies achieved through the electronic forms development methodologies of PACTS^ECM. For example, the mean time for completion of the case plan was 36 minutes using the older methodologies. The mean time using the electronic forms development methodologies of PACTS^ECM was 6 minutes, a per-plan savings of 30 minutes. Simpler forms like the travel permit and Form 7A achieved smaller savings of 5 minutes and 10 minutes respectively.

The group agreed to test five pretrial services forms for this analysis. Those forms are the initial case supervision plan (ICSP), field sheet, Form 14-A Request for Arrest Record, initial chronological record, and PS 7 Reporting Requirements. These forms were selected by the group because of their frequency of use and because they ranged in complexity from a simple one-page form to the more elaborate multi-page case plan.

That testing demonstrated a wide range of average efficiencies achieved through the electronic forms development methodologies of PACTS^ECM. For example, the mean time for completion of the ICSP was 38 minutes using the older methodologies. The mean time using

the electronic forms development methodologies of PACTS^ECM was 9 minutes, a per-plan savings of 29 minutes. Simpler forms like the field sheet and initial chronological record achieved smaller savings of 9 minutes and 7 minutes respectively.

The testing described above demonstrates the potential savings that can be achieved in probation and pretrial services offices when applying PACTS^ECM methodologies. Because we could not test every form used by officers, we generalized from the testing done in the Western District of Texas. The following table contains efficiency improvement estimates based on the number of cases handled in the system annually multiplied by the average number of forms per case multiplied by a conservative estimate based on our testing of 3 minutes saved per form. Those efficiencies are then given a dollar amount by multiplying the hourly rate of staff, in an effort to demonstrate the potential real efficiencies that can be achieved.

Table 22.1 presents a *very conservative* estimate of the savings that can be realized through the implementation of PACTS^ECM. The testing we conducted, which showed substantially more savings than we present, was artificially optimistic in favor of the current methodologies. In real life, staff time would be spent assembling the pieces of information necessary to complete the various forms. In PACTS^ECM all that basic information will be assembled instantaneously by the system.

Our analysis investigated the effects of three levels of hypothetical improvement in general efficiency.

- *Low improvement* is defined as a 1 percent improvement in overall efficiency of probation and pretrial services officers, and a 2 percent improvement in overall efficiency of support staff.
- *Medium improvement* is defined as a 2 percent improvement in overall efficiency of probation and pretrial services officers, and a 5 percent improvement in overall efficiency of support staff.
- *High improvement* is defined as a 5 percent improvement in overall efficiency of probation and pretrial services officers, and a 10 percent improvement in overall efficiency of support staff.

Table 22.1   Increased Efficiency Through Electronic Forms Development

	Number of Cases	Average Number of Forms	Total Forms Per Year	Savings Per Form	Minutes Saved Per Form	Hours Saved	Hourly Rate of PSO	Costs Avoided
PRETRIAL SERVICES								
Investigation Cases	63,497	5.10	323,835	3	971,504	16,192	$30	485,752
Supervision Cases	30,502	10.04	306,240	3	918,720	15,312	$30	459,360
TOTAL		15.14	630,075		1,890,224	31,504	$30	945,112
PROBATION								
PSI/PSIG	49,826	10.80	538,121	3	1,614,362	26,906	$30	807,181
Supervision Cases	88,966	13.16	1,170,348	3	3,511,043	58,517	$30	1,755,522

For example, a 1 percent increase in general efficiency among all officers, and exclusive of any efficiencies gained among support personnel, would yield a net savings (cost avoidance) of $3.12 million. For clerical staff a 2 percent increase in general efficiency would yield an annual cost avoidance of $1.56 million. The combined cost avoidance of officers and support staff would yield an annual cost avoidance of $4.68 million.

This analysis used the most conservative parameters; we have included the more optimistic figures here for the purposes of illustration, and because we believe the assumptions of 1 percent for officers and 2 percent for support personnel to be quite conservative (see Table 22.2).

Note that *cost avoidance* is not equivalent to *cost saving*s. The cost avoidance due to increased officer and clerical efficiency will free those resources to perform other mission-critical aspects of their job. Thus, because the personnel will remain on staff, the benefits described in this document attributable to PACTS[ECM] are not actual savings to the judiciary. Rather, they demonstrate the costs avoided in freeing the probation and pretrial services community from their heavily paper-based environment. The benefit—quantified herein as cost avoidance—accrues not to the bottom line on a budgeting statement, but to the community that the federal judiciary serves. That community avoids the costs of inefficient and cumbersome manual procedures, and

increases the effectiveness and perhaps also the range and scope of services provided by probation and pretrial services: ensuring the public safety, monitoring and supervising defendants and offenders, ensuring that conditions are met, and that violations are dealt with speedily.

BUSINESS PROCESS CHANGES

Achieving the benefits of PACTS[ECM] requires more than just installing software and conducting training. It requires a commitment from the chief probation and/or pretrial services officer to change local processes to take advantage of the functionality provided. The introduction of a new computer system into a work environment generally causes some disruption to day-to-day operations. Staff must learn new screens and commands, workflow may need to be changed, conversion of data and customized features from the old system is usually time consuming. All of this must happen while the office continues to accomplish its primary mission. In order to mitigate some of this disruption, tasks designed to ease the transition for data quality staff have been included in the PACTS[ECM] Project Plan. Most of these tasks are covered in two sections of the project plan, Business Processes and Training and Support.

Although PACTS[ECM] will replace the legacy case management systems in each of the federal courts' probation and pretrial services offices, each probation/pretrial services office has the

Table 22.2   Savings Due to Increases in General Efficiency

OFFICER EFFICIENCY

Hours Saved Per Year	Average Hourly Rate	Savings Per Officer Per Year	Number of Officers	Total Savings Per Year
20.8	$30	$624	5,000	$3,120,000
41.6	$30	$1,248	5,000	$6,240,000
104.0	$30	$3,120	5,000	$15,600,000

SUPPORT STAFF EFFICIENCY

Hours Saved Per Year	Average Hourly Rate	Savings Per Clerical Per Year	Total Number of Clerks	Savings Per Year
41.6	$15	$624	2,500	1,560,000
104.0	$15	$1,560	2,500	3,900,000
208.0	$15	$3,120	2,500	7,800,000

TOTAL EFFICIENCY IMPROVEMENT

Rate Officer	Clerical	Total	
Low	$3,120,000	$1,560,000	$4,680,000
Medium	$6,240,000	$,900,000	$10,140,000
High	$15,600,000	$7,800,000	$23,400,000

flexibility to decide how the system will be integrated into the office's work processes. Three basic options are available, with unlimited local variance among them possible: 1) traditional data entry model; 2) officer-centric model; or 3) hybrid model combining both approaches, as shown in Table 22.3.

The choice of the implementation strategy is a management decision that will directly affect the business processes and workflow within the office. To assist management in making this decision, a Business Process Workgroup could be formed to document current business processes in a manner that is easy for managers to review, understand, and modify. Once current processes have been documented and reviewed and the business process model has been chosen, the Business Process Workgroup can prepare the office to begin day-to-day operations using PACTS[ECM] with the process model chosen for that district.

Depending on the model chosen and the degree of change from the current model, the district will have to re-engineer business processes to ensure a smooth transition. For example, having officers enter data will introduce more error into the data entry process. Therefore, management needs to create or modify the district's data quality assurance plan and procedures to reflect the new workflow. That quality assurance program would need to compare entered data to source documents, look for common errors, and report back to staff who make errors on those errors so that staff can become aware of them and avoid similar errors in the future.

The simple fact that the current process is changed could cause the district to establish procedures that are not now necessary. For example, opening up the data entry function could introduce the possibility that cases get lost before they get entered. This has obvious negative implications for workload credit for the office. Therefore, it may be necessary to validate and make any necessary adjustments to new work processes after the PACTS[ECM] system has been implemented to ensure against this type of problem.

Table 22.3   Range of PACTS^{ECM} Business Process Model Options

Traditional Data Entry Model	Some Officers	Hybrid	All Officers	Officer-Centric
• New system replaces current case management system using Traditional Data Entry Model		• Most client records created and maintained by administrative staff		• Officers create client records
• Administrative Staff enter data		• Test group of officers selected to create and maintain client case record information		• Data entered and maintained by officers
• Data quality analysts maintain data integrity		• Data quality analysts and test group of officers share data maintenance responsibilities		• Data quality analysts and officers share data maintenance responsibilities

One final obvious area that will clearly need to be reviewed *encompasses several areas including all local forms, reports, and applications. For example, it may be necessary to modify data collection forms to reflect the new screens and to accommodate the new workflow. The district should also work with the systems staff to determine the need for existing locally developed reports and applications. This analysis should look carefully for any duplication of effort between PACTS^{ECM} and the local system which preceded it.*

THE FUTURE

The probation and pretrial services user community has long desired and sought support for the development of automated functionality that empowers officers in the community. That desire first manifested itself in the Mobile Computing project, which tested the idea of using laptops to provide that functionality. That project demonstrated the value of technology when the officer was away from the office. However, it also demonstrated the limitations of bulky laptop computers in providing that functionality. The expanded use and functionality of personal digital assistants or handheld computers has raised the probation and pretrial service community's interest in meeting their needs through these devices. As PACTS^{ECM} begins implementation, the District Court Technology Panel and Chiefs Advisory Group believe strongly that the Community Technology initiative is the most important need of officers on the street. The objective of this project is to provide probation and pretrial services officers with the automated functionality they need to more efficiently perform the duties required of them by law in the community. The project will focus on using this technology in five critical areas: pretrial services supervision of defendants; post-conviction supervision of offenders; presentence investigations; pretrial services investigations; and safety of officers in the community. Federal probation and pretrial services officers are required to investigate and supervise defendants/offenders as ordered by the court. Those functions require officers to leave the courthouse and go into the community. Therefore those officers are "remote knowledge workers" requiring electronic access to case-specific data from a variety of remote locations. Moreover, the primary concern of the judiciary and officers in the community is the personal safety of officers in the field. Those two needs combined create the need for officers in the community to have handheld computers.

Another potential source of integration is with kiosk technology. The kiosk could collect a live biometric measurement of offenders' hand geometry or fingerprints or one of several other options to verify that the offender is the one interacting with the kiosk. Then, the screen would prompt the probationer to answer a series of questions (in English or Spanish) previously determined by the probation officer, including current address, phone number and other information. The kiosks could also collect fine and restitution payments. Once the electronic reporting session is complete, the system issues a receipt to the probationer. Over time, a detailed history of the degree of compliance is collected on each offender. The system identifies those who are non-compliant, and for whom the probation officer may need to take some direct action.

The future of technology in the field of community corrections is only limited by one's ability to conceive effective uses for the ever-growing waves of technology to the field of community corrections. Harnessing that potential while eliminating those technologies that are more toy than useful tool is the secret to success in these initiatives. However, having a "state of the market" case management system is the first and most essential step in implementing these various technologies in a community corrections system. With the implementation of PACTS^ECM the federal probation and pretrial services system is poised to move forward on a solid foundation.

23

REHABILITATING FELONY DRUG OFFENDERS THROUGH JOB DEVELOPMENT

A Look Into a Prosecutor-Led Diversion Program

HUNG-EN SUNG

Drug offenders develop chronic dependence on the drug economy for subsistence. Brooklyn's Drug Treatment Alternative-to-Prison program seeks to correct this problem by diverting drug-addicted felons into residential treatment with strong educational and vocational training components and by providing job counseling and placement to program graduates through the job developer and a business advisory council. Encouraging preliminary process and outcome statistics indicate that crime reduction through employment can be successful. But structural changes in the market and ex-offenders' psychological inadequacies need to be addressed for the success to be sustained over time.

Governmental attempts to strengthen criminal offenders' ties to the world of legitimate work have been around since the early 1960s (Thompson, Sviridoff, & McElroy, 1981). It is believed that by helping offenders to adopt a more conventional and productive lifestyle, publicly funded programs can successfully reduce recidivism. Both theoretical formulations and empirical evidence during the past 20 years have supported this crime prevention approach. On one hand, economists maintain that the development of human capital (the sum of marketable skills, abilities, and knowledge) leads to gainful employment, which

Sung, H-E. (2001, June). Rehabilitating Felony Drug Offenders Through Job Development: A Look Into a Prosecutor-Led Diversion Program. *The Prison Journal, 81*(2), 271-286. Used with permission.

produces earnings exceeding the economic returns that street criminal activities can generate (Becker, 1968; Ehrlich, 1979). Crime becomes unlikely among those who are well educated and well trained because they are attractive to employers, well paid, and thus have a stake in conformity. On the other hand, research findings suggest that criminal recidivism is determined more by postrelease factors such as postincarceration employment than by background characteristics (Klein & Caggiano, 1986).

The vicious cycle of poverty and drugs has wreaked havoc in Brooklyn, New York City. Poverty produces joblessness. Chronic unemployment deprives people of concrete expectations and goals for their lives, while the inability to support their families gives rise to feelings of futility and incompetence. Worse, the failure to participate in the legal labor market increases the incentive to use and sell drugs. Between 1989 and 1998, more than 224,000 adult arrests for felony and misdemeanor drug offenses were made in this borough alone (New York State Division of Criminal Justice Services, 2000). As a matter of fact, the drug economy has become an institutionalized aspect of social organization in Brooklyn's poorest neighborhoods.

Despite the impressive job growth during the 1990s, the unemployment rate in New York City remains above the national average. Not everybody benefited equally from the sustained creation of new jobs. Joblessness is still common in minority and immigrant neighborhoods. Whereas wages have increased dramatically for college graduates who replenish skilled manpower for booming financial and information technology industries, real wages for people without a high school diploma have remained flat. Whites experienced greater real wage increases than any other racial groups. These employment patterns are not new; they have been around since the 1970s (New York City Department of City Planning, 1995).

Recent policy initiatives are fighting the recursive loop of drugs and poverty on different fronts. For instance, in New York City, recipients of public assistance who are identified as substance abusers are mandated both to participate in simulated work programs that combine work with education, training, and job search skills and to enter drug abuse treatment programs (New York City Human Resources Administration, 1999). It is hoped that through this comprehensive strategy, drug-addicted indigents will regain self-reliance and avoid welfare dependency. Inexplicably, the role of human capital enhancement in drug prevention and control has not been properly recognized in professional literature. Educational improvement and job training are only seen as ancillary interventions in the treatment of chronic drug dependence and are rarely provided in most treatment settings, with the clear exception of residential programs (Anglin & Hser, 1990). Many still believe that behavioral-cognitive psychotherapy should be the only common denominator of all criminal justice–mandated drug treatment modalities (Deschenes & Greenwood, 1994), although existing evidence has shown that programs using both behavioral and life skills training produce the largest reduction in recidivism (Garrett, 1985).

In this article, I describe the human and social capital enhancement components of Brooklyn's Drug Treatment Alternative-to-Prison (DTAP) Program. First, a description of the program is provided, with detailed discussion of the strategies designed to improve the educational and vocational potential of participating defendants. Then, preliminary process and outcome statistics are interpreted. The study concludes with some reflections on the possibilities and limitations of reducing criminal recidivism through planned efforts to sever the link between drug abuse and economic marginalization.

DTAP

In October 1990, the Kings County District Attorney's Office (KCDA) created the DTAP on the premise that nonviolent offenders diverted to it would return to society in a better position after treatment to resist drugs and crime than if they had spent a comparable time in prison at nearly twice the cost. DTAP applies legal coercion (i.e., the threat of lengthy incarceration) to motivate drug-addicted defendants to choose and complete treatment.

DTAP targets nonviolent drug felons who commit crimes to support their drug addiction and who face mandatory prison sentences under the New York State's Second-Felony Offender Law.[1] Qualified defendants who are motivated for long-term treatment plead to a felony and undergo 15 to 24 months of rigorous residential treatment. All treatment is delivered in therapeutic communities, which provide structured therapeutic interventions, counseling, educational and vocational programs, on-site medical care, and assistance in finding housing. Phased individual and group counseling and behavioral therapies are standard interventions in a highly regimented setting to address issues of motivation, self-esteem, interpersonal relationships, problem-solving skills, and relapse prevention. The long-term nature of residential treatment also allows it to accommodate different services, including critical life skills training, to attend multiple needs of the individual, not just his or her drug use. To maximize public safety and to keep the legal pressure realistic, an enforcement team mobilizes to apprehend absconders, as soon as they leave the facility without permission, to return them to court for sentencing on the original charges. In contrast, those participants who remain in treatment have their charges dismissed after successful program completion.

DTAP Strategies to Enhance Human and Social Capital

Aware of the fact that a deficit in both human capital (basic education and marketable job skills) and social capital (networking and job market information) is one of the main obstacles to the reintegration of rehabilitated offenders to society, DTAP brings the drug treatment system, the criminal justice system, and the business community together to tackle the problem. Educational and vocational training is made available as part of residential treatment to enhance job skills, whereas deficiencies in job-related connections are remedied by the collaboration between the DTAP job developer and a business advisory council. DTAP seeks to reduce recidivism by increasing its participants' competitiveness in the world of legitimate work and by helping them adopt a more

responsible and productive lifestyle. The value of employment goes beyond monetary compensation; it imposes discipline and structure and enhances one's self-esteem. In the mind of DTAP planners, permanent success in reaching all these objectives is more likely in a civic partnership.

In-treatment educational remedies. In addition to clinical interventions to correct social-psychological problems, treatment providers contract outside educational and job skills–training resources to sustain positive changes in participants' behavior and attitudes through life skills enhancement. The most common educational remedy program is the General Educational Development (GED) preparation course, which is intended to improve participants' ability to read, compute, interpret information, and express themselves on a level comparable to graduating high school seniors. Participants can choose to take the GED exam during or after treatment. One of DTAP's treatment providers, Phoenix House, has its own high school program for youthful clients.

In-treatment vocational training. Vocational training usually consists of 3 to 6 months of off-site programs. Participants who want to benefit from these opportunities are free to enroll in the type of training program of their choosing. DTAP participants have been trained in a large number of job skills. The most popular vocational programs are those providing training in home health care, commercial driving, copying and printing, counseling, auto mechanics, and data entry. Participants also receive job readiness counseling on how to obtain and hold a job, including résumé writing and job-interviewing skills. The goal of the job search counseling is to eliminate job placement obstacles related to the lack of motivation, poor articulation, inappropriate dressing habits, lateness, and negative attitudes participants may harbor toward employment. Some vocational programs offer job placement services during the reentry stage of treatment. Graduates are expected to increase their educational and vocational skills through these training programs, to acquire some hands-on experience in practicing newly learned skills, and to be ready for reentry into the community.

Job developer. KCDA has hired a full-time job developer who works with treatment providers to identify the work histories and skills of graduates and to match them to the demands of the market. The DTAP job developer acts as a liaison with Brooklyn's business community to make specific job referrals and often has to negotiate with employers to ease up on hiring standards to employ disadvantaged DTAP graduates. Because of their dependence on the less stable, low-stratum, service labor market, some DTAP graduates are unable to hold a job for an extended period of time. Therefore, DTAP graduates are encouraged to maintain regular contacts with DTAP's job developer, particularly when they want to return to the labor market after a layoff or to look for a better job.

Business advisory council. This community body is formed by KCDA and composed of dozens of businesses located in the New York City area. Participating businesses work closely with DTAP's developer and identify and develop employment opportunities for DTAP graduates. These collaborative efforts have allowed many DTAP graduates to break into established business organizations.

In most contracted therapeutic communities, educational and vocational programs are optional services, and KCDA's policy is not to interfere with the therapeutic process, which is totally entrusted to the professional hands of the treatment providers. The collaboration among the job developer, business advisory council, and program graduates usually starts at or after treatment completion when the cases have already been officially closed. Therefore, although DTAP participants are informed of these enhancement programs and encouraged to participate, the utilization of these resources is strictly voluntary and does not constitute part of their formal agreement of program participation with KCDA.

In the following sections, I will describe the background characteristics of DTAP graduates and also explore how DTAP resources were utilized and what effects on posttreatment employment and recidivism can be reasonably associated to these efforts.

DATA, SAMPLES, AND METHOD

This is a retrospective, nonexperimental study based on official records. Information on the demographics, employment history, and treatment performance of DTAP graduates was obtained from DTAP administrative databases, whereas the recidivism analysis was based on official arrest data maintained by the New York State Division of Criminal Justice Services.[2] The Division of Criminal Justice Services' data excluded arrests that were sealed as of the date the computer file was prepared for use in this study. In New York, arrests are sealed if the court disposition is a dismissal or the conviction charge is a noncriminal offense, such as disorderly conduct.

Different subsamples of the entire DTAP population were examined throughout this study. The selection of subsamples was entirely based on the availability of information and is explained in each analysis. The secondary and administrative nature of the data did not contain as much pertinent and complete information as more rigorous multivariate tests would require; therefore, findings were derived from either descriptive or bivariate inferential analyses. Readers should keep these limitations in mind when interpreting the results and conclusions.

FINDINGS

Descriptive Statistics on the Background Characteristics of DTAP Graduates

Like most drug-addicted offenders found in the criminal justice system, DTAP participants come from disadvantaged neighborhoods, and most have poor educational credentials and long histories of unemployment or underemployment. Of the 406 DTAP participants who successfully completed treatment as of October 1999, 319 were interviewed right after treatment entry and again at the time of program completion.

Table 23.1 Demographic Characteristics of Drug Treatment Alternative-to-Prison (DTAP) Graduates ($N = 319$)

Variable	Mean	Minimum	Maximum
Age (years)	32	18	58
Self-report juvenile arrests	1	0	10
Official adult arrests	3	2	13
Average treatment length (months)	23	14	5
Average regular drug use (years)	12	1	35

	n	%
Gender		
Male	268	84
Female	51	16
Ethnicity		
Non-Hispanic White	16	5
Black	111	35
Hispanic	192	60
Primary drugs used		
Cocaine	26	8
Crack	89	28
Heroin	175	55
Other	29	9
Employment before DTAP		
Unemployed	252	79
Employed	67	21
Education before DTAP		
No high school diploma or GED	220	69
High school diploma or GED	96	30
College	3	1

NOTE: GED = General Equivalency Development

Table 23.1 displays the background characteristics of the sample. At the time of admission interview, these graduates averaged 32 years of age and 12 years of regular drug use. Of the sample, 84% were male. More than half of the sample (60%) were Hispanics, and about one third (35%) were Afro-Americans. Non-Hispanic Whites were only a small minority (5%). Of the sample, 94% self-reported using drugs on a daily or almost daily basis. Their criminal career averaged one juvenile arrest and three adult arrests. Heroin was the drug of choice most widely used among them, although a breakdown analysis reveals that ethnic subcultures apparently had played a role in determining their choice of primary drugs. Of both White and Hispanic graduates, 65% had used heroin as their primary drug, whereas 65% of Afro-American defendants had mainly been crack or cocaine users.

When it comes to educational and vocational achievements, this group showed clear evidence of socioeconomic disadvantages. Less than 1% of the sample had a college degree at the time of treatment entry as compared to 35% of New York City's adult population (New York City Department of City Planning, 1995). Moreover, 69% of the 319 participants did not even have a high school diploma or GED. Unemployment was almost a universal experience: Only 21% of them worked full-time or part-time during the year prior to DTAP admission. Most of them relied on government assistance, family help, illegal activities, or some combination of these options to support themselves. Forty-six percent were receiving some kind of public assistance the year prior to treatment entry. I need to highlight that overall, female graduates were more vulnerable than male defendants. Although the aggregate educational attainment of females did significantly differ from that of males (71% and 68%, respectively), they suffered much higher unemployment: 95% of female graduates did not hold a legal job prior to admission. In addition to selling drugs, for which they were arrested, 22% of the female group engaged in prostitution to generate income.

Drug use and contacts with the justice system had dominated the lives of DTAP graduates. Going to jail or prison had become a common and devastating experience for them. Data suggest that conditions of life outside prison had been equally impoverished, dangerous, and unsupportive. Drug selling, a crime for which a majority of these graduates were drawn to the justice system, becomes a viable way of life among addicted and unemployed residents of poor neighborhoods. It keeps participating addicts in contact with sources of illegal drugs and provides money without requiring long-term training or sophisticated skills. Unlike other property crimes, such as burglary or theft, whose opportunities decrease during recession or high unemployment periods when more people are confined to homes and fewer goods are in circulation, demands for drugs are extremely inflexible because they are driven by psychological or physiological addictions.

Process Statistics on the Utilization of Resources

Because the grand goal of DTAP's human and social capital enhancement efforts is to increase employment and to reduce recidivism through participants' involvement with services and programs offered at different stages of their treatment process, the utilization of these enhancement resources becomes an important indicator of implementation success.

In-treatment educational remedies. Educational remedy programs target those participants who did not have a high school diploma or GED prior to treatment entry, and these programs are unvaryingly provided in the form of GED preparation courses. Ideally, participants of GED preparation courses are expected to complete the program and take the test while still in treatment because positive learning effects produce the best results immediately after course completion and diminish over time.

Mixed results emerged from the analysis of the 243 graduates who did not have either a high school diploma or a GED before treatment (see Table 23.2). Although 80% of the participants without a high school diploma or a GED enrolled in the GED preparation course, only 26% of them felt ready and took the test during treatment. More important, only half of those who sat for the test successfully passed it. Although some of those who did not take the test or failed the test before treatment completion expressed their desire to take and pass it later on, I do not believe many of them were well-prepared and determined enough to achieve it based on the principle of diminishing chance mentioned earlier. It was not hard to persuade undereducated clients to participate in the GED course, but it was very difficult to equip course enrollees with sufficient technical and psychological preparation to meet the challenge.

Table 23.2   Process Statistics: Utilization of Resources

| | Level of Utilization | |
Program	n	%
GED preparation (*N* = 243)[a]		
Participated	194	80
Participated and took the test	62	26
Participated and passed the test	32	13
Vocational training (*N* = 319)		
Participated	201	63
Participated and completed	156	49
Job developer (*N* = 93)		
Contacted	48	52
Found job	37	40
Business advisory council (*N* = 93)[b]		
Offered employment	4	4

NOTE: GED = General Educational Development
a Only participants without a high school diploma or GED are included.
b Kings County District Attorney's Office after January 1998 started to collect data on graduates who completed treatment. Only those for whom intake and exit information is available are included. The analysis does not include past graduates dismissed before January 1998 who requested and received assistance from the job developer and/or the business advisory council after January 1998.

The perceived unreadiness among those who did not take the test may have been a fairly accurate assessment of the reality; only half of those who dared to take the test passed it. All this suggests that enrollment in a GED preparation course may be useful to refresh knowledge acquired in the past, to learn new academic skills, or to instill disciplined routines in daily life but is not as effective in improving the educational credentials of participants.

In-treatment vocational training. Participation in job skill training programs was widespread, and the completion rate was remarkably high. Of all 319 graduates, 63% had participated in at least one vocational education program and some in two, and 78% of the enrollees (representing 49%

of the entire sample) completed their training before graduation. The modal length of training is 3 months, although many programs lasted 6 months. Participants were overwhelmingly satisfied with the training they received. During the exit interview, program participants were asked how useful the vocational training was to them, and they gave an average rating of 9 on a scale of 1 to 10. Posttreatment employment data reported below suggest that the perceived usefulness of vocational training was well founded.

Job developer. Discussion on the work of the job developer and the business advisory council is based on the observation of 93 graduates who completed DTAP after January 1998 when KCDA started to gather information in these areas. Because 63% of the 319 DTAP graduates already had found employment on their own, through acquaintances or vocational-training schools at the time of or before program completion, the DTAP job developer concentrated his efforts on those who were still unemployed and also on past graduates who had lost their job or were trying to find a better one. Of DTAP graduates, 52% met with the job developer for employment counseling, and 40% were placed on jobs with the assistance of the DTAP job developer. Graduates who consulted with the job developer but did not get employment through him were mostly exploring opportunities for career change, which often required additional training. A few simply disappeared after getting employment and never returned. The data available do not show how many of the successful job placements were for new graduates and how many were for once-employed graduates who had become unemployed. But given that a great number of graduates who were unemployed at the time of treatment completion were not employable for a number of reasons (e.g., disability, full-time school enrollment, childcare), it is reasonable to infer that rather than initiating new graduates in the labor market, the DTAP job developer mainly reconnected past graduates who had become unemployed or underemployed. This safety net seemed to have protected many DTAP graduates from recurrent unemployment.

Business advisory council. Surprisingly, only 4% (4) of the 93 graduates with available data

accepted employment offers from participating members of the business advisory council as either a fresh graduate or a returnee. In a verbal communication, the DTAP job developer attributed this low utilization of the council to two specific circumstances. First, the job developer was an experienced veteran in the field and had already developed an extensive network of useful personal contacts in the local business community before he joined DTAP. Therefore, he mostly relied on his own leads to place DTAP graduates in employment. Second, despite the enthusiasm of the members of the business advisory council in taking DTAP graduates, mismatch often existed between their needs for highly skilled labor and the generally low qualifications of DTAP graduates. This observation agrees with New York's Department of City Planning's (1993) finding that industries that have traditionally hired low-skilled workers have steadily declined after World War II, whereas industries that employ few low-skilled workers have kicked into high gear.

OUTCOME STATISTICS ON POSTTREATMENT EMPLOYMENT AND RECIDIVISM

Results from data analysis clearly show that improved employment rates prevented posttreatment recidivism among DTAP graduates. As of October 1999, 281 (69%) of the 406 graduates were candidates for employment.[3] At the time of the DTAP arrest, only 26% of these 281 employable graduates were working (see Figure 23.1). In contrast, 92% of them were working at the time of data collection in various fields such as food service, commercial driving, building maintenance, construction, office management, security, health care, substance abuse counseling, sales, and retail management. Their earnings ranged from minimum wages to $34,000 per year.

An analysis of 117 employable DTAP graduates, for whom 3-year recidivism data was available, showed that legal employment was associated with decreased rates of recidivism. Of those employable DTAP graduates who were not working at the time of treatment

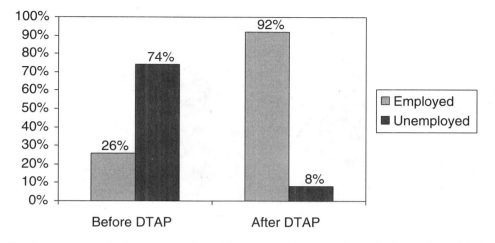

Figure 23.1   Comparison of Preemployment and Postemployment Rates Among Employable Graduates
($N = 281$)

NOTE: DTAP = Drug Treatment Alternative-to-Prison Program

completion, 33% were rearrested during the 3-year posttreatment follow-up period, whereas only 13% of those who were working full-time or part-time were rearrested during the same period of time (see Figure 23.2).

Strengthening criminal offenders' ties to the world of legitimate work must be one of the goals of alternative-to-incarceration programs, because crime is unlikely among those who are well trained and attractive to employers. The DTAP model that relies on contracting providers of residential treatment to develop marketable skills and knowledge among drug-addicted felons proves to be a viable option for crime reduction. Ex-offenders refrain from supplementing periods of low-wage work with periods of criminal activity if they are offered gainful and meaningful employment. Therefore, effective, long-term interventions should include educational and vocational training as well as referral and networking services to end ex-offenders' isolation from the legal labor market.

Discussion

The criminal justice system can make significant contributions toward the reintegration of nonviolent, drug-addicted felons into the legitimate labor market. The DTAP experience has

shown that although this task is not without its predicaments, it can be, and has been, achieved through bold planning, careful implementation, and a continued effort to evaluate and recalibrate the program. Results from the preliminary analysis of DTAP's administrative records reveal that when extant resources in the community are brought together to create a comprehensive network of support for the rehabilitation of addicted offenders, positive outcomes are likely to be found.

Diversion programs for serious or chronic-addicted offenders must be assessed continually and modified as necessary to meet the particular needs of the criminal population. Because outpatient treatment programs are more appropriate for "individuals who are employed or who have extensive social supports" (National Institute on Drug Abuse, 1999, p. 27), long-term residential treatment that focuses on the resocialization of the individual is best suited for the criminal justice population, which is generally less educated and trained and more socially isolated and criminally involved. Making vocational training and job placement counseling available to participants of coerced residential treatment attends to the social, psychological, and financial needs of the individual. In fact, meaningful human and social capital enhancement programs can only be implemented in conjunction with the

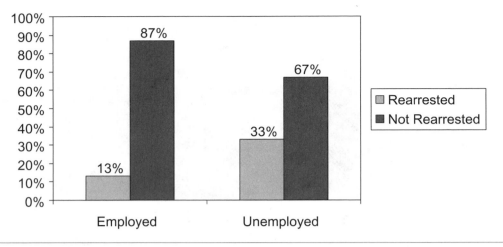

Figure 23.2   Posttreatment Employment and Recidivism (3-year follow-up)

NOTE: DTAP = Drug Treatment Alternative-to-Prison Program

structured and sometimes confrontational environment provided by the therapeutic community. Criminal justice agencies can play a very active role in triggering and maintaining the resocialization process as DTAP has demonstrated.

DTAP participants made extensive use of the educational and vocational opportunities during treatment. Of those without a high school diploma or its equivalent, 80% enrolled in educational remedy courses, while about two thirds of all participants started vocational-training programs. Formal learning and training were evidently integral components of this treatment approach that addresses not only participants' drug dependency and associated medical problems but also deficiencies in self-esteem and life skills. The immediate results of these programs were mixed. Only 16% of GED preparation course enrollees successfully passed the exam, indicating that it is quite difficult to put these relatively aged clients to the level of intellectual and psychological readiness required for this challenge. The limited market value of a GED for a 32-year-old individual with a fragmented employment history might have also contributed to this overall lack of motivation.[4] In contrast, 78% of those graduates who started

vocational training in treatment were able to finish it before treatment completion. It is likely that DTAP participants were more motivated to learn new, marketable skills than to work for a low academic degree, because they expected higher financial returns from the former. The DTAP job developer's work was critical in maintaining an extremely high employment rate among graduates. His valuable experience and effectiveness even greatly reduced the role of the business advisory council in providing jobs to DTAP graduates. Overall, the target clients widely and effectively utilized the human and capital enhancement resources organized by DTAP administrators.

The lack of a control group and of data in this analysis does not allow me to draw definite conclusions about the effect of DTAP's human and social capital enhancement efforts on posttreatment employment and recidivism,[5] nor was it possible to elucidate causal mechanisms that link each of the program components to posttreatment outcomes. Nevertheless, the comparison of the pretreatment and posttreatment employment rates as well as the bivariate correlation between employment and recidivism provided optimistic results. The employment rates among DTAP graduates jumped from the 26% pretreatment level to 92% after treatment completion. There was also indication that graduates who were working at the time of treatment completion were more than 50% less likely to be rearrested during the 3-year follow-up.

DTAP improved employment, which reduced recidivism. Despite these encouraging results, we should refrain from precipitated euphoria. Employment does not automatically reduce the proclivity toward crime. Commitment to work, job stability, and bonding among workers and employers are factors that contribute to ex-offenders' stake in conformity. Lamentably, internal inadequacies and external barriers often undermine the needed commitment to stable employment. Although all DTAP graduates are aware of how desperately they need work, a few would quit or get themselves fired after a short time on the job. It takes time for conventional work ethics to replace street codes of bravura.

But an even greater threat looms over planned efforts to return unemployed or under-employed ex-offenders back to work. Large urban regions, New York City among them, have been moving quickly from a unionized manufacturing economy to a less unionized service economy. As economic restructuring deepens, the less educated and less skilled workforce is systematically displaced to fill cheap, unregulated jobs on the margins of the booming financial and technological economy. The postindustrial market is not well prepared to permanently absorb rehabilitated offenders and only provides them with small contracting or clerical jobs in personal and retail services where employers often ignore regulations governing minimum wages, unemployment insurance, and worker's compensation.

For positive outcomes from alternative-to-incarceration programs, such as DTAP, to last, policy making at local, state, and national levels should strive toward the creation of jobs that offer real opportunities for achieving permanent, economic emancipation. Such policies might include local, targeted tax credits, urban development under neighborhood control, stringent enforcement of affirmative action laws, permanent public sector employment, subsidized on-the-job training, minority contracting and employment guarantees, and full-employment monetary and fiscal policies at the state and national levels. Without support from these macrolevel policies and regulations, the massive restructuring of the American urban economy will quickly dissipate any short-term success gained by local criminal justice efforts in returning drug-addicted offenders to the community.

NOTES

1. Before January 1998, only addicted drug sellers arrested in undercover "buy-and-bust" operations were considered for admission. Since January 1998, offers have also been made to some property offenders, mostly burglars and thieves, who commit criminal activities to support their drug use habit. With a few exceptions, Drug Treatment Alternative-to-Prison Program graduates analyzed in this report were originally charged with criminal sale of controlled substances.

2. Official arrest data were obtained from the New York State Division of Criminal Justice Services under the Non-Disclosure Agreement (No. 14-98).

3. Of the 125 unemployable graduates excluded from the analysis, 40 were physically disabled, 14 were enrolled in school or training programs, 6 were taking care of their children, 17 were detained or incarcerated, 10 had pending criminal justice charges, 15 were voluntarily readmitted to residential treatment, 8 were deceased, and the employment status of 15 was unavailable.

4. In an ad hoc analysis of the 319 graduates not reported in this article, pretreatment education status was found statistically significantly associated with posttreatment employment. Of those with a high school diploma or General Educational Development (GED), 78% were employed at the time of treatment completion, whereas only 62% of those without the same credentials were working ($p < .05$). Nevertheless, when the 243 graduates without a high school diploma or a GED were examined, passing the GED test in treatment did not improve posttreatment employment at all.

5. In a study reported in 1999, the 3-year recidivism of 184 DTAP graduates was compared with that of 215 matched control subjects (Hynes, 1999). Of the comparison group, 47% were rearrested during the follow-up period, whereas only 23% of DTAP graduates were rearrested. This analysis has provided the most solid support to the crime reduction effects of DTAP.

REFERENCES

Anglin, M. D., & Hser, Y. (1990). Treatment of drug abuse. In M. Tonry & J. Q. Wilson (Eds.), *Drugs and crime* (pp. 393-460). Chicago: University of Chicago.

Becker, G. S. (1968). Crime and punishment: An economic approach. *Journal of Political Economy, 76*, 169-217.

Deschenes, E. P., & Greenwood, P. W. (1994). Treating the juvenile drug offender. In D. L. MacKenzie & C. D. Uchida (Eds.), *Drugs and crime: Evaluating public policy initiatives* (pp. 253-280). Thousand Oaks, CA: Sage.

Ehrlich, I. (1979). The economic approach to crime: A preliminary assessment. In S. L. Messinger & E. Bitner (Eds.), *Criminology review yearbook Vol. 1* (pp. 34-36). Beverly Hills, CA: Sage.

Garrett, C. J. (1985). Effects of residential treatment of adjudicated delinquents. *Journal of Research on Crime and Delinquency, 22*(2), 287-308.

Hynes, C. J. (1999). *Drug treatment alternative-to-prison program: The ninth annual report.* Brooklyn, NY: Kings County District Attorney's Office.

Klein, S., & Caggiano, M. N. (1986). *The prevalence, predictability, and policy implications of recidivism.* Santa Monica, CA: RAND.

National Institute on Drug Abuse. (1999). *Principles of drug addiction treatment: A research-based guide.* Bethesda, MD: Author.

New York City Department of City Planning. (1993). *New opportunities for a changing economy.* New York: Author.

New York City Department of City Planning. (1995). *The New York City labor force: 1970, 1980, and 1990.* New York: Author.

New York City Human Resources Administration. (1999). *Welfare to work* [Online]. Available: http://www.ci. nyc.ny.us/html/hra/html/welfare_to_work.html

New York State Division of Criminal Justice Services. (2000). *Selection for criminal justice indicators* [Online]. Available: http://criminaljustice.state.ny.us/crimnet/ojsa/areastat/areast. htm

Thompson, J. W., Sviridoff, M., & McElroy, J. E. (1981). *Employment and crime: A review of theories and research.* New York: Vera Institute of Justice.

PART VI

CORRECTIONAL SYSTEMS AND ALTERNATIVES

24

"ALL THE NEWS THAT'S FIT TO PRINT"

A Content Analysis of the Correctional Debate in the New York Times

MICHAEL WELCH

LISA WEBER

WALTER EDWARDS

cholarship over the past three decades has generated considerable insight into the roles of the media, politicians, and law enforcement officials in constructing images of criminal justice; still, that body of research has rarely ventured into the realm of corrections. Filling this void, we drew a sample of 206 newspaper articles on corrections published in the *New York Times* for the purpose of examining news sources and their quoted statements. Our findings reveal that the *New York Times* relies heavily on political and government sources who—not surprisingly—express support for the prevailing correctional policies and practices. Whereas the *New York Times* also quoted sources critical of the government's

correctional strategies, the dominance of political sources in the press offers evidence of agenda setting in the debate over corrections.

Since the 1970s, scholars have generated significant research documenting various ways in which the media projects official images of crime and criminal justice (Barak, 1988, 1994; Chermak, 1994, 1997; Ericson, Baranek, & Chan, 1987, 1989, 1991; Fishman, 1978; Hall, Critcher, Jefferson, Clarke, & Roberts, 1978; Humphries, 1981; Kidd-Hewitt & Osborne, 1995; Surette, 1992). In particular, those investigations show that mediated information about crime is commonly traced to government sources; namely, political leaders, law enforcement officials, and other state managers. By

Welch, M., Weber, L., & Edwards, W. (2000, September). "All the News That's Fit to Print": A Content Analysis of the Correctional Debate in the *New York Times. The Prison Journal, 81*(2), 245-264. Reprinted with permission.

occupying elevated positions within the hierarchy of credibility, politicians and government officials enjoy the privilege of offering to the media primary—and self-serving—definitions of crime (Becker, 1967, 1973). As a result of this social arrangement, the media afford political and government leaders valuable opportunities for advancing a criminal justice agenda that serves the state's political, ideological, and economic interests by generating public support, legitimizing power, and garnering funds for resources and manpower (Fishman, 1978; Hall et al., 1978; Kasinsky, 1994). The reciprocal relations between the state and the media contribute to the projection of governmental (or official) versions of crime to the public, producing what is *socially thinkable* about the nature of crime and strategies to control it.

Given the pivotal role of politicians in manufacturing the news, critics accuse the media of carrying out propaganda functions for the government's ideological machinery (Herman & Chomsky, 1988; Kappeler, Blumberg, & Potter, 1996; Tunnell, 1992). Similarly, our previous research uncovered important elements of ideology (i.e., beliefs on crime causation and crime control) contained in crime news (Welch, Fenwick, & Roberts, 1997, 1998). Those studies, the first of their kind, were based on a content analysis of experts' quotes appearing in feature articles about crime in four major newspapers (i.e., the *New York Times*, the *Washington Post*, the *Los Angeles Times*, and the *Chicago Tribune*). In sum, we discovered further evidence of journalism's reliance on politicians and law enforcement officials as news sources who in effect serve as primary definers of crime. Compared to intellectuals (i.e., professors and nonacademic researchers) also sourced in those articles, politicians spoke more ideologically about crime causation and crime control insofar as they routinely disavowed the relationship between social conditions and crime.

Applying a similar method of content analysis, this investigation examines newspaper articles devoted to corrections—a widely neglected area in media studies (Marsh, 1989, 1991). While remaining attentive to which issues comprise the corrections agenda, the study tracks chief patterns of news sourcing, particularly among sources supporting the government's correctional strategies vis-à-vis sources opposing such policies. In doing so, we set out to weigh the relative balance of press coverage in the debate over corrections, thus shedding a critical light on agenda setting in criminal justice.

SAMPLE

While being able to generate mass attention to social issues, the media possesses a unique power to shape the nature of debate by defining parameters and selecting key items for public discourse. The debate over corrections presented in the press adheres to that pattern of news production, especially considering the selection process by which certain correctional issues are deemed more newsworthy than others. To further our understanding of how the correctional debate is framed by journalists, we turned our attention to the most circulated and arguably most influential newspaper in the nation.

Ranking first among metropolitan newspapers, the *New York Times* reports a daily circulation figure of 1,074,741, which surpasses its chief competitors, the *Los Angeles Times* (1,050,176), the *Washington Post* (775,894), and the *New York Daily News* (721,256) (*Detroit Free Press*, 1998). Largely due to its vast circulation, the *New York Times* is considered a highly influential newspaper that commands the attention of political, government, and business leaders, as well as the public. Given such influence, the *New York Times* has the capacity to present social issues in ways that dictate the nature of debate on matters of social policy. With this idea in mind, we set out to examine the manner in which corrections is covered by the *New York Times*.

To learn the extent to which correctional issues appeared in the *New York Times*, we administered a computerized literature search using the terms *corrections* and *prisons* as key words in locating appropriate articles. The search was limited to medium (6 to 18 column inches) and long (exceeding 18 column inches) stories to ensure depth of coverage. Between 1992 and 1995, the *New York Times* published 206 articles on corrections, spanning a wide array of 19 issues.[1] Institutional violence ($n = 40$) and correctional programs ($n = 34$)

were the most frequently covered topics, constituting 19% and 17% of all articles, respectively. Other notable correctional issues included health care ($n = 17$, 8%), "get tough" policies ($n = 16$, 8%), and privatization/corrections as industry ($n = 16$, 8%). By comparison, less attention was directed at overcrowding ($n = 10$, 5%), drugs (including the war on drugs) ($n = 9$, 4%), famous (and celebrity) inmates ($n = 9$, 4%), community concerns (including fear) ($n = 8$, 4%), correctional budgets ($n = 8$, 4%), and institutional issues ($n = 8$, 4%). Furthermore, even less coverage was aimed at juveniles ($n = 6$, 3%), officers ($n = 6$, 3%), the death penalty ($n = 4$, 2%), detention in Immigration and Naturalization Service (INS) centers ($n = 4$, 2%), contraband ($n = 3$, 1%), early release ($n = 3$, 1%), women prisoners ($n = 3$, 1%), and prison history ($n = 2$, 1%) (see Table 24.1).

METHOD

A principle objective of the study was to examine in-depth the manner in which the *New York Times* covers corrections; in particular, we set out to evaluate the relative balance of coverage by attending to news sourcing. To achieve this undertaking, we administered a content analysis on the sample of 206 articles by identifying and coding all sources along with their direct quotes and attributed statements. While acknowledging the identity of every source (e.g., politicians, corrections officials, reform advocates), we analyzed the content of their quotes to determine whether they supported or opposed the government's correctional strategies. It should be noted that the research design relied on mutually-exclusive categories dichotomizing support and opposition to the government's correctional strategies.

Interestingly, there were no quotes that could be construed as neutral; perhaps the lack of neutrality stems from the nature of interview questions presented by journalists who force news sources to take firm positions on issues in the correctional debate. Each correctional issue (e.g., institutional violence, correctional programs) was scrutinized according to the classification scheme, thereby permitting us to ascertain whether the debate over corrections in

Table 24.1 Correctional Issues in the *New York Times*, 1992 to 1995: Number of Articles and Percentages ($N = 206$)

Issue (abbreviation)	Number of Articles[a]	Percentage
Violence (VI)	40	19
Programs/ rehabilitation (PR)	34	17
Health care (HC)	17	8
Get tough (GT)	16	8
Privatization/ industry (PI)	16	8
Overcrowding (OC)	10	5
Drugs (DR)	9	4
Famous inmates (FI)	9	4
Budgets (BU)	8	4
Community concerns (CC)	8	4
Institutional issues (II)	8	4
Juveniles (JV)	6	3
Officers (OF)	6	3
Death penalty (DP)	4	2
Immigration and Naturalization Service (IN)	4	2
Contraband (CB)	3	1
Early release (ER)	3	1
Women (WM)	3	1
History (HI)	2	1

NOTE: Figures do not total 100% due to rounding.
a. Totals exceed the number of articles due to cross-listing of overlapping topics.

the *New York Times* appeared balanced or biased. Upon delineating the pattern of sourcing, we delved further into the qualitative content of the articles to extract nuances of news coverage aimed at corrections.

FINDINGS

Comparing various news sources in the correctional debate presented in the *New York Times*, we found that 62% ($n = 593$) of the sources supported the government's correctional strategies whereas 38% ($n = 363$) opposed them. A similar pattern also was evident among the sample of quoted statements: 62% ($n = 1486$) endorsed the government's prison policies whereas

Table 24.2 Support for Government Compared to Opposition, Number of Sources and Quotes per Issue: *NY Times* 1992 to 1995 (*N* = 206)

Issue (abbreviation)	N	Sources				Quotes			
		Support	(%)	Oppose	(%)	Support	(%)	Oppose	(%)
Violence (VI)	40	95	(59)	66	(41)	266	(63)	153	(37)
Programs/ rehabilitation (PR)	34	174	(83)	35	(17)	392	(84)	74	(16)
Health care (HC)	17	36	(56)	28	(44)	100	(59)	69	(41)
Get tough (GT)	16	36	(53)	32	(47)	128	(61)	82	(39)
Privatization/ industry (PI)	16	49	(72)	19	(28)	121	(70)	53	(30)
Overcrowding (OV)	10	22	(44)	28	(56)	49	(44)	63	(56)
Drugs (DR)	9	38	(53)	34	(47)	79	(51)	78	(49)
Famous inmates (FI)	9	11	(32)	23	(68)	34	(29)	82	(71)
Budgets (BU)	8	17	(50)	17	(50)	19	(37)	33	(63)
Community concerns (CC)	8	21	(55)	17	(45)	41	(55)	33	(45)
Institutional issues (II)	8	22	(64)	13	(36)	50	(63)	30	(37)
Juveniles (JV)	6	20	(66)	10	(34)	86	(86)	14	(14)
Officers (OF)	6	13	(59)	9	(41)	30	(47)	24	(53)
Death penalty (DP)	4	9	(60)	6	(40)	21	(50)	21	(50)
Immigration and Naturalization Service (IN)	4	6	(43)	8	(57)	15	(36)	27	(64)
Contraband (CB)	3	10	(59)	7	(41)	16	(57)	12	(43)
Early release (ER)	3	6	(55)	5	(45)	14	(50)	14	(50)
Women (WM)	3	7	(64)	4	(36)	20	(71)	8	(29)
History (HI)	2	1	(33)	2	(67)	5	(27)	13	(73)
Totals	206	593	(62)	363	(38)	1486	(62)	893	(38)

38% (*n* = 893) expressed criticism of them (see Table 24.2). By examining the relative proportion of sources and quotes within each category of issues, Table 24.2 demonstrates that the number of sources (and their quotes) supporting the government's correctional strategies consistently exceeded the sources (and quotes) voicing opposition; this pattern emerged in 15 of the 19 topics studied. On only four correctional issues (i.e., overcrowding, famous inmates, INS, and history) did the sources (and quotes) expressing opposition outnumber those supporting the government's stance on corrections. And on only one issue (i.e., budgets) did the number of sources (and statements) representing both sides of the debate reach parity.

The largest margin of imbalance emerged in articles on programs and rehabilitation (34 articles), a bias favoring the government's position with 83% (*n* = 174) of the sources and 84%

(*n* = 392) of the quotes. The second widest gap was found in the articles on privatization/corrections as industry (16 articles); on that issue, 72% (*n* = 49) of the sources and 74% (*n* = 121) of the quotes supported the government's position. In due course, we shall explain precisely what the government's position is on each of the issues examined, thus allowing us also to interpret more fully the scope of the correctional debate in the *New York Times*. Nevertheless, in the next section we delineate the emergent pattern of sourcing that illuminates the perspectives from which the correctional debate is framed.

Sources and Quotes Supporting the Government's Correctional Strategies

Overall, 23 types of news sources (*N* = 593) supporting the government's correctional strategies were identified, altogether generating 1,486 total quotes. The leading sources in this category are officials in political and government positions (elected and appointed officials, e.g., legislators, attorneys general), accounting for 24% ($n = 144$) of the sources and 22% ($n = 325$) of the quotes. The second and third leading sources supporting are corrections officials (e.g., correctional commissioners, deputy commissioners) (20%, $n = 116$) and correctional managers (e.g., wardens, deputy wardens) (13%, $n = 76$), accounting for 24% ($n = 352$) and 15% ($n = 220$) of the total quotes, respectively. It should be noted that this emerging constellation of sourcing conforms to the government's hierarchy of state managers; indeed, the highest ranking officials were the most often quoted, followed by those officials occupying the lower strata of government power. The dominance of these groups is especially significant because they comprise 57% of the sources and 60% of the quotes endorsing the government's position on corrections. Interestingly, the fourth largest group of sources supporting the state version of corrections consists of inmates, representing 13% ($n = 76$) of all sources and 12% ($n = 174$) of all quotes. The remaining 19 types of news sources clearly remained underrepresented; still, together they contribute substantially to the overall support of the government's position on corrections (comprising 43% of all sources and 40% of all quotes). Other sources include the following: health care professionals ($n = 33$), correctional staff (e.g., officers and various staff members, $n = 26$), prosecutors ($n = 16$), volunteers ($n = 14$), religious personnel (e.g., chaplains and various religious leaders, $n = 12$), judges ($n = 12$), program directors ($n = 10$), reform advocates (e.g., correctional watchdogs, civil liberties activists, $n = 9$), researchers ($n = 7$), professors ($n = 7$), residents ($n = 6$), attorneys representing inmates ($n = 6$), corrections union officials ($n = 6$), relatives of crime victims ($n = 5$), teachers of inmates ($n = 5$), "tough on crime" advocates ($n = 4$), a labor arbitrator ($n = 1$), an attorney representing corrections officers ($n = 1$), and a relative of an inmate ($n = 1$).

Sources and Quotes Opposing the Government's Correctional Strategies

Overall, 24 types of news sources ($N = 363$) opposing the government's correctional

strategies were identified, altogether contributing to 893 total quotes. Perhaps expectedly, the leading critics were reform advocates (e.g., correctional watchdogs, civil liberties activists) who represented 18% ($n = 67$) of all critics, offering 20% ($n = 175$) of all quotes. The second largest group of critics consisted of officials in political and government positions (elected and appointed officials, e.g., legislators, attorneys general), accounting for 15% ($n = 55$) of all sources and 12% ($n = 105$) of all quotes. Inmates ($n = 52$, 14%) emerged as the third most prevalent source of criticism, attributed to 20% ($n = 180$) of all statements, and attorneys representing inmates were the fourth most cited critics ($n = 44$, 12%, offering 95 quotes or 10%).

The government's correctional policies and practices also were called into question by numerous other critics, including 27 professors (7%), 21 corrections union officials (5%), 16 correctional staff members (e.g., officers) (4%), 14 corrections officials (e.g., correctional commissioners, deputy commissioners) (4%), 12 judges (3%), 12 residents (3%), 11 health care professionals (3%), 9 correctional managers (e.g., wardens, deputy wardens) (2%), 5 religious personnel (e.g., chaplains and various religious leaders) (1%), 3 program directors (0.8%), 2 researchers, attorneys representing correctional officers, "tough on crime" advocates, teachers of inmates, and prosecutors (0.5% each), as well as a collector, an auctioneer, a juror, a newspaper editor, and a volunteer (0.2% each).

Interpreting the Nature of the Debate Over Corrections

Reaching beyond a basic quantification of sources and quotes, this study also set out to explore in-depth the content of articles on corrections, drawing more fully on the qualitative aspects of the data. This approach enables us to offer additional interpretations about how the correctional debate is formulated by the *New York Times*, while remaining attentive to the pattern of sourcing. In this section, we concentrate primarily on the five most prevalent correctional issues: institutional violence ($n = 40$ articles), programs and rehabilitation ($n = 34$), health care in corrections ($n = 17$), get-tough campaigns ($n = 16$), and privatization/corrections as industry ($n = 16$).

Institutional violence and riots. As the most common correctional topic, institutional violence (and riots) ($n = 40$) comprises 19% of the articles in the sample. Upon closer scrutiny, however, it should be noted that there are distinct subtopics in this category. For instance, 10 articles were devoted to institutional violence occurring at Rikers Island jail complex (New York City); 8 articles focused on the 1993 prison riot at Lucasville, Ohio; 3 articles covered the questionable deaths in Mississippi jails; 3 articles followed the controversy surrounding prison disturbances that observers claimed were precipitated by the disparity in sentencing for crack (versus powdered) cocaine; finally, 3 articles chronicled the beatings and mistreatment of detainees held by the INS.

Not surprisingly, incidents of institutional violence were covered primarily as *events* (e.g., a particular riot, disturbance, or incident); by comparison, most other articles in our sample were presented as discussions or debates over certain correctional issues. Also prevalent in most of the articles in this category were exposés and other forms of investigative journalism, a type of media coverage rarely found among the other topics, except for a few articles on drugs, contraband, and the INS.

As Table 24.2 illustrates, there is considerable reliance on sources supporting the government's correctional policies in this category (accounting for 59% of the sources and 63% of the quotes). Interestingly, the nature of opposition in several of those articles unveils an otherwise uncommon coalition in the correctional debate. Correctional watchdogs, attorneys representing inmates, prisoners, and correctional officers unions linked institutional violence (or the threat of violence) at Rikers Island to a lack of adequate funding and allocation of resources. Together, those groups opposed New York City's correctional policy that reduced spending for municipal jails. In the words of William H. Booth, chairman of the Board of Correction (a watchdog group), "I'm not one to predict riot, but I know very well when you make cuts like this and take away people's services and rights, they're going to react and you never know how"

(Clines, 1994, p. B1). In response, New York City Mayor Giuliani adamantly defended the government's budgetary cuts for the city's jails, accusing the union of issuing "unfounded warnings" for the purpose of "budget manipulation" (Clines, 1994, p. B8). Stan Israel, president of the Correction Officers Benevolent Association injected further the alarming claim that "gang organization is on the rise" at Rikers Island (Clines, 1994, p. B8). As if in riposte, the Commissioner of Correction, Anthony Schembri, announced after a visit to Rikers Island that "a correctional officer, not an inmate had been arrested in possession of contraband—four razor blades and a sharpened knife blade presumably intended for sale to inmates" (Clines, 1994, p. B8). In projecting the government's interpretation of crime expressed through the media, Surette (1992, pp. 42-44) reminds us that animal metaphors abound in criminal justice rhetoric insofar as the role of police and corrections officers ("sheep dogs") is to protect the public ("sheep") from the criminal element ("wolves") (also see Welch, 1996, 1999; Welch et al., 1997, 1998). The wolf pack metaphor is especially popular in reference to gang violence. Conforming to this pattern of hyperbole, union president Israel referred to Hispanic and Black gangs at Rikers Island as "out-of-control wolfpacks roaming the prison at will and attacking any rival before correction officers can intervene" (Sullivan, 1994, p. B3).

As the debate on this issue suggests, even in articles about institutional violence, a theme of correctional economics is unmistakable. Moreover, political rhetoric and hyperbole is not only issued by the government to justify increased spending on corrections, but also by corrections officers unions who deploy similar tactics to protest budget cuts that affect adversely their membership. Equivocally, corrections officers unions either support or oppose the government's correctional policies depending on how they best serve their interests. It should be noted that the voices of correctional unions were heard early and often in the *New York Times*; union representatives were sourced in 7 of 10 articles about violence at Rikers Island.

Programs and rehabilitation. Although adequate funding for correctional programs has been denied greatly for the past three decades, contrary to the claims of some correctional pundits, the issue of rehabilitation is far from dead (see Clear, 1994; Flanagan, 1996; Welch, 1995, 1996, 1999). As a rough indicator of the degree to which correctional rehabilitation resonates in the mind of the press, our sample features 34 articles (17%) on programs, the second most common topic. Even more to the point, this grouping of articles generated more sources ($n = 209$) and quotes ($n = 466$) than other correctional topics in our sample; undoubtedly, the correctional debate over programs and rehabilitation is very much alive (see Table 24.2).

Consistent with the economic motif already acknowledged, the most common types of programs specified in these articles ($n = 6$) were those related to the promotion of meaningful work and gainful employment as a means of rehabilitation. Similarly, the subtopic of employment also alluded to the correctional value of instilling a work ethic. Perhaps underscoring the unspoken spiritual importance of a work ethic, an equally common subtopic in this section was religion as a source for reform ($n = 6$ articles). The next largest subsets of articles were devoted to the debate over the treatment of sex offenders ($n = 5$ articles) and for substance abusers ($n = 4$ articles). Although the virtues of the vocational, moral, and medical models were fairly well represented in this subsample, the education model, however, remained conspicuously neglected ($n = 3$ articles).

News sources (83%) and their quotes (84%) favoring the government's policies and practices overwhelmingly outnumbered voices of opposition; still, the government's stance on rehabilitation requires clarification. Remarkably, in all 34 articles, the government supported rehabilitation and programs, including those designed for sex offenders. The nature of criticism, however, centered typically on the complaint that correctional programs should be more available and better funded; likewise, many government officials concurred. Interestingly, opposition in the form of advocating harsher punishment rather than rehabilitation was minimal. More significantly, in 24 articles there was not a single news source (or quote) rejecting the government's claim that rehabilitation is an important aspect of its correctional strategy.

Upon reflection, the near absence of opposition on this issue could very well mean that the debate over rehabilitation constitutes a *false polarity*, insofar as it is more universally supported by citizens than politicians realize. Indeed, public opinion polls demonstrate that citizens generally favor correctional rehabilitation, especially in the form of substance abuse treatment (Maguire & Pastore, 1995, pp. 165, 176, 198; also see Cullen, Cullen, & Wozniak, 1988; Flanagan, 1996; Welch 1997a, 1997b; Welsh, Leone, Kinkade, & Pontell, 1991). True, citizens may also demand punishment (Maguire & Pastore, 1995), but contrary to the mind-set of politicians who exploit the crime issue for campaign purposes, rehabilitation and punishment really do not exist as mutually exclusive categories (see Clear, 1994).

Similarly, Wilkins (1991) refers to false polarities in criminal justice by pointing to "just deserts": the idea that offenders ought to receive the punishment that they justly deserve. Elaborating on what he means by a false polarity, Wilkins insists that it is precisely the popularity of "just deserts" that diminishes its value as a principle of jurisprudence, "since no one can really oppose the idea of 'just deserts' due to the absurdity of calling for unjust undeserved punishment" (1991, p. 2). In light of overwhelming public approval for correctional programs—along with the government's endorsement—the debate over rehabilitation often seems superficial, constituting a false polarity. Rather than depicting rehabilitation as a hotly contested issue, the *New York Times* appears to reflect a wide consensus supporting correctional programs.

Health care in corrections. Given the amount of concern and anxiety about health care in the general population, it is of little surprise that medical issues in corrections ranked third among the topics in our sample ($n = 17$, 8%). The most common subtopic in this category was the issue of AIDS/HIV ($n = 8$), followed by tuberculosis (TB) ($n = 4$) and smoking in correctional facilities ($n = 3$)[2]. As Table 24.2 illustrates, news sources (56%) and quotes (59%) supporting the government's strategy for prison health care outnumbered the voices of opposition.

Consistent with the coverage on many of the topics, the theme of correctional economics is clearly evident in articles on health care. Indeed, the government's policies on prison health care were predicated largely by economic initiatives that, in turn, polarized the correctional debate. The government's plan to privatize prison medical services was staunchly opposed by correctional staff unions. Similarly, the government's policy requiring inmates to pay for health care was criticized by reform advocates and attorneys for inmates as well as inmates themselves. Government sources commonly pointed to the rising costs of medical care, including staffing and medication (especially for AIDS/HIV and TB) as reasons to reduce health care expenses.

Issues of correctional debate also included controversial prison management practices, particularly in the realm of AIDS/HIV, TB, and smoking. In 1992, the Erie County, N.Y. jail was sued by Louise K. Nolley, an inmate (charged with passing a check with insufficient funds and forgery) who was subjected to humiliating and improper treatment by correctional staff. The county government approved an institutional policy confining inmates with AIDS to an isolation unit, requiring them to wear plastic gloves while using the typewriter in the prison library and having their belongings labeled with red stickers identifying them as being infected with AIDS. Defending the county's correctional policy on AIDS, jail superintendent John Dray argued that the red stickers were introduced in 1986 when "AIDS became the epidemic that was terrorizing everybody in the business" and they are still necessary "to protect the 13,000 inmates we have in our system every year" (Sullivan, 1992, p. B4). Prisoners and reform advocates challenged the county's correctional policy and practice on AIDS. Eventually the courts intervened on behalf of the prisoners. Judge John T. Curtin of the Federal District Court in Buffalo, N.Y., awarded Nolley $155,000, commenting that, "there is no question that the red-sticker policy was developed, not in response to contagious diseases in general, but specifically in response to the hysteria over H.I.V. and AIDS" (Sullivan, 1992, p. B4).

In its coverage of smoking in prisons, the *New York Times* clearly pointed out that the government's correctional policies and practices included defending itself against inmates suing prisons for illnesses caused by secondhand smoke. In *Helling v. McKinney* (1992), a Nevada prisoner sued the state department of corrections for being housed (in a six-by-eight foot cell) with a cellmate who smoked five packs of cigarettes a day. The U.S. Supreme Court ruled in the case that prisoners who can show that exposure to smoking is a threat to their health may have a constitutional right not to be confined with a chain-smoking cell mate.

On the issue of smoking in prisons, the government's stance is particularly revealing. During the initial stages of litigation involving *Helling v. McKinney*, the Bush Administration as well as 34 states petitioned the Court to rule against the prisoner, insisting that "smoking is wide-spread in society" and that "smoking is a habit that is deeply rooted in our history and experience" (Greenhouse, 1993, p. A8). The Bush Administration further argued that "there was no basis for ruling that exposure to second-hand smoke could even theoretically violate the Eighth Amendment" (Greenhouse, 1993, p. A8). Whereas the *text* (or manifest content) of the debate appears to focus on the issue of health in correctional facilities, the *subtext* (or latent content) seems to be geared toward the Bush administration's (along with 34 state governments') support for the tobacco industry.

Get-tough campaigns. The fourth largest grouping of articles ($n = 16$, 8%) was devoted to get-tough campaigns, a punitive social movement that significantly impacts the form and function of corrections. A principle theme of those articles centered on the retributionist claim that correctional facilities are not tough enough to fulfill their mission of imposing punishment, instilling discipline, and deterring lawbreakers from committing future offenses. Advocates of get-tough campaigns insinuate that corrections coddles inmates and, confounding the issue, imply that high rates of recidivism are the result of institutional conditions that are inadequately punitive. While being driven by a staunch sense of retribution, those campaigns also are inspired by nostalgic sentiments involving criminal justice (Garland, 1990; Jameson, 1991; Simon, 1995; Stauth & Turner, 1988). Get-tough proposals include "three strikes" legislation, the return to hard labor (e.g., breaking rocks), chain gangs, correctional boot camps, the reduction or elimination of institutional amenities (e.g., programs, televisions, weight lifting), and the construction of super-maximum security penitentiaries.

The nostalgic worldview in corrections has emerged as a reaction to the modern and so-called liberal American prison where convicts are afforded certain constitutional protections, allowed to participate in institutional programs, and permitted to keep personal belongings while incarcerated (e.g., civilian clothes, televisions, coffee, cigarettes, snack food). Critics argue that depictions of prisons as country clubs (or "three hots and a cot") are wildly exaggerated, inaccurate, and politically manipulative (Welch, 1997b, 1999). Nostalgia marks a return to harsh images of prison life, similar to those projected in such classic movies as *Cool Hand Luke.* In describing support for the get-tough campaigns in Arizona, Governor Fife Symington waxed his personal nostalgia with movie imagery which includes a sheriff's 400 member "executive posse" patrolling Phoenix in search of law-breakers: "I remember, going way back, watching Tom Mix and Gene Autry and Roy Rogers . . . and I always remember the sheriff swearing in business people on a horse as posse men and saying, 'Go after the horse thieves'" (Mydans, 1995, p. A6; see Koppes & Black, 1987). Sounding the alarm of perceived lawlessness, social disorder, and moral panic, the nostalgic view of corrections expresses a need to regain control of convicts and penal institutions, as well as a return to a simpler society. In doing so, notions of retribution and deterrence are politicized to the extent that the locus of power is personified in criminal justice officials. Sheriff Joseph M. Arpaio, who was elected in 1992 to administer the Maricopa county jail, Arizona, announced to the press, "I want everybody in this country to know that if you commit a crime, you are going into a very bad jail. . . . I want people to say: 'I hate that sheriff. I hate his jails'" (Mydans, 1995, p. A6). Correspondingly, Sheriff Arpaio proudly boasted his beliefs about the virtues of expanding of the criminal justice system, saying, "My whole philosophy is, put more people in jail. We've got a vicious crime problem out there, and the answer is to take

them off the streets and educate them through punishment" (Mydans, 1995, p. A6).

Illuminating the tendency for elected leaders to resort to political hyperbole in justifying the government's financial commitment to increasing the criminal justice apparatus, Georgia Congressman Newt Gingrich proclaimed, "We should build as many stockades as necessary, and as quickly as though this were wartime and people were dying, because they are" (Berk, 1994, p. A9). Likewise, California Governor Pete Wilson blended populism and correctional economics into his personal vernacular of political crime-speak: "There's really no dispute that these reforms will require considerable additional expense for building prisons and operating them. That is an expense, I submit, that the public is willing to pay. We cannot afford not to pay" ("California Governor," 1994, p. A16). Joining the prevailing political mantra to expand the correctional machinery, New York Governor George F. Pataki pleaded to his constituents, "We cannot fail to build new [correctional] facilities if public safety requires it" (Levy, 1995, p. B4).

Whereas the divide between news sources supporting the government's correctional strategy of getting tough (53%) and its opposition (47%) was narrower than others, the total number of statements favored the government (61%; see Table 24.2). Nevertheless, many dissenters offered insightful and reasonable criticisms of get-tough campaigns. In particular, a critic of Alabama's revived chain gangs and rock-breaking ritual pointed to the futility of those practices. Dalmus Davidson, a highway director in Alabama, reported that the Department of Corrections' rock-breaking program would "provide no financial benefit to the state, which already has contracts with quarries to crush rock in various sizes that can be used along highway shoulders and for some road construction" ("Alabama to Make Prisoners," 1995, p. A5). In essence, get-tough campaigns often resort to a form of prison labor that ironically is void of utility and profit, suggesting that such penal exercises take on a purely punitive significance (Foucault, 1979; Welch, 1999).

Privatization and corrections as industry. Correctional economics evident in the growing correctional-industrial complex remained a principle theme in articles on privatization ($n = 9$) and corrections as industry ($n = 6$) (1 additional article focused on the privatization of health care; together these articles comprise 8% of our sample). Overall, 72% of news sources ($n = 49$) and 70% of quotes ($n = 121$) supported the government's correctional strategy advocating both privatization and the use of corrections as local industry. While downsizing the armed services and closing military bases, a move that erodes the economic base of those communities, the government has endeavored to offset local financial loss by installing penal institutions that promise to boost the local economy.

A case in point: In 1992, the federal government announced the closing of Fort Dix, N.J., army base, sending shock waves through a community that had deep financial ties to the military. (In fact, nearly 4,000 civilians worked at Fort Dix, one of the nation's largest employers.) An agreement between the Pentagon and the Federal Bureau of Prisons, however, softened the blow by converting Fort Dix into the federal system's biggest prison (housing 3,200 inmates). An estimated 700 new jobs were created, albeit little compensation for the massive civilian layoffs caused by the closing of the army base. According to Daniel R. Dunne, a spokesman for the Federal Bureau of Prisons, "It's a win-win-win situation for everyone involved. . . . It's an excellent opportunity for the tax-payers. We'd move into a facility that's already established. Both organizations see the importance and economic significance this has and are looking to make it work" (Hanley, 1992, p. 33).

Similarly, in other articles describing the support for corrections as local industry, a Weed, California, city councilwoman retorted, "prisons don't go out of business" ("Residents of Dying," 1994, p. 28) whereas a resident in Appleton, Minnesota, quipped, "It's not a smoke-stack industry. . . . It's a renewable resource" (Terry, 1993, pp. 1,14).

On the other side of the debate, relatively few sources were quoted opposing the government's correctional strategy of privatization and use of prisons to generate revenue ($n = 19$ articles, 28%; $n = 53$ quotes, 30%). Six of the 16 articles did not cite a single source (or statement) of criticism, and in 4 articles, only one critic was quoted. Still, some critics of the government's correctional policies did shed light on the ironic nature of correctional economics contributing to the proliferation of the correctional enterprise.

> Ideally, prisons should be trying to put themselves out of business by doing everything in their power to reduce the recidivism rate. . . . But the very function of a for-profit prison company is to keep people locked up, as long as possible because the companies are paid a certain amount per prisoner, per day. (Jenni Gainsborough, spokeswoman for the National Prison Project of the American Civil Liberties Union [ACLU], quoted in Van Natta, 1995, p. A24)

In a similar vein, Alvin Bronstein, executive director of the National Prison Project of the ACLU, cited criminal justice as one of the few growth industries in the United States, adding, "Corrections today is a gigantic cash machine" (Holmes, 1994, p. 3).

CONCLUSION

Proponents of dominant ideology theory argue that the media's manufacturing of news relies heavily on government sources who, in turn, exploit public access to reinforce state power (Abercrombie, Hill, & Turner, 1980; Hall et al., 1978; Herman & Chomsky, 1988). From a Marxian view of the media, the power of the ruling class extends beyond the ownership and control of the means of material production by exerting influence over the means of mental production. Public pronouncements ratifying the government's criminal justice apparatus, for instance, are mental products transmitted in line with the imperatives of the dominant ideology (see Gramsci, 1971; Larrain, 1983; Marx, 1978; Sahin, 1980).

This research discovered considerable evidence of the politicization of punishment and criminal justice agenda setting. Emulating the hierarchy of state managers, political and government leaders were quoted considerably more often on correctional issues than any other news source, even corrections officials and prison managers. Our findings suggest that media discourse on corrections provides additional opportunities for high-ranking government figures to institutionalize their authoritative position (Becker, 1967, 1973; Chermak, 1997; Hall et al., 1978). In our sample, government sources pronounced repeatedly their agenda to expand the correctional system and to harshen the nature of incarceration, both of which are prominent features of coercive social control. Whereas political posturing on corrections serves to cultivate public support for criminal justice policy, it also functions to legitimize state power through the distribution of punishment (Foucault, 1979; Welch, 1999). By serving as primary news sources, politicians and government officials benefit ideologically, insofar as they determine what is *socially thinkable* about crime and criminal justice; likewise, they also gain materially by promoting their institutional goals and needs (e.g., increased spending on prison construction, maintenance, resources, and personnel) (see Chermak, 1994, 1997; Fishman, 1978; Hall et al., 1978; Humphries, 1981; Kappeler et al., 1996; Kasinsky, 1994; Surette, 1992; Tunnell, 1992).

The politicization of punishment and criminal justice agenda setting in the media have significant implications to correctional policy and practice. The correctional debate presented in the *New York Times*, whose motto is "All the news that's fit to print," concentrates heavily on issues of institutional violence and correctional programs. To a lesser degree, the *New York Times* addresses prison health care, get-tough policies, and privatization/corrections as industry, while neglecting other important correctional issues (e.g., capital punishment, juveniles, and female prisoners). More to the point of agenda setting, which influences correctional policies, those selected issues are introduced more often from the viewpoint of politicians and government officials whose perspectives determine the fate of policy and the flow of funding.

The phenomenon of agenda setting in corrections is highly relevant to criminal justice research, especially considering that popular notions about crime and criminal justice are constructed by the media, often according to the political views of government leaders. Criminal justice scholars entering the public debate over corrections should assess critically what is being reported in the media, remaining attentive to which issues are subject to discussion and which issues are ignored. Equally important, scholars should ascertain which groups serve as the dominant news sources and decipher their recommendations for correctional policy and practice. In doing so, it is crucial that scholars be mindful of the biases among news sources and confront the numerous myths and misconceptions shaping the government's correctional strategies.

In sum, our analysis of newspaper articles on corrections yields findings consistent with previous research documenting the media's dependence on government sources in manufacturing images of crime and criminal justice (Barak, 1994; Chermak, 1994, 1997; Ericson et al., 1987, 1989, 1991; Fishman, 1978; Hall et al., 1978; Humphries, 1981; Kasinsky, 1994; Surette, 1992; Welch, Fenwick, & Roberts, 1997, 1998). However, we also revealed that the press tends to include voices challenging the government's correctional strategies, suggesting perhaps a sense of pluralism. In fact, 38% of the sources (and their quotes) in our sample opposed the government's correctional policies and practices. Still, the degree of diversity evident in this study must be interpreted cautiously. Our examination of news sourcing was confined to the *New York Times*; as a result, the extent of pluralism in our findings may be unique to that newspaper. Additional research is needed to determine whether these findings are representative of other newspapers and media outlets. Despite these limitations, our investigation contributes further to a critical understanding of the social construction of crime news by offering an analysis of corrections: a vital component of criminal justice, though widely neglected by media researchers.

Notes

1. In keeping to our task of evaluating the relative balance of press coverage on corrections in the United States, we eliminated articles on international corrections ($n = 23$), leaving 206 articles in the sample.

2. Newspaper articles on drug and alcohol treatment were assigned to the previous section on programs.

REFERENCES

Abercrombie, N., Hill, S., & Turner, B. S. (1980). *The dominant ideology thesis.* London: Allen and Unwin.

Alabama to make prisoners break rocks. (1995, July 29). *The New York Times*, p. A5.

Barak, G. (1988). Newsmaking criminology: Reflections on the media, intellectuals, and crime. *Justice Quarterly, 5*(4), 565-587.

Barak, G. (1994). *Media, process, and the social construction of crime.* New York: Garland.

Becker, H. S. (1967). Who's side are we on? *Social Problems, 14*, 239-247.

Becker, H. S. (1973). *Outsiders: Studies in the sociology of deviance.* New York: Free Press.

Berk, R. (1994, January). G.O.P. sees crime as a major issue. *The New York Times*, p. A9.

California governor expected to back life terms for repeated felons. (1994, March). *The New York Times*, p. A16.

Chermak, S. (1994). Crime in the news media: A refined understanding of how crimes become news. In G. Barak (Ed.), *Media, process, and the social construction of crime* (pp. 95-130). New York: Garland.

Chermak, S. (1997). The presentation of drugs in the news media: The news sources involved in the construction of social problems. *Justice Quarterly, 14*(4), 687-718.

Clear, T. R. (1994). *Harm in American penology: Offenders, victims, and their communities.* Albany: State University of New York Press.

Clines, F. X. (1994, November 17). Rikers is tense as cuts loom, and official warns of crisis. *New York Times*, p. B1, B8.

Cullen, F. T., Cullen, J. B., & Wozniak, J. F. (1988). Is rehabilitation dead? The myth of the punitive public.

Journal of Criminal Justice, 16, 303-317.

Detroit Free Press. (1998). *100 largest U.S. newspapers* [Online]. Available: http://www.freep. com/jobspage/links/top100.htm

Ericson, R. V., Baranek, P. M., & Chan, J. B. L. (1987). *Visualizing deviance: A study of news organizations.* Toronto, Canada: University of Toronto Press.

Ericson, R. V., Baranek, P. M., & Chan, J. B. L. (1989). *Negotiating control: A study of news sources.* Toronto, Canada: University of Toronto Press.

Ericson, R. V., Baranek, P. M., & Chan, J. B. L. (1991). *Representing order: Crime, law, and justice in the news media.* Toronto, Canada: University of Toronto Press.

Fishman, M. (1978). Crime waves as ideology. *Social Problems, 25*, 531-543.

Flanagan, T. J. (1996). Reform or punish: Americans' views of the correctional system. In T. J. Flanagan & D. R. Longmire (Eds.), *Americans view crime and justice: A national public opinion survey* (pp. 75-92). Thousand Oaks, CA: Sage.

Foucault, M. (1979). *Discipline and punish: The birth of the prison.* New York: Vintage.

Garland, D. (1990). *Punishment and modern society: A study in social theory.* Chicago: University of Chicago Press.

Gramsci, A. (1971). *Selections from the prison notebooks.* New York: International.

Greenhouse, L. (1993, June 19). Court offers inmates a way to escape prison smokers. *The New York Times*, p. A8.

Hall, S., Critcher, C., Jefferson, T., Clarke, J., & Roberts, B. (1978). *Policing the crisis: Mugging, the state and law and order.* New York: Holmes and Meiser.

Hanley, R. (1992, August 30). Fort Dix may become federal prison. *New York Times*, p. 33.

Helling v. McKinney, 61 U.S. LW 3445 (1992).

Herman, E. H., & Chomsky, N. (1988). *Manufacturing consent: The political economy of the mass media.* New York: Pantheon.

Holmes, S. A. (1994, November 6). The boom in jails is locking up lots of loot. *The New York Times*, p. 3.

Humphries, D. (1981). Serious crime, news coverage, and ideology. *Crime and Delinquency, 27*, 191-205.

Jameson, F. (1991). *Postmodernism or the cultural logic of late capitalism.* Durham, NC: Duke University Press.

Kappeler, V. E., Blumberg, M., & Potter, G. W. (1996). *The mythology of crime and criminal justice.* Prospect Heights, IL: Waveland.

Kasinsky, R. G. (1994). Patrolling the facts: Media, cops, and crime. In G. Barak (Ed.), *Media, process, and the social construction of crime* (pp. 203-236). New York: Garland.

Kidd-Hewitt, D., & Osborne, R. (1995). *Crime and the media: The postmodern spectacle.* East Haven, CT: Pluto.

Koppes, C., & Black, G. (1987). *Hollywood goes to war: How politics, profits, and propaganda shaped the World War II movies.* New York: Free Press.

Larrain, J. (1983). *Marxism and ideology.* London: The MacMillan Press.

Levy, C. (1995, December 12). Pataki proposes a ban on parole in violent crimes. *The New York Times*, pp. A1, B6.

Maguire, K., & Pastore, A. L. (1995). *Sourcebook of criminal justice statistics 1994.* Washington, DC: U.S. Department of Justice, Bureau of Justice Statistics, U.S. Government Printing Office.

Maguire, K., & Pastore, A. L. (1996). *Sourcebook of criminal justice statistics 1995*. Washington, DC: U.S. Department of Justice, Bureau of Justice Statistics, U.S. Government Printing Office.

Marsh, H. L. (1989, September). Newspaper crime coverage in the U.S.: 1893-1988. *Criminal Justice Abstracts*, pp. 506-514.

Marsh, H. L. (1991). A comparative analysis of crime coverage in newspapers in the United States and other countries from 1960-1989: A review of the literature. *Journal of Criminal Justice, 19*, 67-79.

Marx, K. (1978). The German ideology. In R. D. Tucker (Ed.), *The Marx Engels reader* (2nd ed., pp. 146-200). New York: Norton.

Mydans, S. (1995, March 4). Taking no prisoners, in a manner of speaking. *The New York Times*, p. A6.

Residents of dying California town see future in a prison. (1994, May 8). *The New York Times*, p. 28.

Sahin, H. (1980). The concept of ideology and mass communication. *Journal of Communication Inquiry, 61*, 3-12.

Simon, J. (1995). They died with their boots on: The boot camp and the limits of modern penality. *Social Justice, 22*(2), 25-48.

Stauth, G., & Turner, B. S. (1988). Nostalgia, postmodernism, and the critique of mass culture. *Theory, Culture, and Society, 52-53*, 509-526.

Sullivan, R. (1992, August 26). Ex-inmate wins award in bias case: Woman with AIDS granted $155,000. *The New York Times*, p. B4.

Sullivan, R. (1994, December 20). 7 stabbings and 5 others hurt in Rikers clash. *New York Times*, p. B3.

Surette, R. (1992). *Media, crime & criminal justice: Images and realities*. Pacific Grove, CA: Brooks/Cole.

Terry, D. (1993, January 3). Town builds a prison and stores its hope there. *The New York Times*, p. 1.

Tunnell, K. (1992). Film at eleven: Recent developments in the commodification of crime. *Sociological Spectrum, 12*, 293-313.

Van Natta, D. (1995, August 12). Despite setbacks, a boom in private prison business. *The New York Times*, p. A24.

Welch, M. (1995). Rehabilitation: Holding its ground in corrections. *Federal Probation: A Journal of Correctional Philosophy and Practice, 59*(4), 3-8.

Welch, M. (1996). *Corrections: A critical approach*. New York: McGraw-Hill.

Welch, M. (1997a). A critical interpretation of correctional bootcamps as normalizing institutions: Discipline, punishment, and the military model. *Journal of Contemporary Criminal Justice, 13*(2), 184-205.

Welch, M. (1997b). The war on drugs and its impact on corrections: Exploring alternative strategies to America's drug crisis. *Journal of Offender and Rehabilitation, 25*(1-2), 43-60.

Welch, M. (1999). *Punishment in America: Social control and the ironies of imprisonment*. Thousand Oaks, CA: Sage.

Welch, M., Fenwick, M., & Roberts, M. (1997). Primary definitions of crime and moral panic: A content analysis of experts' quotes in feature newspaper articles on crime. *Journal of Research in Crime and Delinquency, 34*(4), 474-494.

Welch, M., Fenwick, M., & Roberts, M. (1998). State managers, intellectuals, and the media: A content analysis of ideology in experts' quotes in featured newspaper articles on crime. *Justice Quarterly, 15*(2), 219-241.

Welsh, W., Leone, M. C., Kinkade, P., & Pontell, H. (1991). The politics of jail overcrowding: Public attitudes and official policies. In J. Thompson & G. L. Mays (Eds.), *American jails: Public policy issues* (pp. 131-147). Chicago: Nelson-Hall.

Wilkins, L. T. (1991). *Punishment, crime and market forces*. Brookfield, VT: Dartmouth.

25

PRISON PROGRAMS IN THE UNITED STATES

Preparing Inmates for Release

MARY BOSWORTH

Each year, an estimated 600,000 men and women return to the community following periods of incarceration. Indeed, almost all prisoners (97%) in the United States will one day leave their place of confinement. The skills and attitudes with which these individuals reenter society are paramount to their future success as law-abiding citizens. As a result, most prisons offer programs and training that are designed to help inmates adjust lawfully to life outside the prison walls. This chapter will discuss some of these programs and their history in the federal, California, and New York State penal systems.

OVERVIEW

Until the mid-1970s, most prison systems in the United States and around the world were shaped by a general belief in rehabilitation. Although

specific ideas about how to achieve reform varied, it was generally accepted that a primary purpose of a term of imprisonment was to change an offender for the better.

The first penitentiaries, reflecting views of the time, sought to reform inmates through a combination of religious contemplation and labor. The Walnut Street Jail in Philadelphia, for example, which opened in 1776, held prisoners in solitary confinement, offering them only the Bible as solace. Later, the Auburn Penitentiary in New York State allowed inmates silent congregation during the day to labor together. The few women prisoners at this time worked at tasks like cooking and cleaning, which taught them feminine skills they were thought to lack (Rafter, 1990).

As the 20th century progressed, the federal system and other state departments of correction adopted the so-called "medical model" of rehabilitation. According to this idea, offenders

AUTHOR'S NOTE: Parts of this article appear in Bosworth M., (2002). *The U.S. Federal Prison System,* Thousand Oaks, CA: Sage. Used with permission.

were sick. Women, in particular, were often referred to as "mad" rather than "bad," and men were frequently defined as drug addicts who needed treatment rather than punishment.

Belief in rehabilitation of any sort declined gradually over the 1970s. Following the publication of the critical article "What Works?" (Martinson, 1974), prison administrators began to look for other justifications for their institutions. As a result, most prison systems adopted some version of the "Balanced Model of corrections, which held that rehabilitation was not the paramount goal of incarceration but rather one of several co-equal goals (which also included punishment, deterrence, and incapacitation)" (Roberts, 1994, p. 13). In this strategy, which remains the central tenet of U.S. penal policy today, far less emphasis is placed on reforming individuals. The future success of prisoners is no longer thought to be the central concern of prison administrators. Instead, rehabilitation has become the individual prisoner's responsibility. The prison is merely the facilitator of such change.

Despite the large-scale ideological rejection of rehabilitation by prison administrators, penal institutions still offer many programs designed to help inmates adjust to life after release. This chapter will describe three key areas: work, education, and drug treatment. Comparative information shall be provided about federal prisons and the correctional systems of the states of New York and California.

WORK

Prisoners, as a group, tend to have little legal work experience. Studies have found that, nationwide, up to 40% of all offenders were unemployed or marginally employed prior to arrest and that 83% of probation and parole violators were unemployed at the time of violation. The unemployed may be more likely to engage in criminal activity because of economic necessity and may be frustrated because of their inability to find work.

Since the establishment of penitentiaries in the 19th century, work has constituted a key part of the prison experience. It has always, however, been controversial. Supporters of prison labor believe that it helps manage the inmate population and contributes to lower reoffending rates upon release. In contrast, its critics see inmates as coerced workers. Prison work, in this analysis, is the modern equivalent of slave labor. (On this issue, see, in particular, Burton-Rose, Pens, & Wright, 1998.)

History

In the 1820s, the Auburn Penitentiary in New York State established the model of prison labor that was later adopted by the rest of the states and the federal system. Prisoners in Auburn worked together in silence during the day at various tasks. Unlike the Pennsylvania system, where inmates were isolated to reflect upon their wrongdoings, Auburn enforced a factorylike routine of production.

Elsewhere in the United States, mainly in the southern states, a complex system of convict leasing was introduced following the end of slavery. In this arrangement, the state would loan prisoners to private individuals for a small fee. Such "entrepreneurs" were expected to house and feed the prisoners. In return, the businessmen were entitled to all profit from the prisoners' labor until their sentence was expired. Convict leasing was ultimately abolished, both because of its brutality and because free workers were concerned that prisoners were undercutting their pay and conditions. Nevertheless, as shall be discussed below, private individuals have remained involved in the provision of labor for prison inmates in various ways, usually as representatives of businesses and private industry (Lichtenstein, 2001; Myers, 1998; Oshinsky, 1997; Sellin, 1976).

Overall, 19th- and early 20th-century prison administrators and reformers believed that prison labor offered some chance of salvation for offenders. Convicts, it was hoped, would learn probity and discipline by being set to regular employment during their sentence. At the same time, they would contribute toward their upkeep and even, at the U.S. Penitentiary (USP) at Leavenworth, Kentucky, and at Sing Sing State Penitentiary in New York, help construct the very walls of their place of confinement.

During the 1930s, commitment to prison labor as a tool of reform fell precipitously because of the Depression. Most states and the federal government passed legislation that banned prisoner-made products from interstate commerce. Bucking the trend, however, Congress established the Federal Prison Industries (FPI) on June 23, 1934, "to provide job skills training and employment for inmates serving sentences in the Federal Bureau of Prisons." FPI, given the trade name Unicor in 1978, has remained a key part of prison labor organization in the federal system ever since. Many states, including New York and California, also have prison industries. In addition, unlike the federal system, states may run joint ventures with private corporations. How these issues work in practice shall be discussed below.

Current Practice in the Federal Prison System

These days, the Bureau believes that "meaningful work programs are the most powerful tool prison administrators have in managing the inmate population" (Federal Bureau of Prisons, 2000, p. 27). In federal prisons, unlike other correctional systems, work of some sort is mandatory. The vast majority of federal inmates either take education classes or work to upkeep the prison. Upon arrival, prisoners are offered jobs in various aspects of site maintenance, usually in food services. Following a certain amount of time (usually 90 days), they may shift to another area of prison labor, including grounds maintenance, the prison farm, or work as an orderly. They may also apply for employment in FPI. If they do not wish to work, they must enroll in an education or vocational training program.

Certain prisoners with the lowest security status of "community" may engage in employment that brings them into the public arena, including truck driving and delivery services. Others, as they near the end of their sentence, may be eligible to enroll in a work-release program outside the prison. They will also be encouraged to participate in a work-skills training course and, perhaps, in a mock job fair.

Around 26% of the prison population work for FPI, otherwise known by its trade name Unicor. According to its mission statement, Unicor is meant "to employ and provide skills training to the greatest practicable number of inmates confined within the Federal Bureau of Prisons." Prisoners are employed in one of five areas, roughly defined as "metals," "textiles," "furniture," "electronics," and "graphics/services." Just some of what these areas produce includes "missile cable assemblies, kevlar military helmets, executive office furniture, prescription eye wear, metal prison security doors, military uniforms and data entry of patent and trademark documents" (Unicor, 2000, p. 14).

To avoid competition with private industry, all goods produced in FPI factories are sold to the federal government rather than on the open market. Recently, however, Unicor has been experimenting with subcontractual agreements with other companies who sell to the federal government. In 1999, Unicor also began what the annual report describes as "a pilot program to provide, in the commercial market, services that would otherwise be performed by foreign labor outside the country" (Unicor, 2000, p. 16).

There are other ways in which FPI is tied to the private sector. Unicor points out, for example, that it supports several thousand jobs in the community because of its trade with private companies. According to the most recent annual report in 1999, "FPI spent $423 million purchasing raw materials, supplies, services and equipment from private sector vendors." Similarly, the company argues, "Half of FPI's purchases from private sector vendors were from small, women and minority-owned and disadvantaged businesses" (Unicor, 2000, p. 16).

Unicor is usually represented as the jewel in the crown of the Bureau of Prisons. In official literature, prisoners employed through FPI have lower reconviction rates after release and are more law abiding while incarcerated. They also contribute financially to court-ordered restitution programs through a mandatory arrangement whereby 50% of FPI wages are set aside for this purpose.

Despite this praise, however, Unicor has its critics. First, there are many more applicants to FPI factories than there are jobs because they pay the best wages. This means that there is generally a long waiting list for employment in any facility. Another issue that raises concern for some is the close relationship between Unicor and the Department of Defense. Over 60% of sales from Unicor are made to the U.S. military for a range of services, from uniforms to helicopter cables and wiring used in weaponry. As Jean Luc Levasseur (1998) pointed out, some of the items produced by prisoners inevitably fall into the hands of violent regimes:

> Being contracted to the War Department [sic] means supplying more than just U.S. forces. It means that this military equipment is rerouted by the U.S. to its client states—from Israel to Indonesia—and into the hands of the world's most degenerate and bloodthirsty regimes. In cases like El Salvador, the recipients of the U.S. war supplies used them to kill their own people. (p. 123)

Although prisoners may choose whether they are prepared to support to the military in this way, the financial attraction of Unicor must surely influence their decision. In addition, Levasseur pointed out that prisoners at the second most secure facility in the federal system, USP Marion, have no ability to make up their own minds. Some time in the FPI factory is a prerequisite for transfer out of Marion, and the factory manufactures cables for the Department of Defense.

Finally, ever since it was established, Unicor has been criticized for its inefficiency. In a lengthy and biting assessment of FPI, sociologist Christian Parenti (1999) pointed out that even though Unicor "is guaranteed a labor supply at absurdly low wages, is given direct subsidies, and has a guaranteed market," it is "an economic basket case" (p. 232). Moreover, it is far less cost-efficient than other providers: "Unicor products provided to the Department of Defense, on average, cost 13 percent more than the same goods supplied by private firms. . . . The federal prison monopoly delivers 42 percent of its orders late, compared to an industry-wide average delinquency rate of only 6 percent." Those orders that were filled often delivered poor-quality products. Thus, "A 1993 report found that Unicor wire sold to the military failed at nearly *twice* the rate of the military's next worst supplier" (p. 232).

Current Practice in California and New York

With approximately 160,000 people behind bars, California is equivalent in size to the U.S.

federal prison system. It is also the largest state prison system in the country. New York State holds less than half this number at 70,000. Inmates of the California and New York correctional systems, like those in federal facilities, may be employed in tasks related to the maintenance of the institution, or they may work for the state's industrial program. In California, this program is known as the Prison Industry Authority (PIA). In New York State, it is called Corcraft. In California, some inmates are also employed by private companies in the state's Joint Venture Program.

The California PIA currently employs approximately 7,000 inmates in 23 different facilities. In New York State, Corcraft provides jobs for about 2,200 men and women. As in the federal system, inmates working for PIA and Corcraft produce goods for the government. In fact, all state-run organizations, from the University of California and State University of New York systems to both state capitols, must procure a range of items manufactured by prison industry. From office furniture and uniforms to stationery, many of the items used by government departments are the products of inmate labor. In addition to such physical items, government agencies like the Department of Motor Vehicles (DMV) in New York State employ male and female inmates to answer telephones, providing customer service to the general public (New York State Department of Correctional Services [NYS DCS], 2001a, p. 25).

In contrast to PIA and Corcraft, which may only sell to the government or to individuals who are going to export the goods overseas, California's Joint Venture Programs are designed to produce items for sale to the public. Set up by Proposition 197, which was passed in 1990, the Joint Venture Program aims to recruit businesses to set up operations in state prisons. Companies that participate in this and other similar programs around the country employ prisoners in a variety of tasks, from packaging computer disks to sewing designer T-shirts. According to Christian Parenti (1999), for example, "In San Diego prisoners working for CMT Blues were employed tearing 'made in Honduras' labels off T-shirts and replacing them with labels reading 'made in USA.' Other convicts take reservations for TWA, work at telemarketing and data entry, and slaughter ostriches for export to Europe" (p. 230).

Companies interested in utilizing prison labor receive many tax breaks and other incentives. Often the institution will construct workrooms to their specifications. Most important, there is always a guaranteed labor pool. Companies must, however, pay prisoners minimum wage. And this, as Christian Parenti (1999) observed once more, is often a sticking point: "With wages as low as 40 cents an hour in Honduras, and generous tax breaks to boot, why open a sweatshop inside some bureaucratic hellhole where you have to pay minimum wage?" (p. 235).

Indeed, despite a certain amount of publicity about joint venture programs and prison labor in the 1990s, Parenti (1999) believes that "capital avoids the penitentiary" (p. 233). There is often a lack of space for work. Because prison-made products are popularly regarded as "morally tainted," companies that do utilize prisoner services usually hesitate to publicize the source of their labor. Companies wishing to operate in penal facilities also face numerous problems of restricted flexibility, searches, security, guards, delays for counts, problems of location and delivery, and prisoner resistance (Parenti, 1999, pp. 234–235).

EDUCATION

Like work, education has played an important role in daily prison regimes since the establishment of the first penitentiaries in the 18th century (Roberts, 1971). It was also, at first, tied to ideas of rehabilitation. These days, however, prison classes are justified by a range of alternative viewpoints. According to some authors, for example, learning in prison should be encouraged merely to create more efficient and reliable workers. As Silvia McCollum (1994) pointed out, "Illiterate workers who cannot read instructions, fill in job-related forms, prepare brief reports, or perform work-related math are unnecessary strains on correctional systems that are already carrying heavy resource burdens" (p. 35). For others, education is a way of promoting responsibility and independence and lowering recidivism rates. For many inmates, the benefits of prison classes are more practical; education is a relief from the boredom and frustration of everyday life. It also opens new horizons for those who previously may have had little schooling (Davidson, 1995; Tannenbaum, 2000).

Whatever the justification for education, as shall be made clear below, most prison programs aim to impart only basic literacy and numerical skills. Despite some programs that have much broader goals, prison classes these days are designed predominantly to compensate for inadequate elementary and secondary schooling. Although reading, writing, and arithmetic are undoubtedly crucial skills, the emphasis placed on them appears sometimes to override other, more innovative, programs.

History

The earliest U.S. prisons generally sought to educate prisoners through religious instruction. Pennsylvania's Walnut Street Jail tried to reform and teach inmates by encouraging hard work and religious contemplation. Both activities were conducted in solitude. Over time, however, education outgrew solitary Bible reading, culminating in the introduction of a school in 1798, together with a library of 110 books.

The competing Auburn system that provided the model for most penitentiaries in the federal and state systems was far less supportive of prison classes because of a concern that they might distract inmates from the more important tasks of prison labor (Silva, 1994, pp. 19–21). Nonetheless, by 1870 the National Prison Association, which was the forerunner of the influential American Correctional Association, set out in its Declaration of Principles in 1870 that "education is a vital force in the reformation of fallen men and women" (Conrad, 1981, p. 1). Since then, prison classes have been an entrenched part of the prison experience.

Inmates in the first prisons were taught the basics of reading and writing by prison staff and, if possible, an employable skill that might keep them away from criminal activity upon release. Later, education departments began to have responsibility for law and leisure libraries as well as vocational training and postsecondary schooling options. In most prisons, they also organize sports and recreational activities.

Since the mid-1990s, the legitimacy of prison classes other than basic literacy has been under attack both from within and outside the prison (Silva, 1994, pp. 18–19). Tertiary education has been particularly vulnerable. Although college classes in prison date to the 1920s, academic or postsecondary courses were rarely offered until 1965, when Congress passed Title IV of the Higher Education Act, a major part of which was the Basic Education Opportunity Grants, later renamed Pell grants in honor of Senator Claiborne Pell, the bill's sponsor. This act enabled prisoners, and other low-income people, to afford college education for the first time. This law had a phenomenal impact on prison culture, directly contributing to the growth of prison activism and writing in the 1970s and 1980s.

These days, prisoners are no longer guaranteed the right to a college degree. Pell grants for inmates were abolished in 1994 (Matlin, 1997). Although some college courses in prisons struggle on, most of what is left of a once successful system of prison higher education is college by correspondence for those who can afford to pay for it themselves. Prisoners are limited even here by restrictions on audio and videocassettes and on the number of books they are allowed to have in their cells at any one time. They are also not given access to the Internet. Compounding their difficulties, many universities have established residency requirements for course completion. In contrast to the optimism and passion expressed by many prison educators (Davidson, 1995; Trounstine, 2001), universities seem loath to commit themselves to serving the prison population. Fears of alumni disapproval, faculty absences, and an association of the university degree with offenders are all used as excuses by many institutions from stepping into the breach created by the repeal of the Pell grants.

Vocational Courses, Career Counseling, and Job Fairs

In addition to creating an employable workforce in prison, reduced reoffending rates have always been an important justification for prison education classes. For that reason, vocational courses and apprenticeships are two of the main strategies that education departments pursue to help prisoners prepare for successful release. Certificates or diplomas do not specify that they were earned in a correctional facility.

Like most aspects of imprisonment, the quality and availability of these courses vary enormously. Some facilities offer a variety of choices, from carpentry to cooking. Others, particularly high-security institutions, are much more restricted. Women's prisons also tend to offer fewer vocational options that usually remain committed to traditional feminine stereotypes. Although the situation has improved since legislation governing equal opportunities was applied to prisons, most women's facilities will be more likely to offer classes in hairdressing, cooking, and cleaning than in carpentry, welding, or mechanics (Schram, 1998; Winifred, 1996).

Complementing the vocational training and apprenticeships classes offered, prisons usually run career counseling classes and a variety of self-help programs designed to assist prisoners with career planning and development. As part of this policy, the Federal Bureau of Prisons and some state systems hold mock job fairs to which they invite local employers. However, once again, it seems that this strategy is of limited success, for the fairs are not offered very often. In the federal system, for example, only 14 were held in 1998, in which 550 inmates and 213 volunteers participated (Federal Bureau of Prisons, 1999a).

Prison Libraries

Prison libraries provide an avenue for education in prison outside the classroom. Criminologist Albert Roberts (1971) reports that despite the establishment of a library in the Walnut Street Jail in 1798, most prisons "made no effort to establish libraries until about 1840" (p. 162). Instead, they relied on Bibles and other religious material provided by prison chaplains.

These days, although they are vulnerable to cuts in funding like other aspects of education, libraries are enshrined in most prison systems' rules and regulations. Inmates in special housing units or solitary confinement should also be entitled to library services. To ensure that libraries have an adequate collection, wardens in most systems must provide a variety of reading materials, including but not limited to periodicals, newspapers, fiction, nonfiction, and reference books. In recognition of the cultural diversity of many prison populations, some systems, like the Federal Bureau of Prisons, direct the staff to provide publications in the inmates' native languages. In particular, Spanish-language texts are often available because Latinos and Latinas are the fastest growing inmate population around the country.

According to Virgil Gulker (1973), who established one of the first federal prison libraries in Federal Correctional Institution (FCI) Milan in the early 1970s, the selection of books in any penal facility should reflect the interest and needs of the prisoners. Specifically, he suggested a diverse mix of fiction and nonfiction, with particular emphasis upon political, sociological, and psychological literature. Roberts (1971) affirmed Gulker's vision by suggesting that an institution's book collection should "be geared to all inmates, enabling them to improve their ability to live successfully in these rapidly changing, complex times" (p. 164).

Law Libraries

As part of this task of helping prisoners adjust to the changing world, each prison has a law library in addition to the regular lending library. Prison law libraries are meant to provide inmates with reference materials so that they may prepare their own legal defense. They should also offer some means of copying documents and a typewriter or a word processor for prisoners to write letters and other items.

Prisoners may usually borrow materials from the legal collections to consult in their leisure time. Inmates held in secure custody, in segregation, or in a special housing unit are still entitled access to a law collection. For these inmates, the prison must, as with leisure libraries, either operate satellite law collections in restricted housing units or make copies of books available to inmates confined there.

Inmates generally staff prison law libraries. Over time, these women and men become familiar with prisoner case law. Some, by virtue of this experience, may become known as "jailhouse lawyers," offering legal advice to other prisoners. Criminologists have described at some length the importance and prestige of such individuals in the prison community (Milovanovic, 1988). As prisoners' access to the courts has been increasingly restricted since the 1995 Prison Litigation Reform Act (PLRA), and because few can afford their own attorneys, prison law libraries and well-informed inmates are more urgently needed than ever before.

Recreational Courses

Finally, prison education departments are also responsible for recreation. They organize activities ranging from arts and crafts to intramural sports. Some institutions have a small collection of musical instruments or tools necessary for crafts like needlework, leatherwork, or ceramics. Prisoners are usually allowed to mail home what they make.

Recreational events are also occasionally scheduled, particularly in low-security establishments like prison camps. Events like intramural sporting events or rock concerts break the routine of everyday life and usually recast relations between all concerned. Even so, recreational activities have recently been under attack throughout the U.S. prison system (Wright, 1998, pp. 45–47). Starting with Washington State, for example, many departments of corrections, including the Federal Bureau of Prisons, have been affected by recent legislation banning weight lifting in prison. Fit and muscle-bound inmates, it is suggested, pose a threat to staff. Similarly, in many places electrical musical instruments are now banned. Electric guitars are viewed as potential weapons. They are also described as instruments of aggravation in an enclosed environment that is already noisy. As a result, in most states, worn-out equipment will not be replaced when it breaks.

Current Practice in the Federal Prison System

Currently, little more than one third (35%) of the federal prison population engages in prison education (Federal Bureau of Prisons, 1999a, p. 13). Most of these women and men are enrolled in one of the compulsory classes such as basic literacy, a high school equivalency diploma (GED), or English as a second language (ESL).

The first mandatory literacy program in the Bureau of Prisons was established in 1982. All inmates were required to enroll unless they could demonstrate a sixth-grade level of reading and writing. In 1986 the standard was increased to an eighth-grade level, and in 1991 the current requirement of a high school equivalency (GED) was set.

These days, all promotions in institution jobs above entry level require a GED. Although seemingly a commendable idea, tying education to prison labor so closely places those with little educational experience or those from a foreign or non-English-speaking background in a vulnerable position, rendering them ineligible for many prison jobs.

In addition to basic literacy classes, some federal facilities still offer on-site college classes. Thus, for example, at USP Lewisburg in Pennsylvania, prisoners may study for an associate degree in business from Newport College. Similarly, at FCI Safford in Arizona, professors from Eastern Arizona College run university classes. The problem, however, lies in the limited number of these options. Because inmates may not control their length of stay in any one place, they run the risk of commencing a tertiary studies program in one place, only to find that they cannot continue with it at their next destination.

Current Practice in California and New York

Like the U.S. federal system, both California and New York state offer a range of prison education programs. As in the federal system, however, primary emphasis is placed on general literacy skills. Most prisoners arrive with little experience of education. Consequently, prison classes everywhere tend to focus on Adult Basic Education (ABE), ESL, and training for the GED. Promoting basic reading and writing skills through such courses is thought to reduce reoffending after release (Stayely, 2001).

In New York State, the average reading level of men and women in prison is seventh grade, and inmates' average math skill level is sixth grade. Such low levels may be only partly addressed during incarceration because they reflect the poor quality of education that is available in many communities. Since 2001, prisoners in the New York system have been required to demonstrate proficiency in reading and math at a ninth-grade level because that is the bare minimum thought necessary for passing the GED. As of 2001, approximately 60% of prisoners who enrolled in the GED exam were awarded their certificates, a rate that was slightly higher than the state average (NYS DCS, 2001b, p. 19).

Though all offenders are, in principle, encouraged to earn their GED while incarcerated in New York, those aged below 21 are given the most incentives. Younger offenders who have difficulty with learning, for example, may be eligible for a special education program for those who "have been diagnosed with mental retardation, a learning disability and/or are emotionally disturbed." This program, which is currently available at 15 of the state prisons, is designed to help inmates to participate in mainstream classes (NYS DCS, 2002a, p. 10).

Similarly, offenders aged between 16 and 24 may take advantage of another innovative inmate-centered program based in four of the New York State prisons: Albion, Bedford Hills, Washington, and Greene. At each of these facilities, the Department of Corrections provides "ongoing post-high school services . . . in life, academic and employment skills" (NYS DCS, 2002a, p. 10). Depending on the needs and interests of the prisoner, the program offers training in computer skills and making an employment portfolio and offers additional help in reading or math skills. So far, 1,004 men and women have participated in the training. By concentrating on their needs, prisoners have been able to increase academic test scores and obtain apprenticeships and vocational training and other certificates. Moreover, the Department of Correctional Services (NYS DCS, 2002a) reported that "of the 117 inmates who took the program and have since been released from prison, 94 have been placed in higher education, therapeutic programs and/or employment" (p. 10).

In addition to the regular programming in basic literacy, vocational training, and so on, California offers a unique educational experience for prisoners: Arts-in-Corrections. Under this program, which has existed since 1980, most penal facilities have a state-funded "artistic facilitator" who works with artists in the community to arrange periods of residence.

Judith Tannenbaum (2000), in her book *Disguised as a Poem,* evocatively describes her experience teaching poetry under this scheme at San Quentin. Working with long-term mainline prisoners as well as those on Death Row, Tannenbaum taught poetry for 3 years. To inspire the men to express themselves, she brought in poems and poets from the community. She also published a series of inmate poetry collections.

A primary justification for the Arts-in-Corrections program is that it reduces tension and violence in the inmate community (Cleveland, 1992, p. 80). Programs currently run the gamut of visual, performance, and literary arts. Classes include music, drawing, poetry, drama, creative writing, and so on. Arts-in-Corrections, unlike other education programs, helps break down the barriers between prisons and society during a person's period of confinement. In some prisons, such as Soledad, prisoners in 1983 created murals for the institution and the surrounding community (Cleveland, 1992, p. 83). More typically, as in San Quentin, outside organizations, like theater groups, visit the prison to put on performances inside the facility (Tannebaum, 2000). Finally, some programs produce publications, like a book of poetry, or collected short stories, which are then distributed to the free community.

Though California is one of the few places to run a comprehensive statewide set of programs in arts in its penal facilities, other states are catching on. New York, Massachusetts, and even Texas have some courses of this nature. Rather like the tertiary education in the federal system, however, most of these creative classes remain fragmentary and vulnerable to cuts in funding and punitive public sentiment.

DRUGS AND SUBSTANCE ABUSE TREATMENT IN PRISON

The final means by which prisons try to prevent inmates from reoffending is through the provision of drug treatment programs. Since the 1980s, the United States has aggressively waged a so-called "war on drugs." It is estimated that 80% of state and federal inmates have committed drug offenses, were under the influence of drugs or alcohol at the time of their crime, committed their crime to support their drug use, or have histories of substance use (National Center on Addiction and Substance Abuse, 1998). Under the new sentencing laws, many of these individuals are serving long terms of imprisonment, often for their first offence. This effort to reduce the availability and use of narcotics has had mixed results at best. Though prisons everywhere are bursting at the seams with people sentenced for drug offences, use of illegal substances in the community has diminished little.

Great social and financial burdens are incurred by incarcerating such offenders. Though some of these issues are borne by the community, others must be dealt with by the prison system. As well as coping with overcrowding brought about by the upsurge in the penal population since the declaration of the war on drugs, prisons must also treat those who are imprisoned with addictions. Although most systems offer numerous drug treatment programs and counseling, the evidence of their effectiveness in reducing substance abuse or postrelease offending rates is mixed (Farabee et al., 1999, p. 151).

History

The Federal Bureau of Prisons has led state departments in drug treatment. Specifically, it has offered individualized substance abuse treatment programs since the mid-1960s, following the passing of the Narcotic Rehabilitation Act (NARA). Before NARA, the only drug treatment available was in "narcotics hospitals," where drug-addicted prisoners were forced to undergo rapid drug withdrawal with little medical or psychological support.

NARA programs were unit based and driven by a therapeutic ideal with an emphasis on group therapy. The first NARA unit opened in FCI Danbury, Connecticut, in March 1968 and was quickly followed in 1969 and 1970 by units in FCI Terminal Island, California; Federal Prison Camp (FPC) Alderson, West Virginia; FCI Milan, Michigan; and FCI La Tuna, Texas.

All NARA participants were required to participate in postrelease aftercare, including drug urine tests and community-based counseling programs. NARA graduates generally had lower recidivism rates and less frequent drug use or involvement in drug sales after their release than did nongraduates. Women were more successful in adjusting to life after prison than were men (Murray, 1992, p. 67).

The popularity of unit-based drug treatment was such that, by 1978, there were 33 Drug Abuse Programs (DAPs) in 24 of the federal penal institutions. Residential programs were also opened in many state systems, including New York and California. According to Murray (1992),

> A typical drug treatment unit at that time housed 100 to 125 participants and was staffed by one unit manager, one psychologist, one or two caseworkers, and one or two correctional counselors. Outside consultants (sometimes ex-addicts) and education staff also provided services to the participants. (p. 68)

Like earlier NARA programs, this method of drug treatment was unit based and emphasized group and individual counseling that often continued after release. Inmates involved in the programs also participated in educational, vocational, and recreational classes.

Since the early 1980s, the numbers of prisoners with drug-related needs has made it unfeasible to offer universal unit-based treatment. As a result, drug treatment programs have become more diversified. Although there are specific residential and nonresidential programs in a number of state and federal prisons, drug treatment for most inmates takes the form of education-based courses, individual counseling, or self-help groups run by such organizations as Alcoholics Anonymous and Narcotics Anonymous.

Current Practice in the Federal System

Harsher sentences for drug offenders have had a particularly acute effect on the federal prison system. Currently, 58.3% of the total population is doing time for drug offences, and others are there for drug-related crimes.

According to a 1999 report to Congress, the current Federal Bureau of Prisons strategy of drug treatment "addresses inmate drug abuse by attempting to identify, confront, and alter the attitudes, values, and thinking patterns that lead to criminal and drug-using behavior" (Federal Bureau of Prisons, 1999b, p. 1). To help prisoners most effectively, the Bureau of Prisons differentiates between those who are incarcerated for manufacturing or selling drugs and those who are incarcerated for crimes that were a direct result of their drug use. It is thought that each group requires different counseling and treatment (Federal Bureau of Prisons, 1999b, p. 2). As a result, the Bureau offers three different forms of drug programs through Psychology Services, each of which attempts "to identify, confront and alter the attitudes, values and thinking patterns that led to criminal behavior and drug or alcohol abuse" (Pelissier et al., 2000, p. 5). Currently, drug treatment options in the federal system include a 500-hour residential drug treatment program, a 40-hour drug education program, and a more loosely organized set of counseling and self-help classes known collectively under the title of "nonresidential" drug treatment.

The Residential Drug Abuse Program (RDAP) is the most intensive of the Bureau's drug treatment options. First, the inmate participates in a unit-based program that generally has a capacity for around 100 people. During this time, he or she spends half of each day learning about drug use and the other half of it in ordinary activities like work and education with the general population. Prisoners are screened and assessed at the beginning of the RDAP to work out their treatment orientation. To complete the program, they take a variety of classes, including "Criminal Lifestyle Confrontation," "Cognitive Skills Building," "Relapse Prevention," "Interpersonal Skill Building," and "Wellness," before being returned to the general prison population. Afterwards, they must also participate in 12 months of treatment, meeting with "drug abuse program staff at least once a month for a group activity consisting of relapse prevention planning and a review of treatment techniques learned during the intensive phase of the residential drug abuse program" (Pelissier et al., 2000, p. 4). The residential program even reaches beyond prison. Once an inmate is transferred to a Community Correction Center (CCC), he or she will meet with privately contracted counselors to reaffirm the lessons of the drug treatment program. These sessions may also include other family members.

Residential drug programs are the most celebrated and, apparently, successful, part of the Bureau's current drug policy. According to a recent evaluation, these programs, which last from 9 to 12 months, reduce men's reoffending after 3 years in the community by 16% and women's by 18%. Thirty-six months after their release, men who have successfully completed an RDAP course also are 15% less likely to use drugs on release, and women are 18% less likely to do so.

Because of these findings, the Bureau of Prisons has introduced a series of incentives to encourage prisoners to participate in RDAPs. Some examples of the opportunities available include a small monetary award for successful completion of the program; consideration for placement in a 6-month halfway house; and what are referred to as "tangible benefits," such as shirts, caps, and pens with program logos. The most influential incentive, however, was brought in by the Violent Crime and Law Enforcement Act of 1994, which allows up to a 1-year reduction in sentence from statutory release date. This incentive has obvious attractions, and many prisoners are in favor of it. Others, however, are more critical of this reward. They point out that sentence reductions lead to inconsistent sentencing, in which participants do less time for the same crimes. Critics suggest that this policy unintentionally rewards inmates with drug problems (Pelissier et al., 2000, p. 6).

The 40-hour drug education program is somewhat less intensive than the residential program. It incorporates lectures, movies, written assignments, and group discussion. Participants usually meet twice a week for approximately 10 weeks, covering the reasons for their drug use and abuse, theories of addiction, physical and psychological addiction, defenses, effects of drug abuse on the family, and different types of drugs and their affect on an individual.

Nonresidential drug treatment can include meetings with Alcoholics Anonymous and

Narcotics Anonymous as well as individual and group counseling offered by Psychology Services. Finally, as part of their more general approach to curbing substance abuse, all federal prisons conduct regular random drug tests of all prisoners. Those with outside assignments are tested most frequently. The Bureau's policy appears to have worked. The most recent *Judicial Resource Guide* (Federal Bureau of Prisons, 2000) stated that "the number of positive test results for the random tests continues to be very low for the last few years—1.3% FY95, 0.9% FY96; 1.0% FY97; and 0.9% FY98" (p. 31).

It may be hard to be accepted into either the drug education program or the residential drug treatment program. Unless individuals can prove, through their Pre-Sentence Investigation Report (PSI), that drugs or alcohol played a role in their crime, they may not be eligible for drug treatment. It may still be possible for them to participate in a drug program, but it will be more difficult, requiring documentation from outside and often extensive negotiation with the Psychology Department in the prison.

Current Practice in California and New York

New York State has some of the harshest drug policies in the country. The so-called "Rockefeller Drug Laws," named after the then-state governor, were adopted in 1973 and have caused an exponential growth of the state's prison population. Though there is a current move to rescind some of their harsher components, at the time of this writing, these laws still call for a 15-year prison term for anyone convicted of selling 2 ounces or possessing 4 ounces of narcotics, regardless of the offender's criminal history.

Like the federal system, New York State offers a range of drug treatment programs in its prisons, including therapeutic communities, residential programs, and individual counseling. The oldest of these courses is run at the Mt. McGregor facility, which is housed in a former sanitarium. Established in the 1980s by the prison's Catholic priest, the Alcohol and Substance Abuse Treatment (ASAT) residential program provides a model for the state's other programs. As in the RDAPs of the federal system, inmates in the ASAT program are housed separately from the rest of the population, with each housing unit functioning as a separate therapeutic community. Counseling comes from former drug addicts employed by the prison, as well as outside volunteers (NYS DCS, 2002b, p. 15).

In addition to the ASAT program of Mt. McGregor, the NYS DCS operates a series of Comprehensive Alcohol and Substance Abuse Treatment (CASAT) programs in seven facilities around the state. Established following the 1989 Prison Omnibus Legislation, which called for an expansion of drug treatment services, CASAT programs offer a highly structured course of treatment for inmates inside and outside prison walls. To participate in this program, prisoners must be selected from general population. They are then sent to one of seven annexes to complete Phase I of the treatment, involvement in a therapeutic community with individual and group counseling.

Following successful completion of Phase I, prisoners begin community reintegration, or work release. This part of the program can involve either continuing residential treatment or simply day-reporting at one of 10 participating facilities. The final part of the treatment is release to parole supervision.

According to criminologists Wexler, Falkin, and Lipton (1990), therapeutic communities such as that in Mt. McGregor or those in Phase I of CASAT lower recidivism rates. Comparing another New York State program in the late 1980s, known as "Stay'n Out," that was based on ideas similar to those of other, less structured residential programs and individual counseling, the authors found significant differences. Though all counseling helped, intensive therapeutic communities such as those offered at Mt. McGregor worked best of all.

California also offers numerous forms of drug treatment in prison, but it is currently experimenting with an entirely new approach: diversion. In 2001, the state of California overwhelmingly approved Proposition 36 to divert first and second-time drug offenders from prison to treatment centers. Designed to reduce the state's ballooning penal population, this policy reflects a radical shift in penal policy.

However, its long-term effects are as yet unknown because it is only just now starting to be implemented.

California is often viewed as a bellwether for sentencing policy in the United States. It was here that "three strikes" legislation appeared first in 1994. It was here too that the growth of the penal population that has come to mark almost all states began. If Proposition 36 follows the same pattern, then the situation for drug offenders around the nation may be about to change.

WOMEN

Though it seems from the figures that drugs are an issue for all prisoners, like most things, drug use is, in fact, not spread equally among the penal population. Rather, it varies by age, race, and gender. In fact, unlike other crimes, proportionally more women than men enter prison with a drug problem (Wellisch, Anglin, & Prendergast, 1994). Because the number of women in prisons everywhere has increased, all prison systems have instituted drug treatment programs for female offenders.

According to the National Institute of Corrections (1994), women's programs differ from men's in a number of ways. They are generally less confrontational than men's programs because of a belief that "a central issue for substance abusing women is abusive backgrounds, continued in unhealthy relationships with men" (p. 3). As with most other women's programs, drug education touches on parenting, self-image, pregnancy, and other gender differences.

According to Auherhahn and Leonard (2000), one fact that women's drug programs seem to ignore is that drugs are not always illegally obtained. As these authors pointed out, women have always been overprescribed medication both in prison and in the community. Tranquilizers, like Valium and Xanax, both of which are highly addictive, are commonly prescribed. Women attempting to wean themselves from these legal drugs may encounter many withdrawal symptoms caused by illegal substances. Similarly, they may become addicted to them if they are prescribed to help deal with withdrawing from illegal substances.

CONCLUSION: PREPARING PRISONERS FOR RELEASE

As this chapter has shown, prison systems use a number of different strategies to prepare inmates for release. Since the establishment of the first penitentiaries, work and education have provided the primary means of transformation and reform. More recently, other strategies have been developed, including drug treatment programs.

At first glance, the success of any of these programs seems to be in question because of the high rates of reoffending. With current recidivism rates around 40% nationwide, it seems that few classes, treatments, or skills have much effect on whether people will return to prison. And indeed, people have always wondered about the success of enforced reform. Drug treatment programs in particular are often criticized in this way: How can prisoners realistically choose to change their life when they are mandated by the courts to participate in therapy, counseling, and drug education? Similarly, some authors question whether prison administrators are capable of determining which of the inmates who apply to participate in drug treatment courses will succeed in the program. They ask whether staff members are adequately trained in counseling skills and whether aftercare will be available in the community after inmates' release (Farrabee et al., 1999, p. 152).

In light of the federal employment restrictions on felons brought in by the 1994 Violent Crime Control and Law Enforcement Act, many others query more fundamentally the utility of work and education offerings. What is the point, after all, of helping someone obtain a college degree if he or she will later be blocked from pursuing his or her studies further due to the abolition of Pell and Tap grants? Why too, bother to train people to work if they cannot be gainfully employed in many areas associated with children, home care help, or finance?

Though all of these queries are reasonable and may need to be asked, they do not have to affect the legitimacy of prison programs per se. Indeed, it is unclear whether recidivism rates are an appropriate measure of the success of anything to do with imprisonment. Perhaps, as

Todd Clear (1994) and others demand, it is time that we aim for harm reduction rather than a seemingly unobtainable end to crime. Likewise, whether a person will be able to continue his or her education beyond prison is no reason to roll back opportunities further. The success of programs like Arts-in-Corrections, the satisfaction of those who finally obtain their GED, and the utility of teaching non-English speakers how to read and write in English are all worthy goals in themselves. Add to these other innovative programs around the country, like the AIDS Counseling and Education (ACE) program at New York State's only maximum secure facility for women at Bedford Hills, and it seems clear that we should be offering more, not less. ACE, for example, which began in 1988, trains prisoners to be peer educators so that they can work within the prison and outside once their sentences have ended in the area of HIV education. Providing information and counseling in English and Spanish, this program, which was initiated by prisoners, has achieved national and international renown. It is a model for future practice (for more information about it, see Boudin, Carrero, Clark, & Flournoy, 1999).

Above all, prison programs should draw our attention to three basic aspects of imprisonment. First, they remind us that prisons are still engaged in training prisoners for release. The men and women behind bars are more than just convicts. They will one day leave their place of confinement. Second, the women and men inside our nation's penal institutions are in great need of basic education, training, and assistance. They overwhelmingly come from disenfranchised communities. They have few employable skills. They are addicted to drugs. Third, and perhaps paradoxically, the success of particular programs and the passion expressed by those few prisoners who have been able to take advantage of education, drug treatment, or work options, as well as those practitioners who have worked with them, show that prisoners have much to contribute (Trounstine, 2001).

Reentering the community successfully requires more than just a GED or some math ability; it also requires acknowledgment and trust, from oneself and others, that there is a legitimate place for the former inmate in society. Such a possibility seems best achieved when courses are inmate centered and innovative. Arts-in-Corrections, the classes for New York State young offenders, the ACE program at Bedford Hills, college classes in federal penitentiaries, and therapeutic communities all share a commitment to tapping into prisoners' needs and resources. From the research available, these programs help people change their lives for the better.

REFERENCES

Auherhahn, K., & Dermody Leonard, E. (2000). Docile bodies? Chemical restraints and the female inmate. *Journal of Criminal Law and Criminology, 90,* 599–634.

Bosworth, M. (2002). *The U.S. federal prison system.* Thousand Oaks, CA: Sage.

Boudin, K., Carrero, I., Clark, J., & Flournoy, V. (1999). ACE: A peer education and counseling program meets the needs of incarcerated women with HIV/AIDS issues. *Journal of the Association of Nurses in AIDS Care, 10*(6), 90–98.

Burton-Rose, D., Pens, D., & Wright, P. (Eds.). (1998). *The celling of America: An inside look at the U.S. prison industry.* Monroe, ME: Common Courage.

Clear, T. (1994). *Harm in American penology: Offenders, victims, and their communities.* Albany: SUNY Press.

Cleveland, W. (1992). *Art in other places: Artists at work in America's communities and social institutions.* Westport, CT: Praeger.

Conrad, J. P. (1981). *Adult offender education programs.* Washington, DC: U.S. Department of Justice, National Institute of Justice.

Davidson, H. S. (Ed.). (1995). *Schooling in a "total institution": Critical perspectives on prison education.* Westport, CT: Bergin & Garvey.

Farabee, D., Prendergast, M., Cartier, J., Wexler, H., Knight, K., & Douglas, A. M. (1999). Barriers to implementing effective correctional drug treatment programs. *Prison Journal, 79,* 150–162.

Federal Bureau of Prisons. (1999a). *State of the Bureau 1998.* Washington, DC: Department of Justice.

Federal Bureau of Prisons. (1999b). *Substance abuse treatment programs in the Federal Bureau of Prisons: Report to Congress, as required by the Violent Crime Control and Law Enforcement Act of 1994.* Washington, DC: U.S. Department of Justice.

Federal Bureau of Prisons. (2000). *Judicial resource guide to the Federal Bureau of Prisons.* Washington, DC: U.S. Department of Justice.

Gulker, V. (1973). *Books behind bars.* Metuchen, NJ: Scarecrow.

Levasseur, J. L. (1998). Armed and dangerous. In D. Burton-Rose, D. Pens, & P. Wright (Eds.), *The celling of America: An inside look at the U.S. prison industry* (pp. 122–126). Monroe, ME: Common Courage.

Lichtenstein, A. (2001). The private and the public in penal history. *Punishment and Society, 3,* 189–196.

Martinson, R. (1974). What works? Questions and answers about prison reform. *Public Interest, 35,* 22–54.

Matlin, D. (1997). *Vernooykill Creek: The crisis of prisons in America.* San Diego, CA: San Diego State University Press.

McCollum, S. (1994). Mandatory literacy: Evaluating the Bureau of Prisons' long-standing commitment. *Federal Prisons Journal, 3*(2), 33–36.

Milovanovic, D. (1988). Jailhouse lawyers and jailhouse lawyering. *International Journal of the Sociology of Law, 16,* 455–475.

Murray, D. (1992). Drug abuse treatment programs in the Federal Bureau of Prisons: Initiatives for the 1990s. In C. Leukefeld & F. M. Tims (Eds.), *Drug abuse treatment in prisons and jails.* Rockville, MD: National Institute on Drug Abuse.

Myers, M. (1998). *Race, labor and punishment in the new South.* Columbus: Ohio State University Press.

Narcotic Rehabilitation Act of 1966, 42 U.S.C. § 3412(a) (1970).

National Center on Addiction and Substance Abuse. (1998). *Behind bars: Substance abuse and America's prison population.* New York: Columbia University Press.

National Institute of Corrections. (1994). *Profiles of correctional substance abuse treatment programs: Women and violent youthful offenders.* Longmont, CO: Author.

New York State Department of Correctional Services. (2001a). Bayview. *DOCS Today, 10*(11), 22–25.

New York State Department of Correctional Services. (2001b). Inmate academic bar rises to 9th grade level. *DOCS Today, 10*(8), 19–20.

New York State Department of Correctional Services. (2002a). Feds aid state in education of younger inmates. *DOCS Today, 11*(4), 10.

New York State Department of Correctional Services. (2002b). Mt. McGregor. *DOCS Today, 11*(3), 12–15.

Oshinsky, D. (1997). *"Worse than slavery": Parchmen Farm*

and the ordeal of Jim Crow justice. New York: Free Press.

Parenti, C. (1999). *Lockdown America: Police and prisons in the age of crisis.* New York: Verso.

Pelissier, B., Rhodes, W., Saylor, W., Gaes, G., Camp, S. D., Vanyur, S. D., & Wallace, S. (2000). *TRIAD drug treatment evaluation project: Final year report of three year outcome, executive summary.* Washington, DC: Federal Bureau of Prisons.

Prison Litigation Reform Act of 1995, Pub. L. No. 104-134, §§ 801-810, 110 Stat. 1321 (1996).

Rafter, N. (1990). *Partial justice: Women, prisons, and social control.* New Brunswick, NJ: Transaction.

Roberts, A. (1971). *Sourcebook on prison education: Past, present and future.* Springfield, IL: Charles C Thomas.

Roberts, J. W. (1994). Introduction. In J. W. Roberts (Ed.), *Escaping prison myths: Selected topics in the history of federal corrections.*

Washington, DC: American University Press.

Schram, P. (1998). Stereotypes about vocational programming for female inmates. *Prison Journal, 78,* 244–270.

Sellin, T. (1976). *Slavery and the penal system.* New York: Elsevier.

Silva, W. (1994). A brief history of prison higher education in the United States. In M. Willford. (Ed.), *Higher education in prison: A contradiction in terms?* Phoenix, AZ: Oryx.

Stayely, M. (2001). *Follow-up study of offenders who earned GEDs while incarcerated in DOCS.* Albany: State of New York Department of Correctional Services.

Tannenbaum, J. (2000). *Disguised as a poem: My years teaching poetry at San Quentin.* Boston: Northeastern University Press.

Trounstine, J. (2001). *Shakespeare behind bars: The power of drama in a women's prison.* New York: St Martin's.

Unicor. (2000). *1999 Annual report: Paying dividends to America.* Washington, DC: Federal Prisons Industries, Inc.

Violent Crime Control and Law Enforcement Act of 1994, Pub. L. No. 103-322, 108 Stat. 1796.

Wellisch, J., Anglin, M. D., & Prendergast, M. (1994). Treatment strategies for drug-abusing women offenders. In J. Inciardi (Ed.), *Drug treatment and the criminal justice system.* Thousand Oaks, CA: Sage.

Wexler, H., Falkin, G., & Lipton, D. (1990). Outcome evaluation of a prison therapeutic community for substance abuse treatment. *Criminal Justice and Behavior, 17*(1), 71–92.

Winifred. M. (1996). Vocational and training programs for women in prison. *Corrections Today, 58*(5), 168.

Wright, P. (1998). Prison weight-lifting is a nonsense issue. In D. Burton-Rose, D. Pens, & P. Wright (Eds.), *The celling of America: An inside look at the U.S. prison industry.* Monroe, ME: Common Courage.

26

RESTORATIVE JUSTICE PROGRAMS TO MEET FEMALE OFFENDER NEEDS

Innovation and Advocacy

KATHERINE VAN WORMER

LAURA J. PRAGLIN

Transformation begins when a vision that belongs to one person becomes one that belongs to many.

—Mahatma Gandhi

Current rallying cries for a "justice paradigm shift" from retributive to restorative processes of criminal justice (Zehr, 1995) have set the stage for our current discussion regarding restorative justice programs designed to meet the needs of women offenders. The result of this rallying cry has been dramatic for both the criminal justice system and more theoretical discussions regarding the intricate and murky interrelationships between religious and philosophical concepts such as justice, mercy, forgiveness, and reconciliation.

In this chapter, we will explore several critical issues related to restorative justice. First, we explore briefly its reliance upon native forms of settling disputes, as well as its roots in certain religious traditions that emphasize reparation and reconciliation over retribution. Second, we look at four of its incarnations—community conferencing, family group conferencing, reparative probation, and victim-offender dialogue. We then shift our focus to female offenders and consider ways in which restorative justice model programming can meet their special needs. After a consideration of research findings on program effectiveness, the chapter concludes with a list of guidelines for restorative justice advocacy.

With its roots in the rituals of indigenous populations and traditional religious practices, restorative justice is a three-pronged system of justice, a nonadversarial approach to settling disputes that seeks to offer justice for the individual offender, the victim, and the community. Restorative justice represents a growing international movement with a relatively clear set of values, principles, and guidelines for practice (Umbreit, 1998). Its purpose is to restore the torn fabric of community and wholeness to all those affected by crime, to repair the harm done to the victim and the community, and to make the offender accountable to both.

Humanistic in its treatment of offenders, restorative justice has as its focus the welfare of victims in the aftermath of crime. In bringing criminal and victim together to heal the wounds of violation, the campaign for restorative justice advocates alternative methods to incarceration, such as intensive community supervision. The most popular of the restorative strategies are victim-offender conferencing and community restitution. In many states, representatives of the victims' rights movement have been instrumental in setting up programs in which victims/survivors may confront their violators.

On the international stage, the thrust for a restorative vision has been embraced through the role of the United Nations. In consultation with nongovernmental organizations, the UN is in the process of drawing up formal standards or guidelines for countries to use in restorative justice programming (Van Ness, 2001). The United States, however, has not officially endorsed these procedures.

Worldwide, restorative justice has come a long way since two probation officers first pushed two tentative offenders toward their victims' homes in 1974 in Ontario (Zehr, 1997). The criminal justice literature is almost exclamatory regarding this new development. Restorative justice has variously been called "a new model for a new century" (van Wormer, 2001), "a paradigm shift" (Zehr, 1997), and "a revolution" (Barajas, 1995).

For the offender charged with a crime, the goal is to "beat the rap." The process of adjudication typically involves "copping a plea" to a lesser charge through a deal between prosecutor and defense attorney or participation in the adversarial arena, the modern equivalent of trial by combat. One party wins; the other party loses. The dominant form of justice in most of the Western world is retributive justice. But where is the reconciliation, the mercy, the healing?

We contend that such qualities of reconciliation, mercy, and healing may be preserved alongside accountability in nonadversarial criminal justice approaches found in the restorative justice movement. There are now more than 1,000 such programs operating throughout North America (315) and Europe (707), according to the international survey done by the Center for Restorative Justice and Peacemaking (Umbreit, 2000). Included in this number are many programs operating inside the U.S. prisons. There is also a significant restorative justice focus in the Australian prison system: A search of the keyword *restorative justice* in the Australian Institute of Criminology Database revealed over 200 references alone from January through March of 2002.

TWO RESTORATIVE JUSTICE ADVOCATES

Whether inspired by religion or human rights activism, an unusual coalition of idealistic lawyers, religious leaders, and even conservative victims' advocates has been at the helm of this restorative policy movement. Among these movers and shakers are two known to one of us personally, Linda Harvey and Mary Roche. Members of the helping professions, their work has special relevance for female offenders and victims, the topic to which we will return shortly.

Linda Harvey

"I have a passion for the restorative justice philosophy because it is about the healing of relationships and harm. I believe that I am doing the ministry of reconciliation to which God is calling all of us." So says Linda Harvey (2001), social worker, devoted Catholic, and founder of Transformation House in Lexington, Kentucky. Transformation House provides assistance to victim survivors who are struggling with the emotional devastation of homicide of a family member. Operating under independent auspices, this program is not derived from the criminal justice system. It builds on personal relationships and relies on the goodwill of public officials such as the prison warden. Harvey's vision for promoting reconciliation includes working closely with incarcerated women, especially those who have killed their abusers (L. Harvey, personal communication to K. van Wormer, November 27, 2001). For such women and others at the Pee Wee Valley Correctional Institute, Transformation House provides a series of educational classes and seminars. The dual focus of the classes and seminars is to help women take responsibility for what they have done and to experience catharsis and healing (Transformation House, 2001). During the seminar, a panel of victims and survivors share their stories of grief and loss with the inmates.

Transformation House, for Linda Harvey, was the culmination of a dream. The idea started with a vision to bring Death Row inmates together with their victims' families. In 1999, Harvey brought together Patsy Smith, whose mother had been murdered, with an inmate at the women's prison who had been convicted as an accomplice in the mother's death. The meeting turned out to be a powerful healing experience for both parties, according to the newspaper's account:

> For Smith, the meeting challenged her faith and gave her the chance to live it out. It was a time for facing her sorrow and anger. It was what her mother would have wanted her to do, she said. She was also finally able to hear that her mother had died a few minutes after midnight. (Kimbro, 1999, p. A1)

(For more information on Linda Harvey and other restorative justice pioneers, see the Web site of the International Centre for Justice and Reconciliation at *www.restorativejustice.org.*)

Mary Roche

Mary Roche's husband died in 1993. A young Iowan who left a 2-year-old son, he had been in recovery from alcoholism for 7 years and very involved in AA (Alcoholics Anonymous). So the fact that a drunk driver killed him was ironic. As Mary Roche tells it:

> I very much wanted to have a voice at the trial and express my feelings to him. People wouldn't feel that way, but to me I needed to see him, to talk to him. He needed to hear it from me.
>
> I called the Department of Corrections. They said, "You're the victim, you can't do that." A few years later in therapy, I realized I could run into him at Kwik Star after his release. So I became motivated and [again] called the Department of Corrections. This time I was referred to Betty Brown in Des Moines, who is now the administrator of restorative justice for the state. Betty met with Robert Clay (who had killed my husband) at the prison in Fort Madison. At first he said there was no way. We both went through months of preparation. This was a serious crime, so the meeting was about dialogue, not restitution.
>
> "What do you want?" This was the question I worked on. I wanted him to make amends, to make amends so that he would not drink because of what he had done. I wrote out questions I wanted the answers to so when we met we'd know each other's responses. Arrangements were made through the prison system. The warden was suspicious, however. "I have no idea why you want to do this," he said. The meeting lasted three and a half hours. I was seated first. The security guards looked on. When Robert walked in, I had a strong physical reaction. Betty was with me and Robert's counselor with him. Robert's counselor had not wanted him to participate, so Betty had to speak with her. After a period of time, we were dialoguing. I played a tape of my son, Sam, now 6, during this meeting. Was it cathartic? Yes. And did Robert open up? Prior to the meeting, Robert was very quiet. This is what his counselor said; he expressed no emotions. This meeting had an effect on him; now he's smiling for the first time. One day a woman called me to speak on a victim impact panel. I started with MADD (Mothers Against Drunk Drivers). I found this had an impact on offenders, especially juvenile offenders. Then, 2 years ago, I was hired because of my mental health training [Roche has an MA in mental health counseling] to become a victim liaison for the Department of Corrections in Waterloo, Iowa.

Today Mary Roche carries the message to students and church groups. She offers seminars on restorative justice for members of the community and conducts classes at the women's prison in Mitchellville.

PRINCIPLES OF RESTORATIVE JUSTICE

From the grassroots level to state and local headquarters, restorative justice is rapidly gaining momentum within the United States. In other countries, such as New Zealand, Australia, and South Africa, restorative justice practices have become institutionalized nationally. Victim-offender conferencing (sometimes called mediation) is the oldest and most widely used expression of restorative justice, with more than 1,300 programs in 18 countries (Umbreit, Coates, & Vos, 2001).

To gauge the level of academic interest in the restorative process, we conducted a computer search of *Criminal Justice Abstracts* from January 1, 1977, through April 10, 2002. The search yielded 185 abstracts of books and articles on the subject. This high number of listings effectively validates Turpin's (1999) claim that restorative justice has become one of the most discussed topics in criminal justice today. A comparable search of *Social Work Abstracts* from its online inception in January 1977 through April 2002 provided a total of 3 references. This numerical discrepancy is understandable to the extent that restorative justice is an outgrowth of or alternative to the criminal justice process. Philosophically, however, the results are surprising. Given its stress on empowerment, reconciliation, social justice, seeking the good in people, and advocacy for social change, restorative justice is highly compatible with the teachings and values of social work (for a summary of the values, see the National Association of Social Work [NASW] Code of Ethics, 1999).

THE FOUR MODELS

Four models of restorative conferencing find their way into our current discussion concerning effective programs for women offenders. These include community conferencing (and circle sentencing), family group conferences, community reparation, and victim-offender conferencing. Although derived from Bazemore and Umbreit's (2001) typology, which was developed within the juvenile justice context, the models may easily be adapted to the needs of adult offenders as well. All four models take place within a community-based context and seek, in a nonadversarial setting, to bring victims, offenders, family, and community members together so that they may come to terms with the dimensions of pain and violation caused by the offender's actions.

Restorative conferencing addresses offenses committed while stressing accountability. At the same time, it tries to repair injuries to victims and to communities in which these crimes take place. Whether these conferences occur before, during, or after adjudication, they promote education and transformation within a context of respect and healing. These models are neither mutually exclusive nor complete in and of themselves. They can be combined or adapted depending upon the specific situation at hand. It is, in fact, critical that the implementation of such models be context specific to promote compassionate healing and to do no harm. Following from this, one must also be aware of evaluative processes and outcomes (Umbreit & Coates, 1992), including the nature of the harm done to the victim, personality characteristics of the offender, and the existence of any relationship between victim and offender (Peachey, 1992).

Community Conferencing and Circle Sentencing

The first of the restorative processes is community conferencing. Community conferences make it possible for victims, offenders, and community members to meet one another to resolve issues raised by the offender's trespass (Swart, 2000). A particularly promising form of community conferencing is termed circle sentencing. Historically, sentencing circles may be found in U.S. and Canadian Native American cultures. These circles were adopted by the criminal justice system in the 1980s as First Nations peoples of the Yukon and local criminal justice officials endeavored to build more constructive ties between the criminal justice system and the grassroots community. In 1991, Judge Barry Stuart of the Yukon Territorial Court introduced circle sentencing to empower the community to participate in the justice process (Parker, 2001). One of the most promising developments in sentencing circles, as indicated by Parker, is the Hollow Water First Nations Community Holistic Healing Circle, which simultaneously addresses harm created by the offender, healing the victim, and restoring community goodwill. Circles have been developed most extensively in the western provinces of Canada. They have also experienced a resurgence in modern times among American Indian tribes: for example, in the Navaho courts. Today, circles increasingly may be found in more mainstream criminal justice settings. In Minnesota, for example, circles are used in a variety of ways for a variety of crimes and in varied settings (Bazemore & Umbreit, 2001).

How does a sentencing circle work? As Parker (2001) described the process, participants in healing or sentencing circles typically speak out while passing around a "talking piece." Separate healing circles are initially held for the victim and offender. After the healing circles meet, a sentencing circle (with feedback from family, community, and the justice system) determines a course of action. Other circles then follow up to monitor compliance, whether that involves, for example, restitution or community service (Bazemore & Umbreit, 2001; McCold, 2000). Though few studies have been done on the effectiveness of sentencing circles, those few studies do show generally positive results, despite some concerns over the duration of the process and the need to better prepare circle participants.

Family Group Conferences

The second model of restorative conferencing grows out of both aboriginal and feminist approaches and practice concerns stemming from the international women's and children's rights movements of the late 1980s and beyond. Evoking the family group decision-making model to try to stop family violence, the family group conference made its mainstream criminal justice debut in New Zealand in 1989. It also made a stage appearance about the same time in England and Oregon. This model is currently being tested in Newfoundland and Labrador, as well as in communities in New Zealand, Austria, England and Wales, Canada, and the United States (Hudson, Galaway, Morris, & Maxwell, 1996). Currently, family group conferences are used in many countries as a preferred sentencing and restorative justice forum for youthful offenders. Despite differences among jurisdictions, one common theme is overriding. This is the fact, as noted by Pennell and Burford (1994), that family group conferences are more likely than traditional forms of dispute resolution to give effective voice to those who are traditionally disadvantaged.

Community Reparation

Reparative Probation

Often referred to as the "Vermont model" of reparative probation, this form of restorative conferencing can be implemented more quickly within existing structures and processes of the criminal justice system (Dooley, 1995). Vermont's radical restructuring of its corrections philosophy and practices stems from influences of the communitarian movement and of personalist philosophy generally (Hudson & Galaway, 1990; Thorvaldson, 1990). This model involves a "reparative programs" track designed for offenders who commit nonviolent offenses and who are considered at low risk for reoffense. The "reparative programs" track mandates that the offender make reparations to both the victim(s) and to the community. A reparative probation program, moreover, directly engages the community in sentencing and monitoring offenders and depends heavily upon small-scale community-based committees to deal with minor crimes (Sinkinson & Broderick, 1998). Reparative agreements are made between perpetrators and these community representatives, while citizen volunteers furnish social support to facilitate victim and community reparation.

Restitution

Restitution, the most popular model of restorative justice, enables in-kind or actual return of what has been lost. Restitution is best viewed within the larger context of "making amends." This includes the provision that the offender offer a sincere apology and promise to change the behavior that caused the initial injury. Ideally, restitution also encourages a move beyond "balancing the scales" toward a true spirit of generosity and forgiveness on both sides (Van Ness & Strong, 2002). Victims identify and then seek redress for their losses, whether identified as material (damage to property), personal (bodily or psychic harm), or communal (injury to the quality of life of an entire community). In many cases, offenders can offer individuals, families, or communities compensation for such losses. Yet in cases of homicide, for example, what is lost can in no way be restored. Using a community justice focus, Ventura County, California, juvenile justice officials have implemented a successful program that emphasizes restitution as the key element of the reparative process. To compensate for material losses, for instance, youthful offenders in the South Oxnard Challenge Project offer various financial types of financial or in-kind restitution to victims. The latter have included needed projects for victims, whether individuals or neighborhoods, such as painting a victim's garage, cleaning up graffiti, or working in an office reception area (Karp, Lane, & Turner, 2002).

Victim-Offender Conferencing

The fourth and final model of restorative conferencing to which we shall refer is victim-offender conferencing. As suggested by its name, this model, introduced formally into the criminal justice arena in the 1980s, encourages one-on-one victim-offender reconciliation facilitated through a mediator. Key issues still to be resolved include the utilization of co-mediators; the nature and duration of follow-up victim-offender meetings; the use of victim-offender mediation for more serious and violent offenses; and the implementation of victim-offender mediation in multicultural and prison settings. Successful victim-offender programs appear to have been implemented in Ontario, Canada; Valparaiso, Indiana; Minneapolis, Minnesota; Quincy, Massachusetts; and Batavia, New York,

as well as in Germany (Bannenberg, 2000; Umbreit, 1985). The verdict is still out, but initial outcomes research does indicate considerable satisfaction among both victims and offenders, although the question of recidivism looms. How such programs can begin to be viewed with utmost seriousness as a viable alternative to traditional retributive justice procedures by attorneys and judges remains to be seen.

FEMALE OFFENDER SPECIAL NEEDS

First, let us look at a description of the needs of adult offenders—both male and female—from the point of view of the restorative justice school. Mackey and Shadle (1998) compared the victim's need for healing through the offender's admission of guilt with the offender's need to heal the brokenness in him- or herself: "To be transformed, the offender needs to hear experiences of pain; understand the impact of the crime; [and] develop compassion" (p. 31).

Like many discussions in the field of restorative and criminal justice, however, such thinking assumes a generic rather than gendered quality, and the special needs of girls and women are not taken into account. When we treat female offenders generically, we often confuse equality with sameness. Gender-blind treatment of girls and women in the criminal justice system subjects them to discipline designed for antisocial men, without allowances either for the role of motherhood or for a history of personal victimization. Addiction and dependency on drug-using and often violent men is another female-specific theme. To help meet the special needs of female offenders, van Wormer (2001) has introduced a strengths-restorative approach, which joins a strength-based feminist perspective with a restorative framework.

The finding of distinct gender differences in pathways to lawbreaking is a focus of contemporary studies on female offenders (Belknap & Holsinger, 1997; Chesney-Lind, 1997). Drawing on the empirical finding that there is a disproportionately high rate of multiple victimization for female compared to male offenders, Chesney-Lind describes the pathway that often leads a girl, desperate to escape sexual and physical abuse at home, to run away, seek solace in drugs and "bad company," and survive on the streets through prostitution. Some end up in prison, mostly as a result of incarceration secondary to drug involvement. In their path, they have victimized others as well—their children and family members, and sometimes strangers—through theft and robbery.

So are restorative justice strategies appropriate for such offenders, given their own histories of victimization and low self-esteem? We now turn to some examples.

INNOVATIVE PROGRAMMING

As Umbreit and Carey (1995) indicated, restorative justice does not involve a specific program but rather a way of thinking. It involves a reorientation of how we think about crime and justice (Zehr, 1997). It involves a call for a new paradigm to heed the cries of victims who feel abused or neglected within the present criminal justice process and to heed the cries of offenders and their parents and children as society shows its ugliest face—the face of vengeance and untempered power.

Within the diverse array of interventions that fall under the category of restorative justice, we do find principles in common: restorative justice condemns the criminal act, not the actor; holds offenders accountable to the victims and to themselves; involves all parties to the crime and their respective support systems; and offers a way for offenders to make amends and, in many cases, be reintegrated into the community.

In this section, we will explore strategies designed to meet the special needs of girls and women in conflict with the law. Central to these strategies is a focus on developing healthy relationships within the context of prior familial abuse and/or neglect. There are few reports of female-focused restorative initiatives for offenders, although there is some attention in the literature to the impact of restorative conferencing techniques on female victims, especially victims of violent crime, and of the need for sensitivity to power imbalances. From conferences and personal networking, however, we have discovered several gender-sensitive restorative justice programs.

Restorative justice concepts focus not only on criminal offenses but also on preventative measures such as education and community awareness. Restorative justice, in its focus on repairing the harm done to people and communities, above all emphasizes communication and reconciliation. The aforementioned healing circle, for instance, a form of interactive conferencing used in the Toronto and Waterloo, Canada, school systems, seems ideally suited to the needs of fighting and squabbling teenage girls. Facilitated by a social worker, the healing circle brings offenders and their victims face to face, forcing them to listen to each other and to grasp the impact of their behavior—such as threats, rumor mongering, teasing, and holding grudges. As indicated in the *Toronto Star,* the use of the circle has been highly effective in healing wounds and reducing animosities all the way around ("Healing Circle," 2001).

Looking within the juvenile justice system, Miriam DiBiase (2000) made the case for gender-specific treatment programs to meet girls' developmental needs. Programs such as the Alternative Rehabilitative Communities program at Harrisburg, Pennsylvania, address personal and romantic relationships, abuse issues, victimization, self-esteem, and sexual responsibility. The programs also emphasize victim empathy and community restitution. Because many of the girls have been victimized themselves, they can often come to appreciate the impact of their behavior on others and the need to move beyond excusing themselves from responsible behaviors due to their own histories of victimization. Minnesota in particular has infused restorative justice strategies within its Department of Corrections and has also developed gender-specific programming within its juvenile and adult institutions. The Minnesota Department of Corrections furthermore employs restorative justice planners to train people at the county level for diversionary conferencing (Pranis, 1999), emphasizing above all a spirit of dialogue and healing.

AMICUS in Minneapolis is an exemplary program for troubled girls, given its unique combination of gender-specific concepts with the principles of restorative justice (Goodenough, 2000). AMICUS has admirably faced the challenge of trying to counter what the girls, hardened by their experiences with the criminal justice system, have learned: Don't trust anyone; don't look your offenses straight in the eye; the victim is out to get you. Instead, they are now asked to sit in a circle with victims, family members (their own and the victims') and their supportive probation officer and to trust that healing and truth will emerge from the circle. Although the individuals level with the offender about how her behavior has caused them harm, a spirit of empathy, dialogue, and healing prevails. Meanwhile, the AMICUS boys' counterpart, JUMP! (Juvenile United with Mentors and Peers), maintains a similar spirit in its outreach to serious and chronic juvenile male offenders who link up with adult mentors, reconnect with family and community, and hopefully develop positive peer relationships.

ADULT CORRECTIONS

Within prison walls, some powerful encounters are taking place between offenders and their victims. Encounters occur, however, only after the facilitator has provided a great deal of preparation for both parties. Often, as at the Kentucky Correctional Institute for Women, the starting point for offenders is participation in the Impact of Crime on Victims seminar. Yet what is unique here is that the focus is upon inmates convicted of murder. Preliminary reports from some of the inmates who attended the seminar indicate a reduction in anger and more acceptance of responsibility (Tereshkova, 2000). If inmates later desire a meeting with the homicide victim's family, facilitators act through local victim advocates or through the church to gauge the survivors' desire for such an encounter. If family members so wish, volunteers prepare them for a face-to-face meeting.

Out of such sessions, there will be most likely an apology by the offender. In rare cases, forgiveness from the victim's family may result, although it can never be forced.

Whether involved or not in personal meetings with those they have harmed, female inmates are often especially plagued by feelings of grief, loss, and self-denigration. One of the particularly meaningful classes offered by Transformation House professionals and volunteers to inmates deals with shame and self-forgiveness. Among the questions pondered in small group discussion are these:

- How hard are you on yourself?
- What qualities or behaviors do you think are your "gifts to the world"?
- What are some ways you can nurture your growth and well-being? (interview with Linda Harvey, January 12, 2002)

In New Zealand and Canada, indigenously based programs are directed toward the needs of Native women in confinement. The Sycamore Tree Project in New Zealand follows the principle of restorative justice in helping female inmates move toward reintegration into the community (J. V. Wilmerding to K. van Wormer, personal communication, July 7, 1999). The Tree of Creation project introduced into the women's prison in northwestern Canada focuses on the spiritual process of healing (Lambert, 1998). The symbol used for this nature-based program is a large model of a tree, the branches of which take female inmates from the past to the present. The particular focus of that exchange is on healthful practices to prevent the spread of HIV. (Though a small percentage of the Canadian population, First Nations women constitute 20% of federally incarcerated women; a much larger proportion is found in the western provinces; Faith, 2000).

To provide women with the continuum of care they need as they leave prison, transitional programs are essential. Karlene Faith (2000) of the Centre for Restorative Justice of Canada's Simon Fraser University described the work of the support group Strength in Sisterhood. Members of this grassroots group of activists befriend women who are newly paroled to help them get reestablished in the community. Many mentors from this group are themselves former prisoners. The only real way to strengthen the women's bonds with the community, as Faith indicates, is to invest in these types of less expensive and nonpenal community resources.

A LOOK AT PROGRAM EFFECTIVENESS

Zehr (1995) has constructed a theoretical yardstick for measuring the effectiveness of restorative justice practices. The five questions to be considered are:

- Do victims experience justice?
- Do offenders experience justice?
- Is the victim-offender relationship addressed?
- Are community concerns being taken into account?
- Is the future being addressed? (pp. 230–231)

Yet as Goodenough (2000) asked, "How can we measure the impact of an apology that was ten years in coming, or of an offender visiting a grave site, or of a circle of loving eyes focused on one girl who thought she had no one?" The best we can do is to provide subjective documentation of the success or failure of witnessed interventions, or, for more objective measures, collect empirical survey data from the offender-victim mediation efforts. For purposes of generalization, key variables such as gender, race, and class should be delineated.

Given the recent acceptance of restorative strategies in the criminal justice system, there is much research yet to be done that would provide further justification for the proliferation of restorative justice sanctions. In her comprehensive review of empirical data on restorative intervention for juveniles, Schiff (1998) indicated that the kind of data we need on long-term effectiveness of victim impact panels and family group conferencing is lacking. Research on victim-offender mediation has been more forthcoming. Results consistently indicate a positive impact on offenders in terms of both compliance rates and the belief by offenders surveyed that they were held appropriately accountable for their actions. Much more is known about restitution compliance due to standardized record keeping, and success here is widely reported. These interventions, however, are not always tied in to a restorative model.

In their meta-analysis of relevant outcome assessments, Presser and Van Voorhis (2002) called for systematic assessment of recidivism measures as well as follow-up measures of change in attitude as a result of specific interventions. Sophisticated experimental designs are needed so that we can know what works and under what circumstances.

Sophisticated designs must include an experimental and control group and random assignment of subjects to each. McGarrell (2001) conducted one such recent study that included a measure of gender differences. A team of trained observers attended 157 restorative conferences involving youthful offenders in Indianapolis. Approximately one third of the young offenders were female. Findings indicated that almost all the conferences involved apologies by the youth; completion rates of the total program were significantly higher in the experimental than in the control group; and the rearrest rates were significantly lower. Most significantly for our purposes, the difference in outcome between the comparison groups was greater for females than for males, although there were no racial differences in outcome. Victims and family members alike later reported that they were impressed with the treatment they received in the conferences.

The most extensive and comprehensive studies of the impact of restorative justice interventions to date have been conducted at the Center for Restorative Justice and Peacemaking in Minnesota under the leadership of Mark Umbreit (1996). Data were collected from four Canadian victim-offender mediation programs on 4,445 offenders, mostly adults, who were referred to the programs. Approximately 75% of the victims and the offenders who were interviewed reported satisfaction with the process. Ninety percent from both groups felt they had been treated fairly. A comparison group, who did not participate in mediation, were significantly less satisfied with the handling of their case.

Umbreit's summary (1998) reported that for victims, the possibility of receiving restitution appeared to initially motivate them to enter the mediation process. After mediation, however, they reported that meeting the offender and being able to talk about what happened was more satisfying than receiving restitution. Even in cases of extreme violence, victims and offenders often highlighted their participation in mediated dialogue as a powerful and transformative experience that helped them express personal pain, let go of their hatred, and, finally, heal. In a study of arranged and monitored meetings between incarcerated inmates and victims in British Columbia, victims reported seeing the offender for the first time as a person rather than a monster and thus felt less fear and more at peace. Offenders, conversely, felt more empathy regarding the victim's feelings. Similar results have emerged from large-scale victim-offender conferencing in Vermont, where the restorative justice model has been institutionalized (Marks, 1999), as well as from a Canadian government evaluation of a restorative project in Winnipeg (Bonta, Wallace-Capretta, & Rooney, 1998).

What do these research results tell us about the effectiveness of such programming for women? From comprehensive reviews in the rehabilitation literature, we can learn which types of interventions are effective with female offenders. Dowden and Andrews (1999) conducted a meta-analysis of the rehabilitation literature on treatment effectiveness for female offenders and singled out a number of key variables correlated with a reduction in the recidivism rate. Human service programs emerged as highly effective; these related to inmates' psychological needs such as targeting of issues of victimization and low self-esteem, cognitive work, and intensity of treatment for high-risk offenders with a history of criminality. What we learn from this analysis of the treatment literature is that interventions directed toward the needs of female offenders are significantly positive, especially if they include work in the areas of past trauma and in cognitions (e.g., attitude and irrational thinking). Such an approach is consistent with gender-based restorative justice conferencing programs, although more specific research related to specialized restorative justice interventions is needed to definitively assert these claims.

ADVOCACY

Vision and leadership are two key ingredients singled out by Umbreit and Carey (1995) for effective restorative justice planning. In an article directed at corrections administrators seeking to move from a standard criminal justice model to a restorative model, the authors recounted the journey toward innovative systemic change in one community corrections department in Minnesota. The starting point was a change in thinking from an offender-only concern to a three-dimensional conceptualization involving victims, offenders, and the community. This involved staff education through a variety of presentations and workshops, given that collaboration with staff and administration was critical to the formation of an effective vision and final action plan. Bazemore and Pranis (1997) similarly emphasized partnerships between managers and direct service staff to overcome the stumbling blocks often encountered in inherently rigid bureaucratic structures. Other guidelines for advocacy proposed by Bazemore and Pranis include

- Seeking input from those most affected by crime: crime victims and their advocates
- Addressing crime victims' issues first
- Being careful not to equate restorative justice with victim services so that offender treatment/rehabilitation will not be overlooked
- Involving judges, prosecutors, police, and the public in order to legitimize the program, ensure its effectiveness, and prevent its marginalization
- Involving the community in the process: for example, in projects to repair the damage of the offense

Among corrections administrators, it appears that broadcasting successful strategies implemented in one state can help effect change somewhere else. For example, the policy of requiring inmate restitution (like that of mandatory arrests of batterers) spreads from one jurisdiction to another. State and national agencies can carry out pilot programs to demonstrate applications of principles (Pranis, 1999), and often grant money is available to encourage initiatives and to hire personnel and to set up such experimental programs. The Kentucky Department of Justice, for example, awarded Transformation House a small grant for their victim-offender conferencing initiative.

Often such ideas are spread through grassroots organizing. In Minnesota, several community groups and a nonprofit criminal justice agency sponsored a conference on restorative justice and introduced the idea to established criminal justice practitioners. Statewide conferences and involvement from key leaders from all parts of the corrections system followed (Pranis, 1996). Today, the Minnesota Department of Corrections Initiative provides technical assistance throughout the state for designing applications related to the restorative vision.

We can take a lesson from the victim impact panels first initiated by MADD (Mothers Against Drunk Driving). To change attitudes about drunk driving so that it could be regarded as serious and preventable, panels of victims' families provided firsthand testimony on the pain and suffering caused by car crashes. The strategy spread and is now used in prison and jail settings as well as in treatment programs.

Advocacy on behalf of female offenders and victims can be especially effective given the strength of the advocacy groups. Faith (2000) largely credited a Canadian feminist task force with the regionalizing and feminizing of women's prisons in Canada. Pranis (1999) offered specific guidelines on steps to take in building community support for restorative justice.

Advocacy on behalf of an individual or a community is an obligation at the very core of the social work profession and is in fact spelled out in the *NASW Code of Ethics* (1999). Students of criminal justice would therefore do well to refer to one of the many social work books on policy practice and advocacy (see, e.g., Jansson, 1999; Schneider & Lester, 2000). Social work students and practitioners, conversely, would do well to turn their attention to some of the dynamic and innovative work being done by criminal justice planners.

To summarize, effective strategies for restorative justice advocacy are as follows: to embark on cost-effective analyses of ongoing programs; to engage in special outreach efforts to victim/witness assistance groups to dispel any initial skepticism; to unite with progressives in the field of criminal justice as well as natural allies at the grassroots level for educational efforts; to lobby legislators for funding of state and local pilot projects for certain designated categories of offenders; and, finally, to build community support with outreach to minority groups, especially native populations, in order to promote a restorative framework.

Conclusion

Because crime is a violation of people within a community, the community should be involved in helping to repair the harm. Whereas the usual criminal justice approach focuses almost solely on the individual offender and the offense to the state, restorative justice is three-dimensional. Whether the restorative rituals follow the standard court procedures or are a separate entity, the interconnectedness among victim, offender, and community is integral to the healing process of all three.

Restorative justice, in short, has much to offer female offenders of any given age. In this alternative method of dispensing justice, the importance of human relationships, the opportunity to speak one's own voice, expressing feelings, listening, support system involvement, and consensus-style decision making are the very qualities central to gender-specific programming for female offenders. Empowerment, likewise, is integral to both restorative and gender-specific initiatives.

From grassroots activity to the highest level of government, change is in the wind. With healing the wounds of crime as our goal, there is no end to the imaginative possibilities for advancing social justice and bringing people together for the common good. There is a groundswell of support out there, voiced by a diverse group of dedicated professionals and volunteers striving to bring us a form of justice more interested in helping than in processing people and more interested in restoration than in retribution.

References

Bannenberg, B. (2000). Victim-offender mediation in Germany. In *Victim-offender mediation in Europe: Making restorative justice work. First Conference of the European Forum for Victim-Offender Mediation and Restorative Justice, Leuven, Belgium, 1999* (pp. 251–279). Leuven, Belgium: Leuven University Press.

Barajas, E. (1995). *Moving toward community justice: Topics in community corrections.* Washington, DC: U.S. Department of Justice, Federal Bureau of Prisons.

Bazemore, G., & Pranis, K. (1997). Hazards along the way: Practitioners should stay true to the principles behind restorative justice. *Corrections Today, 59*(7), 84–89.

Bazemore, G., & Umbreit, M. (2001). *A comparison of four restorative conferencing models.* U.S. Department of Justice, Office of Juvenile Justice and Delinquency Prevention. Washington, DC: U.S. Department of Justice, National Institute of Corrections.

Belknap, J., & Holsinger, K. (1997). Understanding incarcerated girls: The results of a focus. *Prison Journal, 77,* 381–405.

Bonta, J., Wallace-Capretta, S., & Rooney, J. (1998). *Restorative justice: An evaluation of the restorative resolution project* (Rep. No. 1998–05). Ottawa, Canada: Office of Solicitor General.

Chesney-Lind, M. (1997). *The female offender: Girls, women and crime.* Thousand Oaks, CA: Sage.

DiBiase, M. (2000). Psychology and justice working together: Addressing the needs of female juvenile offenders. *Healing Magazine, 5*(2), 4–10.

Dowden, C., & Andrews, D. A. (1999). What works for female offenders: A meta-analytic review. *Crime and Delinquency, 45,* 438–443.

Faith, K. (2000). Reflections on inside/out organizing. *Social Justice 27*(3), 158–168.

Goodenough, K. (2000). *AMICUS: Restorative justice for girls.* Unpublished manuscript.

Harvey, L. (2001). Leading edge: Linda Harvey. Retrieved November 25, 2001, from the International Centre for Justice and Reconciliation Web Site: www.restorative-justice.org/rj3.

Healing circle shows offenders their human toll. (2001, November 10). *Toronto Star,* p. NE4.

Hudson, J., & Galaway, B. (1990). Restitution program models with adult offenders. In B. Galaway & J. Hudson (Eds.), *Criminal justice, restitution, and reconciliation* (pp. 165–175). Monsey, NY: Criminal Justice Press.

Hudson, J., Galaway, B., Morris, A., & Maxwell, G. (1996). Introduction. In J. Hudson, B. Galaway, A. Morris, & G. Maxwell (Eds.), *Family group conferences: Perspectives on policy and practice* (pp. 1–16). Monsey, NY: Criminal Justice Press.

Jansson, B. S. (1999). *Becoming an effective policy advocate.* Belmont, CA: Wadsworth.

Karp, D. R., Lane, J., & Turner, S. (2002). Ventura County and the theory of community justice. In D. Karp & T. Clear (Eds.), *What is community justice: Case studies of restorative justice and community supervision.* Thousand Oaks, CA: Sage.

Kimbro, P. L. (1999, May 30). A profound meeting, a healing bond. *Lexington Herald-Leader,* pp. A1, A16.

Lambert, D. (1998, June 20). *Tree of creation: Aboriginal HIV strategy.* Poster presentation at the annual meeting of the Canadian Association of Social Workers, Edmonton, Alberta.

Mackey, V., & Shadle, C. (1998). *Justice or "just deserts"? An adult study of the restorative justice approach.* Louisville, KY: Presbyterian Criminal Justice Program.

Marks, A. (1999, September 8). Instead of jail, criminals face victims. *Christian Science Monitor,* pp. 1, 4.

McCold, P. (2000, April). *Overview of mediation, conferencing and circles.* Paper presented to 10th United Nations Congress on Crime Prevention and Treatment of Offenders. International

Institute for Restorative Practices, Vienna.

McGarrell, E. F. (2001). *Restorative justice conferences as an early response to young offenders* (Office of Juvenile Justice and Delinquency Prevention Bulletin). Washington, DC: U.S. Department of Justice.

National Association of Social Work. (1999). *Code of ethics.* Washington, DC: NASW Press.

Parker, L. (2001). "Circles." Retrieved November 10, 2001, from the International Centre for Justice and Reconciliation Web site: www.restorativejustice.org

Peachey, D. E. (1992). Restitution, reconciliation, retribution: Identifying the forms of justice people desire. In H. Messmer & H. Otto (Eds.), *Restorative justice on trial: Pitfalls and potentials of victim-offender mediation: International research perspectives* (pp. 551–557). Dordrecht, the Netherlands: Kluwer.

Pennell, J., & Burford, G. (1994). Widening the circle: Family group decision making. *Journal of Child and Youth Care, 9*(1), 1–11.

Pranis, K. (1996). A state initiative toward restorative justice: The Minnesota experience. In B. Galaway & J. Hudson (Eds.), *Restorative justice: International perspectives* (pp. 493–504). Monsey, NY: Criminal Justice Press.

Pranis, K. (1999). Building community support for restorative justice: Principles and strategies. Retrieved November 25, 2001, from the International Centre for Justice and Reconciliation Web site: www.restorativejustice.org/rj3/A.

Presser, L., & Van Voorhis, P. (2002). Values and evaluation: Assessing processes and outcomes of restorative justice programs. *Crime and Delinquency, 48*(1), 162–187.

Schiff, M. F. (1998). Restorative justice interventions for juvenile offenders: A research agenda for the next decade. *Western Criminology Review, 1*(1). Retrieved November 2001 from the Western Criminology Review Web site: http://wcr.sonoma.edu/vlnl/schiff.html.

Schneider, R. L., & Lester, L. (2000). *Social work advocacy.* Belmont, CA: Wadsworth.

Sinkinson, H. D., & Broderick, J. J. (1998). A case study of restorative justice: The Vermont reparative probation program. In L. Walgrave (Ed.), *Restorative justice for juveniles: Potentialities, risks, and problems.* Leuven, Belgium: Leuven University Press.

Swart, S. (2000). Restorative processes: Mediation, conferencing, and circles. Retrieved November 10, 2001, from the International Centre for Justice and Reconciliation Web site: www.restorative-justice.org.

Tereshkova, Z. (2000, March 15). Lexington-based group brings inmates, their victims together for mutual healing. *Lexington Herald-Leader,* p. A1.

Thorvaldson, S. (1990). Restitution and victim participation in sentencing: A comparison of two models. In B. Galaway & J. Hudson (Eds.), *Criminal justice, restitution, and reconciliation* (pp. 23–36). Monsey, NY: Criminal Justice Press.

Transformation House (2001). L. Harvey. Retrieved November 25, 2001, from the International Centre for Justice and Reconciliation Web site: www.restorativejustice.org/rj3

Turpin, J. (1999). Restorative justice challenges corrections. *Corrections Today, 61*(6), 60–63.

Umbreit, M. (1985). *Victim offender mediation: Conflict resolution and restitution.* Valparaiso, IN: PACT Institute of Justice.

Umbreit, M. (1996). Restorative justice through mediation: The impact of programs in four Canadian provinces. In B. Galaway & J. Hudson (Eds.), *Restorative justice: International perspectives* (pp. 373–385). Monsey, NY: Criminal Justice Press.

Umbreit, M. (1998). Restorative justice through victim-offender mediation: A multi-site assessment. *Western Criminology Review, 1*(1), 24–30.

Umbreit, M. (2000). *National survey of victim-offender mediation programs in the United States.* St. Paul, MN: Center for Restorative Justice and Peacemaking.

Umbreit, M., & Carey, M. (1995). Restorative justice: Implications for organizational change. *Federal Probation, 59*(1), 47–54.

Umbreit, M., & Coates, R. (1992). *Victim offender mediation: An analysis of programs in four states of the U.S.* Minneapolis: Minnesota Citizens Council on Crime and Justice.

Umbreit, M., Coates, R., & Vos, B. (2001). The impact of victim-offender mediation: Two decades of research. *Federal Probation, 65*(3), 29–36.

Van Ness, D. (2001, December). UN expert meeting recommends declaration of basic principles on restorative justice. Retrieved December 11, 2001, from the International Centre for Justice and Reconciliation Web site: www.restorativejustice.org.

Van Ness, D., & Strong, K. H. (2002). *Restoring justice* (2nd ed.). Cincinnati, OH: Anderson.

Van Wormer, K. (2001). *Counseling female offenders and victims: A strengths-restorative approach.* New York: Springer.

Zehr, H. (1995). *Changing lenses: A new focus for crime and justice.* Waterloo, Ontario: Herald.

Zehr, H. (1997). Restorative justice: The concept. *Corrections Today, 59*(7), 68–71.

27

WHEN PRISONERS RETURN TO THE COMMUNITY

Political, Economic, and Social Consequences

JOAN PETERSILIA

State prisons admitted about 591,000 people in 1999 and released almost the same number. If Federal prisoners and young people released from secure juvenile facilities are added to that number, nearly 600,000 inmates arrive yearly on the doorsteps of communities nationwide.[1]

Virtually no systematic, comprehensive attention has been paid by policymakers to dealing with people after release, an issue termed "prisoner reentry."[2] Failure to address the issue may well backfire, and gains in crime reduction may erode if the cumulative impact of tens of thousands of returning felons on families, crime victims, and communities is not considered.

Inmates have always been released from prison, and officials have long struggled with helping them succeed. But the current situation is different. The numbers of returning offenders dwarf anything known before, the needs of released inmates are greater, and corrections has retained few rehabilitation programs.

A number of unfortunate collateral consequences are likely, including increases in child abuse, family violence, the spread of infectious diseases, homelessness, and community disorganization. As victim advocates are well aware, the implications for public safety and risk management are major factors in reentry. For large numbers of people in some communities, incarceration is becoming almost a normal experience. The phenomenon may affect the socialization of young people, the power of prison sentences to deter, and the future trajectory of crime rates and crime victimization.

Petersilia, J. (2001, June). When Prisoners Return to the Community: Political, Economic, and Social Consequences. *Federal Probation,* pp. 3–8. Used with permission.

PAROLE: MANAGING
MORE PEOPLE LESS WELL

Changes in sentencing practices, coupled with a decrease in prison rehabilitation programs, have placed new demands on parole. Support and funding have declined, resulting in dangerously high caseloads. Parolees sometimes abscond from supervision; more than half of all parolees are rearrested.[3]

Determinate Sentencing
Means Automatic Release

Parole has changed dramatically since the mid-1970s. At that time, most inmates served open-ended, indeterminate terms, and a parole board had wide discretion to either release them or keep them behind bars. In principle, offenders were paroled only if they were rehabilitated and had ties to the community—such as a family or a job. This made release a privilege to be earned. If inmates violated parole, they could be returned to prison to serve the balance of their term—a strong disincentive to commit crime.

Today, indeterminate sentencing and discretionary release have been replaced in 14 states with determinate sentencing and automatic release.[4] Offenders receive fixed terms when initially sentenced and are released at the end of their prison term, usually with credits for good time. For example, in California, where more than 125,000 prisoners are released yearly, there is no parole board to ask whether the inmate is ready for release, since he or she must be released once his or her term has been served. After release, most California offenders are subject to 1 year of parole supervision. Generally, a parolee must be released to the county where he or she lived before entering prison. Since the vast majority of offenders come from economically disadvantaged, culturally isolated, inner-city neighborhoods, they return there upon release.

Indeterminate sentencing lost credibility in part because it is discretionary. Research revealed that there were wide disparities in sentencing when the characteristics of the crime and the offender were taken into account and that sentencing was influenced by the offender's race, socioeconomic status, and place of conviction. But most corrections officials believe some power to individualize sentences is necessary, since it is a way to take into account changes in behavior or conditions that occur during incarceration. Imprisonment can cause psychological breakdown, depression, or other mental illness or can reveal previously unrecognized personal problems. When this is discovered, the parole board can adjust release dates.

MORE PAROLEES
HAVE UNMET NEEDS

The states and the Federal Government have allocated increasing shares of their budgets to building and operating prisons. California, for example, has built 21 prisons since the mid-1980s, and its corrections budget grew from 2 percent of the state's general fund in 1981 to nearly 8 percent in 2000. There are similar patterns nationwide, with spending on prisons the fastest growing budget item in nearly every state in the 1990s.

Increased dollars have funded operating costs for more prisons, but not more rehabilitation. Fewer programs, and lack of incentives to participate, mean fewer inmates leave prison having addressed their work, education, and substance abuse problems. Yet sentences for drug offending are the major reason for increases in prison admissions since 1980.[5] In-prison substance abuse programs are expanding but are often minimal. The Office of National Drug Control Policy reported that 70 to 85 percent of state prisoners need treatment; however, just 13 percent receive it while incarcerated.[6]

Mental illness is another growing issue. As a result of deinstitutionalization, more mentally ill people are sent to prison and jail than in the past. Nearly 1 in 5 inmates in U.S. prisons reports having a mental illness.[7] Confinement in overcrowded prisons and in larger, "super max" prisons can cause serious psychological problems, since prisoners in such institutions spend many hours in solitary or in segregated housing. The longer the time in isolation, the greater the likelihood of depression and heightened anxiety.[8]

Gang activity, a major factor in many prisons, has implications for in-prison and postprison behavior. The existence of gangs and the related racial tension mean that inmates tend to be more preoccupied with finding a safe niche than with long-term self-improvement. Gang conflicts that start (or continue) in prison also continue in the community after gang members are released. One observer of this phenomenon has noted, "There is an awful lot of potential rage coming out of prison to haunt our future."[9] If these needs remain unmet, there will be effects not just for returning inmates, but for community members who are at risk for further crime victimization.

Parole Supervision Replaces Services

Eighty percent of returning prisoners are released on parole and assigned to a parole officer. The remaining 20 percent (about 100,000 in 1998), including some who have committed the most serious offenses, will "max out" (serve their full sentence) and leave prison with no postcustody supervision. This means offenders who are presumably the least willing to enter rehabilitative programs are often not subject to parole supervision and receive no services.

For parolees, the parole officer plays a vital role. He or she enforces the conditions of release, including the prohibition on drug use and on associating with known criminals, and the requirement that the offender find and keep a job. Parole officers also provide crime victims with information about the offender's whereabouts, conditions of parole, and other issues affecting victim safety.

Despite the essential work of parole officers, their numbers have not kept pace with demand. In the 1970s, one agent ordinarily was assigned 45 parolees. Today caseloads of 70 are common. Most parolees are supervised on "regular" rather than intensive caseloads, which means less than two 15-minute, face-to-face contacts per month.[10] Parole supervision costs about $2,200 per parolee per year, compared to 10 times that much per prisoner. The current arrangements do not permit much monitoring. Parole agents in California reportedly lost track of about one-fifth of the parolees they were assigned to in 1999.[11] Nationally, about 9 percent of all parolees have absconded.[12]

Most Parolees Return to Prison

People released from prison remain largely uneducated and unskilled and usually have little in the way of a solid family support system. To these deficits are added the unalterable fact of their prison record. Not surprisingly, most parolees fail and do so quickly: Most rearrests occur in the first 6 months after release.

Fully two-thirds of all parolees are rearrested within 3 years. The numbers are so high that parole failures account for a growing proportion of all new prison admissions. In 1980, they constituted 17 percent of all admissions, but they now make up 35 percent.[13]

COLLATERAL CONSEQUENCES

Recycling parolees in and out of families and communities has a number of adverse effects. It is detrimental to community cohesion, employment prospects and economic well being, participation in the democratic process, family stability and childhood development, and mental and physical health and can exacerbate such problems as homelessness.[14]

Community Cohesion and Social Disorganization

The social characteristics of neighborhoods—particularly poverty and residential instability—influence the level of crime. There are "tipping points" beyond which communities can no longer favorably influence residents' behavior. Norms start to change, disorder and incivility increase, out-migration follows, and crime and violence increase.[15]

Sociologist Elijah Anderson explains the breakdown of cohesion in socially disorganized communities and how returning prisoners play a role in that process and are affected by it. Moral authority increasingly is vested in "street smart" young men for whom drugs and crime are a way of life. Attitudes, behaviors, and lessons learned in prison are transmitted to free society. He concludes that as "family caretakers and role models disappear or decline in influence, and as unemployment and poverty become more persistent, the community, particularly its children, becomes vulnerable to a variety of social ills, including crime, drugs, family disorganization, generalized demoralization and unemployment."[16]

Prison gangs have growing influence in inner-city communities. Sociologist Joan Moore notes that because prisons are violent and dangerous places, new inmates seek protection and connections. Many find both in gangs. Inevitably, gang loyalties are exported to the neighborhoods when inmates are released. "In California . . ." she commented, "I don't think the gangs would continue existing as they are without the prison scene."[17] She warned that as more young people are incarcerated earlier in their criminal career, more will come out of prison with hostile attitudes and will exert strong negative influences on the neighborhoods to which they return.

Researchers explored similar effects by looking at crime rates in Tallahassee 1 year after offenders who had been sent to prison from there had returned to that community. Rather than reducing crime, releasing offenders in 1996 led to an increase the following year, even after other factors were taken into account.[18] One explanation focuses on individuals—offenders "make up for lost time" by resuming their criminal careers with renewed energy. But the researchers who studied Tallahassee focus on the destabilizing effect of releasing large numbers of parolees. They argue that "coerced mobility," like voluntary mobility, is a type of "people-churning" that inhibits integration and promotes isolation and anonymity—factors associated with increased crime.

Work and Economic Well-Being

The majority of inmates leave prison with no savings, no immediate entitlement to unemployment benefits, and few job prospects. One year after release, as many as 60 percent of former inmates are not employed in the legitimate labor market. The loss of much of the country's industrial base, once the major source of jobs in inner-city communities, has left few opportunities for parolees who live there. Employers are increasingly reluctant to hire ex-offenders. A recent survey in five major U.S. cities revealed that 65 percent of all employers said they would

not knowingly hire an ex-offender (regardless of the offense), and 30 to 40 percent said they had checked the criminal records of their most recent hires.[19] It is possible, however, that current low unemployment may cause employers to reevaluate ex-offenders.

Unemployment is closely correlated with drug and alcohol abuse. Losing a job has similar effects. It can lead to substance abuse, which in turn is related to child abuse and family violence. Moreover, prisoners who have no income because they have no job are unlikely to be able to meet court-ordered restitution owed to their victims.

The "get tough" movement of the 1980s increased employment restrictions on parolees. In California, for example, they are barred from the law, real estate, medicine, nursing, physical therapy, and education. Colorado prohibits them from becoming dentists, engineers, nurses, pharmacists, physicians, or real estate agents. Parolees are not barred from all jobs, but the list of proscribed professions suggests a contradictory approach.[20] The states spend millions of dollars to rehabilitate offenders, convincing them they need to find legitimate employment, but then frustrate what was accomplished by barring them from many kinds of jobs.

Underemployment of ex-felons has even broader economic implications. One reason the U.S. unemployment rate is so low is that 2 million mainly low- and unskilled workers—precisely those unlikely to find work in a high-tech economy—are in prison or jail and thus not part of the labor force. If they were included, the unemployment rate would be 2 percent higher than it is now.[21] Recycling ex-offenders into the job market with reduced job prospects will increase unemployment in the long run.

There are, however, a number of organizations that help ex-offenders find employment. Prominent among them is the Chicago-based Safer Foundation, which offers a full range of services, including job counseling and placement, education and life skills training, and emergency housing. Since its establishment in 1972, the foundation has helped more than 40,000 participants find jobs; nearly two-thirds have stayed on the job for at least 30 days.[22]

Family Matters

More than 1.5 million children in the United States have parents in prison.[23] Among incarcerated men, more than half are fathers of minor children. For women inmates the percentage is larger—about two-thirds have minor children. On average, women inmates have two dependent children.[24] Although women constitute only about 7 percent of the U.S. prison population, their incarceration rates are increasing faster than those of men, so the number of children whose mothers are incarcerated will rise proportionately.

Little is known about the effects of a parent's incarceration on childhood development, but it is likely to be significant. When mothers are incarcerated, their children are usually cared for by grandparents or other relatives or placed in foster care. Roughly half these children do not see their mothers the entire time they are in prison. The vast majority of imprisoned mothers, however, expect to resume their parenting role and live with their children after release, although it is uncertain how many actually do.[25]

Mothers released from prison encounter difficulties finding housing, employment, and such services as childcare. Children of incarcerated and released parents often become confused, unhappy, and socially stigmatized. The frequent outcome is school-related difficulties, low self-esteem, aggressive behavior, and general emotional dysfunction. If their parents are negative role models, children fail to develop positive attitudes toward work and responsibility. They are five times more likely to serve time in prison when they become adults than children whose parents are not incarcerated.[26]

There are no data on parolees' involvement in family violence, but it may be significant. Risk factors for child abuse and neglect include parental poverty, unemployment, substance abuse, low self-esteem, and ill health—attributes common among parolees. Concentrated poverty and social disorganization increase the likelihood of child abuse and neglect and other problems related to life after prison, and these in turn are risk factors for other kinds of crime and violence.

Mental and Physical Health

Prisoners have significantly more medical and mental health problems than the general population, because they often live as transients or in crowded conditions, tend to be economically disadvantaged, and have high rates of substance abuse, including intravenous drug use. In prison, people aged 50 are commonly considered old, in part because the health of the average 50-year-old prisoner approximates that of the average 60-year-old person in the free community. While in prison, inmates have state-provided health care, but upon release most cannot easily obtain health care. In recent years, escalating health care costs, high incarceration rates and, in particular, the appearance of HIV and AIDS have made the health care of prisoners and soon-to-be-released prisoners a major policy and public health issue, one whose complexity can only be intimated here.

Inmates are particularly prone to spread disease (especially such conditions as tuberculosis, hepatitis, and HIV), and thus pose public health risks.[27] In New York City, a major multidrug-resistant form of TB emerged in 1989, with 80 percent of the cases traced to jails and prisons. By 1991, New York's Rikers Island Jail had one of the highest TB rates in the Nation. In Los Angeles, a meningitis outbreak in the county jail spread to surrounding neighborhoods.

At year-end 1997, 2.1 percent of all State and Federal prison inmates were infected with HIV, a rate five times higher than in the general population.[28] Public health experts predict the rate will continue to climb, and eventually HIV will manifest itself on the street, particularly as more drug offenders, many of whom use drugs intravenously and share needles or trade sex for drugs, are incarcerated.[29]

As noted before, larger numbers of mentally ill inmates are imprisoned—and released—than in the past. Even when mental health services are available, many people who are mentally ill fail to use them because they fear being institutionalized, deny their condition, or distrust the mental health system.

POLITICAL ALIENATION

As of 1998, an estimated 3.9 million Americans were permanently unable to vote because they had been convicted of a felony. Of these, 1.4 million were African American men—13 percent of all black men. Assuming incarceration rates increase, the numbers of incarcerated black men will also increase. A young black man aged 16 in 1996 had a 29-percent chance of spending time in prison at some time in his life. The comparable figure for white men was 4 percent.[30]

Some observers may see the disenfranchisement of felons as an acceptable part of the penalty for crime. Nevertheless, denying large segments of the minority population the right to vote is likely to cause further alienation. Disillusionment with the political process also erodes citizens' feeling of engagement and makes them less willing to participate in local political activities and to exert informal social control in their community.

HOUSING AND HOMELESSNESS

The most recently available figures indicate there are about 230,000 homeless people in the United States. The number is surely higher now, as many cities report a shortage of affordable housing. In the late 1980s, an estimated one-fourth of homeless people had served prison sentences. In California, 10 percent of all parolees are homeless, but in urban areas such as San Francisco and Los Angeles, the rate is as high as 30 to 50 percent.[31]

The presence of transients and vagrants, and the panhandling they sometimes engage in, increase citizens' fears, ultimately increasing crime and violence. Crime often becomes worse when people are afraid to go out on streets defaced by graffiti or frequented by transients and loitering youths. Fearful citizens eventually yield control of the streets to people who are not intimidated by the signs of decay and who often are those who created the problem. A vicious cycle then begins. Criminologists James Q. Wilson and George L. Kelling famously illustrated the phenomenon by describing how a single broken window can influence crime rates.

If the first broken window is not repaired, people who like to break windows may assume no one cares and break more. As "broken windows" spreads—as homelessness, prostitution, graffiti, and panhandling—businesses and law-abiding citizens move out and disorder escalates, leading to more serious crime.[32]

RETHINKING PAROLE

Government officials voice growing concern about the problems posed by prisoner reentry. Attorney General Janet Reno called it "one of the most pressing problems we face as a nation."[33] In response, several jurisdictions throughout the country have launched a new approach to the public safety challenge posed by released offenders. In a project sponsored by the U.S. Department of Justice, eight jurisdictions are serving as pilot sites of the Reentry Partnerships Initiative, whose goal is better risk management via enhanced surveillance, risk and needs assessment, and prerelease planning. The Department's new Reentry Courts Initiative, with nine sites participating, is based on the drug court model and taps the court's authority to use sanctions and incentives to help released offenders remain crime free.

The usefulness of initiatives like these depends to a great extent on their grounding in scientifically sound analysis and debate. It is safe to say that parole has received less research attention in recent years than any other part of corrections.[34] I have spent many years working on probation effectiveness but know of no similar body of knowledge on parole effectiveness. Without better information, the public is unlikely to permit corrections officials to invest in rehabilitation and job training for parolees. With better information, it might be possible to persuade voters and elected officials to shift from solely punitive sentencing and corrections policies toward those that balance incapacitation, rehabilitation, and just punishment.

Revisiting the Parole Board

The eclipse of discretionary parole release also needs to be reconsidered. In 1977, more than 70 percent of all prisoners in the United

States were released after appearing before a parole board, but 20 years later that figure had declined to less than 30 percent. Parole was abolished in many states because it symbolized the alleged leniency of a system in which hardened criminals were "let out" early. If parole were abolished, politicians argued, parole boards could not release offenders early, and inmates would serve longer terms. However, this has not happened. A recent study of inmates released in states that had abolished parole showed they served 7 months less than inmates released in states with discretionary parole.[35] Similar experiences in Florida, Connecticut, and Colorado caused those states to reinstate discretionary parole after discovering that abolition meant shorter terms served.

Parole experts have long held that the public is misinformed when it labels parole as lenient. By exercising discretion, parole boards can single out the more violent and dangerous offenders for longer incarceration. When states abolish parole or reduce the discretion of parole authorities, they replace a rational, controlled system of "earned" release for selected inmates with "automatic" release for nearly all inmates.[36] No-parole systems sound tough but remove a gatekeeping role that can protect victims and communities.

Parole boards can demand that released inmates receive drug treatment, and research shows that coerced treatment is as successful as voluntary participation.[37] If parole boards also require a plan for the released offender to secure a job and a place to live in the community, the added benefit is to refocus prison staff and corrections budgets on transition planning.

Involving victims in parole hearings has been one of the major changes in parole in recent years. Ninety percent of parole boards now provide victims with information about the parole process, and 70 percent allow victims to attend the parole hearing.[38] Parole boards also can meet personally with the crime victim. Meeting victims' needs is a further argument for reinstating parole.

Perhaps most important, when information about the offense and the offender has been gathered and prison behavior observed, parole boards can reconsider the tentative release date. More than 90 percent of offenders in the United States are sentenced because they plead guilty, not as the result of a trial. Without a trial, there is little opportunity to fully air the circumstances of the crime or the risks posed by the offender. A parole board can revisit the case to discover how extensive the victim's injuries were and whether a gun was involved. The board is able to do so even though the offense to which the offender pled, by definition, involved no weapon. As one observer commented on this power of the parole board, "In a system which incorporates discretionary parole, the system gets a second chance to make sure it is doing the right thing."[39] Again, this can make a difference for crime victims.

Toward a Balanced System

Ironically, no-parole systems also significantly undercut postrelease supervision. When parole boards have no authority to decide who will be released, they are compelled to supervise a parolee population consisting of more serious offenders and not one of their own choosing. Parole officers believe it is impossible to elicit cooperation from offenders when the offenders know they will be released, whether or not they comply with certain conditions. And because of prison crowding, some states (for example, Oregon and Washington) no longer allow parolees to be returned to prison for technical violations. Field supervision of parolees tends to be undervalued and, eventually, underfunded and understaffed.

No one would argue for a return to the unfettered discretion that parole boards exercised in the 1960s. That led to unwarranted disparities. Parole release decisions must incorporate explicit standards and due process protections. Parole guidelines, used in many states, create uniformity in parole decisions and can be used to objectively weigh factors known to be associated with recidivism. Rather than entitle inmates to release at the end of a fixed time, guidelines specify when the offender becomes eligible for release.

The question of who should be responsible for parole release decisions is also worth rethinking. In most states, the chair and all members of the parole board are appointed by the Governor. In two-thirds of the states, there are no professional qualifications for parole board membership. While this may increase public accountability of parole boards, it also makes them vulnerable to political pressure. Ohio is an example of an alternative approach. There, parole board members are appointed by the director of the state's department of corrections, serve in civil service positions, and have an extensive background in criminal justice.

THE PUBLIC POLICY CHALLENGE

Parole supervision and release raise complex issues and deserve more attention. Nearly 700,000 parolees are "doing time" on the streets. Most have been released to parole systems that provide few services and impose conditions that almost guarantee failure. Monitoring systems are becoming more sophisticated, and public tolerance for failure is decreasing. All this contributes to the rising tide of parolees who are returning to prison. As the numbers increase, they put pressure on the states to build more prisons and, in turn, siphon funds from rehabilitation programs that might help offenders stay out of prison. Parolees will then continue to receive fewer services to deal with underlying problems, ensuring that recidivism rates and returns to prison (not to speak of crime victimization) remain high and public support for parole remains low.

This presents formidable challenges for policymakers. The public will not support community-based sanctions until they have been shown to "work," and they will not have an opportunity to work without sufficient funding and research. But funding is being cut, as California's situation exemplifies. In 1997, spending on parole services was cut 44 percent, causing caseloads to nearly double. When caseloads increase, services decline, and even parolees who are motivated to change have little opportunity to do so.

In 2001, there will likely be more than 2 million people in jail and prison in this country and more people on parole than ever before. If parole revocation trends continue, more than half the people entering prison that year will be parole failures. Given the increasing human and financial costs of prison—and all the collateral consequences parolees create for their families, victims, and communities—investing in effective reentry programs may be one of the best investments we make.

NOTES

1. Calculations based on data provided by the Bureau of Justice Statistics, U.S. Department of Justice.

2. The issue was addressed in another paper in this series. See Travis, Jeremy, But They All Come Back: Rethinking Prisoner Reentry, Research in Brief— Sentencing & Corrections: Issues for the 21st Century, Washington, DC: U.S. Department of Justice, National Institute of Justice, May 2000, NCJ 181413.

3. Petersilia, Joan, "Parole and Prisoner Reentry in the United States," in Prisons, ed. Michael Tonry and Joan Petersilia (Crime and Justice: A Review of Research, Volume 26), Chicago: University of Chicago Press, 1999: 512.

4. Tonry, Michael, Reconsidering Indeterminate and Structured Sentencing, Research in Brief— Sentencing & Corrections: Issues for the 21st Century, Washington, DC: U.S. Department of Justice, National Institute of Justice, September 1999, NCJ 175722.

5. Blumstein, Alfred, and Allen J. Beck, "Population Growth in U.S. Prisons, 1980-1996," in Prisons, ed. Michael Tonry and Joan Petersilia (Crime and Justice: A Review of Research, Volume 26), Chicago: University of Chicago Press, 1999: 20-22.

6. Byrne, Candice, Jonathan Faley, Lesley Flaim, Francisco Piuol, and Jill Schmidtlein, Drug Treatment in the Criminal Justice System, Washington, DC: Executive Office of the President, Office of National Drug Control Policy, August 1998, NCJ 170012.

7. Ditton, Paula M., Mental Health and Treatment of Inmates and Probationers, Washington, DC: U.S. Department of Justice, Bureau of Justice Statistics, July 1999, NCJ 174463.

8. Liebling, Allison, "Prison Suicide and Prisoner Coping," in Prisons, ed. Michael Tonry and Joan Petersilia (Crime and Justice: A Review of Research, Volume 26), Chicago: University of Chicago Press, 1999: 283-359.

9. Abramsky, Sasha, "When They Get Out," Atlantic Monthly 283 (6) (June 1999): 30-36.

10. Petersilia, "Parole and Prisoner Reentry in the United States," 505.

11. Petersilia, Joan, "Challenges of Prisoner Reentry and Parole in California," CPRC Brief 12 (3) (June 2000), Berkeley, CA: California Policy Research Center.

12. Los Angeles Times, quoted in "State Agencies Lost Track of Parolees," Santa Barbara News Press, August 27, 1999: B9; and Beck, Allen, Bureau of Justice Statistics, personal communication, March 28, 2000.

13. Snell, Tracy L., Correctional Populations in the United States, 1993, Washington, DC: U.S. Department of Justice, Bureau of Justice Statistics, October 1995, NCJ 156241: 12; and Beck, Allen, and Christopher Mumola, Prisoners in 1998, Bulletin,

Washington, DC: U.S. Department of Justice, Bureau of Justice Statistics, August 1999, NCJ 175687: 12.

14. Hagan, John, and Ronit Dinovitzer, "Collateral Consequences of Imprisonment for Children, Communities, and Prisoners," in Prisons, ed. Michael Tonry and Joan Petersilia (Crime and Justice: A Review of Research, Volume 26), Chicago: University of Chicago Press, 1999: 121-162. This discussion focuses primarily on the consequences for prisoners. For crime victims, of course, the consequences of imprisonment and reentry are equally serious, because victim safety is put at risk when offenders are released.

15. Wilson, William Julius, The Truly Disadvantaged: The Inner City, the Underclass, and Public Policy, Chicago: University of Chicago Press, 1987.

16. Anderson, Elijah, Streetwise: Race, Class, and Change in an Urban Community, Chicago: University of Chicago Press, 1990: 4.

17. Moore, Joan, "Bearing the Burden: How Incarceration Weakens Inner-City Communities" (paper presented at the Conference on the Unintended Consequences of Incarceration, New York, January 1996), 73.

18. Rose, Dina, Todd Clear, and Kristen Scully, "Coercive Mobility and Crime: Incarceration and Social Disorganization" (paper presented at the American Society of Criminology meeting, Toronto, November 1999).

19. Holzer, H., What Employers Want: Job Prospects for Less-educated Workers, New York: Russell Sage, 1996.

20. Simon, Jonathan, Poor Discipline: Parole and the Social Control of the Underclass, 1890-1990, Chicago: University of Chicago Press, 1993.

21. Western, Bruce, and Katherine Beckett, "How Unregulated Is the U.S. Labor Market? The Penal System as a Labor Market Institution," American Journal of Sociology 104 (4) (January 1999): 1030-1060.

22. Finn, Peter, Chicago's Safer Foundation: A Road Back for Ex-Offenders, Program Focus, Washington, DC: National Institute of Justice, National Institute of Corrections, and Office of Correctional Education (U.S. Department of Education), June 1998, NCJ 167575.

23. Hagan and Dinovitzer, "Collateral Consequences of Imprisonment for Children, Communities, and Prisoners,"137.

24. Snell, Tracy L., and Danielle C. Morton, Women in Prison (Survey of State Prison Inmates, 1991), Special Report, Washington, DC: U.S. Department of Justice, Bureau of Justice Statistics, March 1994, NCJ 145321: 6.

25. Bloom, Barbara, and David Steinhart, Why Punish the Children: A Reappraisal of the Children of Incarcerated Mothers in America, San Francisco: National Council on Crime and Delinquency, 1993.

26. Beck, Allen, Darrell Gilliard, Lawrence Greenfeld, Caroline Harlow, Thomas Hester, Louis Jankowski, Danielle Morton, Tracy Snell, and James Stephen, Survey of State Prison Inmates, 1991, Washington, DC: U.S. Department of Justice, Bureau of Justice Statistics, March 1993, NCJ 136949: 9.

27. McDonald, Douglas C., "Medical Care in Prisons," in Prisons, ed. Michael Tonry and Joan Petersilia (Crime and Justice: A Review of Research, Volume 26), Chicago: University of Chicago Press, 1999: 427-478.

28. Maruschak, Laura M., HIV in Prisons 1997, Bulletin, Washington, DC: U.S. Department of Justice, Bureau of Justice Statistics, November 1999, NCJ 178284: 1.

29. May, John P., "Feeding a Public Health Epidemic," in Building Violence: How America's Rush to Incarcerate Creates More Violence, ed. J.P. May and K.R. Pitts, Thousand Oaks, CA: Sage Publications, 2000.

30. Bonczar, Thomas P., and Allen J. Beck, Lifetime Likelihood of Going to State or Federal Prison, Special Report, Washington, DC: U.S. Department of Justice, Bureau of Justice Statistics, March 1997, NCJ 160092: 2; and Mauer, Marc, The Race to Incarcerate, New York: The New Press, 1999.

31. Beyond Bars: Correctional Reforms to Lower Prison Costs and Reduce Crimes, Sacramento, CA: Little Hoover Commission, 1998.

32. Wilson, James Q., and George L. Kelling, "Broken Windows," Atlantic Monthly (March 1982): 29-38.

33. Reno, Janet, "Remarks of the Honorable Janet Reno on Reentry Court Initiative," John Jay College of Criminal Justice, New York, February 10, 2000. Retrieved May 19, 2000, from the U.S. Department of Justice database, on the World Wide Web: http://www.usdoj.gov/ag/speeches/2000/doc2.htm.

34. A congressionally mandated study of crime prevention programs included just one evaluation of parole among hundreds of recent studies examined. See Sherman, Lawrence, Denise Gottfredson, Doris MacKenzie, John Eck, Peter Reuter, and Shawn Bushway, Preventing Crime: What Works, What Doesn't, What's Promising. College Park, MD: University of Maryland, 1997.

35. Stivers, Connie, "Impacts of Discretionary Parole Release on Length of Sentence Served and Recidivism," unpublished paper, Irvine, CA: School of Social Ecology, University of California, Irvine, 2000.

36. Burke, Peggy B., Abolishing Parole: Why The Emperor Has No Clothes, Lexington, KY: American Probation and Parole Association, 1995.

37. Anglin, Douglas M., and Yih-Ing Hser, "Treatment of Drug Abuse," in Drugs and Crime, ed. Michael Tonry and James Q. Wilson, (Crime and Justice: A Review of Research, Volume 13), Chicago: University of Chicago Press, 1990.

38. Rhine, Edward E., "Parole Boards," in The Encyclopedia of American Prisons, ed. Marilyn McShane and Frank Williams, New York: Garland, 1996.

39. Burke, Abolishing Parole: 7.

Name Index

SUBJECT INDEX

ABOUT THE EDITOR

Albert R. Roberts, PhD, BCETS, DACFE, is Professor of Criminal Justice (former Chairperson) and Director of Faculty and Curriculum Development of the Administration of Justice and Interdisciplinary Criminal Justice Programs in the Faculty of Arts and Sciences, Livingston College Campus at Rutgers, the State University of New Jersey in Piscataway. He was the recipient of the Richard W. Laity Outstanding Academic Leadership Award presented in May, 2002, by Rutgers Council of AAUP Chapters, American Association of University Professors. Dr. Roberts has been a tenured professor at Rutgers since 1989. Previously, he taught at Indiana University, Seton Hall University, Brooklyn College of CUNY, and the University of New Haven. He has approximately 29 years full-time university teaching experience including 15 years of administrative experience (10 years as Chair, and 5 years as a Program Director). He received an MA degree (major in sociology; minor in counseling) from the Graduate Faculty of Long Island University in 1967 and a DSW in 1978 (which became a PhD in 1981) from the School of Social Work and Community Planning at the University of Maryland in Baltimore. Prior to his doctoral studies in social work, he completed a 40-credit doctoral minor and specialization in criminology and criminal justice at the University of Maryland in College Park under Dr. Peter P. Lejins. He is the founding Editor-in-Chief of the *Brief Treatment and Crisis Intervention* journal (Oxford University Press). He is an Editorial Advisor to Oxford's Professional Book Division in New York City and currently serves on the editorial boards of five professional journals. He recently edited a special issue of the *Brief Treatment and Crisis Intervention* journal titled "Stress, Crisis, and Trauma Intervention Strategies in the Aftermath of the September 11th Disasters."

Dr. Roberts is a member of the Board of Scientific and Professional Advisors and a Board Certified Expert in Traumatic Stress for the American Academy of Experts in Traumatic Stress. He is also a Diplomate of the American College of Forensic Examiners. He is the founding and current editor of the 36-volume Springer Series on Social Work (1980 to present) and the 7-volume Springer Family Violence Series. He is the author, coauthor, or editor of approximately 150 scholarly publications, including numerous peer-reviewed journal articles and book chapters and 24 books. His recent books and articles include *Handbook of Domestic Violence Intervention Strategies*; *Social Workers' Desk Reference* (coedited by Gilbert J. Greene), *Crisis Intervention Handbook: Assessment, Treatment and Research* (2nd ed.); *Juvenile Justice: Policies, Programs and Services* (3rd ed., forthcoming), and *Battered Women and Their Families: Intervention Strategies and Treatment Approaches* (2nd ed.).

Dr. Roberts's recent current projects include continuing to teach courses on crisis intervention, domestic violence, introduction to criminal justice, research methods, program evaluation, victimology and victim assistance, and juvenile justice at Rutgers University; directing the 18-credit Certificate Program in Criminal Justice Policies and Practices at Livingston College of Rutgers University; training crisis intervention workers and clinical supervisors in crisis assessment and crisis intervention strategies; and training police officers and administrators in domestic violence policies and crisis intervention. He is a lifetime member of the Academy of Criminal Justice Sciences (ACJS), has been a member of the National Association of Social Workers (NASW) since 1974, and has been listed in *Who's Who in America* since 1992. He is the faculty sponsor to Rutgers' Sigma Alpha Kappa chapter of the Alpha Phi Sigma National Criminal Justice Honor Society (1991–present).

About the Contributors

Gaylene S. Armstrong, PhD, is Assistant Professor in Administration of Justice at Arizona State University West. She received her PhD in criminology and criminal justice from the University of Maryland. Her primary research interests include corrections, juvenile delinquency, and violence prevention programs.

Mark Blumberg, PhD, is Professor of Criminal Justice at Central Missouri State University in Warrensburg, Missouri. He received his bachelor's and master's degrees from the University of Kansas, Lawrence, and an additional master's and doctoral degree from the State University of New York at Albany. He has authored numerous book chapters and journal articles on police use of deadly force, the impact of AIDS on the criminal justice system, and other issues. His work has appeared in such publications as the *Criminal Law Bulletin, American Journal of Police, Justice Professional Crime and Delinquency, Criminal Justice Policy Review,* and *Prison Journal.* He is also the editor of *AIDS: The Impact on the Criminal Justice System* and the coauthor of *The Mythology of Crime and Criminal Justice.*

Mary Bosworth, PhD, is Assistant Professor of Sociology at Wesleyan University. Her research interests include prisons, race, and gender. She is the author of *Engendering Resistance: Agency and Power in Women's Prisons* (1999) and *The U.S. Federal Prison System* (2002).

Beau Breslin, PhD, is Assistant Professor of Government at Skidmore College. He teaches courses on constitutional law, judicial process, and civil liberties. His research interests include constitutional theory and the interplay between the First Amendment right to free expression and the country's most valued social and political institutions. He has also published on subjects ranging from capital punishment to restorative justice.

Timothy P. Cadigan, MA, earned his degree in Criminal Justice from Rutgers University. He is currently Chief, Program Technology and Analysis Branch, Administrative Office of the U.S. Courts, and Executive Editor, *Federal Probation Quarterly.* Previously, he was PACTS Project Manager, Federal Corrections and Supervision Division, Administrative Office of the U.S. Courts.

Norman B. Cetuk, MMH, received his master's degree from Drew University and is currently enrolled in Drew University's doctorate program, where his studies have focused on critical incident stress and stress management for the emergency service worker. He has been a police officer for the past 29 years, working at the municipal and county levels. He is currently a Detective Lieutenant with the Somerset County Prosecutors Office, New Jersey, and is the Advanced Training Administrator for Advanced Police Training and Continuing Education. He has been assigned to the Child Abuse and Sex Crimes Unit and has been a supervisor of the Major Crimes Investigation Unit, Forensic and Crime Scene Investigation Unit, and the Fire-Arson Investigation Unit. He is an adjunct faculty member at Centenary College and Raritan

Valley Community College, and also teaches and lectures throughout the United States on crime scene investigation, death investigation, crime scene processing, and crisis negotiations. He has been a member of the Somerset County Emergency Response Team since 1987 and for the past 9 years has been the Coordinator of the county's 20-member Crisis Negotiation Unit. He is on the Board of Directors for the New Jersey Crisis Negotiation Association and has been a representative to the National Association of Crisis Negotiation Associations.

Meda Chesney-Lind, PhD, is Professor of Women's Studies at the University of Hawaii at Manoa. She has served as Vice President of the American Society of Criminology and president of the Western Society of Criminology. Nationally recognized for her work on women and crime, her books include *Girls, Delinquency and Juvenile Justice* (1992), which was awarded the American Society of Criminology's Michael J. Hindelang Award for the "outstanding contribution to criminology"; *The Female Offender: Girls, Women and Crime*; and *Female Gangs in America*. She received the Bruce Smith, Sr. Award "for outstanding contributions to Criminal Justice" from the Academy of Criminal Justice Sciences in 2001. She received the Herbert Block Award for service to the society and the profession from the American Society of Criminology. She also received the Donald Cressey Award from the National Council on Crime and Delinquency for "outstanding contributions to the field of criminology," the Founders Award of the Western Society of Criminology for "significant improvement of the quality of justice," and the University of Hawaii Board of Regent's Medal for "Excellence in Research." In Hawaii, she has served as Principal Investigator of a long-standing project on Hawaii's youth gang problem funded by the State of Hawaii Office of Youth Services. She has more recently also received funding to conduct research on the unique problems of girls at risk of becoming delinquent from the Office of Juvenile Justice and Delinquency Prevention.

Walter Edwards, BA, earned his degree in the Administration of Justice from Rutgers University. He is a former research assistant in the department of Administrative Justice at Rutgers University.

James Alan Fox, PhD, is Professor at Northeastern University. He specializes in serial/mass murder, juvenile violence and statistical methods, having published 15 books and dozens of articles on these topics. His many publications include *Mass Murder: America's Growing Menace* (1991), *Multiple Murder: Patterns of Serial and Mass Murder* (1998), *Homicide Trends in the United States* (1999), and *Dead Lines: Essays in Murder and Mayhem* (2001).

As a leading authority on homicide, Fox appears regularly on national television and in newspapers around the country and has been profiled in several media outlets, including a two-part cover story in *USA Today*. He frequently gives lectures and expert testimony, including 11 appearances before Congress and briefings to President Clinton, Vice President Al Gore, Attorney General Janet Reno and Princess Anne of Great Britain. Finally, Fox is a research fellow with the Bureau of Justice Statistics.

Lisa A. Frisch, MA, is Director of Criminal Justice Policy and Training for the New York State Office for the Prevention of Domestic Violence, a position she has held since 1992. Prior to earning a master's degree in Criminal Justice and completing coursework for a PhD from the State University of New York at Albany, she was a campus police officer and probation officer. This was long before domestic violence was a recognized social and criminal justice problem. In 1985, as a graduate student intern with the then Governor's Commission on Domestic Violence (now the New York State Office for the

Prevention of Domestic Violence), she helped coordinate a public hearing at Bedford Hills Correctional Facility, where she helped prepare battered women who were incarcerated for killing their abusers to testify about their negative experiences trying to get help from the system before they were compelled to kill or be killed. All the years of graduate school did not compare to what was learned from working with those courageous women. Since that time, she has seen tremendous change as well as steadfast resistance to that change while directing statewide domestic violence policy, training, and research initiatives for criminal justice professionals. This experience, which necessitates working closely with victim advocates, courts, police, prosecutors, probation and parole, and federal, state, and local governmental and political entities, has been invaluable in showing—firsthand—the pros and cons involved in striving toward the goal of criminalization.

Kathleen M. Heide, PhD, is Professor of Criminology at the University of South Florida, Tampa. She received her BA from Vassar College in psychology and her MA and PhD in criminal justice from the University at Albany, State University of New York. She is an internationally recognized consultant on homicide and family violence. Her extensive publications include two widely acclaimed books on juvenile homicide, *Why Kids Kill Parents: Child Abuse and Adolescent Homicide* and *Young Killers: The Challenge of Juvenile Homicide.* She is a licensed psychotherapist and a court-appointed expert in matters relating to homicide, sexual battery, children, and families. Her work has been featured in popular magazines such as *Newsweek, U.S. News and World Reports,* and *Psychology Today.* She has appeared as an expert on many nationally syndicated talk shows, including *Larry King Live, Geraldo, Sally Jesse Raphael*, and *Maury Povich.*

Vincent E. Henry, PhD, who recently retired from the New York Police Department following a 21-year police career, is Associate Professor in the Department of Criminal Justice and Sociology at Pace University. He earned M Phil and PhD degrees in criminal justice from the Graduate Center of the City University of New York (John Jay) and BA and MS degrees in criminal justice from Long Island University. The first American police officer to be named a Fulbright Scholar, he held a wide variety of uniformed and plainclothes patrol, undercover decoy, investigative, and supervisory assignments in the New York City Police Department (NYPD). As Commanding Officer of the Police Commissioner's Office of Management Analysis and Planning's Special Projects Unit from 1991 to 2000, he had an integral role in researching, developing, and implementing policy initiatives throughout the agency, particularly those related to the Compstat process and the NYPD's reengineering. He is the author of *The Compstat Paradigm: Management Accountability in Policing, Business and the Public Sector.* In the foreword to that book, former NYPD Police Commissioner William J. Bratton wrote that "as somebody who was 'there' from the beginning, Vincent Henry has been able to create the most accurate accounting yet published of the 'revolution' that occurred within the NYPD beginning in 1994 and the impact of the 'Compstat Paradigm' on not only that department, but on policing and criminal justice systems in the United States and abroad." Henry also holds the American Society for Industrial Security's Certified Protection Professional (CPP) credential and has been a consultant to police agencies and private security entities in the United States, Australia, and Japan. His most recent book, tentatively entitled *Death Work: Police and the Psychology of Death,* is in press.

Sean David Hill, PhD, has been Research Associate at the Criminal Justice Center of Sam Houston State University since January, 2000. His primary responsibilities include coordinating student internships, writing proposals for grants and contracts, continuing research on violent extremist groups, and teaching courses on terrorism, criminal investigation, and comparative criminal justice. From 1997 to 1999, he was employed by the Office of International Criminal Justice at the University of Illinois in Chicago. His primary responsibilities included conducting research in China and Vietnam on international drug trafficking, maintaining a database of open source information on terrorist organizations, and liaison and escort for senior foreign law enforcement officials and international visiting scholars. He received his PhD and MA in Criminal Justice from Sam Houston State University. His BA in Criminal Justice and Psychology is from the University of Illinois at Chicago. Dr. Hill has also conducted research for the Bureau of Alcohol, Tobacco and Firearms, and the Cook County, Illinois Investigative Services. He served in the United States Marine Corps from 1993-1997 in the Fleet Antiterrorist Security Team Company, and the Marine Corps Security Force Battalion.

Mary S. Jackson, PhD, is Professor at East Carolina University in Criminal Justice Studies, School of Social Work and Criminal Justice Studies, Greenville, North Carolina. She received her doctorate from the Mandel School of Applied Social Sciences at Case Western Reserve University, in Cleveland, Ohio. She is coauthor of *Delinquency and Justice* (2003) and the author of *Policing in a Multicultural Society* (1996). Her scholarly journal publications focus on juvenile delinquents, substance abuse, gang resistance programs, and policing.

Michael Jacobson, PhD, is Professor of Sociology and Criminology at the Graduate Center of the City University of New York and the John Jay College of Criminal Justice. He was recently awarded a Soros Senior Justice Fellowship for 2001 to write a book on strategies for reducing prison use through parole reform. The book, titled *Downsizing Prisons: Crime Control for the 21st Century,* will be forthcoming in 2003. Prior to this, Jacobson was the New York City Corrections Commissioner from 1995 through 1997 and was the New York City Commissioner of Probation from 1992 to 1996. Before being appointed Probation Commissioner, he worked for 9 years at the New York City Office of Management and Budget, where he was a Deputy Budget Director.

Jacobson was also Adjunct Assistant Professor at the Wagner School at NYU from 1985 to 1997, teaching courses on governmental budgeting, public policy analysis, and organization theory. He has a PhD in sociology from the CUNY Graduate Center.

David R. Karp, PhD, is Assistant Professor of Sociology at Skidmore College in Saratoga Springs, New York, where he teaches courses in criminology and criminal justice. He conducts research on community-based responses to crime and has given workshops on restorative justice and community justice nationally. Currently, he is engaged in a research study evaluating Vermont's community reparative probation boards and is a member of the New York State Community Justice Forum. He is the author of more than 30 academic articles and technical reports and a trilogy of books on community justice. The most recent is *What Is Community Justice? Case Studies of Restorative Justice and Community Supervision* (coedited with Todd Clear). He received a BA from theUniversity of California at Berkeley, and a PhD in sociology from the University of Washington.

Betsy Wright Kreisel, PhD, is Assistant Professor in the Department of Criminal Justice at Central Missouri State University. She received her PhD from the University of Nebraska at Omaha. Kreisel is the author of several publications in the area of police perceptions, citizen complaints of police misconduct, and police use of excess force.

Karel Kurst-Swanger, PhD, is Assistant Professor in the Department of Public Justice at the State University of New York at Oswego. Her research and teaching interests include criminal justice, human services, and public policy. Prior to her appointment at SUNY at Oswego, she served as the Director of Continuing Studies for the School of Education and Human Development, SUNY at Binghamton where she developed specialized continuing education programs for practitioners in criminal justice and human services. Kurst-Swanger is the Book Review Editor of Brief Treatment and Crisis Intervention. She also served as the Executive Director of the Crime Victims Assistance Center, Inc. of Broome County, New York and the Victim Services Coordinator of the Victim Assistance Unit of the Rochester Police Department, Rochester, New York. She continues to serve as a planning consultant to numerous private and public human service/criminal justice agencies. Her first book, *Violence in the Home: Multidisciplinary Perspectives,* coauthored with Jacqueline Petcosky, is due to be released in 2003. She received her PhD from the State University of New York (SUNY) at Binghamton and holds an MSEd and BS from SUNY at Brockport.

Jack Levin, PhD, is the Irving and Betty Brudnick Professor of Sociology and Criminology and director of the Brudnick Center on Conflict and Violence at Northeastern University, where he teaches courses in prejudice and violence, criminal homicide, and social psychology.

He has authored or coauthored 23 books, including *Mass Murder: America's Growing Menace* (1985), *Overkill: Mass Murder and Serial Killing Exposed* (1994), *Killer on Campus* (1996), *The Will to Kill: Making Sense of Senseless Murder* (2000), *Hate Crimes: The Rising Tide of Bigotry and Bloodshed* (2001), and *The Violence of Hate* (2001). Levin has also published more than 150 articles in professional journals and newspapers, such as the *New York Times, Boston Globe, Dallas Morning News, Philadelphia Inquirer, Christian Science Monitor, Chicago Tribune*, and *USA Today*. He appears frequently on national television programs, including *48 Hours, 20/20, Dateline NBC, The Today Show, Oprah, Rivera Live, Larry King Live,* and all network newscasts. Levin was honored by the Massachusetts Council for Advancement and Support of Education as its "Professor of the Year." In November 1997, he spoke at the White House Conference on hate crimes.

Charles Lindner is Professor in the Department of Law, Police Science and Criminal Justice Administration at the John Jay College of Criminal Justice. He has JD and MSW degrees from Fordham University and has written extensively on probation and corrections. Prior to becoming an academic 23 years ago, he was the Executive Assistant to the New York City Probation Commissioner.

Doris Layton MacKenzie, PhD, is Professor in the Department of Criminology and Criminal Justice at the University of Maryland and Director of the Evaluation Research Group. Prior to this position, she earned her doctorate from Pennsylvania State University, was on the faculty of the Louisiana State University where she was honored as a "Researcher of Distinction," and was awarded a Visiting Scientist position at the National Institute of Justice. She has an extensive publication record on such topics as examining what works to reduce crime in the community, inmate adjustment to prison, the impact of intermediate sanctions on recidivism, long-term offenders, methods of predicting prison populations, self-report criminal activities of probationers and boot camp prisons. She directed funded research projects on the topics of "Multi-Site Study of Correctional Boot Camps," "Descriptive Study of Female Boot Camps," "Probationer Compliance with Conditions of Supervision" and "The National Study of Juvenile Correctional Institutions" and "What Works in Corrections."

Daniel P. Mears, PhD, is a Research Associate in The Urban Institute's Justice Policy Center. His research focuses on the causes of crime and effective ways to prevent and intervene with crime and justice problems. He has conducted research on delinquency, juvenile and criminal justice programs and policies, domestic violence, immigration and crime, correctional forecasting, and drug treatment in prisons. Recent publications include articles in *Criminal Justice and Behavior, Criminology, Journal of Research in Crime and Delinquency, Law and Society Review,* and *Sociological Perspectives.*

Scott K. Okamoto, PhD, is Assistant Professor in the School of Social Work at Arizona State University, Main Campus. He received his PhD in social welfare from the University of Hawai`i at Manoa in May 2000. His research areas include child and adolescent mental health, juvenile delinquency, practitioner issues, and barriers to effective practice with high-risk youth. Two of his publications in 2001 were "Instrument Development: Practitioner Fear of Youthful Clients: An Instrument Development and Validation Study," in *Social Work Research,* and, with Meda Chesney-Lind, "Gender Matters: Patterns in Girls' Delinquency and Gender Responsive Programming," in *Journal of Forensic Psychology Practice.* He has practice experience working with children, adolescents, and families in residential, shelter-based, and school-based programs.

Joan Petersilia, PhD, is Professor of Criminology, Law, and Society, School of Social Ecology, University of California at Irvine, and former Director of the Rand Corporation's Criminal Justice Program to which she now acts as a consultant. She is a past president of the California Association for Criminal Justice Research, and past President of the American Society of Criminology. In 1994, she was the recipient of the American Society of Criminology's August Vollmer Award for her distinguished contributions to criminal justice policy. Petersilia has numerous research publications on juvenile and adult corrections, police performance and practices, intensive probation supervision and other intermediate community based sanctions.

Laura J. Praglin, PhD, is Assistant Professor of Social Work at the University of Northern Iowa. Her interests lie in the area of cultural diversity and the religious/spiritual dimensions of restorative justice.

Gina Pisano–Robertiello, PhD, is Assistant Professor in the Department of Criminal Justice at Seton Hall University in South Orange, New Jersey, a position she has held since 1996. She received her BS in administration of justice from Rutgers University (New Brunswick, New Jersey) in 1991 and her MA and PhD in Criminal Justice from Rutgers University-School of Criminal Justice in 1993 and 2000, respectively. She regularly teaches on research methodology, police, victimology, violent crime, criminological theory, and community experience. Most of her research efforts are in two major areas: perceptions of police and victims of domestic violence. She is a member of the Academic Review Committee, Bergen County, New Jersey, Sheriff's Office and was appointed Honorary Deputy Sheriff in Bergen County, New Jersey, in 1999.

Joseph Ryan, PhD, is Chair and Professor of the Department of Criminal Justice and Sociology at Pace University in New York and is a national expert on community policing and police management-related issues. He is a 25-year veteran of the New York City Police Department and was their expert on community policing and violence, especially as it relates to spousal, child, and elder abuse, and was a former Visiting Fellow with the National Institute of Justice. He received his PhD in sociology from Fordham University, and his master's and bachelor's degrees from John Jay College of Criminal Justice. He has conducted extensive research in a variety of crime-related areas. Most notable is the recent landmark National Evaluation of Title I of the 1994 Crime Act as it relates to the hiring of the 100,000 community police officers.

Margaret Ryniker, JD, is Assistant Professor and Student Advisement Coordinator in the Public Justice Department at the State University of New York (SUNY) at Oswego. She currently teaches core criminal justice courses and elective courses in the areas of children's and women's rights. She is a former Children's Protective Services case-worker who specialized in sexual abuse cases. She holds a Juris Doctorate from Syracuse University College of Law and has practiced law since 1979. She served briefly on an Assigned Counsel Panel and clerked for a Surrogate Court judge. She has maintained a general private practice since 1981. Currently she specializes in adoption law. In Spring 2002, she and Dr. Celia Sgroi had an article, "Preparing for the Real World: A Prelude to a Practicum Experience," published in the *Journal of Criminal Justice Education.*

Lori Koester Scott, M.C., is the Administrator of Sex Offender Supervision for Maricopa County Adult Probation in Phoenix, Arizona. She received her B.S. in Sociology from St. Louis University and a Master's in Counseling from Arizona State University. She began her specialization in this field in 1981 as a sex offender therapist in the Arizona State Prison. As a trainer and consultant to probation departments throughout Arizona and other states, she is presently researching the recidivism factors of probationary sex offenders revoked to prison.

Donald J. Sears is a former New Jersey police officer as well as criminal investigator for the Public Defender Service, Washington, DC. He holds a B.S. in criminology from the University of Tampa, and a J.D. from Rutgers University Law School, Newark, New Jersey. As an Assistant County Prosecutor, he obtained extensive criminal litiga-tion experience in both trial and appellate courts. He also acted as a police legal advi-sor, lecturing on the use of deadly force and drunk driving laws. He is currently a practicing attorney with the law firm of Busch and Busch, specializing in criminal defense cases. He is also Adjunct Professor/Visiting Lecturer at Rutgers University in the Administration of Justice Department, New Brunswick, New Jersey. He has taught courses on issues confronting police officers as well as on the criminal justice system. He is the author of *To Kill Again: The Motivation and Development of Serial Murder* (1991). He is also a member of the Society for Police and Criminal Psychology.

Cassia C. Spohn, PhD, is Professor of Criminal Justice at the University of Nebraska at Omaha, where she holds a Kayser Professorship. She is the coauthor of two books: *The Color of Justice: Race, Ethnicity, and Crime in America* (with Sam Walker and Miriam DeLone) and *Rape Law Reform: A Grassroots Movement and Its Impact* (with Julie Horney). She has published a number of articles examining prosecutors' charging decisions in sexual assault cases and exploring the effect of race/ethnicity on charging and sentencing decisions. Her current research interests include the effect of race and gender on court processing decisions, victim characteristics and case outcomes in sexual assault cases, judicial decision making, sentencing of drug offenders, and the deterrent effect of imprisonment. In 1999, she was awarded the University of Nebraska Outstanding Research and Creative Activity Award.

Hung-En Sung, PhD, is the Research Director for the Drug Treatment Alternative-to-Prison at the Kings County District Attorney's Office in Brooklyn, New York. He received his PhD in criminal justice from the State University of New York at Albany, and has published on migration and crime, drug use and treatment, and policing.

Katherine van Wormer, PhD, grew up in New Orleans and has worked in Northern Ireland and Norway. She has an MSSW from the University of Tennessee-Nashville and a PhD in sociology from the University of Georgia. Active in various civil rights struggles and peace activism, she has recently written in the areas of harm reduction and restorative justice. She is the author of six books, including *Social Welfare: A World View*; *Counseling Female Offenders and Victims: A Strengths-Restorative Approach*; and *Addiction Treatment: A Strengths Perspective* (coauthored with Diane Rae Davis). She is currently a professor of social work at the University of Northern Iowa, Cedar Falls.

Richard H. Ward, PhD, is Dean and Director of the Criminal Justice center at Sam Houston State University in Huntsville, Texas. Prior to this he served as Executive Director of the Office of International Criminal Justice (OICJ) at the University of Illinois at Chicago (UIC) (1994-1999), Associate Chancellor (1994-1998), and as Vice Chancellor for Administration (1977-1994). At the same time, Dr. Ward held an academic position as Professor of Criminal Justice in the Department of Criminal Justice of the College of Liberal Arts and Sciences, and was named Professor Emeritus upon his retirement from the University of Illinois in 1999. He is the author of numerous books and articles in criminal justice, and has directed research projects in excess of $4 million. Prior to his appointment at UIC, Dr. Ward served as Vice President of John Jay College, City University of New York, where he also held the positions of Dean of Students, Dean of Graduate Studies, and professor of criminal justice during an eight-year period (1970-1977). Dr. Ward also served eight years with the New York City Police Department as a detective and youth investigator. (1962-1970), and with the U.S. Marine Corps (1957-1961). Dr. Ward holds his doctorate from the University of California at Berkeley and undergraduate degrees from the City University of New York.

In addition to more than 100 professional presentations, Dr. Ward has visited more than 50 countries, frequently as a consultant or trainer, and has lived and taught in China and Egypt, also serving with U.S. Marine Corps in the Philippines and the Caribbean. In 1994 he was awarded the Friendship Medal by the government of China for his work there, and over the years has received numerous awards and commendations.

Lisa Weber, MA, earned her degree from the School of Justice Studies at American University in Washington, D.C. She is a former research assistant at Rutgers University.

Michael Welch is Associate Professor in the Administration of Justice Program at Rutgers University, New Brunswick, New Jersey. Previously an Associate Professor at St. John's University in Queens, New York, he also has correctional experience at the federal, state, and local levels. He received his Ph.D. in sociology from the University of North Texas, Denton. Welch's research interests include corrections and social control. He has published numerous book chapters and articles that have appeared in *American Journal of Criminal Justice, Journal of Crime and Justice, Dialectical Anthropology,* and *Journal of Offender Counseling, Services, and Rehabilitation*. He is also the author of *Corrections: A Critical Approach*.

David B. Wilson, PhD, is Assistant Professor in the Administration of Justice program at George Mason University in Manassas, Virginia. Prior to 2001, he was Assistant Research Professor at the University of Maryland, Department of Criminology, and a Research Associate at Vanderbilt University. He earned his PhD in Applied Social Psychology from Claremont Graduate School, and is the author of numerous research articles on meta-analysis, juvenile justice, substance abuse, school-based interventions, boot camps, and sex offender treatment.